"A sweeping tale of despots and democrats, philosophers and philanderers, idealists and iconoclasts. . . . Winik is one of the nation's leading historians." —*The Baltimore Sun*

"No one ties together great historical themes with as much flair and erudition as Jay Winik. He's a scholar with a twinkle in his eye. . . . He never forgets that he's telling a story about human beings, complete with dreams, frustrations, flaws, and inconsistencies." —*The Chicago Tribune*

"With grace and insight and sweep, Jay Winik has given us a marvelous account of an epoch that fundamentally shaped the way we live now. *The Great Upheaval* is great history, vividly told." —Jon Meacham

"This is important work. . . . Winik deserves praise for his efforts to integrate U.S. history into the broader story. . . . He has a real gift for narrative history." —*Foreign Affairs*

"Winik concentrates his jeweler's eye on the [world] of the Founding Fathers. . . . He is a master at character depiction and dramatic narration. The book has the cliff-hanging pacing of fictional adventure." —*BookPage*

"A wild ride through the final decade of the eighteenth century. It makes the era come alive in an opera of sex, violence, and great ideas. . . . Winik is a master at portraying great leaders in crisis." —*Financial Times Magazine*

"Only a masterful writer could shape such a stirring narrative from such a wide-ranging field of original research. Jay Winik's *The Great Upheaval* is a terrific work that will endure for years to come." —Doris Kearns Goodwin

"Winik knows how to tell a gripping story. . . . A sweeping account." —*The Washington Post Book World*

"History at its finest." —*The Dallas Morning News*

THE GREAT UPHEAVAL

George Washington and the Marquis de Lafayette at Valley Forge

THE GREAT UPHEAVAL

✧

*America and the Birth of
the Modern World, 1788–1800*

JAY WINIK

HARPER  PERENNIAL

NEW YORK • LONDON • TORONTO • SYDNEY • NEW DELHI • AUCKLAND

HARPER ⬤ PERENNIAL

A hardcover edition of this book was published in 2007 by Harper,
an imprint of HarperCollins Publishers.

P.S.™ is a trademark of HarperCollins Publishers.

HarperCollins books may be purchased for educational, business, or sales
promotional use. For information please write: Special Markets Department,
HarperCollins Publishers, 10 East 53rd Street, New York, NY 10022.

FIRST HARPER PERENNIAL EDITION PUBLISHED 2008.

Designed by Ellen Cipiriano

Library of Congress Cataloging-in-Publication Data is available upon request.

ISBN 978-0-06-008314-4 (pbk.)

08 09 10 11 12 NMSG/RRD 10 9 8 7 6 5 4 3 2 1

To my family, Lyric, Nathaniel, and Evan "B.C."
—forever and always

Contents

✧

PART I: THE PROMISE OF A NEW AGE

PART II: TURMOIL

PART III: TERROR

PART IV: A WORLD TRANSFORMED

Maps

Introduction

◆

It is the age that gave birth to the modern world. It is also arguably one of the most significant eras in all of human history. It is, too, comprised of a galaxy of humankind's greatest thinkers and doers, not to mention the most idealistic generation ever witnessed, people bound and determined to alter the very course of civilization—for good and for ill.

Over a mere ten years, a great dynasty would be toppled; a savage world war, stretching from Europe to the Middle East and the Caribbean, would be fought; and the most civilized country in the world would be plunged into chaos. Revolution, liberalism, democracy, republicanism, nationalism—all would irretrievably rise in this period and all would formally take root. But, just as profoundly, so too would authoritarianism, becoming firmly implanted in the world's largest empire and sealing the end of "the great experiment" that sought to provide enlightened monarchical rule. Meanwhile, in another troubled pocket of the globe, Muslims and Christians would bitterly collide in Holy War. And the world's newest republic, itself tested by three separate rebellions and bitter discord, would almost miraculously survive a tumultuous infancy.

As the 1790s began, America entered the most critical period in its lifespan, Russia towered as one of the great imperial powers of Europe, and France fell into a monumental revolution. But contrary to the way conventional histories like to tell it, none of these remarkable events occured in isolation.

∞

THE WORLD THEN was far more interconnected than we realize. Because they were separated by vast distances and slow communication, it is often assumed that America, and the world at large, emerged virtually independent from each other in the 1790's. But this is a peculiar form of chauvinism of the twenty-first century, an era marked by e-mail, faxes, BlackBerrys, and cell phones; it is also grossly untrue. The late eighteenth century was stitched together in a myriad of ways almost unimaginable to the modern mind. From the French salons in Paris to the young American capital in Philadelphia, from the luxury of St. Petersburg to candlelight dinners in Monticello and Mount Vernon, from the bustle of London and market stalls of Warsaw to the mysteries of the seraglio in Constantinople and the steppes of the Crimea, great nations and leaders were acutely conscious of one another. And year after year, they watched one another, learned from one another, and reacted to one another.

France carefully followed the stunning events in the youthful America—indeed they helped give birth to the French Revolution and inspired the Declaration of the Rights of Man; Russia, in turn, scrupulously monitored the onset of the French Revolution, itself one of the most important turning points in all of recorded history; however, after the Bastille was taken and the tumult began, it led a once-reformist Catherine the Great to irrevocably reverse course and raise an iron fist against dissidents at home, to amputate a whole country, Poland, lest the "Jacobin madness" spread within her empire, and eventually to lunge toward France itself. And America alternately trembled and thrilled to the European spectacle. Born partly out of French largesse—President George Washington prominently kept a portrait of King Louis XVI in his offices—the newly formed United States stubbornly fought to persevere through its own precarious beginning. But increasingly sucked into the vortex of a manichean struggle over who would be the master of Europe, America had to somehow forge a government while confronting the twin threats of foreign intrigue and home-grown insurrection, which no less than the later nightmare of the Civil War threatened to fracture the upstart republic. And as the furies were unleashed by the French Revolution, and the continent was drenched in its own blood, America all but became a nation at war; then, much to the distaste of the Founders, it split into two political parties, warring factions divided both over who would be the true heirs to the American Revolution as well as the great global contest unleashed by the French Revolution. And as in France itself, it too faced the specter of being overwhelmed by rebellion, sliding into dictatorship, or falling prey to unrestrained anarchy.

Consider also the fluidity of the age, the sort that would be almost unheard of today. With insurmountable idealism, political figures and ardent rebels, forward-looking humanists and towering intellectuals, all freely crossed and recrossed borders, switched allegiances, spoke in foreign tongues, and fought for foreign causes with great relish, often making revolution not once, but twice. Thomas Paine, whose great pamphlet *Common Sense* roused a fledgling America in 1776, became one of the period's most fascinating international revolutionaries: He traveled to France, joined the revolutionary Assembly, and was almost beheaded by the guillotine; his life was saved less than twenty-four hours before he was to meet the blade, and he was later released due in part to pleas by James Monroe. At just nineteen, the Marquis de Lafayette left the glamorous world of the French court to fight with the American revolutionaries, where he became George Washington's beloved protégé; he then sought to reenact the American experiment in monarchical France, only to be forced to flee for his life from the very revolution he had helped launch. The great naval hero John Paul Jones thrashed the British on the high seas and was subsequently recruited in Paris by Empress Catherine for her jihad against the Islamic Ottoman Empire—ironically, the meeting was brokered by, of all people, Thomas Jefferson; at the same time, another Jefferson friend, Poland's Thaddeus Kosciuszko, fought with the Americans and helped conceive of West Point, later returning to Poland to lead a national rebellion against Russia. As for Catherine? After having earlier helped the cause of American independence, one of history's forgotten stories, she then dedicated herself to destroying the very idea of republicanism.

In a symbolic passing of the torch from the Old World to the New, Voltaire, France's immortal Enlightenment philosophe, embraced the rustic republican Benjamin Franklin in Paris, yet simultaneously worshipped Catherine—"She has the soul of Brutus," he enthused, "with the charms of Cleopatra"; meanwhile Franklin and Jefferson—and George Washington too—were ardently studied by Russian reformers Nikolai Ivanovich Novikov and Alexander Radishchev. Even Thomas Jefferson advised the budding French revolutionaries—in truth, he fervently believed that the survival of America's republicanism hinged on the success of France's revolutionary crusade—while another signer of the Declaration of Independence, Gouverneur Morris, counseled the doomed Louis XVI. And Talleyrand, the notorious French foreign minister, would spend the bloody days of the Terror in exile in America, only to, upon his return, brazenly threaten the safety of the young United States by "ravaging America's coasts" and fomenting "secession" west of the Appalachians.

The Americans, the French, the Russians, as well as the Ottomans, the Poles, and the British, were all part of one grand, interwoven tapestry. And it is these relationships and interrelationships, as much as any one country alone, that laid the foundations for the world we know today. Yet too often the story of these connections is left out, or is seen from only one side or one perspective. We do not see the world moving in concert. But with each passing year, it did. Thus, to appreciate the full magnitude of the remarkable American success in solidifying the United States in this decisive period, including how it managed its own rebellions and tensions that nearly split the still embryonic nation apart, one has to see it in the context of the larger setting and the world—just as the Founders did. To appreciate and understand the enormity—and ultimately the tragedy—of the French Revolution, one has to see how it unfolded in relationship to events on the other side of the Atlantic, in America, as well as to the East, in the Russian colossus. To grasp how the Francophile and liberal Catherine the Great started out as an ambitious reformer much like her progressive European and American brethren, and yet ended up as a repressive despot, one has to see the vast and rapidly changing world as she did too. In each case, the critical interplay is profoundly illuminating and indispensable.

In the end, this founding moment of our modern world is not simply a story of America or France or Russia, or for that matter, any one nation. It is a global story, and it is an extraordinary tale.

∞

It began in 1787 with boundless promise. America's independence sparked the first great contest over liberty. The people of Europe glimpsed a virgin political civilization on the farthest precinct of human existence, a wondrous republic on the other side of the world. Where, they wondered, was their paradise? While kings blithely reviewed fleets and attended state dinners, enlightened intellectuals and ordinary citizens alike asked this not simply in Versailles and in Paris, as their own historic revolution began, but in England and in Ireland, in Germany and in Italy, in Belgium and in Switzerland and Poland and Russia. The problem, however, was that the French revolutionaries, having promised a nirvana in words so magnificent that they still touch us today, lapsed into tyranny and expansionism, and the rest of Europe's rulers, forming a mighty armed coalition, turned reactionary. Indeed, once the French Revolution made the struggle for liberty a global one, the crowned heads of Europe and Russia thought they could ignore the United States, just as they thought their generals could stamp out its rebellious offspring, the French Revolution. They were wrong.

Yet if France's Revolution eventually stalled under foreign military pressure and internal despotism, and ancient Russia's budding liberalism was strangled it its crib, why didn't America's falter? Actually, it almost did. Under George Washington and John Adams, it confronted its own harrowing insurrections and civil unrest, the Shays, Whiskey, and Fries Rebellions. Against a backdrop of turmoil and terror, it survived tumultuous political splits, cries of sedition at home and meddling in its internal affairs by outside powers, and almost irresolvable divisions over its sister revolution in Paris. It groped its way through conflict or near conflict with the predatory giants of the day, a political crisis with Britain and a fearful quasi-war with France, not to mention the horrifying example of Russia dismantling a constitutional Poland in the east as well as threatening the cradle of the Islamic empire in the south. Still, despite all this, not only did America refuse to become a graveyard for the superpowers of the age, it resisted the temptation that enticed a utopian France: to treat its written constitution as a scrap of paper to be shunted aside, rather than as a sacred contract.

Of course, no less than in France or Russia, all was not perfect. Far from it. For one thing, America tragically retained slavery, thus sullying its beginnings and inexorably sowing the seeds for a terrible civil war. Nor did it reconcile the tensions between states' rights and federal rights. And much like the great powers of the day, fueled by the turmoil of the era, it too engaged in politics of paranoia, often vilifying the opposition as traitors, legalizing the ruthless Alien and Sedition Acts, and writing the Virginia and Kentucky resolutions, which all but endorsed secession. Yet in the end, notwithstanding the storm of widespread dissent and the growing specter of cataclysmic war, America never succumbed to organized violence, wanton repression, or institutionalized murder, which of course happened in cosmopolitan France and sparkling Russia. Instead, it chose a radical new path, of politics, thereby sowing the initial seeds of modern democracy. In doing so, it bequeathed to us the beginnings of a very different world.

BUT THIS IS more than simply a story of conflicting ideologies and uncommon circumstance. For here, in this timespan, are some of history's most dramatic moments: the surprisingly poignant beheadings of a tragic king and a desperate queen; the sudden challenge to America's larger-than-life president by a contemptuous French envoy whipping Americans into a revolutionary fervor, as well as by his own secretary of state; and the unparalleled dueling voices of Paine and Edmund Burke, the Enlightenment and the Terror, Lafayette and Robespierre, not to mention Jefferson and Alexander Hamilton, each struggling to shape the forces of modern politics. Here as well is the horrific carnage wrought by the war between the Ottoman Muslims and Russian Christians in a bloodstained arc from the Crimea to the Balkans, and the terrifying specter of French revolutionaries murderously turning on one another—as the American ambassador in Paris chillingly reported back to George Washington, "In the groves of the Revolution, at the end of every vista, you see nothing but gallows"—and looming concerns that the Americans might do the same. Here also are the historic military battles that more often than not dwarfed the savagery of the American Civil War and World War I, battles like Valmy, Ishmail, Ochakov, Praga, the Pyramids, Alexandria, and the Arcole. Here too in civilized France is a level of man-made cruelty that presaged Hitler's terrifying death machine in the Holocaust: the systematic drowning of priests in Nantes; the deliberate, if not heartrending, massacres of elderly men, young women, and little children in Savenoy; the guillotine that worked overtime in Paris; and the fearful suggestions that gas or poisoned drinking water might speed the job of destroying the Revolution's enemies. And here finally is the unremitting clash of presidents and prime ministers, tsars and sultans, cabinet officers and kings, and of ordinary citizens, all fighting desperately for the ideals they believed in, whether divinely inspired autocracy or man-made democracy, whether constitutional republicanism or Allah's law.

Finally, this is a comparative story of men and women and of the choices they made that would transform the globe. Indeed, how to fully appreciate, and understand, the wisdom and vision and courage of the man who proved to be the age's master spirit, George Washington—though few knew it at the time—than to see him against his reigning peers, a confused and dispirited Louis XVI, or such giants as a charismatic but aggressive Catherine the

Great, or an increasingly dictatorial Napoleon? Or how to comprehend the greatest generation of talent in American history: the visionary Hamilton, the sublime Jefferson, the iconoclastic Adams, and the sober Madison, than to see them in relation to the great revolutionary spirits of France, like the aristocratic Mirabeau, the fulminating Marat, the audacious Danton, and the intense Robespierre—or, for that matter, the dashing Polish hero Kosciuszko, or the inimitable Russian brain trust Prince Grigory Potemkin? Or how better to grasp the way that the Americans successfully absorbed their disputes into a political process, rather than with the guillotine and firing squads, than to see them in relation to France's Terror, or to the court intrigues and machinations of Catherine's empire?

As such, if this story begins with the most glittering civilization in the world, monarchical France, and is bookended by the other domineering juggernaut of the period, tsarist Russia, it ultimately ends with, and in the long lens of history is defined by, that most improbable of nations: America.

∞

THOUGH THE CHAPTERS of this book are organized by accounts of America, France, and Russia, which serve as the frames through which to see the larger age, the story is actually one continuous, interlocking narrative, unfolding much as the protagonists of the day themselves saw it; thus, American events are woven into France or Russia, or Russia and France into America, or France and America into Russia. For example, from the American side, one can't fully appreciate the frightful divisions fostered by the rise of political parties or the subsequent "democratic societies" so feared by George Washington, without seeing them in the context of a French Revolution where Jacobins, Girondins, and monarchists were squaring off against one another with increasing ferocity and bloodshed. One can't sufficiently understand a Whiskey Rebellion that if unchecked potentially threatened to split America East and West, or devolve into a civil war or a French Revolution in miniature—and where insurgents sang the French revolutionary song "Ça Ira" while carrying mock gallows and toasting Robespierre—without seeing it in relationship to the beheading of Louis XVI and the mounting Terror in Paris—or, for that matter, compared to the Pugachev Rebellion in Russia, which threatened to upend Catherine the Great's dominions.

Similarly, one can't appreciate the tensions fostered by a remarkable free press that nonetheless peddled gossip, rumors, and all sorts of scandal—and even wished for an "early death" of George Washington—without viewing it in relation to the mayhem and chilling violence inspired by the likes of Jean-Paul Marat's journal in Paris or Catherine's clamping down on Enlightenment journals she once so admired. And of course, one can't adequately fathom the Alien and Sedition Acts in America, or its quasi-war with France, or American fears of foreign invasion without seeing them in relation to Napoleon's armies that were, in Mallet du Pan's inimitable words, devouring Europe "leaf by leaf," and that were spreading their might as far as Austria and as wide as the Middle East—or for that matter, against the backdrop of Marshal Suvorov's "tidal wave" of Russian armies laying siege to Islam, pouring over the mountains of Italy, and literally wiping Poland off the face of the map. For the Americans, this last example, of course, was perhaps the most nerve-racking: Poland's rebels were grandly inspired by the American and French revolutionary experiences, and this "partition spirit of the times" (the *Pennsylvania Evening Post*'s words), served as a harsh warning to the young American republic about the perils of military weakness and internal chaos, let alone the fragility of their own enterprise.

But it was not just civil unrest and military battles and matters of governance that defined these nations in this formative period. Each wrestled with deep moral and philosophical questions as well—with slavery being paramount among them. In arguably the most searing debate in the new American republic, what began as an effort by a dying Benjamin Franklin to embrace liberty for all Americans and put slavery on the road to extinction, only ended up legitimizing this "peculiar institution." Catherine the Great too had large plans—her initial goal was to reform and then eventually extinguish servitude. But chastened first by the Pugachev Rebellion and then frightened by the maelstrom of events in France, she instead resorted to greater repression, and actually increased the number of serfs as well as their load. By the end of her reign, the institution scarcely differed from chattel slavery in America; on any given day, one might see in the *Moscow Gazette*: "For sale, a girl of sixteen, of good character, and a secondhand carriage, hardly used." And, in one of the era's most fascinating twists, it was bloody, revolutionary France that ended up abolishing slavery. In France, blacks too would stunningly become known as "citizens."

Even the very transfer of power at the decade's close would become a telling emblem of this remarkable era. The French Revolution would end not with a stirring election or republican democracy, but in the resurrection of monarchical-style rule and a coup d'état: The thirty-year-old general Napoleon Bonaparte would come to power at the end of a gun. Russia's secession would be equally tumultuous: In an extraordinary moment of palace intrigue, Catherine's son would, against her own wishes, become tsar, and in yet another twist, he would be murdered in a plot led by her twenty-four-year-old grandson, Alexander, the empress's favorite. Ironically, it was Alexander who would later thwart the grand army of Napoleon, thus spelling the great general's downfall. Only America, this provincial, marginal state, would escape the fate of these other nations, and here the contrast is stunning. George Washington would step down, one of the heroic events of the era; as England's King George III had once earlier commented, "If he does that, sir, he will be the greatest man in the world!" In a country already hotly divided, John Adams would subsequently become the second president, and by 1800, the young republic would do something quite unique in the annals of history: peacefully transfer power from one political party to the next, made more astounding by the fact that it was the tensest election in the nation's history, requiring thirty-six ballots in the House of Representatives and nearly precipitating civil war.

∞

IN TRUTH, THE world has never seen anything quite like this epoch and likely never will again, a period marked by cold, dark battlefields but also by lofty debates, by a clash not just of ideas and ideologies but of cultures and religions, and by a cast of characters unequaled in history. It is crucial to see the world as it unfurled before their eyes, to feel the fears as they felt them, to experience their hopes and aspirations and uncertainties as they did, and most of all, to witness this tumultuous, interconnected world that month after month was rapidly emerging.

But more than just re-creating the interwoven stories of these nations during this decisive, founding decade, to see these nations and peoples together is to grasp the very forces that have made us modern. To contemplate the 1790s, an era of republics and revolution, of despotism and democracy, of authoritarianism and religious war and the cold machinery of totalitarianism too, is to begin to understand the tides that have swept the globe and battered nations, large and small, ever since. America, France, and Russia from 1787 to 1800 are most assuredly of their time—but their history, their founding, their latter-day echoes loudly resonate for us as well.

For out of this crucible was born not just their world, but ours.

Jay Winik
Chevy Chase, Maryland

THE GREAT UPHEAVAL

The Orangery at Versailles

Prelude

AT THE THRESHOLD

H E HAD SURVIVED MORE than seventy summers of wet, fly-infested heat. And more than seventy winters too. He had outlasted infection and epidemics. He was almost as hard a man as the New England stone that marked his fields. He is believed to have once beaten a servant so badly that the man died from the wounds; he fought with his neighbors, even stole from them, and, in his last year of life, failed to defend his wife when she was accused of being a witch.

Barely a month later, he also stood accused. He spent his final summer being moved, in tandem with his wife, from stifling jail to stifling jail, first in Salem, then in Ipswich, then in Boston, and finally back home to Salem again; a single dungeon could not begin to house all of the colony's suspected conjurers. In each town or city, his family had to pay for his food and drink and even the wood for his fires, as well as fees to the jailers. But when it came time for him to answer the charges of witchcraft and be tried before a special court, which ultimately sent nineteen souls to face the noose on a barren slope near Salem Village known as Gallows Hill, he refused.

On the morning of September 17, in 1692, Giles Corey, a landed farmer and recent church member, was stripped naked. The last voluntary act of his life was to lower his quivering flesh to the ground and wait as a wooden board was placed upon his chest. Then, in full view of his neighbors, Salem's sheriff called for rocks to be piled upon the board. A friend of Corey's and even members of the court pleaded with him to agree to a trial. In turn, he pleaded for more weight to bring a speedy end to his suffering. It took two long days for the slow increase of rocks upon the board to crush the breath from his lungs and halt the pulse of blood from his heart. As he lay dying, in a final, painful indignity, the sheriff drew his cane and forced Corey's flaccid tongue back into his mouth.

One of the Salem magistrates, Samuel Sewall, laconically wrote in his diary, "About noon, at Salem, Giles Corey was press'd to death for standing mute."

But Giles Corey's brutal execution was not some aberrant punishment concocted on the fringe of the Europeanized world, where, a mere seventy miles away, Indian war parties were murdering colonial farmers in their fields. No, here in this budding paradise of liberty, Corey died by means of a centuries old, rather conventional European punishment, popularized by Henry IV and named the *peine forte et dure*, and employed as a standard sentence for those who refused to submit to the state's will. In such ways, big and small, did the hand of the old reach out to touch the new.

∞

THE WANING DECADES of the seventeenth century remained, for a majority of mankind, the bleakest of times. Kings ruled at the pleasure of the Almighty; all others did what they were told to do. For millions of subjects eking out a meager existence and shackled in ignorance, life was fleeting, dirty, and cheap. Public opinion scarcely existed, even as a meaningful concept, and the weak endured what they must, forcibly grateful for what little they had. Once the sun dipped over the horizon, the world—or at least the limited world they knew—was enveloped by a vast swath of darkness. Fires crackled and smoked, precious tallow candles smoldered, but for the most part life paused. And whatever secret dreams they held—or lingering agonies they grappled with—were subservient to the bare exigencies of brute survival.

Only half of infants made it past their first year; in the British colonial city of Boston, a quarter of all newborns died before they reached a mere seven days. Middle age was one's thirties and few lived beyond forty. Most were haunted by disease. Throughout much of the century, pestilence was one of the greatest killers—plagues and epidemics, cholera, smallpox, or microscopic fleas nesting in the coats of rodents, each leaving behind fresh graves. So did an array of mysterious ailments, like the "sweating sickness" that devastated England. Indeed, it is a remarkable fact that in the seventeenth century Europe's population actually declined. Nor was this just due to dark reigns of pestilence. The simplest of infections were perilous. And with depressing regularity failed harvests and famine swept the continent, which only added to death's grisly toll. The years of hunger were terrible: Peasants might be forced to sell all that they owned, and cannibalism was not unknown. There were even tales of starving men who tore hanged bodies from the gallows, frantic to eat the raw flesh while it was still warm. Rampant filth and negligible public sanitation also raised the butcher's bill—mosquitoes harbored malaria, lice ferried typhus, and heaps of animal dung lured storms of flies in the streets. The New World was hardly better. On top of the same scourges, hunger, bitter cold, and crippling heat, many settlers lived in fear of Indian raids and heard lurid tales of men, women, and children being hacked to death by silent, efficient war parties as unlucky colonists gathered to chat on a porch or checked on a crop in the fields.

On European soil, it was also an age of horrid cruelty. As the historian J. H. Plumb once wrote: "life was cheap." Torture was universally employed for all manner of crimes, from speaking ill of the king to stealing a tradesman's wages or even loaves of bread. Rarely was there mercy. To be sure, one might be hanged, drawn and quartered—that was simple. But ordinary criminals and political dissidents alike were routinely beheaded, burned, or broken alive on the wheel. Or slowly crushed by the infamous *peine forte et dure*. Or they were subjected to the rack and the rope. Or the knout. Treason, but not only treason, often yielded more creative methods. Rapists, for one, were castrated. Counterfeiters were punished by pouring molten metal down their throats. Curiously, those speaking ill of the sovereign were, to a point, more fortunate: They simply had their tongues ripped out.

News was scarce, gossip and storytelling supplied most information. And people of all stations believed that their physical world was suffused with the comings and goings of spirit beings. Illness or health, privation or plenty were the work of otherworldly forces. Even some of the century's leading scientific minds were not immune: Men like Francis Bacon believed in witchcraft and the curative power of south-facing windows; Robert Boyle, who initiated the beginnings of modern chemistry, wanted legions of miners to be interviewed about their encounters with subterranean demons; and Sir Isaac Newton devoted more time to his studies of the occult than to the physics principles that won him lasting fame.

Whether one was standing in Salem, Massachusetts, or in Paris on the banks of the river Seine, let alone amid the cold snows of Moscow, the future, as much as anyone could see of it, appeared unchanging and immutable, an era of perpetual illness and barbarism, abduction and subservience, superstition and famine. This was all the more ironic because the day was fast approaching that would yield to an era of teachers and lawgivers, builders and administrators, philosophers and learned men, and enlightened rulers as well. Ironic too because a subtle and powerful new spirit would soon be rising in the world, a mighty political storm of revolution, republicanism, democracy. And ironic because in many ways, more so than at any time in human history, tucked inside a few corners of seventeenth-century Europe and especially the New World, there was already an increasingly forward-looking community filled with uncommon radiance and energy. Nowhere was this more evident than in the supreme monarchy of the day.

Between the people and the sovereign a bright line was drawn. The royalty of the age was invested in glory, bathed in mystique, and clothed in magical powers. To be king was not simply to be lord of men, a host of great feasts, but a giver of rings, of gold, of landed estates. It was to be a builder, a warrior, a patron of the arts, and a later-day version of Apollo. It was to be the very embodiment of civilization, answerable only to God, and in one instance, to be civilization itself. Never was this more the case than in France, the forefront of world civilization, where Louis XIV, his most Christian majesty, reigned, and where, in 1682, one of the greatest of palaces in the world—it was actually a château—was being finished: Versailles.

∞

HIS LEGACY STILL echoes down the years. Once saluted as "the invisible divinity," he is known both by his royal name of Louis XIV and his adopted sobriquet, the "Sun King." For seventy-two years, he presided over some 19 million subjects, and was alternately referred to as *"Louis le Grand," "Le Grand Monarche,"* and *"Le Roi Absolu."* And he was the most influential ruler in the world, looming across every facet of European life, politics, diplomacy, and civilization. Few monarchs in any era have approached his undeniable grandeur; Voltaire compared him to the great Roman emperor Augustus. The symbol of absolute monarchy in the classical age, his very presence was designed to overwhelm; his merest gesture became the subject of endless talk. The Sun King claimed dominion over the center of all Europe; it was his France against which the rest of Europe felt obliged to combine and which set the course for a swath of lesser nations, whether it was the future of the Spanish possessions, the independence of the once-mighty Holland, or the survival of England's parliamentary revolution. And this was just the beginning.

France itself was a land of sharp contradictions. With its fertile soil, it was the wealthiest country on the globe, even if millions still lived in appalling poverty. The aristocracy held sway as in no other era, yet there was also a growing middle class: lawyers, office holders, bureaucrats, and merchants. France was a self-sufficient country, yet the French of this century reached out everywhere: trade in India and Madagascar; the founding of Canada—it was French Canadians along with their Indian allies who tormented the settlers on the wild Massachusetts frontier, sending terrified refugees streaming south to Salem—and farther west, penetrating the Great Lakes and the Mississippi Valley. They established plantations in the West Indies, expanded ancient French commerce with the Levant, and enlarged the French navy, an act that would one day have unforeseen but immense consequences for America and the world.

France was cultural and intellectual center as well, infused with literary giants, like the tragedists Corneille and Racine and the playwright Molière, who ridiculed the newly rich and the foppish aristocrats. There were Descartes, the towering mathematician and scientific thinker, and Bayle, the father of modern skeptics, and Pascal, the renowned scientist who became a profound spokesman for Christianity. But always there was, at the core, the Sun King himself.

Louis was the third king of the Bourbon line, but the eminence of his majesty flowed from his character: his prodigious pride, his unimpeachable self-assurance, and his overarching ego. He was five feet, nine inches, quite tall for that day, and astoundingly handsome too. His face was pitted from smallpox, but he retained his rich brown eyes and captivating gaze, and delighted in wearing high heels to accent his well-muscled calves. The ultimate effect was a robust and "noble" appearance, which made, as one court diarist noted, others "tremble" in his presence. As he grew older and his locks of sleek chestnut hair thinned, the king donned a wig of long black curls that only added, as an observer confessed, to his "majesty."

Genial and ruthless, he was a king almost from the cradle. His own father, Louis XIII, had died in 1642, when Louis was four. Ruling in his name was his mother, Anne of Austria, and her chief minister, Cardinal Mazarin, the brilliant protégé of the great Cardinal Richelieu. But it was not to be a peaceful regency.

∞

WHEN LOUIS WAS just nine, France erupted into revolution, spurred on by discontented nobles and a new rash of stringent government revenue schemes. Known as the Fronde, it was the precursor to the French Revolution: Barricades were thrown up and street fighting broke out in Paris; along with the nobility, Paris's courts also rebelled, seeking to circumscribe the crown. And at the height of this violence, armed bands, led by nobles, roamed the countryside terrorizing the peasants. One night, the boy king himself and his mother were besieged by an angry Paris mob. Surrounded by horrid cries and the crackle of muskets outside his windows, Louis was hastily spirited off to St. Germain, where he spent the night on a bed of straw. It was a frightful moment—and formative. While the uprising was eventually subdued—it took five murderous years—the rebellion made a lasting impression on the Sun King. Scarred by the tumult, he was adamant from then on to be his own man, answerable to no one. And he was equally determined never to be controlled by scheming ministers or thickets of court intrigue, a choice with ominous portents for his heirs to come.

Louis was no great intellect, or so his critics said. But this was deception. True, his formal education was mediocre; true, he was not the wisest man of his time. Neither was he the most gifted. But he always grasped the intricate ingredients of leadership. Overbearing and iron-fisted, with an insatiable appetite for admiration and flattery, he instinctively understood how to gain power, how to magnify power, and ultimately how to embody power itself. He was every inch a king. And after the Fronde and years of religious wars, the bourgeois of France were, unlike the British or later the Americans, eager for a strong monarch. This is what Louis meant to give them.

He ebulliently boasted (or so his enemies said), "*L'état, c'est moi*"—"the state is myself," and set about making precisely that happen. In 1668, twelve miles west of Paris, amid the rolling woodlands of the village of Versailles, he decided to build a palace to embody his own personal majesty. That majesty took fourteen years to make flesh. Some 36,000 men toiled, month after month, year after year, on the platforms that wrapped around the building. They struggled in the muck and heat and darkness, laying formal gardens in the long-fallow ground. They chipped and carved away, chisels in hand, fashioning statues of stone and marble. There were the skilled labor, the slave labor, and the imported labor. And there was the death: The mortality rate was appalling. After dusk came the unearthly creak of wagons carting away the dead, bodies crushed by massive stones or broken in violent falls. Deadly bouts of malaria and other maladies also tore through the barracks, but it was of no matter; the work continued. Meanwhile, 6,000 horses dragged timbers or blocks of stone, until in 1682, the king officially moved into the palace.

To all who saw it or heard tell of it, it was a monument to worldly splendor, the marvel of all Europe and the envy of lesser monarchs everywhere. And it was unique, starting with the fact that it had no ramparts, no moats or walls. Such was the overweening confidence of the Sun King that he could bask in his wealth and glory without the most minimal of defenses to protect himself. Even the château's gate was almost never locked.

The original site had been home to a hunting lodge for Louis XIII. That was then. Now the Sun King's Versailles had a facade a full one-fifth of a mile in length—it was filled with exquisite reception galleries and chambers of state, dance halls and meeting rooms, private suites for the royal family and public rooms for government offices, rows of apartments for the royal guests, alcoves and bustling shops, let alone all that was hidden away from public view, like the ample closets that housed the royal clothing, the kitchens that produced the feasts, and the endless corridors, guardrooms, and hideaways. In decor, it was stunning, from its polished mirrors and rich, gleaming floors, to its magnificent Gobelin tapestries and walls lined with patterned velvet, all overlooking a formal park with fountains and shaded walks. By day, blazing sunlight illuminated great arched ceilings and frescoes of mythological royalty; at night, thousands of candles burned brilliantly in sparkling chandeliers and polished candelabra. And everywhere there was silver: silver chairs, silver tables, silver planting pots, and of course silver table services. Even the giant doors were branded in gold with the symbol of Apollo, a glittering mark of his regal majesty. As if to reinforce this, the great French painter Charles Le Brun himself rendered Louis as God on the vaults of Versailles.

Below the château, the gardens were equally arresting. Where there had once been nothing more than a sandy rise and enveloping woodlands, millions of flowers, bushes, and trees were planted amid emerald swatches of finely clipped hedges and lush green corridors. Terraces gave way to ramps, staircases descended to an intricate display of ponds, lakes, and bubbling fountains, and all were populated by an extravagant menagerie of rare birds, ostriches, and herds of wild animals. On warm spring evenings, indoors and outdoors flowed together in a dazzling array, as the entire court spilled on to the grassy lawns and into the Grand Canal, where a fleet of gondolas awaited. And then there were the orange trees, thousands of them, which the king had imported from as far away as Santo Domingo and which almost miraculously bloomed year round.

One of the most ambitious constructions ever conceived, Versailles rivaled the imperial forums of Rome and the great temple complex at Karnak, Egypt. But it was far more than an unfettered pleasure dome. All across the Continent—and beyond—kings, princes, and potentates dreamed of their own royal colossus, instructing their architects and builders to erect palaces worthy of Louis's royal grandeur, down to the imported carpets, the cut crystal, and the filigreed porcelain. So outside Vienna, the Hapsburgs built Schönbrunn; outside Berlin, at Potsdam, Frederick II built Sans Souci; outside Constantinople, the Ottoman sultan built a summer palace, Sa'adabad; and eventually, in St. Petersburg, Tsar Peter the Great built Peterhof. Well into the next century, wealthy planters in the American South modeled their gardens on Versailles, and even the radiating broad boulevards of Washington, D.C., were crafted by a Frenchman in imitation of the Sun King's magnum opus.

And of course, there was the politics. Behind the gilt and gardens, Versailles teemed with political machinations. Far more than a royal residence, Versailles was in many respects the ultimate public building, the instrument for the king to concentrate all power in the hands of the monarchy—his monarchy—even as France remained a hodgepodge of competing jurisdictions, special privileges, and bureaucratic ineptitude. Here, under the royal eye, Louis housed the great French nobility, the very classes who had started eleven civil wars over the course of forty years and instigated the dreaded Fronde. As political theater, it was genius. As political strategy, it was ingenious. Away from their lands and estates, absorbed into the ritual of the court, the French aristocracy soon became satellites of the king, rather than his rivals or potential regicides. So Louis lured them to his palace, corrupted them with gambling, exhausted them with dissipation, and seduced them with ritual. Under the Sun King, there would be no repeat of the Fronde.

Thus, at Versailles, and at the Sun King's command, there were no idle hours for plotting or excessive intrigue. Louis was a master of overstatement. Everything unfolded in precise and formalized detail around him. From the moment the sun streamed over the horizon straight into *la chambre du roi*—it was no accident that the king's bedroom stood at the exact center of the palace—his daily routine of rising (*lever*), of eating (*dîner*), and of going to bed (*coucher*) became an infinite series of minutely orchestrated, ceremonial acts. He awoke ("sire, it is time"), was massaged with rosewater, and was shaved and dressed by the most exalted of his subjects—six of them for this routine alone. So it was that the landed and titled desperately sought the privilege of removing the royal nightshirt, of carrying the royal clothes, and of presenting the chair for his daily "natural functions" or the royal bed for his lovemaking. When he prayed, when he ate, when he grumbled, and when he laughed, as he strode through the palace or sauntered through the parks, a bevy of lords and nobles flowed in his royal wake, with rigid etiquette dictating who could be present, who could speak, who could hold his sleeve ("right" versus "left"), who could share *la viande du roi* (the king's dinner). And whomever the king chose to be in his presence, the court honored in turn.

Aside from the many feasts, which were themselves immense, and thrice-weekly social evenings, the amusements were endless: There were the tournaments and the tennis matches, the billiard games and the boating parties, and the great balls. And then there were the special occasions: a parade of torchlight suppers in the open air, all-night gambling soirees, and multiday festivals when the entire court dressed up as Persians or Turks, Romans or American Red Indians; Versailles had literally become heaven on earth. Equally titillating was Louis's love life: True, he slipped into his wife's bed twice a month to make love to her. But the rest of the time, as long as she lived—she died in 1683—he was picking a series of lovers: When Louis went to war in 1673, he was accompanied by the queen and two of his mistresses, one of whom was pregnant; while mules carried their vast train of wardrobes, the three ladies rode in the same carriage. Moreover, in matters of politics, Louis was equally willful: He never called the ancient Assembly of Notables or the Estates-General, and he ignored the regional courts. And though it is true that he rarely took action without carefully weighing the advice of his fleet of experts, all major decisions were ultimately made by his eminence: the Sun King at Versailles.

Yet under Louis, France was at its pinnacle, laying claim to the epicenter of the world: Its arts, its courtly manners, its elaborate clothing and elegant speech suffused all of Europe—French was the language of diplomats and the tongue of the most prominent men of letters; the French navy rivaled Britain's and its army was among the finest on the continent; and it boasted great scientific minds. Of course, few decrees were more commanding than those signed by *Rex Christianissimus*, "his most Christian majesty": Louis XIV. In his hands, nothing couldn't be done, nothing couldn't be accomplished, nothing couldn't be dreamed.

But Louis's radiance belies the fact that he was hardly universally loved. While his politeness was without parallel, he was routinely thoughtless, or boorish, or stern, bearing little affection for the common people. His religious beliefs were painfully narrow; he brutally repressed the Protestant Huguenots. He even once decreed that all prostitutes consorting with soldiers within five miles of Versailles should have their noses and ears cut off. And there was the matter of Louis's penchant for armed conflict with his neighbors: To much of the rest of Europe, the Sun King was "a bloodthirsty tiger," a brash Catholic despot.

Indeed, if Versailles were the thunderclap reverberating across Europe, his martial activities were even more earthshaking. Increasing his forces from 100,000 to 400,000, Louis made war a near constant activity of the state. For much of his long reign, he was either at war, preparing for war, or biding his time, and his army soon became the scourge of Europe. He harbored designs against Spain and in 1672 invaded the Dutch provinces; in 1679 his troops infiltrated the dissolving frontiers of the Holy Roman Empire in Alsace and Lorraine; in 1681 his men occupied the city of Strasbourg. Then, in 1683, when the dreaded soldiers of Islam, the Ottoman Turks, moved up the Danube and stormed the gates of Vienna, they did so with Louis's initial tacit support; ironically, Catholic France and the Muslim Ottomans were long-standing allies. And he seized the Rhine and annexed Flanders. As his empire swelled ever more, Europe would repeatedly band together against him: There was the coalition of the kings of Spain and Sweden, the electors of Bavaria, Saxony, and the Palatinate, the Holy Roman Emperor, and the Dutch Republic. That was in 1686. Fifteen years later, a combined Europe would again be forced to act against him: England, Holland, Portugal, as well as the Italian duchy of Savoy; this war would continue for twelve years.

Ultimately, this conflict was a considerable setback for Louis. The continuous war produced only poverty, misery, and criticism at home and led to the loss of valued royal possessions abroad. Belgium was lost, and so were Newfoundland and Nova Scotia (then Arcadia), as well as the disputed American Northwest, the Hudson Bay territory, ceded by treaty to the British. And there was a further, inglorious price to be paid among the people of France themselves: When Louis died in 1715, his body was borne to the Saint-Denis basilica—to the jeers of the teeming populace. For much of his life, he was the subject of excessive flattery, but in death, he was serenaded by indiscriminate abuse.

The king's final years were blighted by domestic tragedy as well. His only surviving legitimate child, the Grand Dauphin, died in 1711. A year later the new dauphin was gone, dead of measles. Almost simultaneously, he lost two of his grandsons. And within days Louis's great-grandson, the next dauphin, died of the same disease. In the Sun King's direct line, there now remained only one great-grandson, a red-cheeked child with brooding eyes; frail and of ill health, he was only two. And he too had measles. But the governesses bolted his doors and refused to allow the doctors to treat him. Miraculously, the child survived.

On his deathbed—Louis was suffering from painful gangrene and his flesh literally seemed to be falling off his bones—the Sun King summoned his surviving great-grandson to his side. As shafts of bright light filtered in through the windows, the Sun King leaned over and whispered prophetically, "My child, you will one day be a great king. Do not imitate my taste for war. Always relate your actions to God." He also added: "I am going away, but the state will always remain."

Louis XV did his duty and listened. He would spend fifty-nine years on the throne. Indeed, together these two men would reign in France for 131 years.

With such power and pedigree, despite nagging wars and even territorial reversals, the grandeur and glory of the French monarchy seemed invincible. So when Louis XVI, the fifth ruler of this distinguished line, assumed the throne, it was believed by all that he—like all of France's rulers since Clovis was crowned king of Francia in 496—would follow in their magnificent wake, arousing awe, testifying to the immensity of their heritage, and trailing endless blazes of glory.

How, it seemed, could it be anything otherwise?

As the splendor of Versailles was rising above the French countryside and the nation basked in the brilliant, if imperfect, glow of its Sun King, some fifteen hundred miles away to the east lay the sleeping colossus of the age—ancient, inert, xenophobic: the old Muscovite state of Russia.

In the West, Russia was seen as backward and foreboding, a view that was not without merit. In countless ways, Russia was primeval, oriented not toward Europe or America but toward the vast heartland of central Asia, its gaze fixed on Persia and China, across long stretches of deserts. The bazaars of Moscow were frequented by Persians, Afghans, Kirghiz, Indians, and Chinese, while traders and artisans peddled an eclectic slice of the Asiatic world: silks, brass and copper goods, tooled leather and bronze, and innumerable objects of hand-carved wood. The city itself was peopled with tattered, itinerant holy men and bearded priests, as well as ruddy, callused peasants in cloth leggings and soldiers in voluminous caftans. Meanwhile, women of the upper classes were secluded and often wore gold-braided veils, men wore beards and skirted garments, and the two sexes never mixed at social functions. Russian customs were uncommonly coarse—basic things like cutlery and toothpicks were unheard of; and drunkenness was so rampant that on feast days, travelers were stunned to see naked men, passed out, who had sold their clothing for a drink. Dwarfs and fools, increasingly out of fashion in the West, still amused the tsar and his retainers. Arithmetic was hardly understood, Arabic numerals were not used, and the calendar dated from the beginning of the world. Astonishingly, even predicting an eclipse was regarded as a form of magic. And clocks, imported by Europeans, elicited wonder and awe.

By all accounts, Moscow, or ancient Muscovy as it was often known, had all but ignored Versailles and London, not to mention the nascent colonies in North America. In this regard, Europe, as seen from Moscow, was in the rear; indeed, the Russian government was so fearful of being contaminated by foreigners that well into the seventeenth century, it was reluctant even to permit foreign embassies.

But Moscow was hardly an itinerant backwater. It was the capital and holy city of the vast Russian Empire—an empire that already spanned from nearly the Pacific Ocean to the northern forests of Poland. And by 1682, as the last stones were being laid at Versailles, and a decade before the nightmare that would engulf Salem, Massachusetts, Russia was increasingly breaking out of isolation, slowly reaching out to the West. For some time now, trade between England and Muscovy had boomed, and there was a bustling German community in Moscow's suburbs. And while the English were building Boston and the Dutch erecting New York, the Russians were stretching out towns—a whole string of settlements—5,000 miles across northern Asia, to the Pacific itself. Meanwhile, Russia's upper classes increasingly intermarried with Europeans.

From a distance, Moscow struck one Western traveler as the most "beautiful city in the world," an urban feast topped by hundreds of gold-crusted domes and a sea of glistening crosses that surmounted the treetops. Unlike the stone and marble of its European counterparts, Moscow was a city hewn from wood; even the streets themselves were planked with timber, not trampled down or paved with stone. Also unlike anything in the Western world was the somber medieval citadel of Russian power, the Kremlin, which imbued the city with an exotic mystery.

Exotic was the word: Even today, the Kremlin evokes unfathomable intrigue. With its massive red walls jutting from the bank of the Moscow River, the Kremlin was not a single building but an entire walled city—*Kreml* literally means "fortress" in Russian—ringed by two rivers and a deep moat. Inside this mighty citadel rose gorgeous cathedrals (three), an astonishing number of chapels (sixteen hundred) and hundreds of houses, as well as government offices, law courts, barracks, bakeries, laundries, stables, and a mighty whitewashed-brick bell tower that rang constantly, often driving Muscovites to distraction. And Moscow had a spiritual dimension rivaled only by Jerusalem and the Vatican: It was the "Third Rome," the center of the Orthodox faith.

Muscovites were an intensely religious people, and most of the city, rich and poor alike, fell under the church's spell. Few had a hold on the Russian mind or imagination as did the *starets*—the man of God. But the true master who loomed over this ancient land was ultimately the tsar, the very portrait of absolute monarchy. "Kaiser" in German, "tsar" in Russian, it was derived from "Caesar," the ancient designation for the great emperors of the Roman Empire. It fit. Remote and inaccessible, the tsar was the autocrat of all Great and Little and White Russia, an august figure enclosed in an aura of semi-divinity. From infancy, Russians were taught to regard him as a godlike creature ("Only God and the tsar know," went one ancient proverb). The tsar was regarded as the father of all his people—the "Batushka"—and in return, his subjects were cast as children; their role was subservience as well as obeisance. Similarly, Russian noblemen did not simply bow, they flattened themselves before the tsar, touching the ground with their foreheads ("we humbly beseech you, we your slaves . . .").

Of course, as with the Caesars, there were titanic struggles. For more than a decade, a horde of claimants and pretenders sought the throne; at one point, the title of tsar was even claimed by a son of the king of Poland. And for three years, Russia was even without a tsar. But then on a cold, windy day in 1613, the first great Romanov tsar was crowned. The dynasty passed from father to son, that is, until 1682. That year, Tsar Alexis, who had a penchant for wearing "tufts of diamonds as big as peas," died unexpectedly, and following custom and decree, the crown passed to his fifteen-year-old son, Feodor, a semi-invalid. Beset by frailties, Feodor reigned only six years. He had no eldest son, and thus no direct heir. As the boyars—the supreme noblemen—of Moscow passed by the bed on which the dead tsar lay, brushing their lips on his cold, limp hand, the bells of Moscow's churches tolled. Then, inside the thick, dark walls of the Kremlin, the boyars feverishly deliberated.

Tradition dictated the throne should pass to the elder of the two surviving candidates, in this case, Feodor's two remaining brothers. Normally, the choice should have been uncontested. But sixteen-year-old Ivan was lame and nearly blind, and could speak only with great difficulty. And the other brother, Peter, was only ten years old. The Grand Patriarch of Moscow turned to the crowd that had anxiously massed outside the palace windows. "To which of the princes do you give the rule?" the patriarch called out.

The crowd cried for Peter. So with their voices echoing ever more loudly, the patriarch decided that Ivan was to be shunted aside in favor of his strapping younger brother. The choice thus made, Moscow's elite sat back content, believing that the crisis had been stanched, and the tumult was over.

They were sorely mistaken.

∞

FOR MORE THAN one hundred years, Moscow's impenetrable Kremlin had been watched over by the Streltsy, armed guardsmen comparable to ancient Rome's Praetorian Guard. Service was a lifelong commitment, the position passing from father to son: They were Russia's first professional soldiers, though they were also at times ill disciplined and unruly—and lately unpaid by the crown. Resentful of foreigners, ignorant of politics, sheathed in prejudice and superstition, these disheveled musketeers were always dangerously ready to wield their power—or to mutiny—against someone challenging their privileged position.

Their revolt began after Feodor's death. Rumors abounded: Feodor had been poisoned, it was whispered. Foreign doctors were behind the assassination, it was said, and Ivan, the rightful heir, had been pushed aside. And the rumormongers claimed that Peter, the new tsar, would punish the Streltsy.

On May 15 the tinder was lit, and the Streltsy struck. Enraged, hundreds marched on the palace banquet hall inside the Kremlin and clamored to see the Tsarevich Ivan. Peter's mother mounted the Red Staircase, and, trembling, presented the two children, the infirm Ivan and her beloved Peter, to quell the mutinous army. But after a brief calm, the rank-and-file Streltsy remained unappeased, erupting into an orgy of violence. The first victim was their own commander. Lifting him above their heads, they threw him from the balustrade onto the pike spears of the rebels below. Impaled, his quivering, blood-smattered body was cut into pieces. And any hope of disciplining the mutineers died with him. The next victim was Artemon Matveev, one of Russia's elder statesmen and a former prime minister. He was hurled from the Red Staircase onto the field of upraised spear points. Within seconds, his body was hacked to pieces. Then the real dying began.

The Streltsy roamed and pillaged the palace, dismembering their victims. Statesmen and nobles were dragged from their private chambers and bloodily massacred. No one was to be spared. The privy councillor and director of foreign affairs were cut to pieces. And young and old alike were targeted: An aged boyar was brutally killed, as was the son of one of the tsar's chief aides. The bodies and pieces of bodies—hands, heads, torsos— were then dragged into Red Square, where corpses were piled one upon the other. To the chilling stampede of feet, the rampage continued the next day, and the next. For three days, the Streltsy picked apart the palace, wandering through its dark rooms and maze of tiny apartments, mindlessly looking for victims. On the third day the bloodthirsty troops issued a fatal ultimatum. Peter's mother was informed that she must give up her own brother, or everyone would be slaughtered.

Fearing she had no choice, Peter's mother surrendered her sibling, an agonizing decision. While his sister wept, he spent his final moments in the palace chapel, taking communion and receiving last rites. He then rose and presented himself, unsteadily clutching a holy icon. Beaten for hours, he too was thrown into a sea of spears held aloft by the muttering mob. Now his hands and feet were cut off. Finally his body was chopped to pieces, the bloody carcass mashed into the filthy mud.

Only then did the killing cease.

The soldiers had come for money, as well as blood, and they got it. They demanded back pay, which they received. They demanded amnesty, and received that, too. They demanded that their victims be designated criminals, and that too was done. And thus the stage was set for a coup. It was Peter's older half-sister who then assumed power as regent of Russia, now ruling in both Peter's and Ivan's names.

The young Peter, much like the young Louis of France, was deeply scarred, and for the rest of his life harbored a deep revulsion for the ancient Russian capital of Moscow. And one day, like the Sun King, he too would extract his revenge. After he became master of Russia at twenty-four—eventually earning for himself the sobriquet of Peter the Great—he would go on to crush the Streltsy: Two hundred would be hanged; others were beheaded with axes; and countless more were tortured, all as evidence of the tsar's wrath. To complete his revenge, he exiled his half-sister, forcing her head to be shaved and sending her to a nunnery, where, like Mary Queen of Scots, she was under guard day and night. She spent six years in her solitary confines, dying at age forty-seven.

An imposing six foot seven, with large black eyes, Peter would also dismiss the glory of the Kremlin. He first rejected its cloistered, xenophobic ways—as a burgeoning young man, his closest companions were a Scot, a German, and a Swiss—then eventually he stripped Moscow of its rank altogether. Instead he fixed his eyes on Europe; his first act was to travel the Continent, a striking repudiation of old Muscovy. As biographer Robert Massie notes, no tsar had set foot outside his dominions for more than six hundred years; indeed, no tsar had ever been seen in the West. But Peter saw Europe, traveling largely incognito and indulging every curiosity, even prying wigs off heads to examine them as he walked down the street. He came back determined to Europeanize the Russian behemoth. When the church protested against his godless, foreign ways, he nationalized it. Not yet content, he then created not merely a great château, like Louis XIV, but an entirely new capital, perched far beyond the arid plains ringing Moscow. For the hallmark of the new Russia, he chose a once obscure fort taken from the Swedes, located on the banks of the Neva at the mouth of the Gulf of Finland; he named it St. Petersburg. In the process, Peter pried his unwieldy empire from its Slav heritage, realigning its very axis from East to West.

St. Petersburg—the original name was the Dutch "Pieterburkh"—was to be a new Rome. Called the Venice of the North, the Babylon of the Snows, St. Petersburg was turned into a European, not a Russian city, and quickly became one of the great urban centers of Europe. Its architecture, its style, its morals and thought were all Western. Society and the court spoke French or German, not Russian; meanwhile, the best clothing and furniture were ordered from Paris. And with its opera and ballet, its great balls and its chamber orchestras, St. Petersburg would come to rival all other European capitals. So too in politics. As the decades passed, it would also come to play a dominant role across the continent. But, even as Russia looked West, one thing would not change—the lawless, often anarchical, violent means by which the tsars were selected.

Before his own demise, and dissatisfied with his rebellious son, Alexis, Peter had him violently tortured; within days, Alexis was dead. Then Peter decreed the right for each tsar to name his own successor. Unlike in France, or England, or the American colonies for that matter, transmission of the supreme power was thus enshrined outside the domain of law, and with no principle of succession, dynastic or otherwise, the empire slipped into an unending struggle of plots, praetorian revolts, and palace rebellions. Many years later it would be said, "All of Russia is one vast madhouse": for decades, and as a quick and violent sequence of tsars ruled after Peter, that certainly seemed to be true. But it was in 1762 that the real changes came, with another majestic occupant of the Romanov throne, who would cast a broad shadow across the remainder of the century and the four corners of the globe. Yet this tsar wasn't a man, or a Romanov, or even a Russian. She was a minor German princess, Sophie of Anhalt-Zerbst—born to a small house of royalty in the hodgepodge territory of the Holy Roman Empire. We know her today by a different name, Ekaterina, or later, as Catherine the Great.

At the age of fourteen she left Prussia to wed Peter the Great's grandson, Peter III—the marriage was arranged. It was a disaster. For eighteen years they lived together, in some ways more as wary companions than man and wife, or duke and duchess. Peter himself was dissolute, often drunk, just as often pitiful, more often than not a bully. And he loathed all things Russian while loving all things German, particularly the Prussians; in the process, he dangerously alienated many of the key forces of

Russian society. But Catherine's position was equally precarious. A convert to Russian Orthodoxy and a transplanted foreigner, she spent her life in risky limbo between the throne and the dungeon. Actually, after Peter was crowned tsar, the dungeon appeared perilously close. One evening the freshly minted tsar openly insulted her at a public dinner, mocking her and shouting *"Dura!"* (fool). Reportedly—the facts are unclear—he then ordered an adjutant to arrest Catherine and pack her off to a monastery. Regardless, Catherine's humiliation was now nearly complete—Peter was already living openly with a mistress, whom he boasted he would soon marry—and her danger was palpable.

What was she to do? For the better part of a century, the power of the Russian throne had been neither elective nor hereditary—it was occupative. Only wits or ruthlessness could dictate who ruled this burgeoning empire. Too often, the strong triumphed and ruled; the weak were invariably overthrown. And when the plotting began in the tiny, inbred world of the Russian court, Catherine was at its center.

Even now, much of the next sixty-four hours remains cloaked in conspiracy and considerable mystery, but this much is known: On June 12, after a scant six months on the throne, Peter left Petersburg for a holiday, taking with him his mistress, his black manservant named Narcissus, and his dog. Catherine, at her summer palace in Mon Plaisir, moved fast. By June 27 all was in readiness. Roused in the dark of night, she knew the coup would succeed today—or never. If it failed, at the age of thirty-three she would face the ax or mount the scaffold. In a deeply symbolic gesture, she changed into the ancient green costume of the Preobrazhensky Guard, the elite soldiers marshaled by Peter the Great to replace the hotheaded Streltsy. Never again would she be German. Or a transplant. Or a foreigner. She then grasped a naked saber in her bare hands and mounted a splendid gray stallion. The effect was electric: *"Vivat! Vivat!"* the men shouted at the guards' barracks. Some 12,000 guards rallied to her side, swearing allegiance to her as Catherine II, while those who resisted were promptly jailed.

She then marched on Petersburg.

Peter, who still controlled the army, hastily sent emissaries to negotiate. Coolly, she rebuffed them, and remarkably, they defected to her. Peter's response sealed his fate: Falling to his knees, the disoriented tsar surrendered, asking that he be allowed "his dog, his fiddle, and his blackamoor." This prompted the steely Prussian ruler Frederick the Great to proclaim: "Peter allowed himself to be driven from the throne as a child is sent to bed." Under guard—three hundred of them—Peter was taken to his estate at Ropsha. Within days he was dead, ostensibly after a wild night of drinking. One of the carousers was the brother of Catherine's lover, a military officer named Grigory Orlov.

Whether Catherine sanctioned the killing is unclear, but she issued a terse statement, eerily reminiscent of dictatorships two centuries hence, in which the palace absurdly insisted that Peter had died of "hemorrhoid colic." Catherine was now a usurper and an adulterous regicide. With the gutters of Russia already awash in blood, even this was scandalous. But on September 22, at the Assumption Cathedral in the heart of the Kremlin, the resplendent thirty-three-year-old former German, bejeweled and wearing an ermine mantle made from 4,000 skins, was anointed "Lady Catherine the Second, Empress and autocrat of all the Russias."

It has been said that "Peter the Great created Russia's body, but Catherine endowed it with a soul." She did, in spades. Catherine would quickly prove to be a gifted politician, a consummate actress, and a visionary statesman. Her reign would attract eminent scholars and poets, philosophers, architects, and artists. She would build the grand Hermitage for her good friend, the esteemed French philosopher Voltaire, in case he ever came to Russia. And she embraced many of the other great philosophes of the age, including Diderot, Locke, Blackstone, Grimm, and Montesquieu, a number of whom she personally befriended. During her rule, she herself became an enlightened legislatrix, a patron of the arts, a playwright and poet and writer of epigrams, satires, histories, and dissertations in her own right. And among her many charms, she had the gift of wit, a sharp tongue, and a certain impishness.

But few leaders would also come to embody the tensions and ferment of the age as Catherine, who would become inextricably intertwined within the tapestry of the two great revolutions germinating in America and France, all with seismic consequences for years to come.

∞

SLOWLY, BUT INEXORABLY, the Old World and its immutable orders, and the New, were on a collision course. Exploration was not only traversing the oceans, but the realms of the human mind in science, music, art, and literature. Such staples of modern life as the telescope and the microscope, the thermometer and the compass, wax candles and streetlights, were all rapturously unveiled. The arts flourished: Vivaldi, Handel, and Bach debuted their music; Molière, Racine, Hobbes, and Locke published their finely honed words; and poets like John Milton captured and educated the aristocratic ear. Freed from blind adherence to ancient shibboleths and religious dogma, scientists hurtled forward into the future, exhuming the minutiae of the natural world. The names of these pioneers are immortal: Descartes plunged into analytic geometry, Boyle calculated the density of gases, and Leeuwenhoek opened up astonishing vistas through the tiny lenses of his microscope. Preeminent among the towering thinkers of the day was Sir Isaac Newton, whose work gave the world the law of universal gravity, not to mention insights into physics and astronomy, chemistry and botany, and even calculus, which soldiers soon employed to better calibrate the deadly trajectory of an artillery shell.

With equal zeal, Europeans were expanding outward across oceans and over the immensity of open spaces, exploring and colonizing, settling and civilizing. Expeditions overcame logistics, disease, epidemics, and great distances. Already, the lion's share of South America and a large chunk of North America were possessions of the Spanish crown. Elsewhere, competition for the world's spoils was stiff: Half a dozen nations had laid claim to swaths of Africa; even the landlocked state of Brandenburg controlled a colony on the continent's Gold Coast. In the harsh, craggy corners of Asia, the foothills of the Himalayas had not yet been reached—they would one day yield the war-torn Afghan frontier—but in the jigsaw of states and principalities that today is known as India, a land of strutting peacocks and ruling rajas, the colors of the English and Portuguese nations had been raised. Meanwhile, pirates cruised to Manila, Palo Condore, China, and the Spice Islands, even as far as Australia.

Without minerals to mine or exotic spices to harvest, the eastern lands of North America seemed a bit of an afterthought. Maps were few and poorly drawn, but still, France and England, the two great European powers, had already established competing colonial outposts. Actually, France's was much larger, stretching from Montreal to Quebec, and the Great Lakes down into the flat plains of today's United States. In fact, while Russia's Peter the Great was dispatching a delegation to observe the fleet of the Knights of Malta, France was laying claim to the Mississippi Valley; in 1699, the mouth of the great river was christened Louisiana, thus exalting the name of his most Christian majesty, Louis XIV.

And there were the English settlements, tightly clustered along the gray Atlantic waters, from Massachusetts to the Carolinas. By the sullen, red glow of twilight at day's end, the map was constantly rearranged. The Dutch New Netherlands—today's New York—had already been ceded to England as a battle prize, divvied up after the second Anglo-Dutch naval war in 1664. And at a time when Versailles was not yet fully finished and St. Petersburg didn't even exist, New York, Philadelphia, and Boston were already thriving English towns with more than 30,000 inhabitants among them; even the venerable college of Harvard had opened its doors as far back as 1636. And the continent was barely explored.

∞

YET THE EUROPEANS increasingly came in droves. Despite the dangers, North America offered a measure of religious freedom for sects like the Puritans, as well as economic opportunity and a fresh start for those in need. So it was that in 1638 one Henry Adams of Somersetshire, England, arrived in Braintree, Massachusetts, as part of the vast Puritan migration, coming to the Massachusetts Bay Colony to escape the long arm of the Church of England and to help build his own godly version of a city on a hill. The Adamses were farmers who supplemented what they could raise in the rocky soil with other trades; Henry thus also made malt for brewing or baking. Two generations later, in 1735, John Adams, a farmer, shoemaker, and church deacon, would welcome a son by the same name, a boy whom he fully expected to follow in the family's now well-established line of "virtuous, independent New England farmers."

Eighteen years after Henry Adams set sail, John Washington, of arguably more noble pedigree, boarded a boat for the colony of Virginia. Records place John's family in northeast England as far back as 1180, where they were among the landed gentry. John's own grandfather, Lawrence, was the master of Sulgrave Manor. Barred by the laws of primogeniture from inheriting his father's estate, Lawrence's fifth son, also Lawrence, attended Oxford and entered the clergy. And here the tale might have ended, were it not for the English Civil War. Accused of being a "malignant royalist," Lawrence Washington died penniless; his wife soon followed him to the grave. Orphaned and just twenty-two, John Washington struck out for the New World. In Virginia he married the daughter of a lieutenant-colonel and received, as a wedding gift, a 700-acre estate on the colony's Northern Neck. It was here, on a now 1,700-acre parcel, that George Washington, a third son, was born in 1732.

Not every colonial was so meticulous about his roots. Thomas Jefferson dryly recorded that his father's family came from Wales, near the mountain of Snowdon, which he characterized as the "highest in Great Britain." On his father's side, Jefferson was at least a fourth-generation colonial—one ancestor may have arrived as early as 1619, making his the earliest of the founding families to arrive in the Americas—while his mother, the daughter of a sea captain, was actually born in England, and claimed a long-standing Scottish and English heritage. James Madison, whose later relatives traced his British family's roots back to Charlemagne and the barons of Runnymede, was himself a fifth-generation Virginian, whose forebearer, John Madison, acquired 600 acres of Virginia Tidewater land in 1653, in return for paying for the passages of twelve immigrants from England. By 1683 those holdings had grown to nineteen hundred acres.

Others were less fortunate in the New World. Alexander Hamilton's maternal grandfather was a French physician named John Faucette, who, during the Huguenot purge, was driven from his homeland by the mighty Sun King, while Hamilton's own father was pushed to the New World by the small, ignoble act of birth order. James Hamilton had grown up in a Scottish castle on a sizable estate, but as a fourth son, he had no realistic hope of inheriting anything. In 1741 he made his way to the West Indies, where he repeatedly failed in business. His love match was not much better, for while Dr. Faucette's daughter, Rachel, was smart and charming, she was also already married. James and Rachel's union nevertheless produced two sons, but the boys were condemned to be known as bastards. The younger one, Alexander, escaped for good in 1772, when he boarded a ship for Boston. He never returned to the West Indies.

Benjamin Franklin was sixty-six years old when Hamilton's ship docked, but he knew something of the vagaries of apprenticeships and birth order—and even bastard sons, having fathered one of his own, William. Franklin's father, Josiah, had arrived in Boston eighty-nine years before, ostensibly as a Puritan refugee, but also as a man looking to provide a passable living for his wife and children. In Oxfordshire, Josiah Franklin was a cloth dyer with no prospects. Thus did the New World beckon. Josiah became a candle and soap maker, a burgeoning profession in Boston. He had seventeen children, thirteen of whom navigated the perilous passage to adulthood. At the age of twelve, Benjamin, the last of his surviving sons, was duly apprenticed to his brother James, twenty-one, who was setting up shop as a printer.

By such circuitous routes did these figures, the virtuosi of the American enlightenment who would transform this slender patchwork of settlements into a nation, gain their formal footing upon the stage. And inside these adolescent colonies and among these men, an unmatched generation of political talent in the annals of the world, would percolate ideas on natural history and global geography, free trade and material prosperity, matters of state and the human condition, the nature of man and the nature of government. In the bustling New World, all were endlessly discussed and debated, in Jefferson's telling phrase, like "Greek colloquia."

∞

BUT NOT YET. It had been said that history was "like a clock, unwinding itself forward according to God's design." However, as the 1700s prepared to dawn in the British colonies of North America, it was hard to perceive the taut, logical engineering of such a divinely inspired design. Even as Versailles was being finished, and the Streltsy revolt was running amok in Russia, a once-thriving corner of North America was in the grips of its own tumult. The Second Indian War was raging on the northeastern frontier, as settlers pleaded with the British crown to intervene on behalf of their beleaguered villages. Meanwhile, less than a hundred miles south, the rule of law was also spontaneously breaking down, but here the genesis was quite different: It was not over politics or economics or dynastic power struggles, but witchcraft. It came to be known as the Salem hysteria.

It didn't happen overnight. It was not uncommon in America, as in Europe, for people to speak of the devil, to warn of the devil, to fear the devil. Others claimed to *see* the devil. It was also a time where there were still pagan nightmares of werewolves and griffins crouched under a full moon, or warlocks feasting on serpents' hearts or witches in disguise. Accordingly, religious dissidents in New England, such as the Quakers, were regularly stripped and examined for witch's marks. For many of these New Englanders, the devil was omnipresent, forever lurking, the haunting master of the witches. In Connecticut, ten people were hanged because of "familiarity with the Devil." Witches were prosecuted in Anglican Virginia, too. So was a "little old woman," suspected of being a witch, in Catholic Maryland; she was cast into the sea to appease a violent storm. But in 1692 Salem suddenly found itself in a grip of accusations, and a sinister brew of trials and inexplicably brutal punishments.

The facts are well known. The hysteria began with two young girls, both of whom suffered fits, screaming and rolling on the floor. The girls were medically examined and questioned closely; they fingered "Tituba" as the source of their trouble—Tituba was a female slave who was part of the household, though different accounts label her as black or Indian. Pressured, Tituba confessed, admitting she was a "servant" of Satan and ranting about hogs and cats and the devil's book "signed by nine in Salem." Thus, the hunt began.

A special court was set up to get to the bottom of the matter—in sorcery cases it was now judged that the ordinary law would not suffice. The proceedings were as outrageous as the accusations. Those who confessed to doing the "Devil's work" were released; ironically, those who refused to plead to crimes they had not committed were judged guilty. Throughout the long, hot summer, the hysteria raged. By early autumn, fourteen women and five men were hanged. Giles Corey was, of course, uniquely pressed to death with heavy stones—the dreaded *peine forte et dure*. Even two dogs were slaughtered. Meanwhile, another 150 people awaited trial in damp, stinking, overcrowded jails; some died there.

It was only in October when the new governor arrived that suddenly the spell was broken. The special court was dissolved. Those under arrest were released. And out of this paroxysm of self-righteousness and superiority against an imagined threat, justice was eventually done: The General Court of Massachusetts passed a motion deploring the acts of the special judges. Members of the jury signed a statement of regret. Indemnities were later granted to families of the victims who had been hanged.

And here Salem becomes a very different tale. In retrospect, what was so striking—at least for this age—was the speed with which the rule of law was restored and also the willingness of the local colonial government to confess its wrongdoings and to make reparations. In cosmopolitan France, where the Sun King's court had its own witchcraft scare, confessions were obtained by torture, trials were dispensed with, and the accused spent the rest of their lives in prison—so why not in rough-hewn America? One partial explanation is that the doctrines of the Enlightenment philosophers, though originating in Europe (and including the eminent Englishman John Locke), had widely penetrated the American colonies, and that Massachusetts itself was growing ever more secular by the year.

Still, there is another partial explanation, one harder to quantify but no less compelling. History is indeed composed of such small incidents, but with the end of this frenzied summer, it seemed as though in British North America there was something new, something strange, and something wonderful taking place.

Even then.

∞

By the mid-1700s this patch of Royal Britain would find itself uncommonly blessed with growing cities, all throbbing with commerce and industry and the sounds and sights of jostling humanity—blacksmiths fresh from the forge plying their wares; farmers in muddy boots making their way to the taverns; pamphleteers hawking their ideas; and newly arrived settlers wrestling with their common lives. And there were the cities, Boston and Philadelphia, New York and New Haven, Williamsburg and Richmond, with their handsome parks, grassy residential squares, and tree-lined avenues, all humming with urban enterprise. Day in and day out, their waters teemed with coastal and oceangoing vessels, ships that with increasing frequency traded not only with the countryside and the mother country across the Atlantic, but with the far corners of Europe, and even Russia, Alaska, and Asia as well. And too, these cities soon became the intellectual spokes of this New World, redolent with the steady cry of politics and the slow building of civic lives, out of which an infant nationalism was to be formed.

Indeed, the first colonists in Virginia, cast in the sturdy English empirical mold of fair-mindedness, were a largely pragmatic people, wedded to the rites of hierarchy and patriarchy, and deeply stepped in tradition. One of those traditions, of course, was governance. A number of the colonies were established not just to further religious freedom but to embody advanced political ideals. As early as 1619 in Virginia they had set up a miniature parliament when they debated "the Dale's code." Within a decade, "sweating and stewing and battling flies," they had already developed a highly impressive legislative machinery, creating the House of Burgesses—the first representative body in the New World. Provincial assemblies elsewhere followed, in Massachusetts, Pennsylvania, and New York, until in one manner or another, across all thirteen colonies these American subjects had exerted a strong measure of self-control over their own lives. And in doing so, the colonies, from the start, seemed to be the perfect mirror of English virtues and vices, promoting startling innovation under the guise of tradition.

Actually, this was just fine with Britain. The American colonies comprised only a small piece of the broader British world. By mid-century, the British Empire was vast and disparate, not to mention decentralized. Thirty-one separate governments operated under the auspices of Westminster, ranging from the kingdom of Ireland to the colonies of Bombay, Madras, and Calcutta, half a globe away and controlled by the East India Company. In 1750 the whole empire itself, having once begun as a tiny island kingdom, now comprised some 15 million people of countless colors and creeds, which though less populous than France or even the Austrian monarchy, was hardly unimpressive. Notably, within this broader tapestry, the distant American mainland from Georgia to Nova Scotia was home to only about 2 million people, about equal to Ireland or Scotland—or Brittany or Bohemia. So it remains a curious twist that the notion of America was at first Britain's idea: It was the British who finally persuaded the American settlers to accept some kind of distinct national identity. Equally telling is the fact that the Albany Plan of Union drafted by Benjamin Franklin in 1754 and commended by British officials was rejected by the colonials themselves. They were still English, through and through.

Things, however, were viewed differently in London. America was but one outpost and by no means the most important one: Indeed, the dominant mood of Britain toward America was the same as its other vast dominions, largely one of complacency, of a smugness in the majesty of its unwritten constitution, and a contentment in its relatively free press and parliamentary system. Summary accounts have never quite trounced the notion of King George III as an imperious tyrant, reigning supreme over Britain and running roughshod over America. Yet nothing could be further from the truth. In Britain, Parliament was as dominant as any monarch in continental Europe, in fact, even more so. It had the power, as one journalist sarcastically jibed, to "do all things except change a man into a woman."

But the ebb and flow of political ideas is rarely ever static, and this was especially the case here. Now came the undercurrents of discontent, first among radicals in England, then among the reformers, then across the ocean, six weeks and 3,000 miles away. In America, Parliament and the king and his "friends" were "corrupt," it was said. "Britain," blasted another critic, "was to Americans what Caesar was to Rome." Or as Thomas Paine would later memorably write: "Freedom hath been hunted around the globe."

Even so, most Americans, like Benjamin Franklin and George Washington, remained deeply loyal to the crown. Hindsight has curiously obscured the first draft of Jefferson's Declaration of Independence; it shouldn't have, for it remains an object lesson for all colonial powers. "We might have been," wrote Jefferson, "a great and free people together." The phrase was later deleted by his colleagues, but the sentiment was real. Franklin, for one, agreed. He spent years fervently seeking to keep the British empire together and remarked balefully that George III was "the best of kings." Indeed, for years, Franklin and most Americans could not imagine that they would one day be forced to choose between their king and their native soil.

But that time was coming.

∞

However subtly, the upstart Americans would in part be pushed by the unlikeliest of sources: the philosophes of Europe. Among educated thinkers and writers of the day, the past was rapidly coming to be regarded as a time of barbarism and darkness. And with each new year, scarcely had there been an age so skeptical toward tradition, so confident in the powers of human reason and science, so firmly convinced of the regularity and harmony of nature, and so deeply imbued with the sense of civilization's advance and progress. The evidence of this was all around. The literate public greatly expanded. Newspapers and magazines multiplied, as did coffee houses and reading rooms. There was considerable demand for dictionaries, for encyclopedias, for surveys of all fields of knowledge. Ideas and people now freely mingled, and literary salons came into vogue. And the bourgeoisie, as well as many of the elites and nobles, not to mention forward-looking kings and empresses of the day—they were called "enlightened despots"—were becoming at once educated but also reflective. As the 1700s unfolded before them, their thoughts turned to the question of governance and the human condition. And mostly to this basic notion—that the existing state of society could be improved. Indeed, in many regards, the final area of exploration and settlement in the eighteenth century was not geographic or scientific, but philosophical and political. We know it as the Age of Enlightenment.

Following on the heels of Bacon and Descartes, Bayle and Spinoza, and above all, the venerated Englishmen Locke and Newton, the most famous of all philosophes were the French trio, Montesquieu, Voltaire, and Rousseau.

Yet trio is almost a misnomer. Profound, diverse, idiosyncratic, and filled with reforming zeal, each man differed considerably from the others. But in turn, each was lauded as a literary genius. Together they helped spawn a distinguished list of followers—d'Alembert, Buffon, Turgot, Helvetius, Baron Grimm, and Diderot, and as the historian Bernard Bailyn has pointed out, Franklin, Jefferson, Hamilton, Adams, and Madison too. And together, their powerful conclusions spawned a belief, one that radiated outward, radically and exponentially, in Europe, east and west, off to Russia, and far away to the colonies in North America, which would set minds ablaze and soon help kindle two great revolutions, not to mention a counterrevolution. The crux of this belief eschewed an order based on the direct will of God and the fixed nature of the universe. Instead, it focused a bright light on man-made law and man-made authority. The philosophes spoke of a rightly ordered government that best guaranteed social welfare. They talked of political liberty and intellectual freedom. And just as often as not, of equality: equality not only for commoners, but equality before the law, equality in the payment of taxes, and equality for the middle classes to more easily rise to positions of honor or privilege. And finally, through these philosophes, the theories of liberty and parliamentary government became matters for general discussion as well.

Montesquieu, who was a favorite of Russia's Catherine the Great and who (like Locke) also cast a long shadow over the constitutional debates of the American Founding Fathers in 1787, loomed large in this drama. He was a man of blunt queries, pointed observations, and stark declarations. Twice a baron, a landed aristocrat, a manorial lord of southern France and a judge in parlement, he was part of the resurgent nobility that had followed the Sun King's death and continued through the eighteenth century. As with his brethren, he was a miasma of contradictions. An aristocrat by temperament, he spoke in italics and exclamation points. An anti-absolutist by philosophy, who believed that the known world was constructed of fixed, observable laws, he was a touch reactionary—he admired the "barbarous" Middle Ages. The first to forcefully articulate a system of checks and balances to limit government and guarantee freedom for the people, he believed in the innate superiority of men over women.

Invariably though, he was referred to as "the great" Montesquieu. For his part, the American revolutionary patriot Patrick Henry would one day call Montesquieu "the greatest and wisest of men who wrote about the science of government." This stemmed largely from his masterful work, *The*

Spirit of Laws, published in 1748, in which he developed two sweeping ideas: the first, that national governments were best determined according to climate and circumstances. For instance, despotism, he said, was suited to large empires, while democracy was not. By contrast, republics were suitable only for small city states. His other doctrine had much larger applications: the separation and balance of powers. Here he had in mind the English system, which he saw as an ingenious mixture of monarchy, aristocracy, and republicanism—the king, Parliament (with its Houses of Lords and of Commons), and the courts, which together embodied a separation of the functions of the executive, legislature, and judiciary. This doctrine, of course, was well known to British Americans, and would, with key adjustments, famously make its way into the American Constitution.

If Montesquieu was contained, dispassionate, methodical, Voltaire, the wit of Parisian society, was flamboyant and bold. His real name wasn't even Voltaire. Born in 1694 to a solid bourgeois family and christened François-Marie Arouet, Voltaire was a verbal invention, the most famous of all pen names in history. Cutting and sarcastic, piercing and logical, and frequently incorrigible, he was a master of finely honed irony and mockery. While his brilliance is without question, it often got him into trouble: Twice he was arrested, once for a spat with a nobleman, the other time, at the age of twenty-three, for impertinence to the French regent; he spent eleven months in the dreaded Bastille. For all his fame, he was a wanderer. For three years he lived in England, a country he came to admire enormously. He spent two years in Potsdam with his friend Frederick the Great, and later shuttled between Cirey and Paris, Potsdam and Berlin, plus a stay in Belgium. Often ill and exhausted by this peripatetic existence, he finally purchased a manor near the Swiss frontier, where he became (in his words), the "hotel keeper of Europe," amid the constant patter of lofty admirers, favor-seekers, and those in need who journeyed to his door.

Journey they did. Already he had met the likes of Alexander Pope and Jonathan Swift, the philosopher George Berkeley, and British Tory leader Viscount Bolingbroke. Now Boswell came to see him. So did Gibbon and Casanova and Catherine the Great's good friend, Prince de Ligne. His fame stretched worldwide: He courted royalty and they returned the favor; he was an ardent admirer of Russia's two great tsars, Peter the Great, and, of course, Catherine, whom he extolled as the Semiramis. The two corresponded for years.

No less than Montesquieu, Voltaire is hard to pigeonhole. Arguably, he

is the easiest of all the great writers to read; his ideas transcended boundaries and nations. Neither a liberal nor a democrat, he was a crusader for
freedom of thought and fervently preached religious toleration. He hated
bigotry, hated intolerance, and hated superstition (he called it *l'infame!*), yet
he was less concerned with political liberty. Thus, he was an ardent admirer
of enlightened rulers. He worked tirelessly to advance the cause of material
prosperity and railed against religious bigotry, writing the first purely secular conception of world history in his masterful *Essai sur les moeurs (Universal
History).*

Yet he was a curious man. Despite his disgust for the organized church,
he favored Christianity over "heathen" religions like Islam. And he always
supported a strong state. What he most wanted was liberty for the enlightened, the kings, the nobles, the artists and writers, whom he believed would
serve the greatest good. In the end, though, he reflected the temper of the
times, and most intellectuals remained closest to Voltaire—the reigning
king and god of the philosophes.

Even in his final days, Voltaire relished civilization in every form. In
1778, one month from death, Voltaire embraced Benjamin Franklin at Paris's Academie Royale. Here were two of history's great bons vivants and par-
odists, one of the Old World, one of the New, one a patriarch of Europe,
the other a republican from America. Gaunt and frail, Voltaire entered.
There was Franklin, dressed in his trademark simplicity: plain yellow coat,
no wig, bespectacled with no other adornments. Every seat, it seemed, was
taken. Expectations among Voltaire's disciples rose. Franklin was then the
most famous American in the world; Voltaire, who at the age of eighty-four
had spanned the age of classicism right up to the revolutionary era, was the
most famous man of letters in all of Europe. Among the raucous crowd,
America's John Adams was present to witness the spectacle. The people
became restless. Suddenly, out came the clamor: *"Il faut s'embrasser à la française."* Just as suddenly, to great applause, the two men took each other's
hands, embraced, and kissed each other's cheeks. Words were spoken. They
needn't have been. The effect was electric. As Adams recorded it (and so
did France's eminent philosopher Condorcet), the cry immediately spread
throughout the kingdom, into the papers, across Europe, that it was "Solon
who embraced Sophocles," thus evoking the specter of the two great Greek
philosophers, one known for his laws, the other for his literature. And the
applause to this embrace was a mere prelude. Though no one realized it
yet—how could they?—it was a symbolic passing of the torch as well.

But among the renowned philosophes, there was one more giant on the stage of ideas: Jean-Jacques Rousseau.

Of the three, Rousseau was the odd man out. He was not French, but Swiss; not Catholic, but Protestant; and not a noble or member of the middle class, but of lower-class origin. Where Montesquieu and Voltaire literally captured the room around them, Rousseau was always the little man, the outsider. Neglected as a child, a runaway as a teenager, he eked out a meager living through odd jobs, such as copying music; not until age forty did he have any success as a writer. Everything about his life smacked of tragedy. He had five children, each of whom he consigned to an orphanage. He never had any social status, never had any money, he was pitifully maladroit, and all his life he felt he could trust no one. A puzzle of complexes, he was haunted by nightmares, fears, and neuroses; he was almost surely paranoid. Few were close to him, and none—including perhaps Rousseau himself—understood the chaotic forces within him. Everything about him, his social life, his personal life, his financial life, even his inner life, reeked of failure, save one thing: his ideas. But they were not to be underestimated, and he may well have been the most profound writer of the social condition of his age. He was almost certainly one of the most intoxicating, and his influence perhaps the most powerful in the long run.

Rousseau believed that the best traits of human character—kindness, unselfishness, honesty, and true understanding—were in fact products of nature. Where Montesquieu and Voltaire delved into the cold particulars of government and rational constructs of political liberty, Rousseau embraced raw emotion, the quick strike of intuition, and the transcendent effect of introspection. His works rested not on critical thought, but on spontaneous feeling. Montesquieu was moved by the rational; Voltaire by clarity; Rousseau by mystical insights. Thus he became the "man of feeling," the "child of nature," and the prophet of the coming age of romanticism.

In his seminal work, *The Social Contract*, in 1762, he held that the evils of society produced the badness of men, and that an improved society could produce good men. In his view, organized society rested not upon a political contract, but a social contract. The underlying notion was radical: Government, in Rousseau's construction, was secondary. So were kings, officials, and elected representatives, all of whom were mere agents of a sovereign people. This was, he declared, the "General Will." Ironically, he discussed little of the mechanisms of government; nor did he have much admiration for parliamentary institutions. Instead he reached for something deeper, more impressionistic, more mystical: a commonwealth of man in which every person could feel that he or she belonged; "Man is born free," he famously declared, "and everywhere he is in chains." This was not the cloth that the slaveholding Founding Fathers in America would weave, not at first. Nor was it even the primary ingredient at the start of the French Revolution. But this is not to say that his ideas didn't feed into the great dramas about to explode. Indeed Rousseau became the most compelling, systemic theorist of community, nationalism, and, later, democracy. Even Voltaire, who could be unsparing in his criticism, remarked on one of Rousseau's discourses that it made him "feel like going on all fours." No one quite captured the notion of responsible citizenship and intimate participation in public affairs the way Rousseau did.

His work was not a road map for any constitution per se, but its effects were every bit as profound. In France, for instance, in the highest circles, Rousseau inspired a new respect for the common man, a love of common things, and an impulse toward human pity and compassion. Tears were suddenly *en vogue*, and so was a new sense of human "equality." Even Marie Antoinette would be swept away by his ideas, commissioning a little village to be built in the gardens of Versailles, where she played at being an ordinary milkmaid; she also once took the unusual step of visiting Rousseau's grave. And in time Rousseau's work would act like an intellectual virus, contaminating and estranging the French upper classes from their own comfortable way of life.

And so was a crisis for the faithful born. Aristocrats in the king's court began to doubt their own innate preeminence. And this would be Rousseau's direct contribution to the coming storm: the eventual burgeoning of American democracy and the rise of the French Revolution.

Thus the cerebral footings were laid. Voltaire preferred intellectual liberty and enlightened despotism. Montesquieu wanted limited monarchy and a separation of political powers. Rousseau dreamed of an ideal republican commonwealth. How much did these men and their followers accomplish? Were they even conscious of the weight of their own words? Initially, most of the crowned heads of Europe didn't much seem to notice. For a time, neither did many of the elites. But if their trumpets blew an uncertain note, and the walls of Jericho did not fall immediately, this did not mean that these revolutionary ideas were not chipping away at the dynastic structures of the old order or widely penetrating the new. These philosophes unleashed a questioning arguably more profound than when

Martin Luther nailed his theses to the church door. They would capture the ignored voices of millions, and from this point onward, their ideas and their sentiments would gather in strength and volume, forming a mighty wave out in the deep that would come to engulf much of the world. In doing so, they raised doubts, discredited custom, bred skepticism, undermined the comfort and support of tradition, and opened up new vistas of civilization scarcely imagined for a thousand years.

But if their ideas were electrifying, they were still not immediate. And while the seeds of change were soon to fall on ready ground, the first great events would actually take place far away, in a remote, almost insignificant corner of the British empire: the American colonies.

In retrospect, it was one of the greatest events in all of history.

∞

FUELED BY REFORMIST sentiments, troubles mounted in the American colonies, particularly after 1763. Unlike Ireland, America did not lie under the shadow of English guns, but she too was excluded from the spoils system and representation in London. With greater frequency and equally greater fervency, Americans began to argue that it was intolerable for loyal citizens and flourishing cities like Philadelphia and Boston to have no voice in Parliament. The English establishment haughtily told their noisy offspring that neither did Manchester, Birmingham, or Sheffield. And there lay a rough, practical truth: The crown's American subjects could not accrue constitutional rights if they were not also bestowed upon the vast, unrepresented swath in England itself. It was thus ludicrously asserted that a member of Parliament from Old Sarum, who had no actual voters, was equally capable of representing the true interests of the Americans as a man directly selected by the vociferous rabble in Massachusetts. So much for local parliament in America—or representation in England. Or as Blackstone, the towering arbiter of constitutional theory of the eighteenth century, rumbled, "No power on earth can undo" the actions of "this parliament."

But this cut absolutely no ice in America. Nor was the idea of "virtual representation" even wholly accepted by the many Englishmen sympathetic to their American brethren. Sir Robert Walpole himself had once prophetically jibed: "*quiete non muovere*," sleeping dogs would safely lie, if they were well fed first. Just so. One of the most eloquent spokesmen for conciliation was no less than Edmund Burke, among the greatest thinkers of the day.

Even his stentorian voice didn't matter. By now, Britain had boxed itself in; its "perfect constitution" was in fact an illogical shambles. What were the English to do? Give in? That was impossible. Ignore the cries of their subjects? Equally impossible. In the end, they had to choose between stability and empire, and much as they valued both, they would eventually choose stability. For after the bloody Seven Years' War and a century of unrest and experiment, it now seemed clear that even as the British government renewed its interest in governing the colonies, the English establishment was unwilling to police its sprawling empire at all costs. The distinguished historian Edward Gibbon, for one, was "bored" with the whole matter. So the English elite was content to bide its time and remain engrossed in other issues.

Meanwhile, misunderstandings and disagreements with the Americans multiplied. In a tale familiar to colonial subjects throughout time, Americans of all stripes and persuasions began to sense tyranny in every corner and under every bed. They saw a British conspiracy against them, in taxes, in trade, in religion, in politics. Thus the 1765 Stamp Act was a plot against American liberties; so too the 1770 Boston Massacre, which was hardly a massacre at all. So too the "Townshend duties." And so too taxes on tea, and the Quebec Act, and the obliteration of the Massachusetts government. And now Americans, in the coastal towns and in the backwoods, among wealthy land speculators and poor squatty frontiersmen, among even the most loyal, came to fear that their hallowed liberty was in retreat everywhere, that, as one critic trenchantly put it, "the admirable work of ages—the BRITISH CONSTITUTION seems fast tottering into fatal and inevitable ruin."

Exactly what did this mean? As time passed, Americans came to believe that they would have to "keep Britain herself from ruin." To be sure, the liberties they still sought were British in nature, which they saw as one of the finest achievements in human history. And aside from a handful of fanatics or "fools," they did not want independence; far from it. No wonder that George III was equally close to the truth when he maintained that certain Americans were in a conspiracy against *him.* No matter how idealistic or disillusioned Americans were—and they were—it was a small group of single-minded and even ruthless men who were overshadowing the more reticent multitude. It is doubtful that at any time a majority of the colonists actively favored independence; a lion's share of the nation remained neutral; perhaps a quarter was deeply loyal to the crown; and the rest of the

Americans, bewildered and uncertain, were somewhere in between. As one Maryland merchant said bitterly, "[Samuel] Adams with his crew and the haughty Sultans of the South juggled the whole conclave of delegates!"

During the seething decade between the Stamp Act agitation and the Boston Tea Party, many able pens and probing minds set out constitutional solutions for British North America's dilemma. It was the same story—it was too late. So it was here that, in 1775, a proud son of Virginia, Patrick Henry, exhorted rebellion and declared, "give me liberty or give me death." It was here that the expatriate Englishman and brilliant propagandist, the sometime teacher, the twice-dismissed excise officer, the former preacher and onetime grocer, Thomas Paine, in his passionate pamphlet, *Common Sense*, railed magnificently against "the royal brute of Great Britain" and identified the independence of the American colonies with the cause of liberty for all mankind. And it was here that, after the first shot upon the British kingdom had been fired, Thomas Jefferson (one of a committee of five) wrote an electrifying declaration draft, produced in a handful of weeks, which was then adopted almost in its entirety: Its ringing sentence, the preamble to the even more ringing case for revolution, said it all: "We hold these truths to be self-evident, that all men are created equal, that they are endowed by their Creator with certain unalienable rights, that among these are life, liberty, and the pursuit of happiness."

The Declaration of Independence did more than help rouse a fledgling rebellion or serve as a kind of loose national birth certificate. These brilliant constructions would also become the classic statement of America's purpose and enduring idea. And they would ripple outward well beyond America's shores.

And they would do one more thing: They would ignite a revolution and start a civil war.

∞

THE WAR WOULD rip British citizen from British citizen, father from son, family from family. This inconsequential string of settlements in North America would suddenly find itself in the midst of a much larger struggle: another European contest for empire. Suddenly these fanatics, these artful schemers and manipulators, would find their voice.

But even then, outside of a group of influential leaders in Massachusetts and Virginia, most colonists wanted merely to remove the yoke of unjust restraints strangling the genius and capacity of Britain's American subjects. Not a single colony wanted independence; not a single step had been taken by the American Continental Congress that was incompatible with that idea. Even as late as the spring of 1776, they still called themselves the United Colonies and flew the Grand Union flag. Many believed, as Massachusetts governor Thomas Hutchinson cried out, that a war between England and America would be "the most unnatural, most unnecessary war" in history. But, not for the first time, the revolutionaries exceeded the expectations of the day. Once warfare began, the Second Continental Congress proceeded to raise an American army, dispatched a force to Quebec, and entered into overtures with England's sworn enemy, France. And on July 4, 1776—it was actually voted on July 2—the Congress adopted the Declaration of Independence, by which the United States hesitantly and awkwardly assumed a separate, if not yet equal station among the powers of the world.

Few words can capture the boldness of this venture—America not just as one more distant province of the sprawling English empire, but an actual self-governing country. With pluck, daring, iron will, and imagination, not to mention a healthy dose of luck, these upstart Americans would take on the greatest naval power in the world. One by one, the delegates affixed their signatures to the Declaration of Independence. Was this audacity or hubris? Divine vision or madness? The choice was put starkly by the radicals to the moderates: It was now a matter of independence or enslavement. One thing was sure: No one knew who would emerge the hero and who the villain. But the Americans did know this—as the Earl of Manchester once warned Oliver Cromwell as far back as 1644: "If we beat the King nine and ninety nine times, but he is King still. But if the King beats us once, we shall be hanged." This was not mere rhetorical flourish. Just a year earlier, British judges had hanged Irish rebels in the most brutal manner ("You are to [be] hanged by the neck, but not until you are dead; for while you are still living your bodies are to be taken down, your bowels torn out and buried before your faces, your heads then cut off, and your bodies divided into four quarters"). Benjamin Franklin got the message. He was heard to wryly mutter: "Well gentlemen, we must now hang together, or we shall most assuredly hang separately." And as Stephen Hopkins of Rhode Island put it: "My hand trembles, but my heart does not."

From such talk formed the chromosomes of a modern country, the first in the history of the modern world. But it would take much more than simply talk.

∞

THE GREAT POWERS of Europe, including France and Russia, all stepped into the fray. The fate of these revolutionary upstarts would now be decided as much by monarchical diplomacy, secret alliances, dynastic rivalries, and balance of power, as by their own men at arms.

France, having been battered by the British in the Seven Years' War, now saw a chance to score some blows of its own. For two years, Louis XVI, the young new French king, remained ostensibly noninterventionist against his ancient rival. This was mere smoke. He poured munitions into the colonies; nine-tenths of the arms used by the American patriots at the battle of Saratoga alone came from France. And by 1778 the king and his advisors had decided that the insurgents were worth the political risk: Shaking the foundations of Europe, they recognized the Americans as "one people." Nor did they stop there. They then signed an alliance with the rebels and declared war on Great Britain. Spain soon followed, calculating that its overseas empire was more threatened by British supremacy in North America than by the unsettling example of a free American "republic." The Dutch too climbed on board, hoping to recapture the glories of their own aging empire. They propped up the Americans by trading with them secretly through the Dutch West Indies and providing an extensive loan, negotiated by one of America's envoys abroad: John Adams.

Meanwhile, in the months and years ahead, American diplomats shrewdly fanned out across Europe to create alliances and solidify trade relations: Silas Deane, Arthur Lee, and Benjamin Franklin made the perilous journey to France and to Holland; John Jay to Spain; and Francis Dana (with John Adams's son, John Quincy, accompanying him), as far as to Russia. But Britain was hardly standing still. Confronted with the increasing frenzy of "His Majesty's unhappy and deluded people" on the other side of the Atlantic, George III's ministers approached Empress Catherine the Great for Russian assistance. Britain had the best fleet in the world but a negligible army, traditionally resorting to hired mercenaries. By contrast, the Russians had a homogeneous force hardened by war, toughened by the elements, and thoroughly brutal; moreover, they were aided by equally fierce Cossacks, for whom the medieval giving of no quarter was the order of battle. Just a year earlier they had proven their mettle, putting down the massive Pugachev Rebellion, the largest uprising thus far in the century.

King George requested 20,000 disciplined infantry, "completely equipped and ready to embark" as soon as the Baltic navigation was possible in the spring; he also sought to hire Russian naval ships to bolster his own navy. It was a tempting offer, but Catherine refused. (On June 30, 1775, she had already solemnly written to a friend, "I am afraid that I shall live to see America break away from Europe.") Instead, hating any sign of royal weakness, she sermonized her British interlocutors about the indignity of the request because the rebellion had involved "no foreign power." Publicly she wrote to George III, wishing him "good luck," but privately she was far more smug, convinced that George had badly bungled his handling of the rebels and "should be taught a lesson."

Britain was forced to resort to its second choice. The crown approached the German House of Hesse to make use of mercenary Hanoverians—and troops from Majorca and, as it happened, Anhalt-Zerbst as well: Tsarina Catherine's homeland.

But Russia was drawn into the conflict in other, ultimately profound, ways. It is a sign of the quirks of the age that one of the first acts of the Revolutionary War was an ill-considered attack—it was basically piracy— not by the British on the Americans—or vice versa—but by an American "corsair" upon a convoy of eight vessels carrying Russian supplies from Archangel to London. On such dimes can history turn. Since the beginning of the seventeenth century Russia had always furnished the "unknown factor" in the European equation of power; at any time, with dramatic suddenness, Russian armies could be poured in to tip the scales of a crisis. And, as the Americans knew all too well, the fortunes of the colonies were now linked, for the first time, with the maddening fabric of diplomatic maneuver and the ever-shifting European balance of power. However, the initial signs were not encouraging.

In fact, writing to the Danish monarchy, Empress Catherine still flirted with an anti-American alliance. Moreover, she was equally unwilling to acknowledge the existence of the American rabble, fuming about the affronts of these "Sons of Liberty." Even later in the war, when America's envoy to the Court of St. Petersburg, Francis Dana, arrived, he labored in almost total obscurity. Over the course of several arduous years, he would be subject to humiliation, bribery, and endless intrigue; in addition, he would be repeatedly outmaneuvered by his British counterpart. Catherine never once gave an audience to the austere Massachusetts Puritan who had been John Adams's secretary; nor did any member of Catherine's court ever even officially acknowledge or meet with him. Hounded by Russian secret police, Dana was unable even to watch Catherine from a distance or to break through the tangled web of intricate court maneuvering. "I am sick, sick to the heart," Dana jotted down at his inability to make any headway. "It is a pitiful existence." John Adams answered him back ruefully, it *is* "miserable." Dana again responded in desperation: "Do they not see America is Independent?" As far as he could tell, for Catherine, presiding over her vast empire, the remote and troublesome colonies barely "warranted a thought."

But as fate would have it, the tsarina was also supremely fickle. Edgy and ambitious, she distrusted republics and despised insurgents; but even more than that, she craved power on the grand European stage. This would lead to one of history's most curious moves. The consequences would be far-reaching.

Under the guise of protecting "freedom of the seas" and "international law," Catherine brazenly proposed the Neutralité Armée: the "League of Armed Neutrality." The ostensible goal was to halt brash attacks on the high seas by British, Spanish, and American navies and privateers. The reality was quite different. While crises abounded, Catherine had grasped her own opportunity to spearhead an alliance that would include the other great powers in Europe—Sweden, Denmark, Holland, and eventually Prussia, Austria, Portugal, and the Ottoman Turks—against the belligerents, which, in effect, meant England. Notably, the French and Spanish also rallied to the cause. Overnight, Catherine was hailed by the French philosophes for her "Olympian gesture"; she quickly became the "Mediatrix" of Europe, and the heroic "arbiter of two continents." Suddenly the Americans had even higher hopes: John Adams and Francis Dana now lauded Catherine's "idealism" and hailed the empress as "our friend." George Washington referred to her as the "great Potentate of the North." And they were convinced that the great tsarina was "well disposed" toward the American colonies. In Holland, John Adams hurriedly told his compatriot that Russia *must* "send an ambassador to the United States" to acknowledge their sovereignty. "This would," Adams soberly noted, "give peace to all mankind." Dana agreed. St. Petersburg was, Dana fired back, the crucial "diplomatic pivot" on which the American cause depended. Yet he remained rebuffed at every turn.

No recognition was forthcoming, but, by many measures, the empress had already given her gift. In one bold stroke, her Doctrine of Armed Neutrality redressed the balance of global sea power. More than that, the tsarina had isolated Britain diplomatically—the first time this had happened in the eighteenth century—and had curtailed Britain's vaunted maritime fleet while aiding France's. In so doing, she helped bolster the hopes of the beleaguered American rebels fighting for their lives and, in effect, almost inadvertently helped midwife America to independence.

∞

YET THE OBSTACLES for the fledgling American republic still seemed insurmountable. General-in-Chief George Washington repeatedly wrangled over men, over the need for supplies, over money; it nearly ruined him. There were no smashing victories, no great tactical maneuvers to cow the enemy. True, he had made his daring gamble, his hairbreadth victory at Trenton in 1776 ("Victory or Death" was the password for his sentries that night), when he crossed the ice-choked Delaware River during a howling storm. But with a few key exceptions, he lost battle after battle. He suffered the winter agonies of Valley Forge and Morristown. He had disgruntled traitors in his midst, a sizable population loyal to the crown, a Congress repeatedly on the run and scarcely able to govern itself, and a general government that feuded with the state governments. New York was in enemy hands, and so were Savannah and Charleston. There were the disappointments: Spain seized the thinly held territory of West Florida from the British, then sat on its hands. In a deeply humiliating spectacle, Thomas Jefferson, the Virginia governor, fled his own capital. And Washington's strategy of winning by not losing, of defending the nation from conquest by avoiding full-scale battles that imperiled his forces and counterattacking only when such assaults posed little peril, often seemed questionable. At one point, Washington, not daring to risk an all-out assault on the British army, camped at White Plains outside New York and waited for help from the French fleet. The clock ticked. The fleet didn't appear. The weather changed. The men grew haggard and restless. While despair escalated, Washington repeatedly pleaded for French military and naval support, to no avail. And the British occupied Georgia to boot. Without a doubt the task for the rebels was an immense one.

But it was immense for the British, too. In part neutralized and isolated by Catherine's Armed League, in part blunted by the extensive participation of French regiments and the squadrons of the French fleet, the British Empire had its own hands tied. Putting down the rebellion soon became secondary to far greater concerns: Britain's global struggle with France and Spain. And in the American South, its European-trained soldiers had trouble against irregular bands of hard-bitten colonial partisans— guerrillas—men like Thomas Sumpter and Francis Marion, the notorious "Swamp Fox," and Andrew Pickens, who gave as good as they got. Fading in and out of familiar territory, they harassed and mauled the British at historic engagements like Kings Mountain, Cowpens, and then Guilford Courthouse. Almost daily, the British forces felt the agonies and burdens of occupation as they increasingly lost control of the countryside. To many in Westminster, the game no longer seemed worth it. It would later be said by Napoleon, "If you start to take Vienna, then take it." In America, Britain would become unwilling to do so. The impregnable empire proved not to be so impregnable after all; Britain would eventually be persuaded to recognize the independence of the United States.

The crucial battle came along the improbable strip of land known as Yorktown, Virginia, at the mouth of the peninsula between the York and James rivers; here, Washington's patience finally paid off. British forces were hopelessly bottled up by French naval power, and the English found themselves irrevocably outnumbered and outgunned. Washington himself boarded Admiral de Grasse's flagship, the *Ville de Paris*, and literally jumped for joy. Cut off, facing 8,800 Continentals and 7,000 Frenchmen, the British commander gave up. With the bands playing "The World Turned Upside Down," 7,000 British troops marched out of Yorktown and stacked their arms in surrender.

Yorktown was as much a French as an American victory, perhaps, indirectly, even as much a result of the tsarina's League of Armed Neutrality as fierce partisan fighting. No matter. After eight years of war, a provisional peace treaty was signed with Britain in April 1783; by September 3, and after much diplomatic jockeying, the final treaties ending the war were signed in Paris. France, the most powerful monarchy in the world, and America, the rebellious republic, were allies. Even Catherine, far off in the cold winds of St. Petersburg, now watched the events of the rebels with "anxious care." And the United States, shrewdly playing off the mutual fears of the European powers, would go on to stun France and indeed all the continent with its brazen diplomacy—once it was recognized as a sovereign, free, and independent state.

The Declaration of Independence was at last a statement of fact, not a wish.

∞

As THE YEAR 1783 unfolded, from New York to Paris to Philadelphia, from St. Petersburg to Constantinople to Warsaw, the modern world— in bloodshed, in revolution, in a clash of ideologies and cultures and betrayal—was about to emerge. So too was a flowering of democracy, republicanism, nationalism, and totalitarianism. The great empire that had held the North American continent in its spell had been humbled. Another great monarchical empire, France, had redeemed its glitter and grandeur. And a third great power, Russia, had helped to seal their fates. But this moment would not last. Darkness would soon descend upon a devastated, unstable continent; in another troubled pocket of the world, Muslims and Christians would collide in holy war, and a great Russian experiment in liberalism would be irrevocably extinguished; elsewhere, the first flickers of a new liberty, boldly inspired by the Americans, would come under unrelenting assault as well.

When we look back now, the path forward seems obvious. But no one—not the emperors, not the kings, not the prelates or popes or parliamentarians, nor even the new cabinet secretaries or U.S. representatives— could see where this day was leading or could fathom what was lying in wait around the bend. Britain had gambled on war and lost; the Americans had wagered everything and won. The French and Russians had played a dangerous game, outmaneuvering their rivals and assuming that amid the advance of a revolutionary tide, they themselves could remain blissfully immune. They wouldn't.

Yet even as late as 1783, the demise of autocratic monarchies and the rise of the competing ideals of republicanism and democracy seemed imponderable. Hemmed in by predatory empires hoping for it to fail, who could have expected thirteen former colonies to prosper over a vast continent, standing as a kind of utopia for the common man, rather than simply collapse and become the newest ward of a great European power? Or that in France, a wondrous revolution, laden with noble ideas, would lead to the destruction of a throne and the brutal execution of the king of one of Europe's oldest dynasties, and then descend into the Terror and a ghastly war that would envelop the globe? And who could have postulated that in this Age of Enlightenment, absolute, despotic rule could stubbornly triumph over the most basic reformist influences in the Europeanized court of Russia?

And finally, who could have predicted that any of these new, democratic impulses would endure? Before the birth of Christ, the Pythagoreans had tried and failed to create a new type of order and lasting civic life. So did the Stoics, the Manicheans, and the Romans after them. Even the seventeenth-century rationalists and the audacious philosophes of the Enlightenment sought to do much of the same. But their constructs had often fared better in the human mind than in the actual hands of men.

With American independence, however, a great fuse was lit, and everywhere, it seemed, was an awareness of both endings and beginnings. And this ferment would set in motion the rarest of phenomena, a level of seismic change capable of altering the course of civilization. Ultimately, though, what would be destroyed and what would prosper depended on a handful of events and a handful of men and women. It would all be subtly yet powerfully interconnected in one fashion or another. And it began with sad glances and heartbreak, not in the Old World but the New: in New York.

But it also began, perhaps more than anything else, with great hope and even greater promise. It began with the promise of a new age.

PART I

THE PROMISE OF A NEW AGE

President George Washington

CHAPTER I

AMERICA

✦

SOLDIERS MARCHED THAT DAY in Manhattan. For almost as long as anyone could remember, the sight of soldiers had invariably meant the same thing, whether they were French or Russian, Austrian or English, whether they belonged to kings or were battle-hardened mercenaries, whether they moved in great formations or galloped along on horseback. Too often their presence was ominous, signaling that the campaign was beginning and the war was deepening, that the dead would increase and the bloodshed would continue, and the suffering would go on. But today their footsteps were unique, booming out the rites of nationhood. They called out a celebration of victory and the raising of the flag—the American flag. It was November 25, 1783. Evacuation Day in New York City.

As morning broke, the crowds converged and the collective pulse quickened, murmuring with exhilaration. A hundred years later, the city would still remember and celebrate this day. By Manhattan's shores, the last British troops, heads bowed, dour and defeated, were ferried out to transport ships waiting for them in the harbor, then, sails aloft, their gleaming masts disappearing into the distance. For the British there was indescribable sorrow at the loss of their "thirteen beautiful provinces." And there was then, as one man remembered, "a deep stillness." And then pandemonium.

This final corner of occupied territory was now free.

It was precisely one o'clock. The bells of New York, all but silent since the Stamp Act's repeal and languishing for years in storage, now rang, while at the southern tip of the island, the flag, torn down in September 1776, was soon hoisted anew to flutter in the wind. All across the city, young and old alike collected in anticipation, by the corner of Broad and Pearl streets, where a roar of applause would ebb and mount, and over to Bowling Green, where in 1776 the Declaration of Independence was read and patriots had toppled the king's equestrian statue and hacked the gilded crown off his head. Handkerchiefs flapped and gawkers hung out their windows, down past Trinity Church, where desperate Americans had once quietly prayed for deliverance. And now, before a thicket of patriots, scores of battle-tested American troops entered to reclaim the city. Led by General Henry Knox and flanked on one side by a hatless George Washington, mounted upon a brilliant white steed, and by Governor George Clinton on the other, here they came. These were the survivors of Bunker Hill, the heroes who crossed the Delaware, the men who had shivered at Valley Forge, and the victors at Yorktown. They were "ill-clad and weather-beaten," but the people loved them just the same. Marching southward in formation under a velvety sky, the triumphant procession wound past Blue Bells Tavern, where Washington reviewed the pageantry, past half-ruined mansions where errant British flags still flew, and past the moldering earthworks and trenches that dotted the roads, down to the island's edge and the streets to the Battery. Crowds gasped and erupted into shouts of "Hurrah." A thirteen-gun salute exploded into the air, while artists and scribblers converged, ready to record the event for posterity.

At Fort Washington, the password of the day was "PEACE." The eight-year war was over.

The dawn of a new era had begun.

∞

FROM A DISTANCE, one British officer marveled, "The Americans are a curious . . . people; they know how to govern themselves, but nobody else can govern them." Yet the Revolution had been hard on the country. At least 25,000 Americans had died in the conflict—a staggering one percent of the population, a number surpassed only by the ruthless carnage of the Civil War—indeed, one estimate held that as many as 70,000 had perished. And there were the memories. Legions of American soldiers had been held captive aboard British prison ships anchored in the East River, ships that were damp, cold, and reeking from inadequate sanitation. The filth and the lice, the disease and malnutrition, not to mention the gross mistreatment, had carried off an astounding eleven thousand continentals—nearly half of all the deaths in the war itself. And with grim regularity, the bleached skulls and skeletons of the dead would lap up on the shore, bearing silent witness to British atrocities.

In New York, after seven years of British rule and martial law, the city was a shambles, a legacy of the transforming burdens of war. The day's delirium aside, as the sun rose that morning, the vistas were chilling. The city was a patchwork of shanty huts and brick skeletons, remnants of the devastating fire of 1776. The enormity of the reconstruction challenge was overwhelming: In every direction spread weed-choked ruins, rotted-out homes, and vacant lots; and everywhere stood the debris of war. The streets overflowed with trash, squalor, and excrement, and block upon block lay bare and decrepit; New York had even been stripped of its fences and trees—the British troops used them for firewood—while its wharves had been left to rot and sink into the river. No less than Trinity Church was reduced to a blackened hull. Bony cows and pigs scavenged freely, and the people themselves were crammed into a haphazard mass of pitched tents and cramped hovels. Pale-faced and unwashed—disease-ridden too—they existed, in the words of one visitor, "like herrings in a barrel." No wonder New York's future mayor, John Duane, ruefully noted that the city looked as if it "had been inhabited by savages or wild beasts."

And what now? In these early days—or the final ones, it depended upon your perspective of the British crown—the signs were hardly encouraging. For the Tory supporters of the king, the hallowed era of British rule had come to an inglorious end: Powerful businessmen and overseas merchants were without homes; prosperous shipbuilders had been reduced to nothing short of beggars; great politicians appointed by the crown saw their houses rummaged through and their family dynasties abruptly undone. And hordes of English-American children were cast aside by the only world they had ever known. Already, some 60,000 to 80,000 Tories had fled to England or to the safer outposts of Bermuda, the West Indies, and Canada. They knew that for thousands of American "patriots," Tories were little more than hated traitors; they also knew that vengeance, greed, and jingoism made for a lethal cocktail. Sunk in grief, many thus became permanent refugees in foreign lands, clinging vainly to the faint dream of return. Tragically, when the exiles made their way to Britain, more often than not they were viewed as public burdens or social embarrassments, or, in the end, as simply mere bores. "We Americans," one loyalist said gloomily, "are plenty here, and cheap."

For those who remained, the dreaded Armageddon had finally arrived. Gone were the customary sights that had for so long been an integral part of their British lives—the elegant redcoats with their scarlet uniforms and burnished arms who were their defenders, the glory of the king and the glamour of their empire, the clatter of official carriages and the pitched whistles of British naval vessels that were the great empire's protector, and, of course, the long skyline adorned by the Union Jacks fluttering aloft; all had changed, absolutely and inexorably forever.

At the moment of the British exodus, one anxious loyalist said tearfully, "The town now swarms with Americans." And the last loyalists themselves? The wreckage of their lives was soon to be revealed in vivid detail: homes seized and sold at auction; family furniture and precious heirlooms abandoned or outright ransacked; thieves callously picking over their personal effects; and shattered dishes littering the floors of once elegant abodes, everywhere the dishes. Most humiliating were the public notices, formally banning the exiles from ever returning to America—or the laws curtailing their civil and financial rights. And soon would come frightening incidents of revenge: One loyalist, seized by a mob in New London, was strung up by the neck aboard a dockside ship, whipped with a cat-o'-nine-tails, tarred and feathered, and thrown on a boat to New York. In South Carolina, another was hanged by embittered ex-neighbors.

So on that morning the remaining loyalists numbly waited, listening to the haunting sound of American military men marching their way, the thud of enemy feet in the streets, the sharp commands ringing in the air— and the terrible echo of celebratory cannons off in the distance. One New Yorker even observed that the loyalists were now in "a perfect state of madness, drowning, shooting and hanging themselves."

But euphoric Americans took little heed. As the loyalists escaped New York, packing the roads and crowding the wharves, a surge of new residents arrived, doubling the city's population in just two years and quickly turning this restless little seaport into the most vivacious and cosmopolitan society on America's shores.

New Yorkers, indeed all Americans, were already looking ahead.

∞

TWO DAYS AFTER Evacuation Day, George Washington, hugging his artillery commander, gave a tearful farewell to his officers at Fraunces Tavern. "With a heart full of love and gratitude," he told his officers, fighting back his emotions, "I now take leave of you." One of his men who witnessed the scene would recall that he had never seen such a moment "of sorrow and weeping." But more than that, they saw something else quite startling. Washington was sending out a powerful signal: To a man, they were all mere servants of the nation, even as he resisted calls to become a king.

After crossing the Hudson, Washington then rode south through the gathering chill to Annapolis, Maryland, where the Congress was now meeting. Around noon on December 23, 1783, Washington was escorted into the State House, where he met the assembled delegates. He rose and bowed, and with a faint quiver in his hands, proceeded to read his carefully chosen words. "Having now finished the work assigned me . . ." His voice dwindled. He continued: ". . . I retire from the great theatre of action, and bidding an affectionate farewell to this august body . . . I here offer my commission and take my leave." Now his eyes filled. Neither the heartbreaking loss of New York, or the brazen victory at Trenton, nor the winter nightmares of Valley Forge and Morristown, or the decisive liberation at Yorktown, could have prepared him for this moment inside these hushed chambers. The spectators, fighting back their own tears, also grasped the importance of the day, itself replete with symbolism: For once more, Washington was relinquishing his military power, underscoring civilian control in the new republic.

In London, King George III was soberly informed that Washington would resign and turn to private life. His reply is legendary. "If he does that, sir," the king exclaimed, no doubt with a slight tremble to his voice, "he will be the greatest man in the world." From a king who could barely hear the words "United States" uttered in his presence and who would turn his back on Thomas Jefferson, this was a subtle admission packed with historic meaning. American liberty was now not simply a rhetorical chant mouthed to stay the hands of a prevaricating despot or a corrupted parliament, but a reality. And this incipient revolution was, it seemed, not destined solely for Americans, but for peoples the world over, and, at long last, it was coming into full reveal.

In the epicenter of Europe in 1783, France, now the globe's mightiest empire, felt it too.

∞

IT WAS A paradox, to be sure. Even if France's support for the young rebels had far less to do with idealism than with a cynical settling of scores with England, and even if the young country to which the monarchy had helped give birth remained a footnote in its attentions, France's fashionable society felt quite differently. Heroic poems with thirteen stanzas became the rage. So were picnics on the thirteenth of the month, in which thirteen toasts to the Americans were drunk. And so were the hundreds of French nobles who had rushed abroad and risked death so that a young republic might live: the Marquis de Lafayette, who would achieve immortality as George Washington's protégé and nearly lose his life at the battle of Brandywine; Admiral d'Estaing, who would take Newport and almost die in the struggles to take Savannah; and Admiral Rochambeau, who would eschew the lavish comfort of the French court for one last glorious crusade to fight side by side with the Americans.

Indeed, nowhere was the French ardor for American liberty more apparent than in their national veneration of Benjamin Franklin. More than anyone else, for the French, he seemed to embody this young republic with all its unspoiled virtues: America as a place of natural innocence, America as a place of patriotic virtue, America as a land of freedom. Thus, Franklin's likeness was everywhere—on painted porcelain and printed cottons, on snuffboxes and inkwells, not to mention prints and dolls and engraved glass and over mantelpieces. It was even displayed in Versailles itself.

This was a time when, as the Comte de Ségur, another French nobleman who had fought with Washington, would put it: "The American cause seemed our own; we were proud of their victories, we cried at their defeats, we tore down bulletins and read them in all our houses. None of us reflected on the danger that the New World could give to the old."

∞

ACTUALLY, ONE PERSON did. Miles away off to the east, in the vast Russian Empire, Catherine the Great, the tsarina and ruler of the world's largest nation, was much less impressed. And amid the deepening winter snows of St. Petersburg, America's envoy, a baggy-eyed Francis Dana, was holed up for months in L'Hôtel de Paris, in "mortifying seclusion." Dana was the very antithesis of flamboyant—he had the soul of Puritan propriety, with a reputation for judicial temperament and an unfailing devotion to his country. Under direction from the U.S. Congress, he was seeking to establish the "mutual interests" (his words) that the two nations had in "intimate connection" with each other. But where John Adams, Dana's friend and mentor, dined luxuriously with the ambassadors of France and Spain, and Franklin was a household name in Parisian society, in St. Petersburg, Dana was frozen out.

The American Congress, "fired with prospects of future glory," insisted that Dana be properly received—the Americans were ambitious upstarts even then. But Dana came to realize that he was less informed about Catherine the Great's comings and goings, let alone her policies, than every groom and lackey in the palace. "I have grown very tired," he sighed, "of being a limb of that sovereign." A limb yes, yet he was far from ignorant of the complicated machinery of the Russian court. Catherine, he gravely realized, was still reluctant to admit the United States into the councils of Europe, with serious ramifications for American trade and security. To the empress, there was no place in the established chanceries of the Old World and their dynastic alliances for the "pretensions" of America. So even as Dana's health declined—the horrid winters were too much for him—he continued to plug away. He should have. Catherine was the great spoiler in Europe, and she maintained enormous influence. Backed by the towering philosophes of Europe—Voltaire, Diderot, Grimm—she was also seen as treading the clouds of a "world Utopia." And as Dana asked, weren't these now America's goals and interests, too?

It didn't matter. By the time peace was secured between America and Britain in 1783, Catherine, the Autocrat and Empress of All the Russias, had grown tired, bored, restless with the whole American matter—to the extent that she thought about it at all. She had already thwarted Britain and its imperial goals, hoodwinked the Hapsburgs, aroused the supportive passions of the great liberal minds of the day, and kept all of Europe off balance. Now she turned her attention away from the wider continental spoils back to her own borders—to the ancient kingdom in Poland, emboldened by the fresh success of the American rebels—and to the Islamic sultanate in Turkey.

And back in the United States, the Founders were drawing their own lessons from Dana's miserable experience. Alexander Hamilton and James Madison, viewing Europe as "kings balancing straws on their noses," both agreed that America's "true interest" required that they be entangled "as little as possible" with the Old World. For his part, George Washington looked out on the world stage and saw worrisome portents of things to come. If the disparate thirteen states didn't form a single, genuine union, he direly warned their governors, Americans would yet become "the sport of European politics."

But intoxicated by the glow of victory, nobody was listening; Washington's admonitions fell on deaf ears. He had watched American liberty survive the challenges of war. Now could it, Washington wondered, meet the demands of peace? Or would it succumb to a hostile world of kings, monarchs, and emperors, greedily waiting for it to fail?

∞

AT FIRST, THE signs were promising. A slow process of healing began between the patriots and Tories, and a number of loyalists who had not actively resisted the Revolution were able to return, largely unmolested. Moreover, the United States had gained a vast new domain beyond the Appalachians, and America became the undisputed mistress of an immense, rich, and sparsely settled territory. New roads radiated west and south. Studding the landscape with hand-hewn homes and traveling by carriage or horse and buggy, Americans were increasingly everywhere over the vastness of its spaces: from the distant Pennsylvania Alleghenies to the mouth of the Mississippi, from the loamy banks of Nova Scotia to the arid plantations in the Carolinas. The country tentatively began an incipient national art; a thriving, rancorous press; a devotion to education (including for women) as well as the theater; and even new national idioms: Rejecting the sharply drawn class system of the European monarchies, or the prerogative that had once only belonged to the gentry, "Mr." and "Mrs." came into vogue. From out of nowhere, newspapers sprang up, and so did almanacs, tracts, chapbooks, and periodicals. Great colleges were established. And too, major cities were emerging and expanding, in Boston and New York, Philadelphia and Annapolis, Trenton and Williamsburg, Albany and Lancaster, Raleigh and Columbia, and beyond. The country became a magnet for the hopes of immigrants and an inspiration for those who cherished freedom worldwide, from England and France, Poland and Ireland, and even among reformers in Russia. Increasingly sophisticated and even wealthy, Americans were filled with an extraordinary sense of optimism and a glowing sense of destiny. As the country experienced astonishing postwar growth and a breathtaking vibrancy, there was much to celebrate.

But it was fleeting: Even as the country was riding this crest of appro-
bation, powerful forces were tugging at it from within.

It was all deception.

∞

THE FRAGILITY OF America as a Union from its very first days cannot be
exaggerated. Unlike the Old World, America was not born out of ancient
custom or claim, its people stitched together from the shadows of feudal,
marauding bands, emerging as a nation by the time they could primitively
write their own history. Where in most countries a sense of nationhood
spontaneously arose over centuries, the product of generations of common
kinship, common language, common myths, and a shared history, America
was born as an artificial series of states, woven together with the string
of precariously negotiated compacts and agreements, charters and cove-
nants. The country did not arise naturally, as in Europe, or Persia, or China,
but was made, almost abstractly, out of the guns of a Revolutionary War,
and ink and paper, crafted by lawyers and statesmen. Significantly, even
in 1776, the birth certificate to become Americans—the Declaration of
Independence—did not make it a nation. In fact, the very word "nation"
was explicitly dropped from the draft, and all references were instead to the
separate states. Thus, the very heading of the final version of the Declaration of Independence described the document as "The unanimous Declaration of the thirteen united States of America," and the momentous
resolution introduced in the Continental Congress on June 7, 1776, by
Richard Henry Lee and seconded by John Adams, declared: "That these
United Colonies are, and of right ought to be, free and independent States."
As the historian Daniel Boorstin noted, "Independence had not created
one nation but thirteen."

There was more than a measure of truth to this. Like the colonies that
preceded them, these new states were as dramatically different from one
another as they were from England. Each jealously guarded its own independence, its self-rule, and its sovereignty. Each meticulously had gathered
its own army, chartered its own navy, commanded military actions to protect its own interests, and oversaw its own Indian affairs and postal routes.
Each had its own legislatures, its own functioning courts, its own taxes,
and, in time, its own individual constitutions. And too often forgotten is

this simple but overweening fact: Before independence, Americans were both British subjects as well as citizens of Massachusetts or Virginia, New York or Connecticut. After independence, they were no longer Britons, but neither were they Americans—as yet, they had no American country to which to attach their loyalties. And so they remained faithful, proud members of their sovereign states, Massachusetts or Virginia or Rhode Island or New York, all stretching back nearly 175 years, and older in fact than the collective American nation was at the time of the Civil War. Indeed, to the extent that there was an American national identity, it was unexpected, impromptu, an artificial creation of the Revolution—and secondary.

It is revealing that Thomas Jefferson, for one, referred not to America as his "country" but to Virginia; that he described Virginians as "his countrymen" and spoke of the federated Congress as a "foreign legislature"; and that he soberly warned that "a single consolidated government would become the most corrupt government on earth." Patrick Henry agreed. So did George Mason and John Hancock. Nor were they alone in these sentiments. It was precisely these sorts of ruminations, then, that George Washington, back in Mount Vernon, absorbed with dismay.

It was justified. In truth, there was no greater testament to the feeble unity of the country than America's governing body: the Congress. After pathetically wandering from Princeton to Annapolis, from Trenton and then to New York, where it finally settled, the national legislature could scarcely muster a quorum. It remained a weak and wayward instrument, whose members either didn't appear, often had no work when they did appear, or weren't listened to when they actually sought to enact policy. At every step, its actions were marked by temporizing, indifference, hesitation. And governing was impossible: They couldn't tax, they couldn't raise an army to repel invaders—Congress had been forced to sell off its last warship—and they couldn't suppress internal insurrections. And under the Articles of Confederation, all thirteen states had to agree to any amendment to the federal government's powers. By any standard, it was woefully impotent.

Abroad, for all the world to see, the weakness of the federal government was impossible to disguise. There were consequences. England took advantage, contemptuously refusing to withdraw its troops from its forts on American soil in the West, as had been promised in the 1783 treaty, and barring American ships from moving freely about the West Indies, stifling much-needed American trade. If that weren't enough, separatist movements were plotted with the Indians in the borderlands of the Northwest and Virginia. Spain took advantage too, taunting the new nation by prohibiting Western settlers from shipping farm produce from the port of New Orleans; they also supported renegade separatist movements, this time in the Southwest. Even Arab Barbary pirates made a veritable sport of preying on American ships, selling off captured American sailors in the harsh Muslim slavery bazaars of North Africa, and destroying the young country's Mediterranean markets.

Month by month, the American confederacy was increasingly shaky, and month by month, with ever-growing impunity, the thirteen states acted like thirteen independent countries, and squabbling, ill-tempered ones at that. New York laid onerous import duties on simple rowboats crossing with produce from New Jersey; it taxed lumber from Connecticut too. Pennsylvania followed suit (indeed, Pennsylvania and Connecticut literally waged a twenty-year war over land). So did Massachusetts, which was selling goods with inflated prices to Connecticut and New Hampshire. Rhode Island tried to stick out-of-state creditors with its debts, as did Maryland. And the inhabitants of Kentucky and the newly formed state of Franklin were threatening to arm 10,000 men to settle the question of navigation on the Mississippi.

Soon, between the swelling debt, shrinking money supply, and dwindling trade, the flush of prosperity was snuffed out. Seamstresses, shoemakers, and other craftsmen and artisans were suddenly without work; the shipbuilding industry collapsed too. As would quickly follow in France, there were now dreadful visions of the poor rising up against the well-to-do. The situation became dire: There was no common trade policy, no real foreign policy, barely any domestic policy. And there was the debt, accumulated during the war. The young nation had borrowed millions from France and Holland to finance the Revolution, which it had scant hope of being able to repay. No wonder James Madison luridly warned about this "flagrant" and "present anarchy." Or that John Sullivan referred to the confederacy as "a Monster with thirteen heads!" In fact, the future president of the Congress, Nathaniel Gorham, openly worried that the clashes between New York and its neighbors would erupt into civil war.

And in 1786, with the formal peace with Britain a mere three years old, New England delegates were suddenly talking about disunion and separate confederacies—James Monroe warned Thomas Jefferson that the "northern provinces" were planning to form "their own nation," while

Northerners would soon fear that the South would "form alliances in Europe," and then "all will be irrevocably lost."

The question now lingered: Could this fragile confederacy withstand the accumulated strains between haves and have-nots, between states' rights and federal powers, between the country's divided nature and divided philosophy? One of the original signers of the Declaration of Independence, Gouverneur Morris, put it this way: "The fate of America was suspended by a hair."

More than ever, this was the case when in the fall of 1786 delegates from all the states were invited to a meeting in Annapolis to discuss the crumbling Articles of Confederation and the mounting chaos spreading across the land—and only a paltry twelve delegates from just five states showed up.

Then came the thunderclap that rattled across the frail Confederation: Shays's Rebellion.

∞

WHILE THE ANNAPOLIS conference foundered, rural turmoil was flickering in western Massachusetts.

It happened like this. Thousands of farmers, wrestling with the gritty particulars of their common lives, were unable to pay their debts. When they weren't imprisoned, and ever-greater numbers were, many had lost their farms and were now asking themselves if this was the liberty for which they had fought. And they did more than just ask. Taking their cue straight from the Revolutionary War, they resolved they would prevail against the state's onerous new taxation scheme much as they had confronted another far-off body, the British Parliament.

So in the tense, early days of 1786, disgruntled farmers grabbed pitchforks and guns, donned their old Continental Army uniforms, and started assembling companies of men—and they drilled. But this was not some rogue, backwoods undertaking. Actually, the chief architect of the rebellion was a much respected officer who had served bravely on the front lines of the Revolutionary War, Daniel Shays. To his critics, Shays was an impatient hothead with a short fuse. But he was also a former captain in the Continental Army, had fought at Bunker Hill, marched at Saratoga, and put his life on the line at Stony Point. With his firm-set lip and charismatic bearing—he soon became a local folk hero—he was, by all accounts, a man displaying a passionate zeal for reform. Living day-to-day in constant fear of losing his small Massachusetts farm, he was also prepared to shake up the edifice of the country. Moreover, he was in good company, aided by like-minded men such as former Continental major Luke Day.

Their voices edged with frustration and anger, the Shays protest movement quickly reached into other Massachusetts counties, and found ardent sympathizers in neighboring New Hampshire, Connecticut, and Rhode Island as well. Then the violence began. When judges arrived at their county courts, furious protesters greeted them with clenched fists and violent insults; ashen-faced, the judges swiftly fled. The Massachusetts governor, James Bowdin, reacted swiftly, prohibiting "unlawful assemblies" and summoning the militia to scatter the mobs. In doing so, he was supported by Samuel Adams, cousin of John and one of the nation's most admired revolutionaries. But if such company was reassuring, it was also, at the time, calamitous. Suddenly the Shaysites were even more determined to fight back: Wearing sprigs of hemlocks in their hats, thus evoking the spirit of 1776, they threatened to overthrow the Boston government. Just as suddenly the image of a Revolutionary War redux was conjured up, now reenacted as an American civil war. The horror—and surprise—was not lost on Massachusetts's desperate governor. He suspended habeas corpus and appealed to the national Congress for help.

It was a futile gesture. No assistance was coming. Bankrupt, the Congress instead asked the states to come up with $530,000 for an army. Twelve of the thirteen "united states" rejected the request. Virginia alone provided aid, and it was hardly enough. Beyond that, Massachusetts was unable to muster a single soldier. In dismay, a Virginia delegate begged George Washington to ride for Massachusetts to use his immense influence to dispel the rebellion. Washington testily fired back, "Influence is no government."

Meanwhile, the extremism grew, and converts in the unlikeliest of places were found. Ominously, many members of the militia broke ranks and also joined the insurgency. Then Shays's Rebellion began to spread. From New Jersey to South Carolina, uprisings swept the countryside. In York, Pennsylvania, an agitated mob refused to allow the sheriff to auction off cattle confiscated for taxes. Maryland's Charles County courthouse was forcibly shut by angry rabble. In South Carolina, judges fled the Camden courthouse under a cloud of destruction and plunder. In Virginia, a fulminating mob torched the King William County courthouse, terrifying local residents and burning all the tax records. With each action, the Shaysites grew progressively more brazen and angrier at the power and privilege arrayed against them: unresponsive judges, unelected tax collectors, and an uncaring government, joining hands to wrest from them the few things they cherished most dearly—their farms and their livelihoods. If this could happen, then what next in this new country? Tyranny? Monarchy? Indignant and emboldened, the Shaysites, in a move hauntingly reminiscent of what would soon happen in the historic storming of France's Bastille, decided to strike. Their target: the federal arsenal at Springfield, which held 15,000 muskets and a number of cannons.

The indispensable step to waging a full-blown insurrection.

∞

SLOWLY AT FIRST, then quickly, the clamor grew. This was rapidly becoming a crisis of unconscionable proportions. Time was now running out, even as the Congress sulked and complained, but was unable to act. And for George Washington, the unspeakable seemed to be playing out before his very eyes: the undoing of the republic. He felt not only a sense of desolation but a sickening sense of déjà vu. During the course of the Revolutionary War, he had hammered together an army in his own image, waged titanic battles against the world's mightiest empire, beaten back despair and desertion and a mutinous rebellion, and held a fledgling nation together. And now this. He was despondent: "There are combustibles in every state," he wrote frantically to James Madison. To another friend he pointedly asked: "Who besides a Tory could have foreseen or a Briton predicted such a situation?"

Who indeed. Massachusetts now scrambled for answers. Forced to solicit money from wealthy private citizens, the governor, taking no chances, raised a volunteer militia of 4,400 men, with former general Benjamin Lincoln in command: As it was, Lincoln was an inspired choice—a distinguished soldier in the Revolutionary War, he had been charged with receiving Lord Cornwallis's sword at Yorktown; later, he had served as secretary of war. Now Lincoln wasted no time. The two groups of men—the ex-revolutionaries and their rebellious offspring—were quickly locked in a sprint toward the Springfield arsenal. Yet on January 25, 1787, the militia unexpectedly came into a bit of desperately needed luck. Supported by cannon salvos and 1,000 troops at the arsenal, they managed to hold it; the Shaysites were checked, momentarily at least. The rebels fell back. Forging ahead through the snow, Lincoln's militia raced to catch up to them; they did. A number of rebels were forced to surrender.

It became a sullen test of wills. Defiantly retreating, Shays looked like anything but a man ready to quit. The worry now was that he would continue to slip through the militia's hands, or that the fighting would mushroom into a cauldron of slaughter. But the militia, led by Lincoln, refused to stand idly by. Early that February, in the midst of the day's slow-growing chaos, Lincoln, ripping a page from Washington's bold Trenton campaign during the war, marched his forces for thirteen straight hours under the evening's shroud through a bitter cold and surprised the rebel camp, forcing additional surrenders. But once again, Shays nimbly escaped, fleeing to neighboring Vermont. While the disorder was largely quashed, the campaign dangerously lingered: a bloody skirmish with state militia near Sheffield, Massachusetts, left one hundred men dead or wounded—more than the American casualties at the final battle for Yorktown. Meanwhile, embattled pockets of Shaysites, as their disgruntled brethren would in France, continued a guerrilla war through the uncertain spring of 1787; leaving a vast corridor of fear and destruction, they burned factories, kidnapped judges, seized merchants, and ransacked stores. More than that, Shays's many sympathizers were actually victorious at the election polls in early 1787, prompting one newspaper to moan that now "sedition itself [can] make laws." And when General Lincoln sought permission to pursue the rebels anywhere in the United States, he failed; a number of states flatly wouldn't allow it.

As the swirl of extremism continued, and the country watched the grim spectacle of heightening civil unrest, even war, Washington pleaded: "Let us look to our national character and to things beyond the present moment." Heeding his words, the handful of men gathered at Annapolis issued a call for another convention: for all the states to meet in Philadelphia in May 1787. The goal of the new meeting would be sharply limited in nature; with a mixture of hope and awkward expectation, they would gather for the "sole and express purpose" of revising the Articles of Confederation, the document that laid out the terms of the increasingly fractious American union.

But fate would intervene and this convention would become one of the most extraordinary moments in the history of human governance. Once and for all, the Americans would demonstrate that they were no longer comfortable in the claustrophobic world of their fathers.

∞

ON MAY 17, 1787, the first week of the convention brought a steady patter of rain and didn't muster a quorum. Only seven states were initially represented, and the meeting would not officially convene for another week. But Alexander Hamilton summed up the thoughts of many of the delegates, tantalizingly observing that the convention would "decide forever the fate of republican government." Overstatement? Perhaps. But what was clear from the outset was that the delegates would ask—and answer—whether the United States was susceptible to "one government," or whether the individual states would have "a separate existence connected only by leagues." And what was equally clear was Benjamin Franklin's warning: Of the convention, he deadpanned, "If it does not do good, it must do harm." That was certainly true.

Meeting in the second floor of the gray East Room in the Redbrick State House, where the Continental Congress had signed the Declaration of Independence, the delegates unanimously elected George Washington as president of the convention. His commanding presence—he said little, he didn't have to—was a powerful reminder to everyone that they had gathered to create a more effective, if not stronger, federal government. Of the fifty-five men who eventually attended, representing twelve states—Rhode Island glibly boycotted the convention—forty-two had served in the Continental Congress, thirty in Washington's army. The majority were lawyers, prosperous, educated, and, of course, white; surprisingly, nearly a dozen of them had been born or educated abroad. Back in Paris, when then-ambassador Thomas Jefferson read a list of their names, he rhapsodized that they were "an assembly of demi-gods." Of these, the most dramatic presence was an enfeebled Franklin. Outside of Washington, Franklin was arguably the most famous American in the world. At eighty-one years old, his body was wracked by gout and tortured by agonizing kidney stones, and he was so weak that he was often reduced to having others read his prepared statements. Taciturn yet charismatic, squinting through his trademark bifocals, he was nonetheless an unfailing presence at the convention: He boasted that it was "the most august and respectable assembly" he ever attended. And he showed up for every session.

The convention began on an audacious note—audacious because they had come there as republicans, dedicated to representative government. But to encourage uninhibited debate, it was decided that all the meetings would be held in the strictest of secrecy. Journalists were forbidden to attend, and so were spectators. At all times, guards remained posted by the bolted doors, blinds were invariably drawn and windows locked to ensure confidentiality, and delegates, sworn to privacy, remained tight-lipped to the outside world. Though the convention had a secretary, Major William Jackson, none of the proceedings of the "Committee of the Whole" would be formally transcribed or printed; in fact, James Madison's own voluminous notes would not appear until many decades later. It was all very daring, it was gutsy, and it worked. And far from simply tinkering with the Articles of Confederation, the delegates would—in effect—become an extralegal body, overhauling the articles altogether. It would be one of the boldest gambits in history. Asked how the convention could defend its unrepublican actions, George Washington slyly quipped: "The event is in the hand of God."

God would be needed. From the outset, the convention reflected a pointed truce in the unending tug of war between those who believed in the necessity of a functioning union and those who feared the encroaching despotism of a national government and clung to states' rights; between fervent slaveholders and ardent abolitionists; between those who worried that small states would be smothered by the political juggernaut of the big states and those who worried about the tyranny of the minority; and between radicals who wanted a clear repudiation of the past and traditionalists who wanted only to tinker at the margins. The reality was that after more than five generations of laying down roots in America, the exigencies of life had only sharpened the old insistence upon demands for personal liberty and equality. Now the Americans found themselves arguing both for the historic and chartered rights that they had once enjoyed as Englishmen, and for the timeless and universal rights of man.

It was complicated, it was messy, and somehow everything came together. Politicians all, the delegates compromised repeatedly. But not at first. In its plan, introduced early on, Virginia called for a bicameral legislature, proportional representation, and a seven-year, one-person executive—measures favored by large states; New Jersey countered with a plan calling for one chamber and an executive council that could be removed by the states—a direction favored by the smaller states. They wrangled back and forth, with little end in sight. By July 1 they were hopelessly deadlocked. More than a few caught a whiff of disaster. The ever-observant Franklin warned, "Our detractors believed that we are here only for the purpose of cutting one another's throats."

He might have noted that many of the collaborators apparently wanted to do the same. Even the most enterprising minds at the convention dramatically differed: Madison wanted a federal government to have veto power over state laws, while Franklin wanted a unicameral legislature and an executive council instead of a president, fearing it would otherwise end in "monarchy." The cantankerous Elbridge Gerry wanted a "three man presidency," with each executive representing a separate section of America, while Alexander Hamilton thought the president should serve for life. Such was the tenor of the day. We are "on the verge of dissolution," Luther Martin reported back home to Maryland, "scarcely held together by the strength of a hair." Indeed, it was a pensive moment, but the strand held. On July 16, weathering the heat, the incessant buzzing of packs of flies, and their differences, they reached a grand bargain—the so-called Connecticut Compromise: There would be equal representation in the Senate and proportional representation in the House of Representatives.

But there were still more troubles ahead. While Hamilton, one of the most spirited voices for a strong federal government, stayed, two of New York's delegates quit the convention in a huff and dashed home, warning that the "large state men" were constructing a tyranny that rivaled King George III's. This was one knot the delegates weren't sure how to untie. Indeed, the convention's thoughts about the presidency were a chaotic jumble. Were they creating a monarch? (Maybe.) Or a despot? (As they feared.) In their defense, it was one of their most vexing tasks—to find a substitute for a king. No less than John Adams considered hereditary rule foreordained and declared, "Our ship must ultimately land on that shore."

In the colonies, though governors represented the majesty of the crown, it had long been felt that the chief executive should be the dutiful servant to the legislature; moreover, it was equally agreed that the English system—before it went awry—of a parliament representing the people remained a persuasive model. And who didn't remember the atrocities of George III? Or flinch at the excesses of an enlightened despot like Catherine the Great or the militarism of a Louis XIV? On the other hand, many of the delegates thought that in a world of predatory monarchs and expansionist kingdoms, this was a perilous habit that had to be broken—otherwise the country could fall prey to powerful and intriguing European powers. America thus could not afford to be the subject of inertia or unending deadlock;

it needed energy. With his pink cheeks and sandy hair, Hamilton put it best: "It is a miracle that we [are] now here exercising our tranquil and free deliberations," he maintained. "It would be madness to trust to future miracles." In the end, the delegates seemed to agree. Along these lines, the convention flipped its fear on its head and found an argument made by Gouverneur Morris of Pennsylvania to be the most coherent vision. Neither mincing words nor wrapping himself in vague generalities, Morris averred that the president should serve as the nation's voice, its protector *against* "legislative tyranny." With George Washington in the back of the minds of virtually all the delegates, the convention bought it. The office of president was added, and he would be able to run for relatively long terms—four years each—to boot.

Then, on September 8, with the heat having at last abated, an equally dramatic change was made to the Constitution, another flourish added by Morris. The previous draft, enumerating, "We the people of New Hampshire, Massachusetts . . ." was scrapped. Now Morris put the matter squarely before the delegates, opening with the ringing phrase: "We the people of the United States." With this one critical change, Morris produced an ingenious summary of the goals of many at the convention, if not for much of the entire American Revolution: the United States conceived not merely as a league of states, but as a union in which individuals were citizens of the United States for some purposes, and of their particular states for others. In other words, people, every bit as much as states, composed the federal republic, and the laws of the United States encompassed not merely the sovereign states, but their citizens, the people themselves. This was, in Gordon Wood's words, "one of the most creative moments in the history of political thought."

At long last, the hour had struck. The debates had been intense, petty factionalisms and rivalries had often intruded, but on September 17, 1787, after four months of bruising battles, the convention came to a close. It had been a fast-moving game with plenty of bluffing, but when it was all said and done, there had been no ideological rift, no clear left, no clear right, no clear center; indeed, it was extremely difficult to settle who was a conservative and who was a progressive at the convention; everyone seemed to be both. Together they represented the variousness of the fledgling country. Except for three notable exceptions—the governor of Virginia, Edmund Randolph; Elbridge Gerry of Massachusetts (who grumbled about the idea of asking the "people" to ratify the Constitution); and the formidable

George Mason—one by one the delegates, representing twelve states, rose to add their signatures to the document.

At the end, Benjamin Franklin, deeply touched, took a moment to steady his voice, then theatrically pointed to the painting of the sun on the back of convention president George Washington's high-backed wooden chair. As a swath of milky light filtered in, he gave the chair a reverential gaze and pronounced this telling epitaph on the day's events: "I have often looked at that sun without being able to tell whether it was rising or setting, but now I have the happiness to know it is a rising and not a setting sun."

∞

Nationhood was now beckoning.

How to capture the audacity of it all? It is almost impossible. In the span of sixteen weeks, the Founders had devised a country not forged by thousands of years of shared history and shared dreams, but conceived in the minds of a handful of men over a handful of months. With pluck, vision, imagination, and daring, they had begotten a national idea, a country unlike any since the beginning of time. To be sure, they didn't do this out of whole cloth: The convenient version of history—and more than a bit chauvinistic—ignores the fact that there was historical precedent. James Madison, this small, bookish, balding man, who arguably had the greatest hand in creating the Constitution, had already been sent a trove of books— some two hundred of them—from Paris by Jefferson on government and philosophy, politics and history, to read as the Founders undertook to craft their new document. And for years, among elites and educated classes, the notions of constitutionalism, federalism, and limited government had been endlessly and excitedly discussed. They had been raised as far back as the Middle Ages, and, more recently, had been set forth in such disparate venues as Hungary, the Holy Roman Empire, and even the Parlement of Paris. But in their prevailing form (and even in Montesquieu's philosophy), they were invariably linked with feudalism and aristocracy. Now, however, the Constitution and the Founders had redefined European thought and, in so doing, had made these notions progressive. As Alexander Hamilton eloquently put it, all peoples regardless of their own history could apply it to themselves, because "the sacred rights of man are not to be rummaged for among old parchments or musty records. They are written, as with a sunbeam, in the whole volume of human nature, by the hand of Divinity itself, and can never be erased."

But was that true? On the crucible of the occasion, fateful questions still lingered: What would knit this country together? Was it to be one country and one nation—as most of the Founders hoped—or would it split into several nations, as many Americans thought and even a number of the most ardent nationalists feared? And was it to last in perpetuity, or would this brave and bold experiment in republicanism founder?

At the core of the problem, of course, was the remarkable yet perplexing nature of America. For all its genius—and genius is the right word—the U.S. Constitution was an uncertain guide. Until this point, constitutions were not national codes, but national inheritances. They were not written down, but existed almost intuitively, the ethereal sum of a whole country's charters, statutes, habits, traditions, informal understandings, and declarations. Yet for the Founders, what had started out as an exercise to do little more than revise the Articles of Confederation—a loose system designed for the exigencies of the Revolutionary War—had instead produced a far more audacious gamble, an entirely new body of laws. Nowhere on the planet had anything quite like it been devised, and actually set down on paper: a central government with authority to tax and maintain an army, dividing its power with sovereign state governments, which of course was precisely one of the very things over which the British Empire and America had split. And at the same time, it created a republican system with its powers scrupulously divided among a president, a House of Representatives, a Senate, and a Supreme Court. But—and it is a big but—when it came to articulating America as one nation, the men in Philadelphia flinched.

The word "nation" or "national" appeared nowhere in the Constitution. Unable to reconcile the gnawing tensions between the proponents of the states, later called the anti-Federalists, and the proponents of the new federal authority that would come into being, the Federalists, the Founders resorted to that more ambiguous phrase, which we know today as "the United States." (Even the use of the word "federal," or "foederal," as it was more often written, was actually meant to describe a relationship resting largely in good faith, *foedus* in Latin being the cognate of *fides*, faith.) When it was all done, an elated Washington recognized not just the historic import but also the precariousness of the whole enterprise; it was, he maintained, "little short of a miracle" that delegates from "so many States" should have united to form a national government.

A national government, yes, but did it form a nation? Or a sustainable Union? That still fell a little short of anyone's miracle. Indeed, in the architecture of nationhood, the United States had achieved something quite remarkable: As one historian has put it, "Americans erected their constitutional roof before they put up national walls." The result? With characteristic anachronism, and the benefit of this ingenious contrivance, Americans had a country before they truly had their nation. It was deft rhetoric, but could it hold? Thus, the question lingered: In the ensuing years, would this collection of independent but united states be able to escape the price for such genius—a price to be paid in hate and blood and more revolution?

That, as yet, was unknowable.

∞

NOR DID THE studied ambiguities of the Constitution deal with another fundamental issue that Americans would increasingly wrestle with—the fact that there was no clause in the Constitution that established the United States' perpetuity. Whereas the Articles of Confederation contended the Union "shall be perpetual," this provision, even in draft form (Charles Pinckney's handiwork), was never even brought before the general body for consideration. And where the Constitution was stunningly framed in the name of "We the People," Article VII declared that it would be ratified "between the states." To the extent there was a discussion about the perpetuity of the Union, there was no consensus. No less than James Madison would write: "each state . . . is considered a sovereign body independent of all others, and only to be bound by its own voluntary act. In this relation then the new Constitution will . . . be a *federal* and not a *national* constitution."

What did Americans make of all this? Praying for the best, the Founding Fathers left the question of perpetuity to posterity, and the most common perception of the Union—as opposed to a loose league or confederation—was best articulated by Washington: America was "worth a fair and full experiment." But here, the examples from the rest of the world were not encouraging. Throughout history, with bewildering rapidity, republics came and went, ever ready to disintegrate into petty, squabbling minor or fragmented entities, or chronically vulnerable to anarchy or violence or their more autocratic and ravenous neighbors. Seen through their eyes, the hard facts were that republics were not just precarious and contingent, but transitory, withering through dissension, failing through conquest, or dying through despotism. Across history's long checkered span, for republics, then, the odds were not good, and, in fact, they never had been.

The Greek example, in Hamilton's immortal words, was "disgust[ing]." After a reign of splendor, the Roman republic fell ignominiously. There were a few republics in Europe, but none of them was large—and none had acquitted itself particularly well either. Not the United Netherlands, which collapsed that same year, in 1787; not the Polish diet, which would soon cease to exist altogether; not the incoherent jumble of Swiss or Italian city-states; not Santo Domingo, which would disintegrate in another few years hence; not Sweden, which dwindled from a constitutional monarchy to an autocratic state; and not the Latin American wars for independence from the Spanish Empire, which from a few vice royalties would produce not one or two, but some twenty-two separate nations.

Tradition was also against the Founders. They were boldly obliged to repudiate the political axiom that had behind it the domineering authority of Montesquieu. He had already pronounced that a republic could function only across a small territory, and the thirteen original states together—even some of them individually—were already considered too large. Patrick Henry was hardly alone when, in 1788, he warned, "Our government cannot reign over so extensive a country as this, without absolute despotism." Geography was also against them. As Americans trekked farther south and traveled farther west, the wilderness and the rivers and the mountains created breathtaking differences: differences in lifestyles, differences in culture, differences in economics, and differences in political outlook. Far from changing after the Constitution, this dizzying pattern would actually accelerate. Not long after the ink was dry, the English delegate Caleb Whitfood was asked by a Frenchman what he thought of the new thirteen United States; his reply is unforgettable. "Yes," he hissed through clenched teeth, "and they will all speak English."

The Founders were, of course, aware of all this, and more than that, they were haunted by it. So as they affixed their signatures to the parchment, bold or hesitant, and the great national debate on the Constitution's ratification was about to begin, these unsettled and unanswered questions would be put to the test. What would be saved and what would be lost would, in great part, now depend upon one man, who in his mind's eye was already urgently riding back home, out of the whirlwind of politics and into his steady refuge from the coming storm—to the bliss of his spreading shade trees, the comforting view of his river below, the quiet satisfaction of his annual crops, and his steady brace against the young nation's challenges ahead: the repose of retirement.

∞

ONE MAN. As the full import of the Constitution's signing—and the upcoming ratification debate—began to take shape, the time of reckoning for George Washington had arrived. If ever there was a man who could seemingly defy the daunting odds arrayed against his infant country, who could keep this republic together, who could meld it by force of will, it was the aging general. This moment has been a long time coming, decades in fact, and it had been an often tortuous journey. In the service of his country, he had endured sneers and calumny when the situation seemed hopeless to nearly everyone except himself; he had hurled his men into battle and the face of certain butchery, and just as badly, engaged in humiliating retreat after retreat, again and again, to somehow keep his pitiful army together. He had embodied an icy resolve to see the war through when the voices of despair were at their greatest, and had begged and pleaded for men, for resources, for time, from a feckless Congress that couldn't or wouldn't deliver. And even in the grimmest hours, when defeat seemed certain, or triumph impossible, or when the sheer costs of rebellion seemed no longer worth it, the inscrutable Washington set his jaw and somehow always saw victory.

Washington's place in our hearts is secure. So is his place in the annals of history. It is impossible to talk about the birth of America and its tenuous survival without talking about Washington. There was always something different, something special, something grand about this man. We almost blithely take his outsize stature for granted, forgetting the fact that he lived in a world ripe with domineering and exceedingly colorful personalities, leaders of immense appetites and accomplishment on the global stage, and men and women of great tragedy—George III and Catherine the Great; Maria Theresa and Edmund Burke, William Pitt and Emperor Joseph; Prince Potemkin, Minister Talleyrand, the young Napoleon, and, of course, America's godfather, his majesty Louis XVI; all were his contemporaries. That Washington somehow seems to soar above them all, a hero looming quietly on a different plane, is remarkable. He wasn't a moral icon of the age—others, like Rousseau, held that appellation. Nor was he the greatest warrior of the day—other generals were infinitely more expert to fit that bill. Nor was he the most brilliant leader of the era, or the most ruthless, or the most experienced—he was none of those things either.

And as is the case with so many historic figures, it is futile to expect consistency or clarity in his genius. Speaking in florid, often didactic sentences, he is the least quotable of the Founding Fathers. At the same time, he rarely indulged, publicly or privately, in excessive sentimentality, and almost never doubted or second-guessed himself. He disciplined his deeply felt emotions—mainly his temper, which was legendary—or repressed them. He was laconic, a master of understatement, yet he also had a sublimely romantic quality about him, a true tragedian who understood his appeal to his followers as well as his enemies. He was a man driven by the crassest of emotions for most of his life—ambition and self-interest—yet was also fired by a deep-seated conviction and belief in his home country that was second to none among his peers, arguably not matched by Jefferson, not by Franklin, not Hamilton, not Adams, not Madison. And ultimately it was these qualities that served him so steadfastly when his young country was confronted with its mightiest challenges—its rebellion, its birth, its survival—and that enabled him somehow to rise to the occasion. Somehow. That is the word. By the force of a sheer charisma that we can scarcely define, and an enigmatic presence we can barely understand, even to this day, he persevered. When fate and luck betrayed him, as they did frequently, he roused himself to become a figure who could give heroic life to what the United States was, and, just as important, to what it might become. In doing so, he inspired his men, his army, and his country. It is a supreme irony that this republican system built so assiduously on checks and balances, with its written constitution and competing institutions of power, should seemingly rise or fall on one man, when it was supposed to be the other way around.

But it did.

∞

NONE OF THIS was apparent early on. Not in the beginning of his life; not in the middle. For much of his life, Washington harbored dreams of being a Virginia gentleman, struggling to retain status and honor in the face of narrowing opportunity. Little in his family background suggested he would amount to much. True, the family pedigree was solid: the Washingtons, or "Wessingtons," as they were then called, could boast a family tree stretching back five centuries in Britain. But while Virginia was filled with great family dynasties, the Randolphs and the Carters, the Byrds and the Fitzhughs and the Lees, families that for decades had served as counselors and burgesses, as able aides to the king and great planters, the Washingtons, by comparison, were of the middling gentry, not common, but far from exalted. They lived in the rolling hills of Virginia; his father, Augustine, died when George was only eleven, leaving the family 10,000 acres, but stretched out over seven parcels and primarily for his two oldest sons, along with forty-nine slaves. The Washington family fortunes, not all that much to begin with, quickly declined.

Washington's mother, Mary, sought to make do; it wasn't easy. When he was fourteen his mother almost sent him to sea—he needed a vocation—but her half-brother commented that British ships were floating hells; so it was decided that George would instead become a land surveyor. He was unremarkable as a boy. He had little formal learning (in a sense, curiously reminiscent of Abraham Lincoln), spelled badly, and knew nothing of syntax. And he was insecure, terribly rustic, and culturally impoverished. But he worked hard to improve himself: He read novels like *Don Quixote* and *Tristram Shandy*, developed handsome handwriting, picked up a smattering of mathematics and science, and became a good draftsman. And he sought opportunity wherever he could, the first break coming from within his own family. His half-brother Lawrence evolved into his substitute father and more or less adopted Washington when he was fourteen. English-educated, debonair, and affluent, Lawrence invited George to his great estate on the Potomac River, Mount Vernon. And it was here, through Lawrence, that Washington met the Fairfaxes, one of the grandest families of the Virginia colony. Beyond the world of horse races and fox hunts, dances and girls, which absorbed so much of Virginia's young elite, Lawrence and Colonel William Fairfax, who became his brother-in-law, provided a primer for young George in the exemplary qualities of a Virginia gentleman: duty and honor, dignity and virtue, public as well as personal. A quick study, Washington learned these lessons well, and, in bits and pieces, then as a whole, they would stick with him for the remainder of his life.

All was not smooth sailing, though. Tragedy struck. Stricken with tuberculosis, Lawrence wasted away before his grief-stricken family. But he was as generous to George in death as in life, bequeathing him his grand clapboard mansion on the Potomac. Suddenly George, this young man of meager education and modest background, was able to become a member of the Virginia aristocracy, to enjoy expensive clothes, fine wines, and other pleasures of the upper echelons of Virginia society. Physically, he fit the bill. He grew to be a big man, six feet two, seemingly dwarfing all those around him. The noted doctor and longtime friend of John Adams, Benjamin Rush, slyly quipped: "There is not a King in Europe that would not look like a valet de chambre by his side." He had enormous hands and large feet, a shock of striking auburn hair, ruddy cheeks, and a high forehead; he also had wide hips and surprisingly narrow shoulders—likely he wore a corset until about age five, as was the European fashion—and he used his massive physique, large for his day, to develop a powerful, imposing presence, which he frequently employed to sway men, from recalcitrant soldiers to squabbling politicians. He often relied on brute physical strength to enforce his will, and could easily throw stones, or a "heavy iron bar," an immense distance, a gift he liked to demonstrate to impress anyone who might be watching. He was also fastidious about his appearance: He never wore a wig (he thought in unflattering), but he meticulously dressed and powdered his hair, securing it with a neat velvet ribbon. And decorum always mattered: As a young man he once insisted upon packing nine shirts, six linen waistcoats, seven caps, six collars, and four neck cloths just to venture into the woods.

Indeed, Washington was riddled with quirks and paradoxes, on the surface at least. We think of him as stiff, and in later years he was, but as a young man, he freely displayed a good sense of humor, and was always quick to laugh or join in a good joke. And well into the Revolution he was an enthusiastic dancer. Nor did he mind a glass of liquor—preferably wine or rum—or a good game of billiards or whist, or cockfights, or for that matter, attractive young girls. We think of him as the picture of robust health, but he suffered from ailment after ailment—malaria, dysentery, pain from deteriorating teeth (he initially broke his teeth cracking nuts and replaced them with false ones made from hippopotamus ivory), anthrax poisoning (or so it was believed), the "bloody flux," pulmonary illness, fevers, debilitating sore throats, ague, rheumatism, pneumonia, a sore back, violent headaches, and a large tumor in his thigh; he almost died in 1790, a scant two years into his presidency. We see him as handsome—and he was—but his cheeks were pitted with scars from smallpox. And we think of him as intuitively, almost effortlessly, rising above the fray, but here again, this was the result of long study and much practice.

Virtually everything about him was self-made: As a young man, after reading a little pamphlet, *The Rules of Conduct and Politeness*, he committed to memory exactly how to dress, how to comport himself in the presence of "the mighty," what manners to display at the table, how to show generosity and prudence, when to display pity for the guilty, and, above all, how to meld modesty with moderation. To improve his manners further, he copied out these 110 maxims, which would in one shape or form guide him for the rest of his life. Thus "If you cough, sneeze, sigh or yawn, do it not loudly, but privately..." "Associate yourself with men of good quality if you esteem your own reputation." "Take no salt or cut bread with knife greasy." In time, he shrewdly learned how to ingratiate himself with the first families of Virginia, less through his wit or banter—he had neither—but through the quiet manners of a perfect gentleman, while keeping his own counsel and his mouth shut.

Increasingly, he became the man we know today: dignified, detached, and unapproachable, a man of immense discipline. His eyes could be large and brooding, but at the same time, when he was angered, his cold stare was unforgettable. And it has been remarked that behind his soft blue eyes dwelled a boundless ambition, not simply to shine with the first families of Virginia, but to ordain himself with military glory. That too drove him—all his life.

He was—again the word is somehow—fated for destiny. In 1754, at the age of twenty-two, leading a force of Virginia volunteers and Indians on a scouting expedition, he stumbled into a skirmish with a French detachment. To this day, the circumstances are cloudy as to who shot first and how the hostilities broke out. What is not in doubt is that the fight became a bloodbath, an orgy of insensate assassination and massacre. "I ordered my men to fire," Washington would later dryly report. This was understatement. As the Frenchmen fell, some dead, some wounded, Washington's Iroquois Indians attacked savagely with their tomahawks. But the French commander, Lieutenant de Jumonville, was still alive. Through a translator, he informed Washington's men that he had come in peace, bearing a diplomatic message from his most Christian majesty, King Louis XV. With an unsteady hand, Jumonville held out the letter. The interpreter read it. Suddenly an Indian leaned over the commander, and exclaimed: "*Tu n'es pas encore mort, mon père*" (Thou art not dead yet, my father). The slaughter resumed. One of the Indians sank his hatchet into the man's skull, and before Washington could stop the mayhem, thirteen more French had been brutally killed. They were scalped and stripped; one man was decapitated, his head then impaled on a stake. This incident, *l'affaire Jumonville,* sparked massive French retaliation and the outbreak of the Seven Years' War.

Even Voltaire commented on the bloodletting: "A cannon shot fired in America would give the signal that set Europe in a blaze." Actually, it was a hatchet. Horace Walpole put it more trenchantly: "The volley fired by a young Virginian in the backwoods of America set the world on fire." It did indeed. Washington, who since boyhood had always dreamed of glory, suddenly awoke to find himself world famous. Even King George III now knew his name. And how did Washington react to the slaughter of these defenseless men? He uncharacteristically boasted: "I heard the bullets whistle, and, believe me, there is something charming to their sound."

Soon, he retired to a planter's life in Mount Vernon. He married well—Martha Dandridge Custis, a widow. The match was a wise and enduring one, though in countless ways the two were opposites: Where he was tough, she was sweet; where he evinced cold steel, she was quite gentle; where he towered over other men, she was only five feet tall and plump. But she became his best friend and closest confidante, and would be invaluable to him during the Revolutionary War. And she was very rich. Now Washington, the master of Mount Vernon, was no longer a minor planter but a major landowner, and he was suddenly welcome in the drawing rooms of the Old Dominion's finest families. At home, he was able to warm his feet by the oversize fireplaces he redesigned or stroll over imported European carpets that he had ordered, or sign off on architectural drawings to expand his estate from a handsomely rendered desk. Ordering busts of his military heroes, Julius Caesar and Frederick the Great, he lived in style, with thirteen house servants, plus an assortment of carpenters and handymen, and one hundred slaves. From 1768 on, over just seven years, he entertained a staggering 2,000 guests, all in lavish style.

As he rose in the ranks of Virginia's social strata, he shrewdly reinvented his persona, doing all the things a proper English gentleman (and the Virginians who mimicked them) would be expected to do. He bred horses—fish too—set up a handsome library and ordered 500 bookplates from London, imprinted with his coat of arms. He also kept hounds—Old Harry, Pilot, Pompey, Mopsey, Duchess, Vulcan, Rover, Truman, Jupiter, June, and Truelove. While he and Martha had no children together—he was almost certainly sterile—he doted on his two stepchildren, buying them fine English toys: "A Turnbridge tea set," "a neat book," "a bird on bellows," "a neat dressed wax baby," "a Prussian Dragoon," and a "turnabout parrot"; sadly, though, neither child would live much beyond the first blush of adulthood. He occupied himself with his estate, experimenting with agriculture—he smartly shifted from tobacco to grains and fibers—amassed a fortune (estimated at $1 million), and moved into politics. He became an outstanding member of the House of Burgesses, a vestryman, a justice of the peace for Fairfax County, a warden of the Pohick Church, and, eventually, a delegate to the First and Second Continental Congresses.

But beneath the surface, one senses from afar that Washington, always the picture of moderation, always the paragon of restraint, was dissatisfied. While he had achieved considerable glory on the battlefield and stature in Virginia politics, he also appears to have felt the years slipping by, that he was growing old and worn, all the while sensing that greater things awaited. But what? As an Englishman with stature in a great empire? Or as a rebel? This is the tougher question, and for Washington, the answer is both important and complicated.

By all accounts, he was the unlikeliest of revolutionaries. He was a man with more English blood in his veins than the Hanoverian German King George III, a man whose estate, Mount Vernon, was named in honor of a British admiral who had commanded his half-brother's expedition in a colonial war against Cartagena, a man so Anglophile that he ordered his suits from London and wrote instructions to the tailor to select the fabric, the color, and the cut in keeping with the latest British fashions. He fervently believed in the traditional world of the British upper class; he equally reveled in the "protector-king" role of British imperial dominion in North America; and he preened in the hidebound traditions that the American colonies were integral parts of an empire that was destined to rule the world. He was fond of using the word "empire," and was proud of England's; ironically, he once tried to reorganize the first Virginia regiment on the model of British regulars. This was no republican or radical from Massachusetts: His instincts were aristocratic, and in time he became regal. Affecting the dignified mien of a country squire, he even dismissed the American custom, becoming increasingly commonplace throughout the eighteenth century, of shaking hands. Instead he bowed. And his soldiers called him "Your Excellency."

But history has its twists and turns. When his ambitions were confronted with the careless snubs and indignities of the British upper class—among other things, he was denied a regular commission in the king's forces—it transformed him. The what-ifs of world events here are irresistible. Had Washington not been snubbed, his entire life might have been different—the prospect of glorious global service to the king, the surety of continued promotion, and the certainty of increased riches. But tradition was against him. To the Horse Guards, the headquarters of the British army in London, the amateurish colonial army officers were slightly more than nuisances, even a George Washington. And for Washington, it was a gratuitous slap and a slight that would set the course for his loyalties, indeed for much of the remainder of his life.

Too often we forget that in 1775, when King George struck to chastise Boston ("that nest of sedition") and punish Massachusetts with his Coercive Acts, it was Washington who rose to speak, his voice cracking with fury, "Shall we, after this, whine and cry for relief, when we have already tried in vain? Or shall we supinely sit and see one province after another fall prey to despotism?" When the Second Continental Congress met, this laconic, undereducated farmer-planter not only galvanized his countrymen with his immortal reply, but as he stood, decked out in his faded, old blue uniform, they then tapped him to lead the way in crossing the Rubicon—to rebellion.

After the first shots were fired in Lexington and Concord and Bunker Hill, the Second Continental Congress chose the now forty-three-year-old Virginia militia colonel as commander-in-chief. The vote was unanimous. Was it an inspired choice? Actually, it was quite risky. To be sure, Washington looked the part: He always attended Congress in his uniform, was tall and composed, and had a stately air that inspired confidence. As one congressman put it, he was "no harum scarum, ranting, swearing fellow, but sober, steady, and calm." But was this the man to lead the rebellion? That was debatable, and invariably a sign of desperation on the part of the colonials. Compared with the professionally trained generals of the British armada, Washington had relatively little combat experience, and not all of it impressive. Early on in the French and Indian War, he was forced to surrender his outnumbered regiment to the counterattacking French and their Indian allies; in many English circles, his name had actually become synonymous with American military incompetence. True, he had spent five years as a soldier in the war, serving under General Edward Braddock in a bloody and humiliating defeat near Pittsburgh, then in a victorious campaign with the British in the Ohio Valley that drove out the French. But for most of the war, Washington was a much-harassed provincial colonel protecting the

settlers of the mountainous frontier against French and Indian incursions—
he was often ill, too. He also showed himself prone to a series of vices. He
had quarreled with his benefactor, Governor Dinwiddie; had shamelessly
played politics with the House of Burgesses in Williamsburg; and had dis-
sembled with British generals in the hopes of receiving a commission—not
to mention his clumsy and impetuous attack on Jumonville's men, which
helped trigger a struggle fought across the globe. And for the last seventeen
years, he had not been involved in military matters at all.

He was not profound, or overly philosophical, or strikingly spontaneous. Why then didn't he sink into obscurity, as a self-satisfied bungler, an
uneducated braggart unfit for command? The plain fact was, in the face of
disaster, he was never mawkishly self-pitying. Where other fighting men felt
panic in the swirl of conflict and the prospect of defeat, he was uncommonly brave. In the same battle in which Braddock was killed, Washington
had four shots rip through his clothes, and two mounts shot from under
him. He stood in that battle alone and, by the sheer force of his will, helped
avert total disaster. And as time went on, he would demonstrate unrivaled
physical and moral courage, a steadfastness in adversity, and an astonishing
durability in the midst of hardship, ill health, and filthy weather. The secret
to the magic that he cast over his men was elusive but no less real: an antique
sense of honor unfathomable by the standards of our time, and even to the
luminaries of his day. And there was the matter of his uncommon poise,
which endeared him to his officers, and a steely discipline that nourished
the resolve of his men. True, he was not, and never would be, a great orator like Patrick Henry or Richard Henry Lee; in fact, he was actually quite
stilted in his speech. Nor would he be a great thinker or theoretician, like
Thomas Jefferson or James Madison. But the fact was that in a thousand
little ways, it seemed he was fated to lead this rebellion. And it was here, in
these qualities, that he would find his voice—and his calling.

Would he be equal to the challenge? No task seemed more improbable.
The British crown was able to gut France's empire, pluck Spain's as an afterthought, and summarily stanch rebellions in Ireland, Scotland, and England
itself. And with its professionally trained armada of 50,000 troops and
30,000 mercenaries, not to mention its seventy warships—Washington
and his forces had none—it could surely destroy America at will.

But it was in the war that Washington made his indelible mark. While
he struck some of his aides as rigid, inflexible, or, most of all, icy and aloof,
he smartly combined the artistry of an actor, the skill of a trained psychol-

ogist in motivating others, and a penetrating mind that cut to the quick. And there was his volcanic temper, which he employed to great effect. Once, when his men were in full-blown retreat, he shouted hotly: "Are these the men with which I am to defend America!" Another time, at the battle of Monmouth in June 1778, Washington swore "till the leaves shook on the trees," and denounced his men as "cowardly rascals!" Told by his commanding general that the troops wouldn't stand up to British bayonets, he glowered and shot back: "You damned poltroon, you never tried them!"

He improvised brilliantly, even if it meant defying his own Congress: In the fall of 1776, rather than give up New York to the British intact, he allowed it to be set ablaze; 600 buildings, a quarter of the city, were engulfed in the conflagration. And he thought nothing of flirting with death, riding out with his men or even leading them in battle: Once, when his beautiful white charger—a gift given to him in honor of his crossing the Delaware—dropped dead from heat, Washington changed to a chestnut mare, rallied his troops ("Stand fast, *my boys*, and receive the enemy!"), and drove the British from the field. The young Frenchman Lafayette marveled: "Never had I beheld *so superb a man.*"

And Washington never lost heart. After losing the battle of Long Island and failing to hold Manhattan, he was near despair: He had lost substantial territory and most of his army as well; rivals vied to replace him; Congress had "reproach[ed]" him; and the Revolution seemed stillborn. But here was his genius, turning adversity to advantage: he would not, indeed, seemingly would never, entertain surrender. As the Comte de Ségur, the French aristocrat—he would later be France's emissary to Empress Catherine's court—put it, "Any other man but Washington would have failed." He knew when to gamble and when to hang back, when to take chances and when to bide his time, when to strike and when to maneuver. His Christmas crossing of the Delaware (his "counter-stroke"), of course, and stunning victory over exhausted Hessians (*"Der Feind! Der Feind! Heraus!"* they shouted: "The enemy! The enemy! Turn out!") has become the stuff of legends. So was his follow-up victory at Princeton. When it was most needed, American morale surged.

But it was also fleeting. As the war ground on, he lacked troops, lacked resources, lacked naval support. He was accused of "fatal indecision." He watched his army starve in the frigid snows at Valley Forge and Morristown. He lost a good number of the battles he commanded, and had virtually nothing to do with the critical northern campaign that forced the surrender of Gentleman Johnny Burgoyne at Saratoga in 1777. Yet year after year, he never wavered. And he learned fast. What he saw, almost uniquely and alone, was that he could win the war by simply not losing, by exhausting an enemy for whom America, this insignificant outpost on the fringe of the world, was little more than a sideshow. Like the Roman general Fabius Maximus—to whom he was often compared—whose strategy of attrition wore down the invading Carthaginians, Washington resisted the calls of a panicky Congress, and the temptation most generals feel to fight one great and grand climactic battle. Instead he bided his time. Thomas Paine aptly wrote that these were "the times that try men's souls." But unlike the politicians, Washington knew that time tried the enemy's soul as well; interestingly, in this regard he was a far keener student of global politics than his peers in the Continental Congress. Thus, when the indomitable William Pitt had rumbled in the British Parliament: "Americans are the sons, not the bastards, of England!" and "Is this your boasted Peace? Not to sheathe the sword in the scabbard, but in the bowels of your countrymen!" Washington knew this meant something. That something was time. That something meant a divided Britain would someday lose its patience.

Meanwhile, he made tough choices; he recruited men with bounties and other incentives, endorsed harsh punishments for infractions of military discipline, personally lashed runaways, and rarely shrank from confrontation with subordinates. He begged for—and accepted—French support and advice, something it would be hard to see a Napoleon or a Pershing or a Patton ever doing. And he sanctioned a *petite guerre* of irregular bands in the south, which the British deemed "unchristian." No matter. Washington accepted the bloodshed, destruction, and hardship of war with an alacrity that civilians, and even many generals (and in time, the British themselves), found hard to fathom. Once the battle began anew, he fought to win as no one else did. If he knew one thing with any certainty in this war, it was this: that he stood alone. But he was prepared to survive the setbacks, the losses, the constant humiliations, and of course, the heartbreaking prospect of defeat. One cannot help but be struck by his tenacious resolve, his detached yet focused energy, his unwavering determination.

This, of course, paid off in the stunning victory at Yorktown.

∞

AND IT LED to his selection as president of the Constitutional Convention. Even here, he was enigmatic; actually, Washington almost never attended the

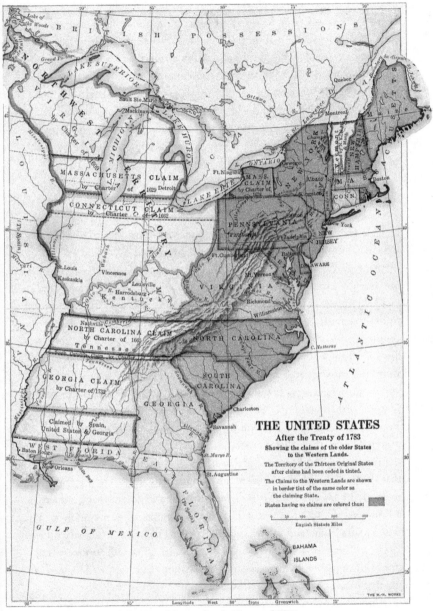

The United States after the Treaty of 1783

meetings. Having promised his countrymen in 1783 that he would permanently retire to his farm, he worried that a return would make his motives suspect. And he feared that if he showed up, it would violate his pledge to withdraw from public life. But he also did not want to show a "dereliction to Republicanism." So he became the convention's president—although he played it coyly: He had no part in drafting its provisions and made only one speech—on the last day. And while he signed the draft Constitution, and privately praised it, he conspicuously refrained from any public statements of support. It didn't matter. The success of the convention, like the success of the war, was credited to him.

Following the convention, a tense ratification process began between the Federalists and the anti-Federalists. Again, Washington's presence loomed large. In the public squares and at town meetings, in the streets of little towns and the big cities, in the remote countrysides of the Appalachian hills and the backwoods and backwaters took place the grandest debate in public history. The pen was the weapon and the debate was played out ardently in the press. And when it was over in 1788—it was nip and tuck; Virginia and New York held out until the bitter end—the Constitution was ratified, and Washington was back in Mount Vernon. But the country was waiting.

Of all the Founders, Washington remains the most difficult for us to wrap our arms around. This is not surprising. He was a man who lived within, possessed by a private vision that he shared with almost no one. Indeed, what so set him apart from the other figures of his day—the other American Founders, the great kings, the fervent revolutionaries and passionate republicans, the brilliant propagandists and the ruthless autocrats—was that he possessed a hidden side that he revealed to no one. But peel back that thick skin, and one could find a dreamer, quite possibly the greatest romantic of his age—which is saying something in an era that produced a Robespierre and a Danton, a Thaddeus Kosciuszko and a Napoleon, a Burke and a Paine and a Richard Price, not to mention a Catherine the Great, an Emperor Joseph, and a Sultan Selim III. How else do we explain one of the most theatrical moments in this nation's history? In 1783, learning that his disgruntled officers were planning a coup—they were actually marching on Congress—Washington dramatically pulled a letter from his jacket; it was a communication from a congressman. He paused, then fished in his pockets for his spectacles. "Gentlemen," he said, his voice laden with emotion, "you will permit me to put on my spectacles, for I have not only grown gray but almost blind in service to my country." The mutinous soldiers were so moved that the coup ended right there.

Driven, singularly willed, quietly but fiercely ambitious, hot-blooded and bold, and uncommonly charismatic, he cherished the military, but he also cherished something even far greater: a sense of history. And virtue. Thus, when a movement among the troops mounted at the end of the war to make him king of America, he rebuked them; it had given him, he said, "painful sensations." Thus, when he learned of the pending mutiny among unpaid and disaffected officers of his army, Washington personally reproached them, saying an insurrection would only "open the floodgates of civil discord" and threaten the very liberties for which they had fought. Thus, when he informed Congress after Evacuation Day that he was relinquishing military power and turning it over to them ("I retire from the great theatre of action"), he was self-consciously imitating the legendary Roman republican leader Cincinnatus, who returned power back to the Senate and retired to his farm. This single action, more than any, was a virtuoso performance: His refusal to seize power that a Caesar, or a Cromwell, or a William of Orange would have eagerly grasped, and, moreover, his refusal even to entertain the notion of accepting power others would have gladly handed over to him, earned him an international prestige and a domestic power that no American, before or since, has surpassed. The old French generals saw him as a great captain because he did not let a superior force destroy his army; the world recognized him as a great man because he did not let his victorious army turn loyalty to him into a military dictatorship of the United States.

When he surrendered his sword to the Congress in 1783, he electrified the world. It did one more thing: As Garry Wills has commented, he perfected the art of gaining power by "giving it away." And ultimately it was not life among Virginia's social elite or military glory where he would make his lasting imprint. It was in leading his nation. By some combination of design and fate, the question of the United States would not let Washington stray from public life, not let him fade into obscurity, not let him rest. As Fisher Ames would put it, "he changed mankind's ideas of political greatness." Even Lord Byron would one day write: Washington had "the all-cloudless glory . . . to free his country."

How would he rise to this great, uncharted task? Some men in history lose winnable wars and other win losing ones. This is one of the mysteries of leadership embodied by Washington. There is another: Lying beneath all his reserve and dignity, beneath the somber face of every sketch and portrait, Washington was actually a strikingly emotional man. And for a man who always carefully controlled his emotions, his passion for his country was always there and insatiable—you could see it in everything he did: in his subtle changes of moods on the battlefield; in the quick flashes of rage and the stare of disapproval at being let down by his men; and in the little twinkle in his eye when pleased. You could see it in his little gestures for his officers, and you could see it in the tears he openly shed as he bade good-bye to his cherished veterans after Evacuation Day, shaking each by the hand, and embracing one of them. But most of all you could see it in his actions when duty called. The Constitutional Convention had made the new chief executive so strong, perhaps so kinglike, precisely because the delegates expected Washington to fill this role. Washington, as always, was concerned for his reputation. He had so much to lose and so little to gain. Nor was this mere coyness or egotism. His hesitations about serving or not serving, his expressions of scruples and qualms, were all part of his strenuous effort to live up to the classical idea of a virtuous leader. Now, with a vengeance, after the Constitution's ratification, he would get the chance to be precisely that. George Washington, half-brother of Lawrence, self-made man and master of Mount Vernon, fatherless husband to Martha and disciple of the great Fairfax family, son of Virginia and triumphant commander-in-chief of the rebel armies, and president of the Constitutional Convention, would in 1789 don a new mantle: first president of the United States. It would be unanimous.

∞

YET FOR ALL the promise of the day, one thing the Constitution, or George Washington, could not provide for the United States was what it most needed: permanence. This was no idle concern. In 1787 and the rush of ensuing months, it would be called into question, for the world—not simply America—suddenly stood on the brink of thundering ferment; of accelerated evolution. Would America become the much vaunted "City on the Hill"—or dwindle into an "American Sparta"? Or would the vessel go aground altogether? In an almost unfathomable display of confidence, Francis Dana, the American emissary in Russia, once amply summed up the aspirations of the Founders, sputtering in frustration: "Does the world not recognize that we are destined to be a great power?"

In truth, to the rest of the world, even as the American Founders boldly harbored utopian visions of the perfect republic, and as their wondrous ideas propounded in the Constitution fanned exuberantly outward, across the oceans, into the courts of the world's great powers, the United States remained little more than "revolting colonies," a fringe republic, an obscure land on the margins of Western civilization. As Bernard Bailyn trenchantly reminds us, America was a "small, unsure, preindustrial borderland" filled with "provincials." One European diplomat, the Comte de Montmorin, summed it up this way: "the united States," he sneered, were "the laughing stock of all the powers." So, while Americans now groped forward to wrestle with their numerous inconsistencies—there were many—and their unresolved problems and seemingly insoluble dilemmas (they were numerous as well), the eyes of the world were not on these rustics. Far from it. Actually, they were fixed on the glitter and grandeur of an expedition now under way halfway across the globe: one of the most remarkable expeditions known to humankind. And then they would rest on the center of the monarchical universe: France.

It was into these twin vortexes that an unsuspecting young republic would be thrust. With such ripples do tidal waves begin.

CHAPTER 2

RUSSIA

✧

IN 1787, AS THE Americans reshaped their government and cast their lots in favor of an uncertain future, the gilded world of the European nobility stood at its summit. As always, fashionable society wore diamonds and furs, lounged in the world of elegant retreats, hunted on magnificent estates, and played billiards and backgammon. Wealthy, worldly, enchanting, and bored, they flaunted love affairs and attended an astonishing procession of seasonal balls, midnight masquerades, and all-night gambling parties; everybody gave one and everybody went. Kings and monarchs continued to flock to one another's courts, hold state dinners, attend the theater, review military parades, and gallantly hoist one another's children onto their laps. But unknown to them, fashionable society, in great numbers, rested one step from the abyss. Behind the polished circle of aristocracy, of ancient dynasties and enlightened despots, there was a wider, all-too-often hidden arena where millions of ordinary souls—the poor, the dirty, and the abused—toiled invisibly, as they had for centuries. But now the omens of upheaval were all around, even if the portents were anything but obvious.

Yet, in early 1787, the chatter of European capitals was consumed by an unprecedented spectacle, taking place not in Paris, or London, or Berlin, not in Vienna or even Rome. Nor was it taking place in the "New World," in Philadelphia, where fifty-five men were preparing to assemble behind closed doors to create a constitution and make a young republic. In fact, it wasn't even being presided over by a man. Amid enormous fanfare and intrigue—did it mean war?—and no small amount of trepidation—would it unleash a clash of civilizations or an unalterable tipping of the European balance of power?—Empress Catherine the Great of Russia, the tsarina who ruled the largest nation on earth, had left St. Petersburg and now stood at the crossroads of the exotic Orient and the splendor of the Occident, smack in the former crown jewel of the Ottoman Empire: the Crimea.

The trip had been years in coming. After more than a decade of bloody war against the Turks, insurrection by the Tatars, and incessant diplomatic maneuvering, the Crimea was now Catherine's. And along its shores lay the glistening prize of warm-water ports and year-round passage through the Black Sea. What Peter the Great had done in the north, capturing the Baltic from the Swedes, building a fleet, and establishing St. Petersburg, and with it, a gateway to Europe itself, Catherine now dreamed of doing in the south on an even grander scale. But in raising an empire, she was prepared to go him one better—for the Crimea was a steppingstone to even more glorious conquest.

During her long, illustrious reign, Catherine had already outlived the giants of Europe: Frederick the Great of Prussia, Augustus III of Saxony-Poland, and Louis XV of France; she had shrewdly outfoxed King George III of England; and she had boldly outmaneuvered Queen Maria Theresa of Austria. And now her mind lay fixed on Tsargrad, the "city of Caesars," the capital of the Ottoman realm—Constantinople, today's Istanbul. Her goal? A feverish scheme that gripped her imagination: the dismemberment of the Muslim Ottoman Empire and the restoration of the old Byzantine Empire, with a Christian emperor on Constantinople's throne—her grandson.

For Catherine, the Crimea was just the start.

∞

HER CRIMEAN EXPEDITION itself was unlike anything seen in history, but not without considerable risk. To go from St. Petersburg to the Crimea, Catherine would have to travel more than a formidable thousand miles over the course of six months. Once having left the safety of the capital, there was the perennial danger of moving through hostile Muslim heartlands, the dreaded specter of assassination, and the possibility of revolt back home. And even under the best of circumstances, bumping and jostling across the interminable, vast landscape of Russia, travel was invariably perilous. At a time when the Appalachian mountain chain was considered the wild country in America, and no one really knew what lay to the west of the Mississippi—there was speculation that the lost tribes of Israel wandered on the immense American plains—so extensive were Catherine's lands that as dusk first settled across their western borders, dawn was already unfolding along the Pacific. Between endless tracts of dark forests and boundless expanses of parched desert and desolate tundra, the winters were bleak and unremitting. It is almost impossible for the Western mind to grasp their full horrors. Winter began as early as October and lasted six surreal months; in St. Petersburg itself breath turned to ice. By the hint of first snow, the sun dipped down at three in the afternoon. Then, in the gathering darkness, blizzards screeched, shaking the doors of the imperial palaces and piling massive snowdrifts along paths and walks. Feathery clouds streamed across the gray skies while Russian peasants huddled with their animals and slept on top of their stoves. And before long, the country was entombed beneath an infinite sea of whiteness: whole villages, entire forests, stretches of fields, and even Russia's mighty rivers. Great masses of quivering pine trees and birch were submerged under deep drifts of snow, while the climate and isolation taxed the mind as well as the body. Hence the profusion of holy men and mystics. Hence the bitter ice-bound seas where few dared set foot. Hence, east of the Urals, in the large empty spaces of windswept Siberia, the temperature dipped to ninety degrees below zero, making sheer survival itself a herculean feat of strength.

No matter. Catherine, one of the most domineering and versatile monarchs of her age—perhaps of any age—was not one to be easily cowed. Indefatigable and determined, passionate and industrious, she had now fixed her eye on her new southern conquests, and beyond that, the prize of the Ottoman Empire, and she was not about to be put off. So as the Ottoman sultan reverently bowed toward Mecca in the mosque of Sultanahmed, on January 7, with an icy wind driving flurries of snow, cannons boomed a salute and the empress's expedition began.

As Catherine's entourage left her palace at Tsarskoye Selo outside Petersburg, it was a dazzling display of imperial splendor scarcely witnessed in the Western world. Her carriage, drawn by ten magnificent horses, was furnished with cushioned benches and tables and was so massive that a man could stand up in it. Escorting her were fourteen other carriages and 180 sledges, which carried hundreds of servants, including a stunning array of imperial subjects: doctors and blackamoors, twenty footmen, thirty washerwomen, and countless silver polishers, cooks, and apothecaries, as well as assorted chamberlains, ladies-in-waiting, and even a master of the horses. "All of society is galloping off," Catherine glibly boasted. It certainly seemed so, but far from a purely Russian affair, the expedition was a microcosm of Europe itself. It quickly evolved into an event of uncommon glamour, a cockpit of cosmopolitanism and audacious diplomacy in which the great issues of the day—politics, war, economic trade, and empire building— would be thrashed out behind closed doors. Beyond the empress's own twenty-two senior courtiers, she was accompanied by the ambassadors of France, Austria, and England; she called them her "pocket ministers." And as the expedition moved deeper south, she would be joined by a multitude of travelers "from all parts of Europe," each of whom would be seduced by Russia's Little Mother and her roving court.

A careful student of the Enlightenment, Catherine had sought out genius wherever she could find it: in England, France, Ireland, Scotland, Germany, Sweden, and even in America. Accordingly, as she journeyed from Petersburg to Kiev, and Kiev to the Caucasus, Catherine's royal orbit would become a rendezvous point for statesmen and scholars, diplomats and philosophers, as she was accompanied by an assemblage of some of the most fascinating personages of the age. There would be Samuel Bentham, the English engineer, builder and inventor, and brother of the famous philosopher, Jeremy; there would be the Venezuelan warrior Francisco de Miranda, the budding father of South American liberation who had fought for the Spanish as far afield as Algiers and Jamaica and who knew George Washington and Alexander Hamilton; and there would be Prince Charles de Nassau-Siegen, a heralded soldier of fortune who was known as the "paladin" of the times (he once tried to set himself up as a king in West Africa and had made love to the Queen of Tahiti). And Catherine had already invited George Washington's beloved protégé, the heroic French nobleman, the Marquis de Lafayette.

Shuttling to her side would also be Charles Joseph, Prince de Ligne, the witty Austrian aristocrat and notorious charmer of Europe, who epitomized how interconnected the eighteenth century had become: He had befriended Frederick the Great; knew the commanding philosophers of the era, Rousseau and Voltaire; and had also met Casanova and Queen Marie Antoinette. As Ligne once quipped wryly: "I'd like to be a foreigner everywhere." An American rounded out the retinue: Lewis Littlepage, the dashing "Virginia gentleman" and a personal friend of Washington's, who was now serving as an aide to Poland's king; he was designated to be Poland's eyes and ears in Catherine's Crimean court. En route, Catherine would spend part of the journey negotiating with the Polish king as well. Stanislas Augustus Poniatowski had waited three months to meet with Catherine; galvanized by America's revolution, he pleaded for Polish reforms and a renewed Polish-Russian alliance. In a mere three hours, she would coldly rebuff him, leaving the king humiliated and the very fate of his nation in doubt. But as it happened, Catherine had other affairs of state and other monarchs to turn to in her fold.

For waiting along the river at the town of Kaidak was Joseph II, the Holy Roman Emperor and older brother of Queen Marie Antoinette of France, who would also link up with her imperial caravan.

The expedition departed precisely at eleven, racing along icy roads lit up with bonfires of pine and birch, which blazed day and night. Heading southwest, by January 29 the party had arrived at the inland port of Kiev, strategically perched along the Dnieper River. Captured from Poland before the reign of Peter the Great, Kiev was considered the mother of all Russian cities and the birthplace of Russian Christianity. Amid its glorious onion domes and great cathedrals, the gathering would remain for three months, until the seasons changed and the river's ice melted. Then, on to the Crimea. Until its departure, the traveling court was a whirlwind of activity: Concerts played daily; there were balls and banquets and all-night dances; and courtiers from across Europe and Asia—including Chinese, Persian, Indian, and Tatar dignitaries—brought urgent dispatches and paid tribute. In all, Catherine received the emissaries of thirty different nations. In the meantime, the magic of the empress's "fairy circle" was continuous. At nightfall, thick clusters of lights blazed brilliantly and dozens of dresses swayed as ladies sank deep in curtsy when Catherine strolled by. By day, plates of cold sturgeon and steamed fish and caviar were demolished. And the flavor of the expedition was as much East as West: If there was a quadrille, then there was a mazurka; if there was a hussar in full-dress uniform, then there was a Cossack in a scarlet tunic. A Polish noblewoman, racing to catch a view of the empress, put it this way: "If one closes one's eyes, one imagines oneself in Paris." The worldly French ambassador, the Comte de Ségur, who had fought with the Americans in the Revolutionary War, offered a different slant, marveling that this remarkable caravan was a "magical theater" that mixed "antiquity with modern times, civilization with barbarism."

In mid-April, the first murmurs of spring came. The ice began to crack, and by April 22 the thaw was fully under way. Artillery salvos boomed, signaling that Catherine's voyage, now on board ships, could begin anew. It was one of the most luxurious fleets in history, filled with all the trappings of palace life, a rigid protocol handed down from a bygone era. To guard the empress, produce this paradise, steer its vessels, scrub its floors, make its beds, pour its wine, polish its silver, serve its banquets, play its music, and bathe and dress its imperial guests required the ministrations of thousands of human hands. There were seven elegant imperial galleys in the fleet, manned by an army of servants: some 3,000 oarsmen, crew, and guards, all attended by eighty other ships. Each barge was outfitted with gleaming floors and polished halls, and ornamented with gold and mahogany, crystal, velvet, and silk; even the oars were painted. For all the guests, there was a communal drawing room, a library, a music room, and a shade canopy on deck. The lavish bedroom suites were decorated with Chinese silk, the beds themselves with "taffety," and graceful wooden writing tables adorned the studies. There were, too, lavatories with their own water supply, a supreme rarity for the day. Orchestrated by Catherine's most trusted aide, Prince Grigory Potemkin, this armada was as dazzling and fantastic as anything on the planet.

Said Prince de Ligne: it was like "Cleopatra's fleet." For his part, another European aristocrat, the Prince of Nassau, called the gathering "one of the most unique things ever seen."

∞

THE TRIP ITSELF was often a chaotic, tangled skein. But soon the Dnieper waters were interrupted by falls and the expedition continued overland, resting each night in lavishly outfitted tents. Nearly a breathtaking 1,000 miles from Moscow, the landscape now changed dramatically, as Catherine approached the infamous black-earth Ukrainian steppe, the gateway to the Crimea itself. Here, only the head and shoulders of a man on horseback were visible above the long expanse of tall, rangy grass. This was the land of the rugged Crimean Tatars, Islamic offspring of the old Mongol conquerors and longtime subjects of the Ottoman sultan, who had lived for centuries tucked away among the steep rugged cliffs of the Crimean peninsula. For vast stretches, under a ruthless sky, there was no habitation, no sign of life, only the flatness of Tatar deserts and great heat; it was, as one observer noted, as though they had "left the world of men." One evening, underneath a shimmering sky, Emperor Joseph took a stroll with Ségur, the French ambassador. Off in the distance, a caravan of camels formed an eerie silhouette against the horizon. "What a strange journey," the emperor crackled. "Who would have expected to see me, with Catherine II and the ministers of France and England, wandering in the desert of the Tatars?" He paused dramatically: "It is a completely new page of history."

"It's more like a page from Arabian nights," Ségur dreamily added.

In truth, however, this new page of history had firm roots in Europe's tumultuous past. On May 12, with huge herds of sheep and horses grazing in the background, Catherine and her compatriots had passed underneath an arch marked prophetically: "This is the road to Byzantium."

It was. Once they crossed through the Perekop Straits, the entrance of the northern Crimea itself soon came into view. After several more hours, the flat, arid prairies of the Ukrainian steppe became lushly dramatic. This was the Crimean peninsula in all its splendor. In every direction, the views were spectacular: Along the southern shores bathed by the Black Sea, jagged mountains gazed down upon the glistening waters. Off in the distance, forests of tall pine graced the upper slopes of the Haila range. And in the valleys and by the sea cliffs below, there was an abundance of verdant gardens and groves, cypresses and orchards, vineyards and pastures. Much as in southern Italy or Spain, medieval fortresses hugged chalky cliffs and deep emerald bays. Each season was mild, and during the Crimean spring a host of fruit trees, shrubs, vines, and wildflowers bloomed in almost miraculous profusion; even tangles of wild strawberries and grapes curled over its slopes. With its shimmering sun and placid sea breezes, its kaleidoscope of colors and enticing odors, it was no wonder the Ottoman Turks claimed this unspoiled wilderness as their "pure and immaculate virgin"—the Sultan's Lake. Strategically, it was crucial as well, for as Catherine knew, whoever controlled the Crimea also controlled the Black Sea. And whoever controlled the Black Sea was poised to strike at Constantinople.

Many had sought—and failed—to attain permanent dominion over this gem: the ancient Greeks, Goths, Huns, Byzantines, Khazars, Karaite Jews, Georgians, Genoese, Armenians, and Tatars. The Tatars were a special case. Ruled by their own khans, the Crimean Tatars were Islamic descendants of the great Genghis Khan, and, despite their subservience to the Ottoman sultans, they maintained a uniquely privileged status, constituting the second family of the Ottoman Empire itself. Indeed, should the

Ottoman sultanate fail, it was widely assumed that the Crimean Tatar khans would succeed them. But for Catherine, the Crimean Tatary had a singular significance as well: For three hundred years, it had been a Mongol citadel that had tormented Russia.

Modern sensibilities tend to obscure this dark epoch in Russian history, but not so for the Russians. Expert archers, the Tatars were bred to kill and trained from infancy to be pitiless; these dreaded warriors turned plunder into an industry. In the 1500s the Tatars were among Europe's most important states, with their empire stretching from Transylvania in the west to Kazan in the east, and in the north approaching Moscow itself. Even in Catherine's day, the khanate ruled from the eastern Kuban steppes to Bessarabia in the west, all precious territory that stood between the Russian Empire and the tantalizing riches of the Black Sea.

Every spring and summer, the Tatars strapped on their bows and arrows and scimitars ("Their country," it was once famously said, "is on the back of a horse.") and plunged deep into Russia, often battering it beyond recognition. This was Russia's misfortune; it was also Russia's humiliation. Year after year, in small groups or immense armies, the Tatars stormed entire Russian and Ukrainian villages. Then the massacres began. Moreover, these massive raids annually captured thousands of Russians and condemned them to be sold in Ottoman slave markets; in 1662 alone, some 20,000 prisoners were chained and whipped and consigned to bondage. And year after year, Russian men were routinely shackled to the oars of galleys across the eastern Mediterranean, while young Russian boys were handed over as tokens from the Crimean khans to the Ottoman sultan. Indeed, so extensive were the hordes of Russian slaves throughout the East that it was once derisively asked if "any inhabitants still remained in Russia" itself.

Whatever anguish the Russian tsars felt—and it was profound—they were largely powerless against this onslaught. Instead, they were forced to pay vast yearly sums as protection money; the khan mockingly referred to it as tribute. Still, the raids did not stop. In fact, twice the Tatars pillaged and torched Moscow itself. By the time they retreated, they had transformed the glory of Muscovy into a reeking slaughterhouse. In another instance, in 1688, the Tatar khan plundered the Ukraine, menaced the cities of Poltava and Kiev, and when he finally withdrew back to the Crimea for the fall, 60,000 bloodied souls staggered behind the khan's dreaded horsemen.

Nearly one hundred years later, Catherine was determined to change this.

∞

HER CHANCE CAME in 1768, when the Ottoman Empire declared war on Russia. This, the first Russo-Turkish War, resulted in an uneasy military stalemate, but Catherine was successful enough that her treaty amputated the Crimea from the Ottoman empire. The Ottomans reluctantly agreed, though they thirsted for revenge. That said, after centuries of Russians being treated like weak and stupid barbarians, Catherine did as well. An independent Crimean Tatary was the first step in the empress's ambitious plan of conquest and dominion. The second step was even more audacious. By 1782, fearing that Russia still stood naked to invasion (in Prince Potemkin's words, the Ottomans could "reach our heart" through the Crimea), Catherine authorized preparations for annexing the region outright—even at the risk of provoking another war with the mighty Ottoman Empire, if not with the great powers of Europe itself.

She also knew that her window was fleeting; in one perfect instant, all Russia's usual antagonists were fatally preoccupied. Britain was still mired in the Revolutionary War with the Americans and the French, Constantinople was wracked by plague, and Austria was, for the moment, amenable to her grand scheme. If there were a time to act, it was then. Lest Catherine waver, Potemkin tantalizingly goaded her with visions of grandeur. He told her, "Imagine the Crimea is yours." And this: "There is no power in Europe that has not participated in the carving up of Asia, Africa, America." And this: "With the Crimea, dominance over the Black Sea will be achieved." And this: "Russia needs paradise." And finally: "You are obliged to raise Russian glory."

The empress agreed. Emboldened by the exhaustion of Turkey's traditional ally, France—it had bankrupted itself financing the American rebels in their struggle against Britain—Catherine now glimpsed that glory. She flamboyantly declared, "When the cake is baked, everyone will want a slice." But Catherine craftily intended to keep the entire prize for herself. She was, however, anything but reckless, hoping to take the Crimea bloodlessly, or at a minimum, that the annexation at least appear to have "the will" of the Crimean people—a pattern that would repeat itself for the next 220 years. "It was," she mendaciously proclaimed, "the love of good order and tranquility that had brought the Russians into the Crimea." Actually, it was the love of empire. But if it was a gamble, it was not a foolish one. The hillsides of the Crimea were still soaked with blood, and the memory of 30,000 Tatars being mercilessly put to the sword by Catherine's armies was still painfully fresh, as was the ghastly image of tens of thousands of Tatars desperately fleeing into exile, dying from cold or starvation as they swarmed across the eastern steppes.

So for Catherine, the region was now ripe for the plucking. With the light of desire in her eyes, she issued the orders to take the Crimea. And once her armies massed along the frontiers for war, her enemies quickly folded. On the anointed day in the ancient Crimean Khanate, 6,000 Tatar tents dotted the Eysk steppe, and these "loyal sons of Mahomet," shuddering before Catherine's arrayed might, stood in their most exquisite robes to pledge their allegiance to an Orthodox empress—perched on a throne more than a thousand miles away. But the empress was taking no chances; she then callously pushed the defeated Tatar khan into neighboring Turkey, where, disgraced, the Ottoman sultanate promptly beheaded him.

Nor did Catherine stop there. Within days, in the Caucasus, her emissaries blustered and bribed the kingdom of Georgia into accepting Russian protection; in the northwest, Russia then took Kuban. Then she twisted the blade further still. In a trance of ardor, Russian generals lunged against the barbarous Muslim mountaineers as far off as Chechnya and Dagestan, and marched into Armenia and the Caspian seashores—today's Azerbaijan.

From afar, Europe shivered at Russia's conquests. Potemkin wrote to Catherine, "You have attached the territories which Alexander [the Great] and Pompey just glanced at . . . order your historians to prepare much ink and much paper." With glee, Catherine shot back, "Upon the envy of Europe, I look quite calmly—let them jest while we do our business." That business was concluded shortly after Benjamin Franklin, John Adams, and John Jay had affixed their names to the peace treaty in Paris, formally ending the War of American Independence. Deprived of French help, on December 28, 1783, the Turks officially acknowledged the loss of the Crimea. Exclaimed one observer, "The Russian state has spread out like ancient Rome." Indeed, Catherine's swollen dominions now stood larger than the entire Roman Empire had at its height.

But there was a price to be paid for this brash expansion. For one thing, Catherine's actions had inadvertently opened the wounds of religious wars that would one day cross borders and carry over into a new age. For another, the Crimea remained a boiling stew of biases, prejudices, and fierce hatreds.

In the years that followed, while annexation was one thing, actually extending Russian control over the region remained an exasperating business. Russian forays into the area known as the Caucasus were especially riddled with troubles—in 1785 a rebellion broke out among a deadly mix of Chechens, Avars, and other tribes. Descending from the mountains, a shadowy leader wrapped in a green cloak and espousing a mystical version of Islam proclaimed a *Ghazavat*, or holy war, against the Russians. With dauntless flair and lightning strikes, this self-anointed "Sheik Mansur" led a coalition of mountain tribesmen that harassed and tormented the Russians with guerrilla warfare, laying the seeds for a conflict that still sputters today.

Yet while the bloody campaigns continued, so did the settlement and building. Across these once empty steppes, the empress constructed shipyards, fortresses, roads—and whole cities. In little more than a decade, under the tireless supervision of Potemkin, about a million settlers would throng to Russia's new Black Sea coast, an event in scale and speed that would be unparalleled until great waves of immigrants began arriving in America in the nineteenth century. And not unlike in the wilds of the American West or the territories of Australia, these settlers would often be a motley collection of paupers and fugitive serfs from Russia, slave laborers and sailors, hussies and workmen. But the tsarina also wooed Albanians, Romanians, Moldavians, Swedes, Corsicans, German Mennonites, Greek colonists, Poles, Jews, and British convicts. Ironically, loyalists from America, those still faithful to the British crown, petitioned her to be welcomed as settlers in 1784. Deemed seditious by the powerful Potemkin, they were singularly refused.

Jealous European visitors claimed that these new outposts were a sham—the infamous "Potemkin villages"—but in truth, the reality is more complicated. While some of the cities were stillborn—slave labor and autocratic fiat can go only so far—many of these early outposts would grow into great industrial and military centers: Odessa, Ekaterinoslav (today's Dniepropetrovsk), Kherson, Nikolaev, and Sebastopol, about which Ségur gushingly told the Empress, "Madam, you have finished in the South what Peter the Great began in the North!"

Throughout the month of May, it was these Potemkin villages that Catherine was unveiling to her vast array of awed and at times envious companions.

∞

IN THE DAYS that followed, the fantastic voyage continued, rumbling past strategic fortresses; past elaborately planted English-style gardens, complete with shady foliage, running water, and pear trees; and past elegant palaces, while fireworks and rockets exploded in the background. Each step along the way, Catherine's procession inspected Potemkin's seemingly wondrous creations. At one point they were accompanied by a regiment of amazons, one hundred "Albanese" women wearing armored breastplates, white turbans, and armed with lances and muskets. At another point, a fleet of magnificent ships in the distance fired their guns in salute. The pageantry culminated with the grandest spectacle of all, 50,000 troops, some dressed as Russian soldiers, others as Swedish, reenacting the historic battle of 1709—Peter the Great's stunning victory over Sweden's Charles XII. "Joy and pride shone in Catherine's eyes," wrote Ségur. "One would have thought the blood of Peter the Great flowed in her veins." Joseph, himself the emperor of a waning empire, marveled at these grandiose projects created by autocratic command, but with one caveat: "We in Germany or France," he declared, "would never have dared undertake what is being done here. Here human life and effort count for nothing . . . The master orders, the slave obeys."

As the imperial carriage proceeded, suddenly there was an earsplitting din of hooves and a great rush of dust. Twelve hundred Tatar cavalry, waving jewel-encrusted pistols and curved daggers, appeared, surrounding the imperial expedition. Was this danger? No. It was an elaborate act of stage and statecraft. To demonstrate her goodwill toward her new imperial subjects, Catherine decided not to be protected by Russian troops; her goal was less to muzzle a rebellious minority than to seduce it. So instead, she made her entrance into Bakhchisarai, the former home of the deposed khan, with an elite guard of native Tatar *murzas*, mounted upon thundering horses and swathed in dark green uniforms adorned with gold. Approaching the Crimean capital, they served as Catherine's personal escort; meanwhile, twelve Tatar boys were designated to be her pages.

The magnitude of this trip now became apparent. In the distance, the travelers could glimpse the skyline of slender minarets. It was the khan's palace. Arresting and vaguely sinister, it housed a city within a city, an expansive compound of castle, seraglio, and mosque. Suddenly the expedition was no longer illusion: For the first time she would see for herself the inner sanctum of a Muslim world that had long been, as Potemkin once feistily put it, the "wart on Catherine's nose." But this was understatement. Her task, even for Catherine, would be a formidable one. Then, as now, her knottiest problem remained: Could she wage war on the central seat of the Islamic world, Constantinople, without inciting the wrath of the Crimean Muslims across her newly acquired provinces?

∞

BUILT BY VAST legions of Ukrainian slaves, the khan's palace was designed to echo the great Ottoman palaces of Constantinople, with a myriad of gates and sparkling fountains beckoning toward his personal residence and harem. With its ancient Greek marbled walls and intricately tiled floors, the palace was awash in Moorish and Arabian influences. With its secret gardens of intertwining vegetation—rose bushes, laurel trees, jasmine vines, and orange trees—it reflected its European underpinnings; Italian architects had helped design it. Meanwhile, alluring smells filled every corner. And with its baths and thick walls, its cushioned divans in the private apartments and the lingering sounds of imams calling the faithful to prayer, it recalled the ancient city of Constantinople itself. So did the sight of veiled women strolling quietly along the gardens and through the courtyards. But now this city belonged to Catherine—and completing every victor's dream, upon her arrival she slept in the vanquished khan's own palace apartments.

In ensuing days, the empress surveyed a scene that resonates to this day, the uneasy coexistence of Muslim and infidel. It was everywhere she went: the juxtaposition of the delicately laced minarets of Islamic mosques side by side with the freshly built onion domes and gilded cupolas of Russian Orthodox churches; the mixing of Tatar men, sinewy and dark complexioned, with light-skinned Russians; the joining of Tatars wearing round black hats and tight white trousers with Slavic generals whose chests sagged with medals. And into this vast melting pot strode the empress. Actually Catherine, an absolute monarch, a woman, a Christian, saw no incongruity between her pledges of peace to the newly annexed Muslim Crimea even as she planned war on the seat of the Islamic empire. In fact, she took special pleasure in assuming the throne of the Tatars who had ravaged Russian provinces for years, and who still beneath their breath called Russians "infidels" and "dogs" and were as disdainful of her religion as of her sex.

Yet while Catherine privately nurtured Russia's historic grievances, publicly she was a shrewd imperialist; for here, tolerance was the watchword. She granted audiences to local imams, according them great deference. She took no umbrage when merchants and craftsmen turned their backs when her imperial carriage passed. And already she had bequeathed ample money to maintain mosques, as well as bestowing a cornucopia of titles and land on local muftis, setting in motion a time-honored tactic, especially among Russian rulers, of coopting conquered peoples. No less than Potemkin himself, who personally governed this southern realm, received courtiers like an Islamic pasha, sprawling on a divan wearing only a pelisse, as though he were the vizier of Damascus, Cairo, or, as Ségur noted, "even Constantinople" itself.

For decades, the contemporary Western world has been mired in debates over whether imperial ventures can last, or if a clash of cultures is inevitable. In the Middle East particularly, the question looms: Do they create only more resentment—or is coexistence possible? On this score, Catherine was not one to be consumed by such sentimentality or self-doubt. And in any case, her policy of tolerance paid off. Rather than rise in revolt, her new Muslim subjects were marked by an air of listless indifference. Instead of fierce chauvinism, there was an almost sleepy impassivity. And she did manage to build some loyalty: At one moment, approaching the Crimean capital, the tsarina's carriage—Emperor Joseph was with her as well—nearly careened down a steep hill. She was boldly rescued by the Tatar horsemen galloping alongside; they almost surely saved her life. Nor was Catherine willing to tempt fate. In the presence of her new subjects, she tended to the littlest of details. One night, at dinner, the Comte de Ségur and Prince de Ligne recounted a story of how they had hidden behind a clump of trees and caught sight of three Muslim women removing their veils; the women, turning their heads upon being caught, covered their faces and began to scream. Hearing this story, Catherine was not amused. In reply, she growled indignantly, "This joke is in very bad taste. You are in the midst of a people who have been conquered by my armies: I want their laws, their religion, their customs and their prejudices to be respected."

That, of course, was only for the Muslims of the Crimea. For the Ottomans, she had a very different view.

It was then, at the outskirts of Sebastopol, that the talk of the expedition ominously turned to war.

∞

ON MAY 22 the empress and Austria's emperor dined elegantly in a gorgeous palace hanging on a mountainous bluff overlooking the sea. The vistas were magnificent. Outside, through French doors, they could see the sweeping horizon—and Catherine's new naval base. Suddenly, in a deeply symbolic moment—symbolic of what was to come, that is—a formidable flotilla of ships and frigates, their varnished masts gleaming in the sunlight, lined up in the placid bay, in battle order, facing the two monarchs. Upon a designated signal, the fleet saluted with all its guns booming. To one of the ambassadors who witnessed it, the message was unmistakable: It proclaimed that the Russian Empire had subdued the south with all its might, and that within thirty hours, Catherine's armies could "plant her flags on the walls of Constantinople."

This was a brazen prediction. It was also fantasy. While the Russian forces looked formidable on parade, they were far from battle-ready. Emperor Joseph realized this too. And despite falling under the full assault of Catherine's charms, he felt suffocated by doubt, wondering if she was exhibiting more zeal than wisdom. This was perhaps predictable. With his handsome features and flashing eyes, the Hapsburg ruler saw himself in the mold of an enlightened monarch; a shrewd scion of generations of European rulers, Joseph also doubted Russia's southern prowess. Thus, when he retired alone with Catherine and Prince Potemkin to discuss the timing of the war against the Ottomans, Joseph urged caution, citing the resistance of France and Prussia. Shaken and furious, the empress was indignant. To her, it was almost incomprehensible that most of the Western states, instead of "supporting the Christian peoples of Europe," should align themselves with the Muslims of Asia; as it was, France's emissary, Ségur, had already delivered a warning to the empress from King Louis XVI. Undaunted, she sneered that the king was too shaky politically to pose much of a threat, even if, as Potemkin put it, France would make "a lot of noise."

Besides, the empress said flippantly to Joseph, "I'm strong enough—it suffices that you won't prevent it."

Indeed, when the question was raised if her ships were equal to the Ottoman ones at Ochakov, Catherine had answered mischievously, "Do you think I dare?" But Potemkin, her commander-in-chief, was more direct: "Everything is there," he bellowed. "All I have to do is say to 100,000 men—March!"

Unwilling to be left out of the taking—his fading empire stretched like a slowly rotting carcass across the eastern boundaries of Europe—Joseph assured her of Austria's support against the Turks. But was his word good? That remained to be seen. Nevertheless, before long, the empress, Potemkin, and a chastened Joseph ravenously made preparations to carve up the Muslim empire.

"Those poor devils, the Turks," Catherine warned. "They had no idea what was in store for them."

∞

IT WOULD NOT be that simple. That night, as a huge spring moon rose and cast its silver light across the shining waters of the Black Sea, hardly did any of them know that on the banks of these very same waters, this very same discussion—holy war or humiliating truce—was also boiling over in Constantinople. Already legions of Ottoman soldiers were descending upon the capital, packing the mosques and coursing through the streets, ready to defend the empire's Black Sea strongholds—all that stood between them and the hated Russians.

In truth, for all her martial bravado, Catherine didn't want an actual war. As with the Crimea and Georgia, she wanted conquest without bloodshed. It could hardly have been otherwise. For despite Potemkin's impressive displays, there was dissension in her high command, and none of her forces was complete or fully prepared; in the Black Sea alone, Russian commanders said they needed another two years before engaging in all-out hostilities. But it was no longer simply Catherine's call to make. Shifting alliances and mutual recriminations were inexorably drawing both sides closer to catastrophe. The maneuvers of Catherine's southern fleet had infuriated the Ottomans, while their support for Sheik Mansur and his marauding band of Chechens galled the Russians. In response to her sly diplomatic overtures across Europe, the Turks repaired their defenses from the Black Sea to the Balkans and ambitiously refurbished their fleets. As Russian agents swarmed over the Balkans, Syria, and Egypt to buy up local alliances, the Turks sought aid from French officers, who happily drew up plans for the sultan's troops to retake the Crimea. And within the walls of Constantinople, her imperial visit to the Crimea further smacked of bullying and aggression. Finally, in the great Kabuki dance of European powers, even England, Prussia, and the Kingdom of Sweden were casting a wary eye over Russian intentions and were inching closer to supporting the Ottomans as well.

Thus a vicious cycle of familiar history was set in motion. Catherine thought her roar could silence the world; it couldn't. And with each week, the brinksmanship and tensions escalated between these two distant cultures. After studying the situation, Catherine sought to play a clever game: Prepare for war, but talk peace. She sent soothing assurances to the sultan; he knew better, and she became stuck in her own net of deception. By June I the empress's envoy to Constantinople, Bulgavkov, was horrified to find the Ottomans wracked with war fever: It was the same story—dark rumors were spread by local imams, xenophobia was at its height, and the sultan's hand was being forced. Meanwhile, Muslim recruits from Asia rushed through the city, headed for Ismail, one of the Turks' key fortresses. Ottoman armies now stood at 300,000, and when Catherine's emissary rejected the sultan's demands for the return of Georgia and acceptance of Turkish consuls in Russian cities—this Catherine could never accept—he was hastily imprisoned in the grim castle of Seven Towers. To Catherine's horror, the result of such disdain for diplomatic protocol was unmistakable: It meant war. Whether unintentional or not, the gauntlet had been thrown.

And war is what Catherine got. Now, in the late summer of 1787, the hour had struck. On August 20 Ottoman ships assaulted two Russian frigates, touching off a six-hour battle. So that Turkish intentions wouldn't be mistaken, the Ottoman grand vizier galloped through the streets of Constantinople under the banner of Muhammad, demanding holy war against the infidels. Out came the fearsome cry—*"Uhra, uhra!"* (Kill, kill!) The prime Ottoman aim: the restoration of the Crimea to the Muslim heartland.

As one of the members of Catherine's dreamlike expedition fatefully wrote, "The performance is over, the curtain has fallen."

∞

IT HAD. AFTER six months, Catherine's expedition returned to the long, white nights of St. Petersburg. Her traveling companions were presented with medals marking the occasion, embossed with a profile of the tsarina on one side and the Crimean itinerary on the other. The inscription noted that the expedition had occurred on the twenty-fifth anniversary of her reign, adding that it was undertaken for "the public good." The tsarina and Emperor Joseph had already parted for the last time. Within three years, he would be dead from tuberculosis. They would never see each other again.

Difficult questions remained. As both sides hardened toward the inevitable Armageddon, had Catherine displayed more enthusiasm than judgment? To be sure, she was guilty of all the familiar sins of great monarchs—

overestimating the power of her armies, assuming the empire that served her would do so without hesitation, and believing that she could, by sheer force of will, personally alter the course of events. This may have been a study in arrogance, but it was not far off the mark. Stubborn and brilliant, she had always been driven by a vision of Russian greatness. Why not now?

She could handle conflagration; indeed, war, or at least brinkmanship, had long been her secret goal. But less palatable were the twists that followed: accounts of Europe's recalcitrance poured in, and its monarchs were dangerously lining up against her. For the empress, once the "arbiter of two continents" during the American Revolutionary War, this was an especially bitter pill to swallow. Nor was it the only one. Europe and the rest of the world seemed to be in the grip of change, a quickening pulse that she was unwilling to acknowledge or unable to fully understand. As France's ambassador, Ségur, had prophetically uttered on the fateful Crimea trip, "The air of our age is that of philosophy and freedom. It is slowly spreading everywhere, and enters palaces as well as huts. It cannot be kept down." And more menacingly, he thundered, "If force is used, as England recently attempted in America, it will only transform into a furious wind."

Actually, one day in her imperial carriage, after having departed the Crimean capital, Catherine felt the first gusts. She overheard a remark made by the English envoy to the French representative that stunned her. The envoy insisted that, all things considered, the American Revolution that had stripped the crown of its thirteen North American colonies would do more good than harm to his country by ending the expense of managing far-off lands. For the tsarina, this was madness. Not that she didn't comprehend the dramatic impact of the American Revolution—a keen student of politics, she grasped all too well the potential global implications of this young, fringe republic. But understanding was one thing, sympathizing another. "I do not know if King George III is of the same opinion as his ministers," she coldly exclaimed the next day, "but I know that if *I* had lost ... but one of the thirteen provinces taken from him, I would have blown my brains with a pistol!"

And when Ségur declared that in distant Paris, King Louis XVI had summoned the fateful Assembly of Notables—the first act of the French Revolution—the empress publicly congratulated him. "Everybody's mind," she said tactfully, "was secretly stirred up by liberal sentiments, the desire for reform." But away from prying eyes, she was more honest, fuming, "we're not impressed." Little did she know that as 1787 came to an end—from Paris to Philadelphia, from London to Kiev, from Hungary to Belgium—these epic transformations would be a mere overture to the great paroxysms to come. Or that by 1790 her ailing and enfeebled ally Joseph, much like his sister Marie Antoinette, would be facing rebellion within his own empire.

Back in St. Petersburg, Catherine uncharacteristically began to feel the strain of war. "I am working like a horse," she anxiously wrote. "My four secretaries are not enough."

It was a remarkable turn of events. Here, in the nation's capital, Russia's "window on Europe," the sights and sounds and grandeur of the great empire were never far away: This time of year the city was drenched in light, up to twenty-two hours a day; by midnight, looking east toward the vast expanse that was Russia and beyond, one could see a pink ribbon marking the start of another dawn. Palace windows were propped open to catch the salty air from the Gulf of Finland, and outside the streets shook with the bustling sounds of imperial carriages, the clank and clatter of assorted vendors, and the sonorous peals of church bells. But behind the thick closed doors of her palace, Catherine, the Empress of All the Russias, ruler of the world's largest and most absolute autocracy, heir to the House of Romanov and keeper of the history of a great state, began privately to waver. Juxtaposed with this northern serenity was the grim reality of what was to come.

Catherine sensed the Ottomans would fight her troops, foot by bloody foot.

So as the weeks passed, despite her legendary singleness of purpose, she often lumbered into her study and sat, uncharacteristically, with her face in her hands, her gaze tired and absent, fixed on nothing. Yet for all her heartaches, she could not bring herself to abandon her expansionist dreams. She knew that the sprawling 2,400 year-old capital of Byzantium was as vital to the world's trade routes as it had been in 85 B.C., when the Roman general Sulla signed a famous treaty with Mithridates VI, King of Pontus, in the ancient city of Dardanus. And there were the calls, of course, of Russia's own past, even harder for her to ignore. In 1711 Catherine's grandest predecessor, Tsar Peter the Great, seeking to avenge humiliations at the hands of the Turkish sultanate, had proclaimed himself the liberator of all Balkan Christians and called on all Catholics, as well as Orthodox believers, to rise up against their Ottoman masters and ensure that "the descendants of the heathen Mohammed were driven out into their old homeland, the sands and the steppes of Arabia." Was this, then, not Russia's unfinished destiny waiting for her to fulfill?

But against this sentiment was another haunting reality. As all-powerful as she was in Russia, she also knew that that failure rarely came without a price. A century earlier, one of Russia's great generals had lost his first campaign against the Tatar khan; the news leaked out, spawning ugly stories. Before he left on his second campaign, an assassin made an attempt on his life. And on the eve of his foray into battle, he discovered a coffin standing outside his door. Affixed to it was a note, announcing that if the next campaign was not far more triumphant than the first, this hard, dark box would "be his home."

∞

THROUGHOUT HER LONG, eventful reign, Catherine had dragged Russia from "semi-Barbarism" into one of the most richly creative cultures in all of Europe. Even her greatest critics, those who derided her as a "a Tartuffe in a skirt" (Pushkin) and criticized her as "debauched," a "bully," and a startling exhibitionist (Frederick the Great once announced, "It is terrible business when the prick and the cunt decide the business of Europe"), or those who spurned Russia as simply "Asiatic pomp blended with European refinement," had to admit she presided over a period of not just empire building, but of glittering intellectual and cultural achievement. This ferment of activity included not just experimentation with new political ideas, but with philosophy and science, music and art, language, education, and architecture.

To be sure, she was responsible for Russia's "disconcerting fickleness" in foreign affairs, as well as its immense appetite for power and conquest. Also to be sure, though the French-speaking court of St. Petersburg was part of the cosmopolitan aristocracy of Europe, portions of the Russian psyche remained painfully medieval, disdainfully referring to "the Peninsula of Europe." And there was no denying that many of the liberal reforms she had once ambitiously harbored as a young ruler had been cast blithely aside. But Russia still had a surprisingly open spirit. Did she not receive the adulation of the leading philosophes of her age, more so than the King of France, or even as much as the republican upstarts in America, the world-famous Benjamin Franklin included? Was she not an ardent patron of the arts? A champion of education, a reformer of ancient legal codes? Did literary salons not flourish in St. Petersburg, rivaling those in London or Paris—or for that matter, did not George Washington, that famed rebel, see her universal dictionary as a step toward promoting "the affinity of nations"? And her supporters were just as surely correct in asserting that Catherine could be witty and vivacious, focused and visionary, an outsize force of nature—and always resolute. Francis Dana, the American diplomat in Russia, and John Adams's good friend, was equally observant in dubbing Catherine "an incalculable planet." But it was Voltaire, one of the Enlightenment's greatest philosophers, who was perhaps most correct in asserting Catherine had "the soul of Brutus with the style of Cleopatra."

Still, there was always that streak of stubbornness that could be her undoing. Dana would also make this telling riposte: "With advancing years, Catherine had herself become a Muscovite—caught in the toils of old European fears. Casting aside the scepter of Europe, the machinery of Leagues and the Mediations, she now listened to the ancient call of eastern conquest."

And night after night in September, as Catherine stood at her desk pacing in her study, receiving dispatches from Potemkin, her commander-in-chief in the Black Sea, the news was hardly reassuring. Was her prince holding up? Was the key Russian stronghold in Kherson safe? Would the Turks attack before Russian forces were ready—or Austria's Joseph still honor his part of the bargain? For the empress, the signs were unsettling. She received the following note from Potemkin: "I am very weak, millions of troubles. . . . Not even a minute's rest, I'm not even sure I can stand it long." She knew anything could still go awry, and in fact it did. In this confused, chaotic war, the first great casualty came not from shots fired by Turks or Russians, but the weather. On September 9 the Black Sea whipped up into a vicious storm, and Catherine's fleet in Sebastopol—the very hub of Russian forces and key to this war—was destroyed. "I've become despondent," Potemkin wrote in explanation, adding soon thereafter: "God strikes us, not the Turks."

Then, suddenly, another ominous development. Potemkin's letters abruptly ceased.

∞

YET EVEN AS Catherine's court was relentlessly fixed on the Ottoman struggle, another equally fateful drama was unfolding back in the West.

With the passing months, France's ambassador, the Comte de Ségur, began anxiously awaiting each dispatch from his own restive nation; so did Catherine herself. Benefactor of the Americans in both money and blood, France was now about to taste its own providential impulse for freedom. As Ségur furiously wrote, "I was far from being the only one whose heart palpitated at the sound of the awakening of liberty, the struggling to shake off the yoke of arbitrary power." And off in Paris and Versailles, that momentous struggle was about to take a decisive turn.

CHAPTER 3

FRANCE

✧

C'est légal, parce que je le veux," King Louis XVI once famously uttered to silence a royal dissenter: "It is legal because it is my will."

In 1787, while Catherine and the Ottoman Empire lurched toward a full-throttle war and the American Founders gathered in Philadelphia, the glory of France towered like a beacon. Across the globe, its language was the tongue of all aristocracies, and it reigned as the most powerful country in all of Europe, if not the world. Militarily, a great war had been won, Britain had been humbled, and France's ally, the rebel Americans, had gained their independence. Economically, France's reach spanned the continent: The French upper-middle classes invested on the stock exchanges in Paris, London, and Amsterdam, and it was estimated that they controlled half the money of all Europe; it was even said that fifty percent of all gold pieces circulating on the continent were French.

But at home, as it had for centuries, French life invariably revolved around the strange and fascinating world of the king. If France was the most advanced country of its day—it was—and the radiant hub of the Enlightenment, Versailles, built originally for the Sun King, remained the core where the cult of the monarch and the glory of the House of Bourbon were celebrated on a daily basis. An isolated, miniature world, it was as artificial and dazzling as anything ever seen on the planet. Every moment of the king's day belonged to the unending routine of court life, from the clothes he wore, to the food he ate, to the royal etiquette dictating his every public utterance, to the time he went to sleep.

At each minute of each hour of each day, like a finely regulated watch, four thousand courtiers regularly promenaded and attended to the sovereign's needs; it was impermissible even for the king's brother to sit in his presence. Beyond the elaborate rooms, even the palace hallways vibrated with the constant bustle of the court. At great state dinners, thousands of candles shimmered and hundreds of dresses swayed as ladies dropped to curtsy when the king walked by. No less than the *bouche du roi*, the "king's kitchen," was spectacular: the thirteen chefs and five assistants of the Grand Pantry, the twenty royal cup bearers, the sixteen hasteners of the royal roast, the four carriers of the royal wine, and an entire coterie of tasters and candle snuffers and salt passers. And the 226 apartments of the château, not to mention its 500 narrow attic rooms, were filled with France's most privileged elite, aristocrats of great rank and women with velvet-lined coaches and diamonds as large as walnuts—in short, those favored by the king himself. Even to be in the garret of Versailles was considered a mark of the highest order, for every man and woman knew that his or her destiny could be changed on a whim by the king and everyone secretly dreamed of being singled out by the all-powerful ruler. A week of excited conversation could be extrapolated from a simple royal nod, a slight smile, or a brief word spoken in an unexpected direction.

So rigid, so widely accepted, was the immutable Versailles protocol— a mysterious blend of the momentous and trivial, the significant and purely formal—that an innocent visitor could easily be misled into thinking that nothing would ever change in the kingdom of France—nothing. And why should it? The people widely accepted the monarchy, and even loved it, looking to its enlightened despotism as an engine of social reform or a source of comfort. Not unlike the massive Russian populace who saw the tsars as divine patriarchs, the French peasants put ample faith in their kings, believing they too could make all things right. It was familiar, it was reassuring, it was divine and patriotic. And century after century, it was theirs.

The coronation of King Louis XVI, in June 1774, governed so precisely by history and tradition, was testament to this. Kings reigned by the grace of God from the instant when their predecessors drew their final breath; coronations were simply public celebrations of what was already established fact. But in 1774, the countryside and even Paris itself had been convulsed by riots over exorbitant bread prices; it was also openly whispered that the elaborate coronation would be a flagrant misuse of public funds. Violence was feared as well, so royal troops were called out as a precaution. And in the highest circles, there were even thoughts of delaying the ceremony itself, as fewer people than ever were expected to witness the historic spectacle. Yet on that luminous morning when the cathedral doors of Rheims swung open to unveil the young Louis, crowned and invested with France's glory, touched by the scepter of Charlemagne, and anointed with the oil of Clovis—each drop believed to have been whisked from heaven by a dove—the assembled masses broke down and wept despite themselves; for that matter, so did the queen.

To great expectations, Louis XVI solemnly swore that day to uphold the peace of the Church, to squelch disorder, guarantee justice, exterminate heretics, and maintain forever the Order of the Holy Spirit. A teenaged Maximilien Robespierre, tears in his eyes, was among those attending in solemn tribute. And three days later, in stifling summer heat, the young king pronounced, *"Dieu te guérit, le Roi te touche"* ("God cures you, the King touches you"), and placed his hands upon 2,400 sufferers from scrofula, the horrifying skin disease that many believed could be cured only by the touch of divinely anointed royalty.

But the old monarchs who had given character to the world were vanishing. And little did the new sovereign or his court realize that the kingdom was perched perilously on the brink of utter ruin. Seduced by its own majesty, entranced by its own ritual, and mesmerized by its own isolation, the monarchy's eyes would be opened too late.

∞

BUT NOT YET. For the aristocracy in 1787, pleasure was the watchword. In opulent surroundings, ladies flaunted their décolletage and shimmering silks, gentlemen of leisure flocked to the hunt, men of fashion read Molière aloud and aristocrats gathered to play silly children's games. The gardens were impossibly green, the food was bountiful and succulent, pleasurable desires were guiltlessly fulfilled. And everywhere one detected the pursuit of happiness—the happiness of play, the happiness of time wasting, the happiness of home, all a celestial paradise on earth. Even America's John Adams observed, "The delights of France are innumerable."

But beyond the frivolity and atmosphere of decadence, imitated and aped by the over-rich courts across Europe—in St. Petersburg and Vienna, Stockholm and Berlin—the French economy, society, and politics were feeling the tug of modernization. Foreign trade was booming, and professionals, scientists, and intellectuals were increasingly provided positions of influence and respect. And too, the life of the average bourgeois was improving: Coffee from the New World was ubiquitous, and café au lait was routinely drunk by the urban working class. As the sparkling shop windows of Paris fueled their aspirations, common people increasingly indulged in what had once been exclusively aristocratic luxuries: extra clothes, foot warmers, even wallpaper. Billiards, cards, and backgammon became the rage. So did showy furniture, such as inlaid writing tables, red leather card tables, and coat stands, all increasingly commonplace. And in the salons and the academies, among the bourgeois and a growing national network of Freemasons, a new spirit of citizenship was arising—based on liberality and justice. Louis XVI himself grasped this new mood and temper. So he gingerly professed the highest respect for public opinion; he smartly chose exceptionally talented and enlightened ministers to direct and oversee the needs of the kingdom; he boldly allowed the philosophes to be used as policy consultants; and he radically experimented with "Physiocratic theory," which supposed that the free market would overcome both the stagnation of the French bureaucracy and the increasingly overextended French state. And he even began to dabble in new systems of political representation and a modernized judicial system.

In turn, across the rest of the world, everyone wanted to see France. By now, Thomas Jefferson was in residence there as America's ambassador, and, after the American Revolutionary War, no less than such prominent figures on the world stage as Britain's William Pitt and William Wilberforce, the great abolitionist, flocked to Paris. So had, already, Catherine the Great's son, Grand Duke Paul—brushing shoulders with Benjamin Franklin along the way. And so had Austria's Holy Roman Emperor Joseph II—twice. As late as the summer of 1788, there was even the fantastic spectacle of three Muslim envoys of Tippoo Sahib, King of Mysore, India, wandering about Paris. The three ambassadors, clad in headdresses, yellow-slippered feet, and green and red leather uniforms, and flanked in the background by a gaudily fantastic array of turbaned slaves, were given an audience with the king; they had come to seek his majesty's protection in curbing England's power.

They were also lavishly entertained: Ferried about in a carriage drawn by six high-stepping white horses, they were given the best seats at the opera and a private viewing of the Versailles gardens.

Indeed, to those who remembered, the season of 1787 seemed especially brilliant. The tall windows of Versailles blazed with light. The streets and shops were filled with bustling crowds. Women did their hair, lounged in their salons, and gossiped deliciously. The fleur-de-lis still flew grandly from naval bases in all points of the empire. And the mood in the capital was one of hope.

But as the Comte de Ségur put it, France could not escape the air of freedom blowing across the rest of the world, nor could it quell the simmering mood in its own country. Day by day, feudalism and privilege were coming under assault, and not simply from below. As one Frenchman—actually, a nobleman—would soon tantalizingly declare: "The hour is striking for us to finally think and talk for ourselves."

Another aide to the king would put it more bluntly: "Your monarchy will be destroyed and the state with it."

∞

BUT WHO COULD have seen it coming? The answer, actually, is very few. France's problem—and it was the problem—is that it paradoxically could not escape the rest of the world. The triumph over Britain was a great feat; but like all wars throughout the ages, it carried unpredictable consequences. True, Louis's birthday was a national holiday in America, a tribute to his enormous contribution, and equally true, his hallowed portrait graced the fledgling country's public spaces. These sentiments were real, but they would also prove to be woefully blinding. Here, Catherine the Great was right. Saddled with debt—the king had sent a staggering $240 million to the colonies in their war for independence—and absorbed in its own domestic affairs, France was increasingly reduced to hollow diplomatic gestures, having, for example, neither the will nor the ability to come to the defense of its traditional ally, the Ottomans, now locked in a fresh conflict with Russia. The government would not even be able to muster 30,000 soldiers to sit on the border with Holland, soon to be consumed by its own antiroyalist revolutionary fervor.

And then there was the slow virus of America, the whiff of liberty emanating from Louis's "honorable war."

Across France, the American Constitution was hailed as a harbinger of new things to come; Jesuit priest turned social critic Abbé Raynal wrote a widely read tract, *La Révolution de l'Amérique*, while eight thousand Frenchmen who had fought alongside the "heroes" of the American struggle returned from the United States with an infectious message: The American rebels had taken charge of their own destiny and cast off the rule of an oppressive king. Suddenly, in the coffee houses and salons, in the bustle of France's great cities and the quiet of its far-off provinces, change seemed possible. Never did this claxon sound more loudly than in one profoundly symbolic incident. Upon his return to France, the Marquis de Lafayette, the young disciple of George Washington and acclaimed hero of "two worlds," was graciously called upon by the queen, Marie Antoinette, a powerful gesture from the royal family. While Lafayette strode out to her carriage, and thunderous fireworks were exploding overhead, the crowd roared in a long ovation. But neither the queen nor the milling throngs heard Lafayette's subsequent observation. Whether said in a hush or a shout, the effect was the same. Unimpressed, he ruminated philosophically that the cost of a lavish court ball could have equipped "a whole regiment" in America. Indeed, increasingly, rituals and grandeur no longer held the people blindly in their spell; and figuratively and literally, Lafayette and countless others were now inhabiting a different world.

The elites of France were ready to fling their caps for independence— even if most of the nation didn't yet know what that meant.

∞

As 1788 DAWNED, the first climate of crisis began to envelop the monarchy. It came in the form of pitiless "acts of God." The winter was the harshest in eighty years: Couriers en route the twelve miles between Versailles and the capital froze to death; in Provence, so did birds, iced to their perches. Meanwhile, wolves prowled in Languedoc and the poor in Ardeche resorted to boiling tree bark for a bitter gruel. Then a hailstorm ripped through central France, devastating 180 miles of fertile terrain—even fruit trees and olive orchards perished by the thousands—and laying waste to a vast swath of land from Normandy to Toulouse. Then came the severe drought, which stunted crops and made a wreckage out of the harvest. While the rich donned furs, slipped on ice skates, or traversed the cold on sledges, the poor suffered and prayed for their very lives. In Versailles itself, there were bonfires at the crossroads and the king often stopped to give alms to the poor. Complicating matters, outside foodstuffs and supplies were hard to come by.

France never had efficient food distribution; year after year, during the height of the summer, grain stores in Paris itself ran dangerously low. But there was now a new and disturbing predicament. Catherine's ongoing war with the Ottoman Turks in the Mediterranean had cut off key supplies for the south of France. Meanwhile, another one of her conflicts, in the Baltic, would only exacerbate the situation further, impeding food sources from Poland and east Prussia. So across all the French countryside, there was no work, there was crushing debt, and, most ominously, there was hunger. And there was turmoil—bread riots broke out, the provinces seethed, and the nation hovered at the edge of financial collapse.

Ironically, France's debt was only half as great as Britain's and less than that of the Dutch Republic. But its tax system was antiquated and arbitrary. The Catholic Church didn't want to pay taxes, nor did the privileged classes of France; thus, while the country itself was actually prosperous, the government treasury was frequently empty. So Louis XVI found himself straddling an uncomfortable middle: Was he foremost a divine monarch, the beloved source of all law and dispenser of privileges? Or was he a patriotic voice of the bourgeoisie, a faithful steward of the public sphere? Or—unthinkably—something altogether different: a citizen, like everyone else?

The first stone of the House of Bourbon was about to crack.

∞

IT WAS AGAINST this atmosphere of gloom that Louis's finance minister, Charles Alexandre de Calonne, proposed radical economic reforms, including a uniform land tax, to replace some of the more outdated measures of the kingdom—indeed, up to now, collectors based their assessments on such imponderables as the mere appearance of wealth. He also sided with the poor, arguing that they should no longer be conscripted for public works projects. To his credit, Louis was persuaded. But always a cautious man, he was extra-cautious here. The king wanted to bypass the Paris Parlement, which was a sovereign court and a powerful political institution unto itself. So he boldly summoned the Assembly of Notables, an archaic body that hadn't met since 1626, to bless his audacious program.

Remarkably, Louis XVI, the Most Christian King of France and

Navarre, who was addressed on bended knee, was appealing for public support.

To a large degree the king's program should have worked, averting the crisis to come. But where the thirteen American colonies fell into convulsions over the perceived unfairness of the British crown's taxation and representation system, France was about to be upended by an effort to modernize its worst elements. The Notables first met on February 22, 1787—by happenstance, it was George Washington's birthday. Lafayette, sensing that great events were about to occur, had by now turned down Catherine's offer to join her in Russia on her grand expedition. But as the Notables bickered and dithered, the young marquis sneered to Thomas Jefferson that they should be dubbed "Not Ables." Yet the Notables kept meeting, and slowly the ferment grew. In truth, while they may have been intensely aristocratic, many were also politically radical; sensing royal weakness, they refused to endorse the king's proposals and instead upped the ante, insisting on concessions and shared control of the government. Such dramatic actions would have horrified Louis's grandfather, Louis XV, not to mention the Sun King himself. But these were different times, and the discussions, in fits and starts, bordered on undermining the very foundations of the monarchy itself— even if no one really quite saw it that way. Then, at least.

Buffeted by the debate, the king, a decent man but neither a democrat nor an autocrat, could not shake his habit of nervous vacillation. He had already earned the love of his people as "the good King Louis." But had he not earned their fealty? Had he not gone further than any of his predecessors? To his contemporaries, Louis could comfortably boast a long list of enlightened deeds: He had freed the last of the serfs on his own lands, forbade the use of torture in criminal examinations, proposed demolishing the hated Bastille and banning the hideous dungeons of Vincennes, and uniquely refused to punish free thought, each stunning in its own right. Notwithstanding the extraordinary American experiment, in the mental world of 1787, notions of popular sovereignty, civil liberty, and equality before the law scarcely existed. It was widely assumed that inequality was a good thing, that it conformed to the hierarchical order established in nature by God himself. In this regard, then, Louis was perhaps a victim of double injustice. But it no longer mattered. The sacred aura of the crown was slowly being tarnished, and events were moving too quickly now.

"It's a delirium," the Swedish count Axel Fersen noted darkly. "Everyone sees himself as a legislator and everyone talks of nothing but progress."

Progress yes, but no one could agree upon what type of progress. Nor would the delirium break. Lacking a public mandate, the Notables were unable to reach their own conclusions. Deadlock ensued. And as the debate became increasingly passionate, Louis, still feeling his way in a vanishing baroque world, was wracked by a deadly cocktail of indecision, apathy, gloomy presentiments, even melancholy. One day he made his way to the queen's chamber and wept in private.

Trying to regain control, the king dismissed his chief advisor, Calonne—the unfortunate Calonne was burned in effigy on the Pont Neuf. As his successor, Louis appointed Lomenie de Brienne, the worldly-wise archbishop of Toulouse and a favorite of the queen's. Brienne tried to push much the same program, this time through the Parlement of Paris. It too balked. So in November 1787 the king made a rare appearance at the Palais de Justice to sway the members; unwittingly, however, he had walked into a trap. Both sides seemed to follow their political scripts until Louis XVI, his patience sorely tested, all but ordered Parlement to accept his edicts. This was perhaps predictable. His days so routinely cannibalized by sycophants and courtiers, Louis was not used to such give-and-take. Nevertheless, after a moment of pausing to measure each other's inflections, Louis's own cousin, the Duc d'Orléans, rose to his feet and defiantly spoke out against the king. To be sure, the duke was fickle and enigmatic, a panderer to popular tastes and a blustering gadfly, and was openly scheming to capture the throne. He also bitterly hated the queen. Yet the rebelliousness in his tone was unmistakable, and by any measure, this display was a staggering act of defiance. The king had come here, in person, to plead his case. And for the first time in the history of the monarchy, a prince had dared question the very fabric of royalty. Actually, intoxicated by his own bravado, the duke's tone grew more querulous. Apoplectic with rage, the king muttered an icy reply and hastily stormed out, betrayed and humiliated by a member of his own royal family, a man who required the king's permission to embark on a simple trip.

Glumly, Louis and his ministers pressed forward in the face of near universal condemnation: Provincial estates now hesitated to act, army officers refused to serve, and independent political clubs were organized. Watching the mounting crisis, Thomas Jefferson wrote that Paris had become "a furnace of politics" and feared that a kind of madness had overtaken the populace, man, woman, and child. His words were prophetic. Soon insubordination toward royal authority itself rapidly began to spread. First came the uprisings: in Besançon, in Pau, in Rennes, followed by Flanders, Midi,

Dauphine, Provence, Languedoc, and Guyenne. Then came a true insurrection, in Grenoble, innocuously called the "Day of Tiles."

It was the first salvo of the French Revolution.

∞

THE AFFAIR BEGAN innocently enough. On June 7, 1788, in Grenoble, an industrious town at the foot of the Alps populated by hemp combers and artisans who made fine gloves that were sold as far away as Philadelphia and Moscow, the government sought to exile recalcitrant magistrates who were defying Louis's royal decree that local parlements be dissolved. Two regiments of the king's soldiers—the Royal-Marine and the Austraise—were sent to convince the officials to go quietly. But word of the king's heavy hand soon spread. Stalls and shops quickly shut, and a mob streamed into the center of town. Then, over the din of speeches and insults, the mob decided to take matters into its own hands. One group tightened to protect the parlementarians, by force if necessary. Another group fanned out to bar the city's gates and hold back the arrival of military reinforcements. A third was dispatched to the governor's house. But up to this moment, there had been no direct confrontation with the king's troops. A fragile peace still held.

Not for long. Absent direct instruction from Versailles, the commander of the garrison faced a series of agonizing decisions. Should he unleash his troops into the streets to suppress the crowd—or merely seek to restrain it? Should they be fully armed—or hold fire? Would wanton force be hotly criticized—or halfhearted measures make things worse? As with the British before them, facing down their fellow subjects in the American colonies, as with so many imperial powers later, as with even the inchoate American militia confronting the Shays rebels, the decisions were agonizing.

The soldiers held back, a fateful error: They had infuriated the mob, but had not intimidated it. Suddenly, from the rooftops, the troops were pelted with a rain of thick slate tiles. Emboldened, the mob moved forward. In a moment of horror, one regiment panicked; taunted beyond endurance, they opened fire. The bullets smacked into the thigh of a twelve-year-old boy, who soon died, as did a hatter. Crimson blotches stained the ground. A middle-aged journeyman was then fatally bayoneted in the face by a royal trooper. Now the frenzy built. The blood-drenched clothes of the dead were hoisted in the streets, while the cathedral bells tolled for peasants in the countryside to join the melee. They did. The defiant mob turned next to the governor's house and angrily sacked it. The governor's furniture was hauled into the streets and a bonfire was lit, ignited by the cheering, triumphant mob.

It was a moment that would haunt the realm again and again in the years that followed. Royal authority had totally dissolved, and the vaunted king's military was impotent in the face of urban riot. But the day became a profound turning point for another reason. While the victims numbered only a handful, their deaths shattered the ancient, legendary belief that the king, the *père nourricier* (the father provider), and the people were irretrievably one. Reflexive loyalty to the crown was unraveling.

Myth has a significance of its own. And henceforth, the myth of Grenoble as the "cradle of the Revolution" grew.

∞

AT VERSAILLES, LOUIS XVI was horrified when he heard the news. How could this happen in France? By all the rules as he knew them, his men, as representatives of the crown, should have been treated with great deference. Yet the clumsy actions of his army only seemed like premeditated cruelty; just as worrisome was the utter chaos. His ministers met hurriedly to consider the problem. But Louis's own inner struggle was almost irresolvable. For nearly two centuries, a delicate balance had prevailed, reassuring the French people that the prerequisites of the crown were compatible with the welfare of the people. And this was a nation that for some time had embraced him with such warmth. No more. The two great ideals, of the Bourbon dynasty and the rights of commoners, were drifting dangerously apart, and the king was foundering in the gulf between them. It was at this point, on August 8, 1788, that a desperate Louis issued a historic call and stunned the nation. Bowing to growing sentiment, he summoned the Estates-General—the assembly of the clergy, the nobility, and the populace—to meet the following May to advise him on how to deal with the problems of the realm; interestingly, Russia's Catherine the Great had done much the same early on in her reign.

Never before in the memory of living Frenchmen had a king sought the advice of his people; indeed, the Estates-General had not been invoked since 1614. But there was now a seemingly vast frontier separating the royalists and peasantry, the church and the professional classes, the aristocracy and the middle class. Actually, Louis's moves were farsighted and far-reaching, and held great promise for holding the country together; by convening the Estates-General, Louis was inching toward the profound step of transforming the monarchy into a representative regime. The people themselves sensed this too. The proclamation was greeted throughout the country initially with elation: torchlight parades, fireworks, brightly lit windows, and songs preaching love for the king.

It was all to be fleeting. The battle lines were too sharply drawn, and these actions, rather than being taken as forward-looking, were ultimately seen as a sign of weakness, even as mass demonstrations insisted on more. But more what?

∞

THAT WAS THE question for the king. Tumult and intrigue increasingly marked his every step. Like other monarchs under duress, Louis XVI undertook one more move: He again shuffled ministers, recalling Jacques Necker. Neither a Catholic nor a Frenchman, but a Swiss Protestant financier, Necker had run the treasury before, yet had been dismissed in 1781 after some of his spending proposals displeased the queen; at this stage, however, he seemed to be Louis's only hope. It was a calculated risk, calculated in part because the king despised Necker. To Louis, the Genevan was conceited, cynical, and, by his own lights, infallible; he harbored dangerous "democratic" ideas as well. But ever a dabbler in shades of gray, he was also enormously popular with the people, and his appointment was greeted with outbursts of joy.

"I did not want to do it," the king privately confessed. "But . . . I'll do everything he tells me to do and we'll see the result."

Experienced and shrewd, Necker now swept in as the freshly crowned peacemaker. And though Necker was a loyalist, he was no stooge; moreover, he possessed one of the ablest minds in France. While his emergency measures and very presence temporarily rekindled confidence in financial circles, he also quickly saw that the country was in complete turmoil. Especially worrisome was the situation in the capital itself. Tensions stewed and mounted over the winter until April, when a mob set its sights on a prominent wallpaper manufacturer, Jean-Baptiste Reveillon. Rumormongers accused him of cutting wages—Reveillon vehemently denied it, but the mob would not be deterred. Protesters, armed with sticks and swelling into the thousands, marched across the Seine, closed traffic, and carried mock gallows. To avoid another Grenoble, Louis's men swiftly dispatched knots of sentries to guard public buildings, while squadrons of soldiers clattered up and down the boulevards. Meantime, a huge throng massed on rooftops and in windows in anticipation of the looming confrontation.

It wasn't long in coming. Some ten thousand rioters converged outside Reveillon's house, where they were met by the Gardes Françaises. Additional troops were called in, but it was too late. Protesters poured through a gap in the lines, destroying Reveillon's home and setting his factory ablaze. By the time the troops' bullets riddled the protesters, the people were already shrieking, "Death to the rich! Death to the aristocrats!"

The day's horror was compounded by its speed, by how rapidly anarchy had taken over. Reports of the number killed ranged from twenty-five to nine hundred, with at least three hundred injured. More ominous still, as Simon Schama reminds us, was the fact that Reveillon was a modern man, a liberal, an exemplary employer—everything the French were calling for. But something had snapped in the people; their mood was now ugly. By contrast, the Bourbon body politic seemed hopelessly paralyzed.

Worried that the country lay at the brink of civil war, and that the throne itself might be imperiled, Necker seized the reins as prime minister without title. He energetically proselytized the king to mobilize public opinion on the eve of the now all-important opening of the Estates-General. One of his first acts was to sway Louis XVI to align himself not with the nobles or clerics at the Estates-General, but with the Third Estate, the commoners. The king agreed. He fatefully broke with tradition and doubled their number of delegates. But then he temporized, insisting the Third Estate would have more representatives, but no additional votes. Thus they would outnumber, but could not outvote, the other two orders.

Was this a gesture of enlightened statesmanship, or a slippery slope toward suicide? Whatever the answer, it was a brazen roll of the dice. Not unlike the new American House of Representatives, the Estates-General would be decided by votes in local assemblies, the closest approach yet made by France toward universal suffrage. The king also requested that all the electoral assemblies submit to him reports (*cahiers*) of their grievances.

Here, of course, was a classic formula for eventual conflict. The *cahiers* from the assemblies—in total some 615 were presented to the king—demanded trial by jury, privacy of the mails, and extensive reform of the law. But where the nobles clung tightly to "preservation" and "maintenance" of privilege, the Third Estate clamored for a reduction of taxes, protection of farms from injury by royal hunts, and the opening of careers to the middle class regardless of birth. Where the clergy wanted an end to religious toleration and the ceding of exclusive control of education to the church, the Third Estate stonily countered with demands for universal and free education. And where all three estates expressed loyalty and affection for the king, none spoke of the divine right of the monarchy. To the extent there was consensus, it was that the king should share his powers with an elected assembly that would, in effect, render him a constitutional monarch.

There was also one class of people entirely left out of this fateful drama—the workingmen, or, as they would become known, the sansculottes (because they wore simple pants, not the knee breeches of the upper classes)—the laborers who toiled as butchers, bakers, brewers, grocers, cooks, lace makers, hat makers, masons, plasterers, dyers, and blacksmiths, numbering perhaps some 500,000 strong in Paris, but unable to pay a poll tax of 6 livres or more. For these slop carriers and laundresses, who made the manors run and the cities work, their dark rooms were unheated, their clothes were tattered, their poverty was fearsome. And in the antique formula of the estates, where each division even had its own peculiar costume, flowing robes for the clergy, billowy hats for the nobles, and a severe black suit for the Third Estate, they remained invisible and unrepresented, consigned to the shadows. Indeed, as in America, the king and his advisors were looking to the bourgeoisie, not the illiterate and emotional masses, as the engine of reform and savior of the nation. But the sansculottes would not stay silent. They too would be avenged. Their cobblestones would one day shatter gilt mirrors, their dirty feet would one day soil gleaming parquet floors.

By then it would be too late. By then the country would be rushing full tilt into revolution.

∞

ON MAY 4, 1789, dawn broke as a brilliant, radiant day in Versailles. Thousands of people lined the streets, jammed balconies, or even clung from branches of trees as the procession of the Estates-General began. It was the same throughout France—everywhere wild excitement and crowds filling the boulevards; almost magically, a wave of patriotism and hope had swept the country. More than resolving abstract questions of governance, a majority of the people believed the Estates-General would end their hunger and give them a voice. So facades of buildings were decorated with tapestries unveiled only for great occasions; musicians played at street corners; and regiments lined the main avenues.

At precisely ten o'clock, the king and queen drove up in their magnificent coach. *"Vive le Roi!"* the spectators cried. "Long live the king!"

The 621 deputies of the Third Estate, dressed in simple black, swordless, and each carrying a candle, led the march, followed by 285 nobles in satin suits and silk cloaks; then came 308 members of the clergy, draped in black habits and episcopal robes, then came the king's ministers, his family, and then at last Louis XVI himself, wearing magnificent cloth of gold, and Marie Antoinette, shimmering in cloth of silver ("the moon to the King's sun") all accompanied by banners and bands.

The next day, the Estates-General officially opened. The king was met with a deafening ovation as he entered. Then, with confidence and dignity, he rose. His voice steady, he told the delegates that France was nearly bankrupt; moreover, he called on them to help return the nation to solvency. And in a moment poignant with possibilities, he spoke of himself as "the first friend of the people." Necker followed with a three-hour speech—enumerating the dreary particulars of economics, taxation, and financial revitalization. When at last the king rose again, he was greeted anew with wild cheers. On the next day, however, this burst of unity quickly fizzled, as the clergy and nobles separated; both believed they should discuss and vote apart, as the last Estates-General had done 175 years before, and that no laws should be made without the unanimous consent of each order and the king.

The Third Estate, however, had other ambitions.

∞

WHILE THE KING anxiously waited for the ancient body to give its assent to his new taxes, the Third Estate, knowing its acceptance was also needed, plunged not simply into radical oratory but also extensive plotting and political maneuvering. The fire of rhetoric and passion lay with the commoners: The tongue was their weapon of choice, and so was the propaganda pamphlet. While the clergy and nobility met behind closed doors—not unlike the Founders had in 1787—the Third Estate shrewdly allowed the public to attend its sessions. Overnight, the streets of Paris overflowed with journals, pamphlets, placards, and clubs. And theirs was a rousing message: No longer need the masses feel powerless or inferior.

The first dissident fruits quickly appeared. On June 11, the Third Estate invited the other orders to join them in deliberations; they received no reply. But what followed was even more stunning: Led by Abbé Sieyès, the delegates of the Third Estate then began to consider whether to declare themselves the representatives of the French nation—their version of a Declaration of Independence. A year earlier, the mere suggestion of this would have been a heresy of the highest order. Now it was mere overture. The Comte de Mirabeau, an aristocrat elected to a commoner's seat and the lion of literary Paris, objected. His sonorous voice rumbling with emotion, he argued that the Estates-General had been summoned by the king and could be legally dismissed at his will. Watching among the delegates was another eloquent spokesman, a young lawyer, Maximilien Robespierre. Remarkably, Mirabeau was shouted down. Here among the people's designees, fierce arguments followed and members even came to blows. Yet on the next morning, June 17, 1789, as huge crowds expectantly waited, the fateful question was put to the vote: "Shall this meeting declare itself the National Assembly?" The tally was 491 in favor, 89 against. Just six weeks after their convocation, the delegates had declared themselves the National Assembly of France and had committed themselves to establishing a representative government.

The first stage of the revolution had now taken life. Boasted Mirabeau, "We are beginning the history of man."

∞

EVENTS WERE SUDDENLY moving very quickly. Greeted by the Assembly with tears in their eyes, some breakaways from the clerical order also voted to merge with the Third Estate. Unnerved, the nobility adamantly insisted that if the clergy and commoners united, Louis should dismiss the estates. For his part, Necker advised the king to compromise—to give in to some of the demands of the Third Estate, without recognizing it as a National Assembly. This would be, he contended, a cheaper, easier way of ending the turmoil. Increasingly torn between the strident voices of his court and an aggressive opposition, Louis wavered, and split the difference. But far from acting cowardly, he decided he would deal with this defection by personally offering a reform program of his own to the fractious estates. In an extraordinary cataract of events, in those three days, everything would change, inexorably altering the very ground beneath the king's feet.

On the evening of the nineteenth, Louis XVI ordered the Hôtel Plaisirs to be closed to prepare for a joint royal session of the three orders on June 22. So when the deputies of the Third Estate showed up on the twentieth, they found the doors bolted and armed guards posted at the entrance. Outraged—they believed, wrongly, that the king intended to dismiss them—they grew bolder. This was ironic: Far from being a statement of royal policy, it was a mere oversight, the by-product of the endless array of ritual that followed the king wherever he went.

Outside, a heavy rain fell on the assembled deputies. Meanwhile, tempers were flaring. In a quirk of historical fate, Dr. Joseph-Ignace Guillotin, best known for his infamous invention, the guillotine, remembered the existence of a nearby tennis court—the immortal Salle du Jeu de Paume. There, with their arms raised in a dramatic salute, all but one of the deputies signed an oath, pledging "never to separate, and to meet wherever circumstances might require, until a constitution should be firmly established." Told of these events, the haggard king sputtered that it was "merely a phrase." But "the oath of the Tennis Court" was a phrase that would haunt him until the day he died.

Postponed for a night, the royal session opened on June 23. Louis XVI went far—further than any French monarch in history: He recognized the principles of individual freedom, freedom of the press, and equal taxation; he also agreed to end lettres de cachet, internal traffic tolls, and all vestiges of serfdom in France. Dragging, almost overnight, the nation from an absolute despotism into a near semi-constitutional monarchy, it was a program many of the American Founders could have accepted. But by now, the king had lost his chance for any meaningful compromise. Thus, when he also reclaimed the right of absolute monarchy ("If, by a fatality . . . you were to abandon me in this great enterprise, then I will work alone for the welfare of my peoples . . ."), and when he steadfastly declared the deliberations of this new National Assembly as "null, illegal and unconstitutional," he was met with a stinging reply. Upon the king's departure, an aide ordered that the hall be cleared. The dissident delegates now took their most brazen step yet, issuing their own edict. The president of the assembly abruptly rose to say that the assembled nation could never accept such an order, while the Comte de Mirabeau impatiently declared, "We are here by the will of the people and shall not stir from our seats unless forced by bayonets."

Actually, their bluff was called. But when the king's troops tried to enter the room, a band of liberal nobles, including George Washington's protégé, Lafayette, unthinkably drew their swords and barred the entrance. When an aide rushed to the king and asked what should be done, an exhausted Louis rasped, *"Qu'ils restent"*—"Let them stay."

With these fateful words, the guardian of the Bourbon dynasty marked the last day that he would ever speak or act of his own free will.

∞

THAT NIGHT, CROWDS flooded the grounds of Versailles. The court was now in disarray. Ominously for the king, soldiers of the Garde Française disobeyed explicit orders to remain at their posts and were openly fraternizing with the throng of renegade deputies; within days, whole companies would soon desert him and flock to Paris, vowing never to fire on the people. And there came another bitter blow: The Duc d'Orléans led forty-seven nobles to join the new Assembly, where they were met with wild exhilaration. Two days later, however, the king struck back, declaring that the Estates-General formally ceased to exist.

Actually, inside his château, Louis was biding time. After the dissolution of the Estates-General, another restive crowd gathered in the Versailles courtyard, clamoring and shouting for the sovereign. In a disarming gesture, the king and queen themselves appeared. Marie Antoinette was without her famous jewels, her hair long and now flecked with gray—it had been a mere three weeks since the dauphin, her son and the king's heir, had died of tuberculosis. In a gesture laden with pathos, she hoisted her surviving royal son in her arms and showed him to the crowd. After a moment's hush, there were yet more cheers, as the crowd lingered until dusk. For the moment, it seemed the king's situation was secure. Thomas Jefferson wrote home, lamenting that his subsequent letters would undoubtedly lack for such high drama, "This great crisis now being over."

But Jefferson was sorely mistaken. Not so many of the assembled nobles, who, sensing the storm to come, retired to their country homes. Warily, the Marquis de Ferrieres dashed off a letter to his sister, writing darkly, "Everything announces a great revolution. If only there was one [dependable] man, I would not regard things as desperate." Instead, though, the marquis predicted that the Estates-General would be marked by "a banner of blood that will be carried to all parts of Europe." He would not be wrong. The king's determination now—or rather his hope—was to some-how save the monarchy.

Egged on by the queen, his first step was to call up ten regiments, mostly German or Swiss, to rush to his aid; by July 10, six thousand troops had massed around Versailles, and some twenty-five thousand more began taking up positions around Paris, quickly occupying every bridge and all major streets. These measures were not without risk: Amid turmoil and ter-ror, the National Assembly quickly weighed its own response, denouncing "the aristocratic plot" and demanding that the king order the retreat of the troops. Slyly, Louis retorted that they were there to "maintain order," while privately he was ready to strike a decisive blow: to dismiss his finance minis-ter, Necker, and appoint the hard-line Marechal de Broglie minister for war. It was clear by now that the king, torn between extremes, was considering counterrevolution.

Feeling the groundswell of fear and suspicion emanating from the side-walks of Paris—grain shortages made simple bread a near luxury, hunger was rising, and troops even had to accompany grain convoys to protect them from attack—the Assembly, in an exalted mood, made its own risky countermove, proceeding to consider a new constitution. But this was not the steady gathering of fifty-five men in Philadelphia; it was a headstrong, fractious mix of hundreds of souls, and one riven by rumors of an aristo-cratic plot against them. Against this backdrop, the overarching question loomed: What role would there be for the monarch? The volatile Mirabeau again stepped on the stage, pleading with the deputies to keep the king as the nation's champion against anarchy and growing rule by the mob. He portrayed the king as a warmhearted man with laudable intentions (true), who had often been ill-advised by weak or malicious counselors (also quite true). Now he boomed prophetically, "Have [you] men studied, in the his-tory of any people, how revolutions commence and how they are carried out?" And then this: "Have [you] observed by what fatal chain of circum-stances the wisest men are driven beyond the limits of moderation, and by what terrible impulses an enraged people is precipitated into excesses at the very thought of which they would have shuddered?"

In effect, Mirabeau was pointing out that their objective was not destruction of the system, but its rehabilitation, and he was warning that if this infant revolution went awry, like Saturn—like so many revolutions—it could be destined to devour its own. Chastened, the delegates followed his advice. But instead of extending his hand in compromise, Louis foolishly picked this moment to draw a hard, unflinching line. He outraged radicals and liberals alike by sacking Necker on July 11 and exiling him back to Switzerland—an act that only perilously reminded the country, especially the Assembly, of the whims of autocracy.

A showdown between the king and the Assembly was now inevitable.

THE NEXT DAY, an impoverished young scribbler of verse, Camille Desmoulins, mounted a table outside a café by the gardens of the Palais-Royal, eloquently condemning Necker's dismissal and calling upon the people to arm themselves against the alien troops. "The Germans will enter Paris tonight," he warned, "to butcher the inhabitants." If this was demagoguery, it nevertheless deftly stoked the wave of emotion in Paris. *"Aux armes!"* he cried out, brandishing two pistols; "To arms!" thousands of people responded back. Several hours later a scuffle broke out between a regiment and a crowd of demonstrators at the Tuileries; meantime, panic spread from neighborhood to neighborhood, and rioters began lighting fires and looting stores. Expecting an armed offensive, Parisians, now wearing green ribbons as an emblem of liberty and hope, raced around the city in a desperate search for weapons. They broke in and seized the weapons housed in the Hôtel de Ville (City Hall); fortified rebels then coursed through the streets, upholding wax busts of the ousted Necker and the maverick Duc d'Orléans, while great crowds gathered at street corners and on church steps.

A sense of alarm began to speed through the city. Much as the Shaysites had attempted in Massachusetts, two days later the swelling crowd invaded the Hôtel des Invalides, a hospital for aged soldiers, and seized 40,000 muskets and twelve cannons. Unknown to the king, rather than standing fast to quell the uprising, his commander, Baron de Benseval— ostensibly worried that his Swiss troops would join the rioters—withdrew them to the suburbs of Saint Cloud. Suddenly the armed mob was now the master of Paris.

But its newfound power was all too ephermeral without ammunition for its arms; as it happened, the demonstrators still needed gunpowder and bullets. It was at this fateful juncture that the most audacious suggestion yet was made.

To enter the Bastille.

∞

ON THE EAST side of Paris, the massive medieval stronghold seemed to be as old and dark as the feudal system itself. It had been built in 1370, first as a defense against the English, then to incarcerate leading enemies of the crown, usually held by secret orders of the king—the much feared lettres de cachet. Voltaire had once been detained there, so had the Man in the Iron Mask (it was actually velvet), and so had Mirabeau. But under Louis XVI, very few prisoners were locked behind its walls; by 1789, only seven remained, living in relative luxury. The marquis de Sade, who had resided there until early July, had brought his wardrobe, tapestries, and a 133-volume library. Moreover, Louis already intended to abolish the gloomy fortress and perhaps replace it with a park. But in their frenzy, the people knew none of this, or if they did, they didn't care; to them, the mysterious dungeon remained a despised symbol of brutish despotism.

In the half light that preceded the sunrise that morning, the tocsin rang out and the population was still roaming the streets. But to do what? The crowd had no plans to assault the Bastille; with its eight-foot-thick walls, protected by seventy-seven-foot-high towers concealing artillery, and surrounded by an eighty-foot-wide moat, that seemed impossible. Therefore, when they reached it on July 14, their aim was at first modest, or so they believed: simply to ask the governor to let them remove the gunpowder—some 30,000 pounds—and firearms stored within. In that way, they could defend themselves against royal troops if the king should turn his men against them. Thus, two delegates from Paris's new municipal council offered to meet with the governor of the fort to present their list of demands.

The governor was a terrified marquis de Launay. A bureaucrat, not a soldier, he was nervous and indecisive, and most certainly not spoiling for a fight. So as best he could, he pleasantly received the crowd's emissaries, and just as quickly gave his word of honor that he would never open fire unless attacked; he also ordered his 114 soldiers to stand down, as well as to pull his cannons out of sight. He even offered his visitors *déjeuner*—lunch. Momentarily, the standoff held.

But not for long. Around one-thirty, a series of violent acts and random blunders unfolded. While the two sides slowly negotiated, some overanxious workmen found an undefended outer courtyard, broke into a guardhouse, and emerged with axes and a sledgehammer, then scaled a rooftop by the chains of the drawbridge, and cut them. Suddenly the Bastille lay naked to attack. The time for talk was over, and the emboldened assailants wasted no time, flocking into the courtyard, taunting and jeering, "Give us the Bastille!" De Launay ordered them to return. They refused. Inexplicably, shots were fired. But by whom? In the mayhem, it was a pointless question. De Launay's soldiers began firing upon the mob.

The fighting seesawed back and forth, and by three that afternoon, the battle became serious. Wounded men were ferried about on makeshift stretchers, carts of straw were lit on fire, sending blinding smoke into the skies, and the rumble of cannon fire could be heard for hours, even miles beyond the capital. Actually, had the king's men moved decisively, Louis could have swiftly crushed this rebellion. But ensconced at Versailles, Louis knew nothing of what was unfolding. And while royal armies had restored order in the face of popular insurrections for centuries, this time they would blink. The vaunted Swiss troops had deserted, and several companies of French soldiers openly defected to the other side.

Now aided by the king's own soldiers, the attackers brought up five cannons, including one that had belonged to the Sun King himself, and began to chip away at the walls of the Bastille. As the peril grew, an isolated de Launay sought to negotiate in earnest. He ordered the drums to roll and hoisted a white handkerchief, offering to surrender if his men were allowed to march out safely with their arms; it was a reasonable request, but reason was no longer the issue. Coldly rebuffed, the bewildered aristocrat now faced the cool, hard logic of defeat; a siege was impossible and so was a prolonged battle: he had only 114 men, a mere two-day supply of food, and no internal supply of water; the attackers numbered at least 900 to 1,000. Unable to carry out a siege or fight back, cut off from royal instructions emanating from Versailles, he lost his nerve and yielded, dropping the other drawbridge. But this only further emboldened the attackers. And as their apostasy mounted, so did their courage. The crowd flooded inside the prison itself, seizing the vast store of gunpowder, then freeing the remaining prisoners, and then turning on the defending troops with a new fury. By day's end it would be a veritable field of blood: Nearly one hundred would lie dead and seventy-three wounded; among the defenders, another one hundred died, while three lay wounded. De Launay himself was seized as a captive.

Drunk with their success—the fall of the Bastille was a sound heard all over the world—the rebels now marched in triumph to the City Hall.

Along the way, a frenzy built. Swarming through the streets, hundreds of people surrounded the hapless de Launay; he was spat on, shouted at, knocked down, kicked with a torrent of feet, and badly beaten. Eventually he snapped, melodramatically mumbling, "Let me die." With these words, the demons were unleashed. Brutally assassinated, he was impaled by sword thrusts, bled by bayonet jabs, tormented by darting knives, and finally succumbed to a volley of pistol shots—as though that were needed. Even this didn't satiate the crowd's thirst. Inexplicably, his head was then cut off, not quickly, with a sword, but slowly, with a penknife. His dismembered cadaver fell to the ground, while his head was planted on a pike, bobbing and dipping aloft, for all to see as a warning to those who might defy the rebels. Several of his subordinates suffered the same ghastly fate. In triumph, as mutilated bodies were dragged though the streets and corpses swung from streetlamps, the exultant mob began to laugh and sing, a cackling, thundering, roaring refrain, chanted over and over again in an ever-rising crescendo.

Invincible, the "Bastille conquerors" then proceeded through the capital.

∞

As the feverish rampage blanketed Paris, one would have expected the king to be hovering close to total despair. In truth, he knew none of this, or more accurately, understood none of this. Otto von Bismarck would one day say that political genius entailed hearing the hoofbeat of history and then rising to catch the galloping horsemen by the coattails. The problem, of course, was that one may hear the wrong horse or leap after the wrong horsemen. Louis did both.

That night he went to sleep peacefully, content in the view that dissolving the Assembly still remained an option; in his diary, he dryly recorded *rien*—"nothing." Historians are fond of noting that "nothing" meant that Louis had not gone hunting; but it also meant he was in total ignorance of the gravity of the day's events. So was the queen, who that afternoon had openly rejoiced at the arrival of freshly sequestered troops at Versailles. But as the hours ticked by, the euphoria and ignorance that had cloaked the palace began to dissolve. Word of the haunting facts reached Louis's advisers. They were desolate: Royal troops had deserted. The show of strength had failed. The Bastille had fallen and Paris was in full rebellion. Upon the betrayal of the royal troops, it would be impossible to recapture the capital.

And most ominous of all, as they would soon learn, a bounty had been put on the head of the queen.

Finally, at two o'clock in the morning, one of His Majesty's aides eased himself into the royal chamber, itself an almost unheard-of act. Yet the message he carried would shake the insular, sheltered world of the king to its very depths. Louis muttered acidly, "But this is a revolt."

"No, Sire," his aide replied. "It is a revolution!"

∞

THE NEXT MORNING, Paris was in convulsions. With each passing moment, events were rapidly slipping further from Louis's grasp. Without his consent, the famed astronomer Jean Bailly was made mayor of Paris; without his input, Lafayette was chosen to head a new National Guard; and without his sanction, delirious Parisians began to demolish the Bastille, stone by stone. And in perhaps the most damning measure of all, the traditional color of the monarchy, Bourbon white, was replaced by red and blue—the colors of Paris—separated with white; henceforth, the new colors of the Revolution, the new national flag of the empire, would be the tricolor red-white-and-blue cockade.

Desperate and worried, the king hurried to the Assembly, unprotected and flanked neither by his royal retinue nor by any ceremonial guards, where he announced that he had recalled the troops that had been dispersed across Versailles and Paris. Incredibly, the king and the Assembly were now under the same roof—as equals. Mirabeau put a stop to the spontaneous applause that greeted the sovereign: "The people's silence," he said menacingly, "is a lesson for kings." Two days later, Louis followed with another about-face: He fired his new cabinet, which had held office for a mere "Hundred Hours," and recalled Necker, to the delight of the financial community and the populace. To the queen's dread—she feared he would never be allowed to return—the king also bravely made another personal trip to Paris, with the goal of promoting calm. In truth, Louis was concerned too. Before he left, he made his last will and testament, dressed in a simple morning coat, and knelt for prayer in the royal chapel. Well he should have. For when the king arrived in Paris, what he saw must have rattled him to his core.

In a matter of days, the capital had been transformed. The stark facts of revolution had taken over. Everything hinged on the whims of the National Assembly and the rumors of the mob. The city was packed with tens of thousands of sweltering, excited people waving cockades and flags, and waiting impatiently for the king. Events were again moving with breathtaking speed. Led around by the deputies and the city militia, the king was not met as the Gallia et Germania, the customary welcome proclaiming French monarchs as the "Gallic Hercules," but was instead confronted with gritty pugnacity. Once, with the snap of a finger, he could have moved troops and clergy and parlements randomly around like pieces on a chessboard; now, taken through streets lined with armed citizen-guards, he was the one being moved. Once, the thick voice of the sovereign had aroused awe and inspired majesty; now, it seemed the king was reduced to a posturing mascot. And Louis was greeted by Lafayette not in the traditional royal manner, but in civilian dress, as if he were not His Majesty, the King of France and agent of God on this earth, but just another *père de famille*—father of the family. Nor was this merely symbolic. Visiting the Hôtel de Ville—the City Hall—he read his official new designation: "Louis XVI, Father of the French, the King of a Free People."

Struggling to maintain his composure, and helpless to alter the day's events, he signified his acceptance of the new council and, for the first time, acknowledged the legitimacy of the new regime ("Help me in this circumstance," he beseeched, "to ensure the salvation of the state; I expect as much from the National Assembly"). Then, in his most dramatic gesture, he allowed himself to be displayed before the people from a balcony. Affixed to his hat was the new emblem of the "French nation"—the red-white-and-blue tricolor. There was a stillness at first, then excitement and great commotion; a trumpet blew and cannon shot drowned out bursts of applause.

Were the king's sentiments genuine—or were they mere pretense? To be sure, the voice was Louis's, but how about the passion, the commitment, the words? It was unknowable just then, but what is knowable is this: His court had all but ceased to exist, the very symbols of the sovereign's power had all but been wrested from him, and he could no longer guarantee the basic safety of his family. Also knowable is this: Returning to Versailles late that evening, he trudged out of his coach, tearfully embraced his wife and children, and told them, "Happily, no [more] blood has been shed, and I swear that never shall a drop of French blood be spilled by my order."

While historians hotly debate this, the events that would then ensue over the next three years would be ironic—ironic because another side of the real Louis XVI, humble, deeply patriotic, dignified, and regal, would manifest itself. To be sure, evidence would later reveal that the king was more than once two-faced, unpatriotic, and double-dealing; as Jefferson himself would pen to George Washington, "I was much an enemy to monarchy before I came to Europe. I am ten thousand times more so since I have seen what they are." But this is only half truth: for as the crisis deepened, it eventually produced some of Louis's finest moments ever.

In the meantime, on July 16, behind the heavy, closed doors of the palace, a conference of nobles had urgently assembled; this would be their last meeting within the boundaries of ancient royal tradition. They advised the king to take flight immediately, to seek asylum in some provincial capital, perhaps at the heavily fortified fortress at Metz by the Netherlands border, or to Switzerland, Belgium, Austria, or Germany, or some other friendly foreign court; meanwhile, over the next three years, many nobles and princes would end up even farther east, in St. Petersburg. A weary Marie Antoinette, fearing for the royal couple's lives ("I am terrified of everything," she nervously wrote, "I have no more strength left."), strongly supported this proposal, and began gathering her jewels and other small treasures in preparation for the journey. Here, surely, was temptation.

And now the king's fate, it seemed, was measured not in years or weeks or even days, but precious hours. But in the end, given conflicting advice, he was unwilling to flee or abdicate. There were practical considerations as well: With the chain of command evaporating before their eyes, how could the king's security be guaranteed? It was decided instead that the king's younger brother, the Comte d'Artois, taking his wife and mistress with him, and the queen's closest friend and confidante, the Duchess de Polignac, all in disguise, would lead the first group of émigrés out of France.

No one knew what lay next. With tears in his eyes, the king insisted: "Don't lose a single minute," while the queen pronounced, "Adieu! The most tender of friends. The word is terrible to pronounce but it must be said."

The first royal exodus had begun.

Much later, when it was too late, the king would recall that exquisite moment of chance, saying sadly, "I missed my opportunity, I missed it on July 14, and it never came again."

∞

KNOWN BY HIS nation as Louis Seize when he was crowned in 1774, amid great fanfare and even greater hopes, he was alternatively the *"Resurrexit"* of the great King Henry IV, who rescued France from chaos, corruption, bankruptcy, and defeat; the *"bon père du peuple"* ("good father of the people"); "your eminence" (by foreigners); or simply *"le Roi"*—the King.

When his grandfather Louis XV had died, all day long through Versailles and Paris, and all across France as the news was spread, men, women, and children cried, *"Le Roi est mort, vive le Roi!"* By Bourbon tradition, Louis XVI was now the sole legislator, the chief executive, and the supreme court. He held elaborate palaces at Versailles, Marly, La Muette, Meudon, Choisy, St. Hubert, St. Germain, Fontainebleau, Compiègne, and Rambouillet. His era, no less than the majestic Sun King's, was believed to be a renewal of the Golden Age. And for a flickering moment, it was. In a short time, he presided over the rebirth of a great navy, the regeneration of a great empire, and the defeat of a long-standing foe; moreover, the young United States would not exist without him. Hence, his birthday was a national holiday in America, and his portrait graced not only the halls of the Senate and the private walls of George Washington's residence, but countless homes in the fledgling republic as well. And Louis's country reigned as the center of Europe: His court was not as imperious as Prussia's, it was no less calculating than England's, nor was it as byzantine as Russia's. And he ruled over a nation blessed with uncommon natural wealth. Yet he lived in one of the great interstices of history; that would be his curse, and it would be his undoing.

As a result, few monarchs in history are as underappreciated for what they stood for, as lampooned for what they misunderstood, as disgraced for what they lost. As a result, few monarchs have been so overshadowed in history: by their wives, by their lineage, by their royal peers in Europe, by the times they lived in, by the dictators who would follow them, and by the men who would cut short their dynasty and end their life. And few monarchs—as monarchs, that is, as opposed to men—have had so many tragic flaws. Yet few monarchs would also leave this world with such dignity, nobility, and grandeur, the very same qualities that Louis XVI ironically all too often lacked in life. So who was this king, Louis XVI? In truth, he was a curious amalgam of the proudest and darkest sides of the French nation, an incipient blend of the old and the new, sometimes twisting in awkward contradiction, other times sauntering boldly hand-in-hand.

∞

HE WAS BORN Louis-Auguste, duc de Berry, on August 23, 1754. His early childhood was nothing short of desolate. He was sickly, meek, and pathologically shy. Ignored by his parents who preferred his handsome older brother, and taunted by his own family as "Louis the Fat," he was orphaned at the age of ten and was then haughtily ridiculed by his formidable grandfather. He became woefully uncomfortable in fashionable society and equally reclusive; all his waking hours, he was happiest when surrounded by the solitude of his books or the graceful trot of his horses. Physically, he walked with a waddle and was plump even as king, but he grew to be quite tall for his times—five feet ten; and he had soft blue eyes. In his youth, he immersed himself in outdoor play and arts, traits he carried with him all his life: He was an avid hunter and swimmer, shot with uncanny accuracy, loved to fell trees and saw wood, and, more than anything else, was most at home with building and tinkering. His eyesight was bad, but his memory was superb. For that matter, so was his moral compass: modest and loyal, almost to a fault, he always esteemed the Church, and, uncommon not just in a French king but in most kings, was ill at ease with his father's many infidelities.

Reserved, ungainly, indecisive, almost naïve, he was also a man of angelic disposition and distinct intellectual gifts. History has portrayed him as shallow and loutish; upon careful scrutiny, neither rings as quite true. He was a proficient Latin scholar, a keen student of history and science, and from an early age he also steeped himself in the science of government, reading Montesquieu, Rousseau, and Fenelon, a Catholic bishop and author for whom he developed a special liking. In his own way, this Bourbon king would become quite progressive, adopting the view that "the monarch is nothing but the steward of the state's revenues." And unlike the elegantly turned Louis XIV, this reclusive Louis was notoriously indifferent to his appearance: Inert and slothful, he thought nothing of showing up at court with wig powder covering his jacket, in waistcoats without buttons, or with fingers that had turned black from hours at the forge, where he indulged his cherished hobby: the making of locks and keys.

How was this young man, who admired the skills of artisans more than the intricate dynastic maneuvers of kings, who loved to work quietly with his hands more than to receive the acclaim of winning great wars, who preferred the silence of his study to opulent state dinners, to find the courage and pride and strength to lead the future Bourbon dynasty? Few princes seemed less emotionally or temperamentally prepared for the intrigues of Europe's most cynical and slyest court, or the responsibilities of the most civilized nation on the globe. That was the question then. It would always be the question.

A creature of his day whose royal authority owed everything to the past, he was a thoroughly modern man as well—a scientist, a sailor, and an engineer, with an instinctive sense of political duty that prodded inextricably to the future. Yet while these traits were revered in the salons of cosmopolitan Paris, they became a handicap in a great Bourbon king. He temporized, he lacked the judgment necessary for strong leaders, he was too often uncertain and irresolute, and, at times, he was shamefully weak; both his grandfather and the Sun King would have wept at the spectacle of his indecision. And to a degree almost without peer, his fate would be intertwined with the woman who would become his wife, the fifteenth of sixteen children born to Empress Maria Theresa of Austria.

We know her as Marie Antoinette.

Born in Vienna, the Archduchess Maria Antonia was the second youngest of the empress's children. Her lineage and her family ties, and most of all the influence of her mother, would stalk her all her life. Maria Theresa, her powerful mother, promoted her country's power with a zeal and decisiveness that would have impressed the Sun King. She gave birth to her children in front of an open window while signing papers of state; she was also cold and scheming, readily inflicting emotional havoc on her high-spirited youngest daughter. "Your beauty frankly is not very great," the empress insisted in one typical letter to her daughter, remarkably when she was twenty-five and queen of France. "Nor your talents nor your brilliance. You know perfectly well you have neither." Painfully neglected, Maria Antonia became a willful child who spent royal parties ensconced behind her fan, giggling, and wanted always and everywhere to be at play. She was also remarkably uneducated, knowing only a smattering of Italian, and unable to write in German or French correctly. And the court of her youth could not have been more different from Versailles: simple, devoid of ritual, and run with an iron fist by her mother, who quashed intrigues and demanded piety and subservience. Not unsurprisingly, her marriage to Louis-Auguste was arranged; it was to be a union of two great empires: The French and the Austrian, the House of Bourbon and the House of Hapsburg.

When she arrived at Versailles at the age of fourteen, she was equipped with little more than her bewitching charm, which was enormous, her flawless, unflinching face (like "lilies and roses"), her coquettish informal manner, and a radiance marred only by her faintly scornful Hapsburg mouth and a forehead that was too high. Few princesses had been less ready for the most cunning court in all of Europe, or for that matter, for marriage to the Dauphin of France. One marvels at the scene in May 1770, when this tactless young girl, imbued with the heart of a child, signed the register to mark her union with Louis-Auguste; she smudged ink on the page. Marie Antoinette, as she was now called, was either hopelessly nervous—or barely able to sign her own name.

A few hours after the ceremony, she was instructed to get into bed with the maladroit sixteen-year-old, who would become France's future king in a mere four years. To the dismay of the dynastic expectations of the French people, it would take the royal couple an astonishing seven years before they would consummate the marriage. Spies from every corner of Europe, Spain, Prussia, of course Austria, reported on their nuptial predicament; even Emperor Joseph II, Marie Antoinette's older brother, would investigate the matter, in person; in disgust, he reported: "Together, they are two complete fumblers." Actually, in bed the problem was not that they were fumblers. The gentle, clumsy Louis was neither impotent nor disinterested; instead, he had a deformity of the foreskin that made erections painful, but that was curable by a simple incision. Eventually Emperor Joseph persuaded his brother-in-law to have the required surgery, and soon, the "great act" was accomplished. The following spring, Marie Antoinette was pregnant with her first child, a daughter, and then quickly thereafter, with three more children, two of them boys, whom the royal couple doted endlessly upon. Moreover, they had ensured—or so they believed—the perpetuation of the mighty House of Bourbon.

But by then, the cards were already rudely stacked against them.

∞

FOR ONE THING, Marie Antoinette seemed to forever live in a long-distant century of delicate hands and coddled minds, which of course was fantasy. For another, she joined the court at a moment when the French were extremely antagonistic toward Austria, an enmity that stretched back for centuries. The first of many epithets for Marie Antoinette was thus *l'Autrichienne*, a barb made all the more stinging by the fact that it was built around the word *chienne*, or "bitch." For her part, the queen did little to remove its stain. As time went on, she made lethal mistakes: She was impetuous, she was tactless, and she politically stumbled, over and over, through the marbled corridors of the palace. She forgot her own promise that she made as a fourteen-year-old while en route to France, when she declared, "Do not speak German, gentlemen; from today I understand no language but French." If that weren't enough, she later flocked with Anglomaniacs to the newest rage of fashionable Paris—English horse racing. She also foolishly allowed herself to be manipulated by Maria Theresa and later Joseph II into promoting Austrian interests in the halls of Versailles, which further alienated her subjects. Still, she was not without her appealing sides, but even here they were marred. She had a strong benevolent streak and built a loving memorial to Rousseau. But her aristocratic appetites were infinitely stronger: the trips to the opera, the theater, the extravagance of her dress and the love of jewels. Whenever she was floating through the lines of courtiers in the Gallery of Mirrors, or leaning back indifferently on a settee in conversation, or stretching out her delicate white hand to be kissed, she was a prisoner—a prisoner of pleasure. And she was utterly out of touch with the reality outside her many palaces. Did she ever say, "Let them eat cake" to the bread-starved masses? Almost certainly not, but she might as well have anyway.

She and her husband made for a combustible mix: In this sense, maybe Joseph was right—they were fumblers. Where at every step her husband's actions were marked by temporizing, indifference, hesitation, she was actually quite decisive, but here again it worked against her and the king. Time and again, she criticized and even sabotaged a host of talented ministers who might have salvaged the regime's precarious finances: She alternately disparaged Robert Turgot, cursed Alexandre de Calonne, and detested the wildly popular Jacques Necker. And in an age of great transition, where Louis XVI's inchoate Rousseauist leanings might have produced a British-style constitutional monarchy, which a sizable cast of moderate revolutionaries sought in 1789—actually, for a fleeting moment they seemed to come close—Marie Antoinette undercut her king and incited the populace by rejecting any diminution of the royal role. Remarkably, never once, even as the growling discontent of Paris was echoing in the state rooms of Versailles, did she evince a desire to learn about the provinces she ruled, let alone steal an hour and visit one of her subjects. Then, in 1790, a time when France was increasingly a cauldron of smoldering resentments against the royal family, she would impoliticly dismiss the constitution as "monstrous" and a "tissue of absurdities." And where Louis was touchingly sweet and straitlaced, even prim—he was the only monarch in generations to be faithful to his wife—she engaged in a passionate love affair and devoted friendship with Count Axel Fersen, the Swedish nobleman, with Louis's blessing.

She was heedless, trusting, ill at ease with protocol; it seemed that nothing she did turned out right, and all this, of course, affected the king. She never grasped that hers was a court steeped in the use of the subtle pinpricks of flattery and fiendishly well adapted to lead an unknowing queen astray. Thus she was endlessly portrayed as a devious, disloyal foreigner—and this from schemers in her own court. Her intimate friendships with her sympathetic souls and royal favorites, Princess de Lamballe and the Duchess de Polignac, led her to be labeled a debauched hussy, and a lesbian at that. Her model village at the Petit Trianon, a testament to Rousseau; her opulently fashioned palace at St. Cloud; and her extravagant rewards to her favorite minions led her to be seen as looting the national treasury; her epithet, by the 1780s, dangerously became "Madame Deficit." In truth, everything stuck to her: Her involvement in the infamous Diamond Necklace affair, perpetrated by a gang of brazen swindlers (she herself was innocent), left her so thoroughly despised that for a time she could barely leave the palace. Rather than being seen as gently advising the king, she was thought to be undermining the nation by manipulating her husband. Her maelstrom of gambling parties, masked balls, and amusements, rather than adding to the mystery and mystique of the court, and hence its royal power, only left her more isolated. And even her proud reserve was viewed not as the majesty of a great queen, but as proof of her corruption and iniquity.

There was no escaping the criticism. As the revolution took life, she remained a marionette among other marionettes, the queen of rococo perpetually sacrificing responsibility to amusement, the people of France to the nobles of Versailles, the real world to the world of dreams. She herself once said with alarming frankness: "I am terrified of being bored." But a queen who forgets her subjects is taking great risks. And this she did in spades. She fought almost every measure that passed through the Assembly, thereby earning yet another withering epithet: "Madame Veto." Her misjudgments were deadly too. Indifferent to the century's great philosophical movements, lacking insight into the aspirations of the majority of the French people, surrounded by an ill-informed entourage, she saw conspiracies all around the king. Thus, an unreconstructed autocrat, she stirred up opposition in the court opposed to any alliance between the crown and the moderate reformers, and even then remained painfully ignorant of the vast social ills that would soon lead the "howling mob" to embrace a republic and seek vengeance upon the monarchy. But by then, it almost didn't matter.

In 1789 she became the object of an unremitting stream of pornographic and violently sexist tracts produced by a legion of underground pamphleteers: Rumor inflated her into a monster of depravity who presided over sinister orgies in the palace (actually, despite her adultery, she, like Louis, was also quite prudish). Routinely, she was a "she-wolf," a "tigress," a "hyena," a "daughter of Satan." And as the crisis deepened, all the complaints the French had about the queen—her Austrian birth, her coldness, her autocratic leanings, and her distant melancholy air—blended into one single sweeping torrent of hatred. Ironically, both the nobles and the Jacobins peddled the same gossip with the same relish and the same sighs of disgust. Inevitably it seeped over into criticism of the king: according to Babeuf, he was a "donkey," "weak," and "abstinent." The king was mocked for walking with "a waddle." Thomas Jefferson labeled him a "fool." And that was just for starters.

But this simplifies matters. In fact, there was more going on than that. Can one make too much of the queen's follies? Absolutely. Once Marie Antoinette cried out, "My fate is to bring bad luck." And the royal couple did have their own share of bad luck. Four weeks before the fall of the Bastille, their beloved dauphin, their eldest son and heir to the throne of the world's most important monarchy, literally wasted away of "consumption"—tuberculosis—before their very eyes. Louis and Marie Antoinette's agony was real. On April 15 the little prince, a gentle boy with a wonderful face, who was eight, weighed only sixteen pounds fully clothed; for the queen, sitting beside him, unable to help, every cry seemed a sword thrust into the bottom of her heart. She was at his bedside when he died on June 4, just after midnight. The king and queen spent the day in complete seclusion, wracked with grief; then, physically and emotionally drained, both withdrew to the solitude of their country house in Marly. But there was no respite from the whirlwind of political events around them. When told the Third Estate insisted upon his presence, the king, allowed no time to mourn, answered stonily: "Are there no fathers among them?"

Louis XVI's task was an immense one. Long before the Revolution broke out, the country was, one observer noted, "confusion and chaos everywhere." A case in point: The south of France kept Roman law; the north of France kept common and feudal law. The French state itself was a maze of privileges, tax exemptions, immunities, and hereditary office holdings—we would call them special interests today—that was incapable of modernizing itself without a mass abolition of all these special favors granted to one group or another. Something had to give. And the political structure itself was a sheer babble of confusion: There were the old titled families with straitened finances, which the king was supposed to relieve; there were the nobles with their own separate courts who refused to pay taxes; there was the church which saw itself as an entity unto itself; and there were all the royal prodigalities, the pensions, the gifts and the salaries, and the sinecures paid out of revenues drawn from the economic life of the nation, costing France 50 million livres a year—a staggering tenth of the total income of the government. If ever there were a social, administrative, and political system blocked, it was France before 1789.

And finally there were the distinctions between the classes and privileges that were nearly irresolvable. The story is complicated and knotty. What is not is the fact that the hatreds among the classes of status groups in France ran extraordinarily deep: how else to explain the ferocity of the bloodthirsty mobs, the wanton cruelty, the paranoia, the mass death, and eventually the Terror, organized by the state itself. To be sure, violence *would* one day be the Revolution—this is always the danger of utopian movements—just as Mirabeau had predicted. But there were also mighty currents, long pent up, that only a truly representative political structure could have absorbed—but which didn't yet exist anywhere in the European world—that would not lightly go away.

Could this revolution have been nipped in the bud? Or put differently, could Louis rise to the challenge now arrayed against him? It is one of history's great imponderables, but had Louis XVI been a stronger or more effective leader, the answer is quite possibly yes. The crux of the issue—and it was the crux—is that Louis XVI was a reluctant king. When he inherited the throne at the age of nineteen, far from embracing it, the courtiers who had rushed to his apartment found the dauphin on his knees, weeping and praying, "Oh God, protect us! We are too young to rule." The pathos, and the tragedy, is that not until it was far too late did he find his voice and his regal instincts. So, when the Bastille was taken and the crisis had hemorrhaged, Louis found himself nearly paralyzed.

Compare him to the other leaders on the world stage: He wasn't cunning or manipulative like Maria Theresa of Austria or a swaggering iconoclast like Frederick the Great; he didn't have an interior of steel that would enable him to ruthlessly put down the Revolution with as much force as would be needed, as Catherine the Great did with the two-year Pugachev Rebellion, or for that matter, as George Washington would contemplate when America faced its own violent divisions, just several years away. Nor could he bring himself to wage war on his own people, as King George did. And for all the sweetness and goodness and benignity of his nature, and even his harshest critics attested to that, he couldn't decide if he should be like Henry IV, who centuries before would gallantly rescue the people, or like the crafty Louis XIV, who would co-opt them. His executive management was poor, and so were his arts of persuasion, so necessary in any leader, whether a king, emperor, tsar, or elected president. In times of war, rebellion, or crisis, leaders must sometimes be cruel or cold; he was neither. In times of war, rebellion, or crisis, leaders must sometimes be able to hate; he was not much of a hater either. In times of war, rebellion, or crisis, to some extent or another, most leaders must be fanatics: stubborn, steel-willed, driven, secretive yet able to inspire, fastidious, and zealous. They must be prepared to act alone, without encouragement, relying on their own inner resolve. And they must be indifferent to approval, reputation, and even love, cherishing only their own personal sense of honor or vision, which they allow no one else to judge.

But this was not the case with Louis XVI. Yes, reluctant is the word. That is why he fell asleep at important state occasions; it is why he was so often depressed; it is why he lost himself in hunting and gorging on food, as though nothing had changed and as if his kingdom wasn't falling down around him, when of course it was. That is why, having been almost at war with a rebellious aristocracy in 1788 and then embracing the commoners, he abruptly changed his mind, siding with the nobles: In doing so, he lost control over the Estates-General, failed to exert leadership or present a program until it was seemingly too late, and provided no symbol behind which parties could rally. Most importantly, up until this point, he failed to make use of the profound loyalty to him felt by the common people—and make no mistake, even in the darkest moments of 1789, they still felt it—who yearned for nothing more than a king who would stand up for them, as in the days of yore, against the aristocracy of birth and status. Even the Assembly sensed this, too.

The result? The emerging middle-class revolutionaries now feared the nobles more than ever, and believed, apparently with some good reason, that they now had the king in their hands. And after Louis's trip to Paris, after the first exodus of nobles, and upon Louis's return to Versailles, there was no going back. Great leaders—and especially great kings—find ways to overcome obstacles, rather than be thwarted by them. They command, but they may never complain—even when luckless. And somehow they must move on. Moving on now meant decisiveness, one way or the other. It meant being a revolutionary king, or an absolutist king. And as hindsight tells us, it meant realizing that anything in between was a perilous no-man's-land.

But as the year 1789 entered its waning months, and the Revolution fiercely accelerated, Louis was overwhelmingly a melancholy man—barren, empty, defeated. The same even for the outwardly defiant Marie Antoinette. Little did he know that the agony was just beginning. And little did he know that as France now embarked on an unprecedented journey, one that would alternately horrify and arouse all the world, fueled by heavily charged ideas and magnificent, lasting words, he and his wife would also embark on their own personal journeys. Caught up in the sweep of mighty historical forces, they would, to be sure, blunder even further. But it would be near the hour of their deaths, when they were mere citizens, that, without reluctance or hesitation and with no small measure of dignity, they would at last grasp the mantle of monarchical devotion as king and queen of a great nation.

Yet even after the fall of the Bastille, as Louis surveyed his fate, there were still hopeful options. In Russia, Catherine had beaten back her own rebellion, and even if King George had bungled his, he at least was able to cut his losses and keep his head. At the same time, the infant United States had cauterized the mutinous Shays's Rebellion while it was still largely contained in Massachusetts. So it could be done. And if there was hope, it would lie in the loyal subjects in the countryside, who still resounded with cries of "Long live the king!" and who saw Louis as "the new Augustus," the man who would renew the "Age of Gold." For them, his personal popularity was a huge asset, and the Revolution was his to command. And hope also would lie in the ultimate brotherhood on the world stage, his fellow autocrats. True, on one hand, many Americans remained optimistic that the French Revolution would mirror theirs, peacefully evolving into a constitutional monarchy. Thus Thomas Jefferson, exuberant and delighted at the events racing before his eyes, reported back to James Madison of the "astonishing train of events as will be forever memorable in history," while another American observer, the inventor James Ramsey, gleefully noted, "American principles are bursting forth in Every quarter." James Monroe felt much the same and so did a young Andrew Jackson, not to mention the vaunted George Washington himself.

But on the other hand, watching this incipient demon of democracy unfold, the collective royalty of Europe shuddered. "I often think this cannot be the eighteenth century," Queen Charlotte of Britain morosely reflected, "for Ancient History can hardly produce anything more Barbarous and Cruel than our neighbors in France." For his part, comparing the French ferocity with the accounts of the St. Bartholomew's Day Massacre of French Protestant Huguenots two hundred years earlier, Baron Jean Simolin, Catherine's minister in Paris, hotly reported back home in disgust. "The Revolution in France has been carried out," he wrote, "and the royal authority annihilated." He added that there was one crucial difference— this rebellion was fueled by "political, rather than religious fervor." Even William Pitt, in adversarial England, would remark: "To suppose that any nation could be unalterably the enemy of another was weak and childish. It had neither its foundation in the experience of nations nor in the history of man." In fact, at this pivotal juncture, Louis would soon have choices— with his peers abroad. He also knew this: Revolutions in Holland, Geneva, and Poland had been suppressed by foreign intervention. So it could happen in France too.

Soon, the eighteenth-century European order would come to matter, a great deal. And that order was a Hapsburg in Vienna, a Romanov in St. Petersburg, a Hohenzollern in Berlin, and, yes, a Bourbon in Paris. The antique social systems of Europe did not see themselves due for demolition. Far from it.

In the critical months to come, the king would still have room for maneuver, both in his own country and outside. The final question was: Would he seize or squander that opportunity?

∞

THE FIRST INKLING of all that was to come, and the measure of desperation the king would feel, would soon be apparent. By the time the first émigrés fled, an eerie quiet had fallen over the Château de Versailles. The courtiers, *plus royalistes que le Roi* (more royalist than the king), now made themselves unobtrusive; the great drawing rooms and antechambers that had swelled with glittering crowds often fell to a hush. Already there were memories of a world that no longer seemed to exist: Once, the glass doors of the château would be thrown open and the fragrance of the lush gardens would fill the rooms; once, lighted chandeliers blazed in clusters catching the gowns and jewels of the women and the bright uniforms of the nobles; once, cotillion suppers and sumptuous feasts were held; and dancers and lovers amiably strolled in the gardens. Now an age of opulence and crafts-manship, of gilt and beauty, seemed to be dying. There were whispers and there was fear. Indeed, the entire country was suddenly succumbing to an uncontrolled panic. It was called *la Grande Peur*—the "Great Fear"—and it spread with rumors that the nobles and their minions were murdering those who wanted reform. In turn, frenzied peasants rose up in revolt, burning châteaus, looting, and taking up arms.

And nine days after the fall of the Bastille, the Great Fear edged dan-gerously close to Versailles itself. Joseph-François Foulon, one of the minis-ters in Louis's government—he was to have replaced Necker—was brutally murdered by a mob, despite impassioned efforts by the Marquis de Lafay-ette to save him. Hay was stuffed in his mouth and his head stuck on a pike, which was then thrust before his son-in-law with the admonition to "kiss papa." Then his naked remains were strung up and mutilated. Shortly after that, his son-in-law was butchered alive with a knife, piece by piece, while a hysterical mob screamed encouragement.

Caught in his own agony and indecision, the king, still adrift in the memories of the past, was forced to ponder his next moves, even as Ver-sailles' cloistered rituals remained strangely unchanged.

But those too would not last long.

PART II

TURMOIL

Empress Catherine the Great

CHAPTER 4

AMERICA

✧

I T WAS NOT JUST France. Disorder now seemed to stalk the very globe. At the remote edges of her empire, Catherine's war machine was poised to unleash an assault upon the Ottomans. In France, a baffled King Louis XVI was slowly realizing that any abiding faith in the everlasting glory of the realm was slipping. And where did the young American republic fit in? Precariously. Wedged in the south and west between predatory powers, it was haunted by the twin phantoms of immaturity and military weakness; the great European states routinely scorned and ridiculed the enfeebled America, as did the Barbary states of North Africa. Having watched its first form of utopian government collapse, and having erected another, it had—just barely—survived a bruising battle over the ratification of the Constitution; nevertheless, Rhode Island was still holding out, and already the political culture was marked by endless intrigues and rampant dissension. Fearful of a national government boldly consolidating too much power, America also worried about the stalking horses of disunion and civil war. And it well knew that one more wreck, in whatever form, could be fatal to the country.

Yet amid all this, there were promising signs. At home, whether in Charleston or Philadelphia, New York or Boston, there were scores of newly proud, educated, civic-minded citizens; even workmen and artisans were literate and independent. This, of course, was just the sort of soil in which a permanent republic could flourish. And abroad, however tentatively, America was spreading its wings far and wide. The American ship *Empress of China* had even left New York harbor bound for Canton. Reaching the ancient empire, the world's newest nation was wildly greeted as "the New People," and as the Stars and Stripes was raised, the Americans were dubbed the *"Hwa-Chi"*—people of the "flowery Flag." Meanwhile, George Washington kept abreast of the global turmoil. Mindful of Catherine's machinations, he anxiously monitored the great clash to come. "The Russians and Turks," the Marquis de Lafayette reported to him, "are quarreling. The Empress is going to . . . meet the Emperor of Austria." And with even greater interest, if not satisfaction, he followed the rebellion in France. Actually, he would soon hold in his hand the great wrought-iron key to the Bastille, relayed to him through the writer and master pamphleteer Thomas Paine. "That the principles of America opened the Bastille," Paine wrote with jubilee to Washington, "is not to be doubted." President Washington hung the key next to his portrait of Louis XVI.

And in April, Washington had been inaugurated as the young republic's first president.

∞

FOR EIGHT DAYS, as Washington had made his way from Mount Vernon to the temporary home of the federal government, New York, the fledgling nation had been in the grip of celebration. It was a time suffused with hope and symbolism. He left the great brick estates of Virginia and moved north, passing by neat little villages with two-story white wooden houses and spired churches. His carriage crawled under triumphal arches in endless towns, to clanging bells and flower petals strewn in his path, as he moved off through blazing sunshine and rain-streaked skies, heading for Manhattan. At Elizabethtown, New Jersey, he was transported on a ceremonial barge towed by thirteen pilots, across the Hudson River to New York City, accompanied by leaping porpoises and booming cannon. Masses of cheering people came out to see the president-elect, ships ran up their colors, church bells pealed, and when Washington landed at the foot of Wall Street and walked to his house on Cherry Street, tears ran down his cheeks—several times he was seen stopping to wipe his eyes. And that night, while thousands of candles were set aglow in windows all across the city, Washington surrendered himself to a noisy and packed reception, a state dinner given for him by the governor. Shortly thereafter, on a brilliant sunny day, April 30, Washington boarded an elaborate carriage to make the inaugural pilgrimage to Federal Hall, overlooking Wall Street.

Children waved, people cheered, and a band played. The papers screamed with news of the day's events. And from Broad Street to the Battery, from the waterfront to Golden Hill, ten thousand New Yorkers jammed the roadways, climbed trees, and hung out their windows to catch a glimpse of the fifty-seven-year-old Virginian, while on the Hudson River, ship captains held their breath in anticipation and flags snapped in the wind. Wagons ground to a halt underneath thickets of trees in the distance, and the thump and swish of feet could be heard all along Wall Street, as spectators elbowed themselves closer to the president-elect. Finally the crowd caught sight of Washington himself in the presidential coach, richly appointed with velvet, tassels, and a handsome seat; a roar of applause mounted from the throng as he made his way to the appointed spot, and just as quickly suspended as he stood, towering over the other men.

With his vice president, John Adams, standing beside him, and flanked by Federal Hall's neoclassical splendor and a magnificent backdrop of gold stars on a blue background, the presidential party moved onto the splendid second-story balcony. As Washington rose, the crowd tipped forward, to catch every word; when he spoke, they shook their heads in silent affirmation and listened with awe and glee. Washington slowly set his hand on the Bible. "I solemnly swear," his voice rang out, "that I will faithfully execute the office of the President of the United States and will, to the best of my ability, preserve, protect and defend the Constitution of the United States." A glowing sun flooded the entire gathering in radiant light, and all around, the air quickened with anticipation. Then, in a prophetic note of the challenges to come, Washington uncharacteristically departed from the script. Bowing his head, he kissed the Bible—it was resting on a soft crimson cushion—and ended with an emphatic, and spontaneous: "So help me God."

The crowd burst into applause and broke into tears, and chanted back: "Long live George Washington, President of the United States!" From a ship in the Hudson River, an artillery salvo exploded in the wind, and over the voices and above the guns rose the song of massed church bells.

Mustering his energies, the first president left the cheering assembly to read—nervously, as it happened—a brief inaugural address in the Senate chamber. Then he marched up Broadway to St. Paul's chapel, where he solemnly knelt in prayer. It would be needed. He would soon rejoice that it was a "Miracle that there should have been so much unanimity . . . and increasing good will."

Neither would last for long.

∞

MANY YEARS LATER, Abraham Lincoln, after his election, would quip to a gaggle of reporters, "Well boys, your troubles are over now, mine have just begun." These words could easily have been uttered by Washington himself. James Madison, the father of the Constitution, lamented, "We are in a wilderness, without a single step to guide us." Washington himself had his own deep reservations: He felt like "a culprit going to his place of execution," he wrote just before he set off for New York. As usual, his dart had hit its target. He now had to solve complex problems for which precedents and guidelines were virtually nonexistent. He had to produce a national government that until this point was mere theory; create a cabinet with clear lines of authority and responsibility; and strike a happy medium—as if that were easily accomplished—between regal dignity and republican austerity, while melding a still divided country together. And as if he didn't have enough to worry about, somehow, in the midst of all that, he had to define what the presidency was, knowing that the fate of America depended on it; the new federal government, he had already once moaned, "may be compared to a ship between the rocks of Scylla and Charybdis." And finally, he had to forge a balance between the moral imperatives and high sense of purpose enshrined in the young country's two founding moments: the first embodied in 1776, which established American independence and whose heroes exalted the virtues of individual liberty and feared the "excess of power in the rulers" (Benjamin Franklin's words); and the second in 1787, which established American nationhood and whose statesmen clung to an altogether different conception, that of a single, powerful national government, with the executive as the driving engine of policy.

In a sense, nothing had changed but now everything had changed. What had once seemed so clear and compelling—the exigencies of revolution, freedom from a corrupted British monarchy, the imperatives of liberty—now was, and would increasingly become, a matter of debate.

∞

BUT GEORGE WASHINGTON had a vision. At the war's end, he had circulated a letter to the thirteen governors outlining four things America would
need to do to survive, if not to attain its rightful place on the world stage:
The states would have to be consolidated under a vigorous "federal" government; there would have to be timely payment of the staggering debts
left over by the war; an army and navy needed to be created; and finally,
there must be harmony among the people. This was the theory at least; in
truth, the reality was a cyclone of bewilderment and confusion.

For starters, Washington, who knew what he knew (fighting and politicking), and knew what he didn't, was keenly aware that everything about
his administration assumed heightened importance, because each act, each
official gesture, was setting a precedent, not simply for the present, but for
the future as well. So the machinations of these early days were complex,
they were nuanced, and, to later generations, almost wholly perplexing and
prone to being misunderstood. Even the minutiae mattered. There were the
broad philosophical questions: For starters, how should the president be
addressed? Was he to be called "Your Excellency" or "Your Mightiness"
and treated like a king? John Adams thought so, proposing Washington
be referred to as "His Highness, the President of the United States of
America and Protector of their liberties." The Senate debated this. Or
was he a servant of the government? The House thought so, deciding that
Washington should be simply the "President of the United States." The
House version was kept. Then there were the questions of day-to-day government. Should Washington meet regularly with visitors? It was decided
that the office required an aura as much akin to a king as to a homespun
republican. Thus, not unlike Louis XVI, Washington received visitors at
weekly "levees," but he never stayed more than half an hour, never returned
visits, and with monarchical reservation, he bowed to his guests, rather than
shake hands. Of course, there was the matter of dealing with the legislature. So Washington held private dinners limited to eight visitors, appearing in a black velvet coat, yellow gloves, and a dress sword hanging from
his breeches. Then there was the issue of adding grandeur to his office.
So on public occasions, Washington's buff-colored coach was pulled by
six magnificent white steeds, guided by two liveried positions; at the same

time, there was the need to maintain his republican credentials. So Washington took daily walks among the people every afternoon, promptly at two o'clock. And there were the trappings of his planter's heritage—he had seven slaves shipped up from Mount Vernon to assist his white household servants.

And then finally there was the question of his government. Here, the Constitution said little about the executive departments and didn't mention the cabinet at all. The administration was a nebulous concept, a tabula rasa. Washington literally had to invent his government. But what he didn't have to invent were the issues facing the country. The questions loomed large over the infant republic: As the financial state of the new government was especially precarious, who would solve the debt issue? Who was the man to represent the young country to the councils of the world? Who would make America's military might respectable? And who would oversee the laws of state? As for Washington himself, he was strong, but would he be just? Or wise?

At the start, Washington himself saw the presidency as something of a cross between a monarch, a head of state floating above the partisan fray, and the presidency of today. His vice president, John Adams, the short Yankee from Massachusetts with his patch of scrubby white hair and a crackling voice, was seen as more a head of the Senate, rather than an integral part of the administration's decision-making apparatus; his role was negligible. Then there was the cabinet itself. For now, the beating heart of America would not be the Constitution, or the people, but the men who served Washington. The president, who had the gift of superior judgment, chose astutely; arguably, his assembled talent was the finest in this nation's history. For secretary of war, Washington picked his skilled artillery commander, Henry Knox, although this was understatement: The president could well have been his own war secretary. For secretary of state, he chose the current minister to France, the forty-seven-year-old drafter of the Declaration of Independence, Thomas Jefferson. And for the crucial role of secretary of treasury, he asked the man whose name hung on everyone's lips, the financier of the revolution, Robert Morris of Philadelphia. Actually, Morris, on the verge of a financial scandal, turned him down, instead recommending Washington's former aide-de-camp, the thirty-two-year-old principal author of *The Federalist*, Alexander Hamilton, "a man of superior talents." Never in history would a second choice have such a large impact; Hamilton eagerly accepted. For attorney general—which didn't even have a department—he asked the thirty-six-year-old Edmund Randolph, the tall, handsome Virginian. And for chief justice, he nominated New York's John Jay, a signer of the Declaration and one of the peace commissioners in Paris. Conspicuously missing from this list was James Madison, the diminutive Virginian and arguably the most sophisticated and subtle political thinker of the revolutionary generation; he was elected to the House and would serve as both de facto leader of the people's body as well as de facto advisor to the President—that is, when they weren't clashing.

Because national habits and customs had yet to take root and the political institutions were so untested, the upcoming decade would literally be held together by sheer force of personality. Bureaucratic routines and regulations were nonexistent, and to modern eyes, the system seemed incorrigibly haphazard. Since there were only three departments, each secretary wielded considerable power; they *were* the government. Moreover, Washington's cabinet members lacked clearly delineated lines of authority and often worked at cross purposes, frequently interfering with one another's tasks: Hamilton raided Jefferson's and Knox's turf; in turn, Jefferson did the same to Hamilton's, while from their perches Madison, Randolph, and Knox often meddled as well. Administratively, the government seemed to be a confused jumble, but this served Washington just fine. He wanted men of creativity and capability, gifted executives and incisive thinkers, and an exceptional team that could magnify one another's strengths. And in the end, it was the ever-prudent Washington who made the final decisions. Rather than always confer with his full cabinet, he sometimes liked to consult with individual secretaries about matters that concerned them. Still, he wasn't consistent about individual consultation either and at times sought out his favorites, like Hamilton, for diplomatic matters, which naturally offended Jefferson. At other times, Washington didn't bother to confer with anybody and simply made decisions himself. And soon, to the president's consternation, then to his worry, then to his profound dismay, the secretaries would be bitterly fighting for spoils of the office or Washington's ear.

The first major decision of the republic would be made collaboratively, if not contentiously, by Hamilton, Madison, and Jefferson. After that, it would be all-out war.

∞

HAMILTON, MADISON, AND Jefferson. These were the giants of the American enlightenment: they fought the Revolution, crafted the Constitution and Declaration of Independence, and came home to the political arena. Among them, their thoughts are part of the air we breathe and their ideas formed the framework for virtually every political discussion in the 1790s—and for many decades hence. In those early hectic days, they were the father, the son, and the holy spirit of Washington's advisors. The volatile Hamilton was a restless, rawboned immigrant from St. Croix, who lusted for power and fame, and relentlessly pursued his vision of the supremacy of the federal state. Presiding over the largest department—he had thirty-nine employees, compared to five at State—he was a swashbuckler who reveled in debate and was quick to insult. A smooth-shaven Anglophile with almost feminine rosy cheeks, an exceedingly fair complexion, and charm to burn, he was thought handsome by many. He thought so himself. He was well-tailored, cared about the cut of his suits, and fussed about the style of his hair (pulled back). And he ardently studied Jacques Necker, Louis XVI's finance minister, the Bank of England, and Sir Robert Walpole, Britain's first prime minister, and with equal ardor examined the lessons of history: the wars of Charles XII and Peter the Great, the oppression of Roman provinces, and the lessons of modern and ancient leagues. He took them each to heart.

His problem in life was not talent—he had an abundance of that—but temperament. He was everything his opponents said he was: invariably pugnacious in debate (Adams once called him an "insolent coxcomb"), he could be needlessly tactless, or overly provocative, or way too driven, or wildly impetuous, or downright reckless. Curiously, though fascinated by power, the use of power, or the misuse of power, he nonetheless was never one for the give-and-take of politics or compromise—he sought to remake the government in overarching sweeps or by sheer force of his ideas. But they were not to be lightly dismissed, for he was a force of nature, brilliant and tireless, thinking in rapid bursts and with broad sweeping strokes, almost without peer. Accordingly, while lacking Washington's inner sense of decorum (the president, it was rightly remarked, possessed "the gift of taciturnity"), he would supplant Madison and eclipse Jefferson, becoming Washington's favorite son.

This was no easy task. Thomas Jefferson, the former governor of Virginia, was tall and suave, lean and aristocratic. He also had striking reddish hair and captivating cool green eyes, which, despite his often disheveled dress and surprising impulsiveness, gave him an imposing presence; Abigail Adams once described him as having a look "not unlike God." With his lips tightly sealed, his gaze intense, he was simultaneously charismatic while conveying a patrician ease to all around him; Senator William Maclay rightly spoke of Jefferson's "lofty gravity." And where Hamilton was hyperactive, Jefferson was warm and engaging, even unusually shy. Still, many thought him aloof, stiff. And he spoke with a lisp. Detesting controversy, he was guarded and courtly, but this was all deception. His talents were legendary: He was a sphinxlike man of unshakable reticence who also happened to endow the Declaration of Independence with a soaring majesty of prose. His passions matched his words: He hated monarchy, hated the British (those "rich, hectoring, swearing, squabbling, carnivorous animals"), and lived in fear of a strong executive government that would infringe on Congress, the guardian of popular liberty, and worse still, infringe on the states, an even greater trustee of republican freedoms. His view of the good society was unshakable: The working class was corrupt; merchants were corrupt; speculators were corrupt; cities "pestilential"; in the ideal republic, only farmers were dependably good. Though he came from an elite Virginia pedigree, and relished the fineries of life—expensive wines, gourmet foods, classical architecture, beautiful women, the best art, the grandest gardens, the latest inventions—he despised all aristocracy as well. By disposition and design, he was a planter and a philosopher, a spokesman for the devout wisdom of the common man and a protector of republican purity, and an incurable lover of revolution; he even advised Lafayette and the revolutionaries in France. Revealingly, after Shays's Rebellion, he commented dryly that "the tree of liberty must be refreshed from time to time with the blood of patriots and tyrants."

And returning home from Paris to New York in 1790, he immediately understood the moral burdens he would have to shoulder, and the high seriousness of the challenge to come. With a wry smile on his face, he was unfailingly polite and brilliantly catered to those around him, but this too was a smokescreen: He knew how to scheme and squabble, how to flatter and deceive, all the while pouring a honeyed potion in his enemy's ears. Observers who watched him closely saw that he had strong tastes, even stronger views, and was fierce in political combat, even as he waged it from the shadows—or by proxy.

And one of those proxies would eventually be Madison, a close ally and friend of Hamilton's, but with whom Jefferson would ultimately develop an almost mystic bond. John Quincy Adams called their relationship "a phenomenon," adding that it was "like the invisible and mysterious movements of the magnet in the physical world." This did not exaggerate. As time marched on, the two fed off each other brilliantly: Madison rejoiced of Jefferson as one of the "most learned men of the age;" Jefferson felt the same way about Madison. Eight years Jefferson's junior, three years Hamilton's senior, Madison was small, pale, colorless, and unsmiling. If Hamilton was a mercurial risk-taker, and Jefferson quietly elegant, Madison was remarkably introverted, bland, a sober, silent man; thus, where Hamilton favored bright waistcoats, Madison cloaked himself in black. Many found Madison "gloomy," "stiff," "mute," even "repulsive" as well as "little and ordinary." And he had a "croaking" voice. A graduate of Princeton, he was, in truth, an exceptionally alert, incisive, and intricate man. He read voraciously and often slept only four hours a night. He also was frail and had a nervous disposition, which kept him from serving in the military; a fanatical hypochondriac, he apparently suffered some sort of breakdown after college and was gloomily convinced that he would die. Instead he emerged as an ingenious defender of religious liberty, as the master builder of the Constitution, as the other indispensable author of *The Federalist*, and with Hamilton and John Jay an eloquent spokesman for a vigorous federal government. He could be gregarious, but by nature he was a loner who trusted no one—save for Jefferson. His self-derogation was real; so was his apparent simplicity. And there always seemed to be a corner of remorse lodged deep inside, furiously eating away at him.

But Madison possessed a fierce determination and a fanatical sense of conviction; he was one of the most willful and cunning men of his time, really of any time. How else to explain that this diminutive man with a weak, reedy voice fended off the great rhetorician Patrick Henry in the 1788 Virginia debates on ratifying the Constitution? Henry mesmerized the convention delegates with his blistering attack on a Constitution that "squints toward monarchy," but over the course of three long weeks, Madison kept chipping away, methodically marshaling facts, employing logic, composing a brilliant defense of the proposed new government. In the end, Virginia followed his lead and voted for ratification. And no mind was more supple or original than Madison's. As much as any of the Founders, he knew history, knew political science (having helped create it), knew philosophy, and was fond of quoting Voltaire, Locke, or Hume. John Quincy Adams would one day write that it was Madison's good fortune to be "sent into life at a time when the greatest lawgivers of antiquity would have wished to live. How few of the human race have ever enjoyed an opportunity of making an election of government—more than air, soil or climate—for themselves and their children." And it would not be an exaggeration to give Madison the appellation of philosopher king of the 1780s in America, more so than Franklin, more so than Adams, more so than Jefferson.

But paradoxically, this bland, reflective man also proved to be a formidable politician as well—we too often forget that he had served in Congress since 1780, beginning at the young age of twenty-five. It showed. He was filled with grand ideas, but was also meticulous in detail; he was ruthlessly calculating and a virtuoso of political indirection and stealth, always plotting, always planning and strategizing, and, like Jefferson, doing so laconically, shrewdly operating behind the scenes or scheming in the shadows. Alliances were his forte; his lightest touch invariably commanded results. Not unsurprisingly, as the 1790s marched on, he would become dubbed "the Big Knife," then "the general" (Jefferson was the "the generalissimo"), so adept was he at cutting deals.

Hamilton, Jefferson, and Madison. They were what Pelatiah Webster observed was so necessary for the new country: *"grave, wise and faithful men."* Madison provided the architecture for the republic, Hamilton its masonry, and Jefferson its soul and poetry. And now, in the heady early days of 1789 and 1790, with Washington as president and the three working together—none of them believed in political parties, which they feared would lead to "rage," "dissolution," and eventual "ruin" of the republic—it seemed that anything and everything was possible.

Certainly that appeared to be the case for the country's first great collaboration. It began with an idea and ended with a dinner, which in countless ways would determine the outlines of the country's future.

∞

THE FIRST OVERARCHING issue confronting the nation was the tangle of debt: $25 million in state debts, as well as the issue of America's credit— a foreign debt that amounted to a staggering $11.6 million. The total war debt, foreign, national, state, exceeded $76 million. And, as the Founders knew, the growing turmoil in the ancient monarchy of France was vivid testimony to the dangers of letting debt spiral out of control.

At the outset, arresting the state debt was the linchpin. Hamilton's initial proposal was that all states should assume payment of the debt equally—in his phrase, "assumption." His plan was to treat the debt as a national responsibility rather than as belonging to "fourteen separate governments," which would, he contended, only lead to "collision," "confusion," and "interfering regulations." Drawing from the English example, he also proposed a "sinking fund" created from surplus revenues of the Post Office. He was convinced that his plan would also have the endorsement of his old ally, Madison, who had voiced his support for the outlines of assumption during an afternoon walk at the Constitutional Convention. "We were," Hamilton tartly stated, "perfectly agreed." And for some time, the two had uniquely shared a continental perspective.

No longer. By February 1790 the political landscape had changed dramatically; Madison was not speaking as the Constitution maker or coauthor of *The Federalist*, but as the congressman from Virginia, and he stunned Hamilton by sharply reversing course. In his eyes, the choice now lay between his home state of Virginia and republican liberty, and an impersonal, unresponsive general government, and he saw no shame in siding with the Old Dominion. Complaining that Virginia and most other Southern states, including Maryland and North Carolina, had paid off the bulk of their wartime debts, he asserted that "assumption" would unfairly penalize their fiscal rectitude. To Hamilton, however, this position smacked of pure parochialism. But four separate times, between February and July 1790, Madison foiled attempts to enact assumption. He was not without his own arguments: He proposed that all state war debts—both those that had been paid and those that hadn't—be absorbed by the federal government; he called it "descrimination." The effect would be to double the nation's debt; it would, no surprise, also favor Virginia.

Yet Hamilton stuck to his guns. He retorted that the debt was produced by the revolution, whose rewards all Americans equally shared, and thus they should assume collective responsibility. For Hamilton, assumption increasingly was the central issue for the nation; he was proposing not only to resolve the debt, but to create capital and yank America into capitalism. Most importantly, for Hamilton, assumption was the most effective and irrevocable way to tie the states together into a permanent national government—which for Madison was precisely the problem. All through the spring the debate raged, but on April 12, led by Madison, the matter came to an abrupt halt. The House voted Hamilton's assumption plan down (Madison's plan failed as well), and two weeks later, it voted to discontinue all debate on the issue. Hamilton was aghast. The country was split. It smacked of a return to the paralyzing days of 1783–1785 that rent apart the Articles of Confederation. Hamilton knew he had to strike a deal. But what kind?

The "what kind" came in the guise of another make-or-break issue that would indelibly define the character of the new country: the permanent location of the capital. Since the Constitutional Convention, which had envisaged the creation of a federal district (ten miles square) in an unspecified venue, the matter had been the subject of fierce debate, intensive lobbying, and a mist of intrigue. Never was an issue more ripe for conflict: The winning state would be the recipient of enormous wealth, power, and prestige—there would be London and Paris, Berlin and Rome, Vienna and St. Petersburg, and . . . ? Moreover, the debate was a referendum on the very nature of what America should look like. The capital was, in part, an answer to a far deeper set of questions. From the rural woods to the hubbub of the cities, Americans asked themselves: Should the United States be a pastoral paradise, a wondrous agrarian nation of yeoman farmers and great planters? Then a southern state, most likely Virginia or Maryland, was the logical choice. Or should it be a bustling, urban society, favoring the "stockjobbers," the moneyed men, the manufacturers, and the mercantilist and commercial interests? Then the North was the logical choice. Alternatively, should the capital be a great metropolitan center uniting all the political, economic, and cultural energies in one place, as found in Europe? Or should the two be split, much as in Russia, where Moscow was the religious and spiritual center, and St. Petersburg the seat of political power?

Already, sixteen possible sites had been proposed, but each had failed to muster a majority, including Annapolis, Baltimore, Carlisle, Frederick, Germantown, New York, Philadelphia, the Potomac ("Petomek"), Susquehanna, and Trenton. Still, given its centrality between North and South, it was assumed Pennsylvania had the edge. Complicating matters was the fact that there would be a makeshift, temporary capital, most likely New York or Philadelphia, until the permanent capital was ready. But as the debate dragged on, it opened old wounds and created new ones. It divided America between anti-Federalists and Federalists, those favoring sectional interests and those wanting national allegiances, those who clung to agrarian visions of oceanic isolation and those desiring a broad web of commercial international entanglements, and finally, those who believed a weak federal government was the savior of the country versus those who wanted a powerful one. And, of course, it divided Madison and Hamilton.

Now was the time for Hamilton, the quintessential New Yorker, to unsheathe his broadsword again. To sulk or fume or give up would have been unthinkable. Of course, he was not capable of detachment—that wasn't in his nature. Nor should he have been. Someone had to sound the alarm; he did. He favored his home state, but fearing that wasn't in the cards, let it be known he was willing to back Germantown, Maryland, or Trenton, New Jersey. Then came the rumor that Pennsylvania and Virginia had already reached an understanding: Philadelphia would become the temporary capital, and a site on the Potomac ("a wilderness monstrosity") would be designated as the permanent capital. Hamilton was now eager to trade. And anxious. For suddenly he feared that the talks were moving ahead without him.

Against this uncertain backdrop, Hamilton ran into Jefferson one late spring evening.

It was fortuitous. Hamilton, who for all his prodigious energy could fall prey to black melancholy and blue funks, seemed to be in a state of despair. He struck Jefferson as "somber," "haggard," and "dejected." In the evening glow, the treasury secretary spoke darkly of the "secession" and the "separation" of the states. He dramatically painted a picture of a national government that would "burst and vanish." And he movingly made an appeal for Jefferson to employ his influence to salvage his cherished policy of assumption ("a common duty," he pleaded, "should make it a common concern"). For his part, Jefferson, recently back from Paris, was barely in better shape, suffering from a debilitating migraine for weeks. He too was worried about the specter of "secession and dissolution"—from the northern states. Actually, he was deeply suspicious of Hamilton's assumption plan—he feared this was yet another example of the avaricious hand of the unscrupulous "money powers," the sprawling, hydra-headed creature associated with banks, stock markets, and devious speculators, especially in New York, Boston, and the City of London, not to mention unrepublican, un-American attitudes of all kinds—everything he despised. But as secretary of state, Jefferson equally feared that a stalemate over the debt could wreck the union (in his words: "the greatest of all calamities"), with dire repercussions for American credit abroad, leading to its "total extinction."

The two walked for an hour and a half, pacing past the president's door. Jefferson was actually quite sly. He suggested he was a stranger to the whole subject—in truth, he knew the matter in great detail—and he cagily proposed a "friendly discussion" between Hamilton and Madison over dinner at his house the next evening, not unlike the remarkable dinner among political rivals that he had already hosted in France to salvage its revolution. The stage was set. On June 20, 1790, would commence the most famous dinner in American history.

Some of these details, coming from Jefferson, are perhaps apocryphal, but not the main themes. The next evening, Madison agreed not to oppose assumption—which amounted to supporting it—if something were granted in return. "The pill would be a bitter one to the southern states," he asserted. "Something should be done to soothe them." And what was that something? Simple. Philadelphia would be the temporary capital for only ten years—this had already been in the works—but the capital then would be permanently relocated to a site on the Potomac. By the same token, there would be no maneuvering to keep Philadelphia as the nation's home by default. Madison also sought, and received, more favorable treatment for Virginia's wartime accounts. That, so to speak, was the ball game. The prospect had been hooked. Hamilton would now work on the Pennsylvanians, Madison on the House, and the deal would be consecrated.

Here was politicking at its best. Everybody gave and everyone got. And they all agreed to stick it out. A second dinner between Hamilton, Jefferson, and Secretary of War Knox with the Pennsylvanians on June 28 finalized the agreement. On July 10 the House approved the Residence Act, and on July 26 the House narrowly passed the assumption bill. The now ubiquitous Madison, true to form (and to save face in Virginia), voted against the measure, but paved the way for four dissenters to switch their votes, enabling the bill to pass. The scene in the Congress was a piece of superior stage management from the people who had almost—"almost" being the important qualifier here—brought the young country to the brink of catastrophe.

To be sure, there were brusque cries of protest. Agitated New Yorkers hurled solemn epithets at Pennsylvanians, while New York senator Rufus King, hearing the news, burst into tears; and Pennsylvanians in turn cursed their own delegates who had sold them out to the Virginians. Yet by all measures, this was republican statesmanship at work, just as George Washington, just as the Founders, had envisioned. And once the national government prepared to move in the fall to Philadelphia, the heady idealism of the country seemed to be in full bloom. Indeed, in 1787 Hamilton had predicted that the federal government would either "triumph altogether over the state governments . . . [or] the contests . . . will produce a dissolution of the Union." But now a tremulous crisis had been averted. Now, so long as the country flourished, the spirit of forward-looking compromise dominated the day. And now all the participants, as if in chorus, themselves felt this too.

As secretary of state, Jefferson was convinced—then at least—that "whatever enables us to go to war, secures our peace." And for Hamilton, the matter was quite simple: A sound economy generated trade, trade generated wealth, wealth generated power, and power generated security. And "prosperity" at home, he professed (here agreeing with Jefferson), made America "respectable abroad." In the course of this, he was also laying the foundations for a state built on centralized, federal power, and one that could claim to be fully separate and secure from Europe.

And then there was Madison. No wonder Madison, who for months now had shaken his head in opposition and had masterminded an artful campaign from the outset, was overjoyed. He too celebrated the "spirit of accommodation" that had resulted in this great compromise, believing it augured even grander things to come. But had they all been too clever? For amid their triumphalism and rejoicing, amid their belief in America as a kind of miraculous Eden crafted by men, came a warning with ominous, even resounding portents. Portents because it appeared in the form of a petition to Congress—the truest voice of the people, as intended by the Founders. Resounding because it would bear the stamp of approval of no less than one of the most famous men on the planet and one of the legendary founding spirits of the Republic: a dying Benjamin Franklin. And ominous because it threatened to drive a stake into the heart of the new republic by addressing, in full public glare, the one issue that even the Founders brazenly wouldn't articulate outright: slavery.

This stunning debate would culminate in one of the most momentous decisions in the nation's lifespan. It would speak to the intractable dilemma of crafting a union; and it would say much of what America was and what it would become. And what was clear as the debate began on that unusually cold March afternoon was this: America may well have been an Eden. But suddenly, lurking quietly, the snake had reared its head.

∞

THE HOUR'S BITTER chill did nothing to stanch the blistering debate erupting in the young Congress that day. Spectators watched tensely from the galleries, while members of the House of Representatives nervously scrawled notes before leaping to their feet to urgently rebut their colleagues—if, in fact, they were even permitted to be heard. Slavery was, uniquely, the one issue where there was hot debate over the actual right to speak. "I am certain," a South Carolinian bellowed, "that even talk of abolition will sound an alarm, and blow the trumpet of sedition in the Southern states!" A murmur could be heard in the background, as other South Carolinians, Georgians, and eventually Virginians, nodded their heads and shouted in agreement. Rising among the cramped desks, a Georgian boomed rhetorically, "Is it good policy now to bring forward business which is likely to light the flame of civil discord?" to which he added portentously, "For the people of the South will resist one tyranny as soon as another!" In desperation, Northern members countered any Southern moves to silence the debate, citing not only the enlightened support of the antislavery societies, but "the general tide of public feeling." No matter. Before the gavel sounded and the session closed, a South Carolinian representative laid down a final apocalyptic warning about trespassing on Southern "rights."

"Let me remind men who expect a general emancipation by law," he chillingly insisted, "that this would never be submitted to the Southern States without civil war!"

∞

IT WAS MORE than jarring. Since the ratification of the Constitution, existential concerns had lingered about whether the federal government would endanger or protect individual liberties, and whether it would reaffirm or destroy the state governments. So it was no small measure of reassurance when New Jersey's Elias Boudinet promptly declared to the House that the new Congress had indeed accepted all the fifty petitions, memorials, and addresses presented to the chamber by an impressive array of American citizens, rich and poor, powerful and powerless alike. As such, though the much-feared specter of national despotism might still be hovering across the land, for now it seemed to have stopped at the doors of the people's tribune, the Congress.

But did it? For then came three antislavery petitions. Like a train wreck, they arrived unexpectedly in February 1790, and in so doing, openly questioned whether the grand experiment of representative government could flourish intact in a union also dedicated to slavery. Remarkably, this was the first time Congress had ever broached the subject in public. It unfolded

like this: Even as Madison and Hamilton had been jousting over his funding scheme, Quakers from New York and Pennsylvania had submitted a
petition ostensibly to abolish the overseas slave trade, while a third group,
the Pennsylvania Society for Promoting the Abolition of Slavery, led by
the eighty-four-year-old Benjamin Franklin, filed a more expansive petition
seeking to abolish the very institution of slavery itself. The implications
were immediate. And here was a timeworn recipe for conflict: Legislators
were faced with the unconscionable dilemma of adhering to the requirements of a republican government—free debate among the people's representatives—while a good portion of that government, slaveholders,
stubbornly maintained that there could be no further debate on slavery; in a
word, that it should be muzzled altogether.

On the surface, to modern lights at least, the debate seemed innocent
enough: The petitioners were admonishing Congress simply to turn "its
serious attention" to the African slave trade; to enlist Congress's support
in dealing with the "evil of slavery"; and to implore Congress to "step to
the verge of its powers" and ultimately address "our oppressed brethren."
But the South, which contained ninety percent of the Union's slaves in
1790, would hear none of it, and the delegates from the Deep South were
unmistakable in their intentions; sparing no words, they demanded that
those proffering the antislavery petitions should be gagged, smothered, and
ejected from Congress. James Jackson of Georgia, making mocking faces
toward the Quakers in the galleries, thundered that the memorialists should
"mind their own business." William Loughton Smith, the South Carolinian
federalist, his voice dripping with disgust, declared that this slavery "business" be dropped at once. Adenenus Burke went so far as to accuse the
memorialists of "meddling," while Jackson then demanded to know why
they were even allowed to be introduced at all. And Maryland's Michael
Jennifer Stone testily sniffed that Congress might as well consider petitions
on "honesty," "prudence," or any other sundry topic of "general interest."
Meanwhile, Burke of South Carolina averred that all spectators should be
removed from the galleries, lest their minds be corrupted by such wanton
blasphemy.

The debate became increasingly acrimonious. Presaging the arguments
that the South would make over the next eighty years, the slavery proponents presented an affirmative defense as well: Jackson growled that the
Bible itself approved slavery. "Search . . . that book," he angrily challenged;

Smith, for his part, contended that abolition would only make a bondsman's lot harsher and an overseer's lash more ruthless. And to serve general notice that antislavery was a forbidden topic, Thomas Tudor Tucker threatened that the debate could lead to "civil war!" while Smith then proclaimed that it was the "duty" of slavery proponents to oppose the memorials by "every means possible"—even if it meant disunion, and even as the Southerners fumed against the "eastern states" and their "fanatics." Roused by these inflammatory words, the Senate acted with extreme dispatch. After one day of debate on February 15, they refused to accept the petitions—and called on the House to do the same. Senator William Maclay of Pennsylvania, who supported the antislavery cause, later remembered the seething, discordant atmosphere; sadly, he also recalled being almost utterly abandoned by his Northern colleagues.

In the House, however, the memorials were accepted, but then quietly referred to a House committee. Madison, ever the master strategist, and who had just ushered through the Bill of Rights, actually harbored sympathies for the abolitionists. But this didn't stop him from privately telling Edmund Randolph, "the true policy of the Southern members" should be to "let the affair proceed with as little notice as possible." In other words, let the issue be ignored and simply wither away.

The question was: How had it come to this point?

∞

How indeed. In a moment of candor, William Loughton Smith had his answer. "We took each other with our mutual bad habits and respective evils, for better for worse," he once admitted. "The northern states adopted us with our slaves, and we adopted them with their Quakers." But in truth, the reality was far more complex, rooted in economics, history, and the evolution of humankind to that date. And it began almost quite by accident, roughly two centuries earlier.

∞

"About the last of August came in a dutch man of warre that sold us twenty Negars," John Rolfe, a Virginia colonist and soon to be husband of the Indian princess Pocahontas, recorded laconically in 1619. The first blacks to enter an English settlement in the New World—as indentured servants who could theoretically be freed in five years—their arrival marked the start of American slavery.

The "Negars" were landed at Old Point Comfort, a sandy wedge that divided the James River from the broad stretch of the Chesapeake Bay. Ironically, this English colony had set up the first General Assembly in North America, thus ushering the country down two totally different and contradictory paths: the rise of representative institutions and the eventual use of chattel slavery. In ensuing years, American slave ships from the Yankee ports of New England—the later birthplace of the American abolitionist movement—would become a common sight on the Atlantic Ocean. Slaves, bought in Africa for five pounds sterling, brought from thirty to ninety pounds in the West Indies, a hearty sum that laid the footings for many a New England fortune. The colonies of Spain, Portugal, France, Holland, and Britain similarly became strongholds of slavery, and the institution flourished in the American North until 1780, when Pennsylvania enacted the New World's first gradual abolition law and Massachusetts's highest court abolished bondage, boldly declaring it null and void. Other Northern states followed suit, banishing slavery or at least debating it—even in New York and New Jersey, where slavery was almost as deeply imbedded as in the South. It was hoped that the Southern states would do likewise, especially in the Lower South. But here the greatest tragedy unfolded, at first unwittingly, even reluctantly, until over the course of 171 years slavery became so intertwined and intermixed in the fabric of the South that any assault on slavery was to be seen as an assault not simply on the lifeblood of the Southern economy, but on Southern institutions, Southern values, and the Southern way of life.

Even back then.

Quietly in the beginning, however, then more noisily, over the course of the eighteenth century the ferment over slavery only grew. As many Africans were introduced into the United States during the first twenty years of the 1700s as during the entire time since the 1620s, and by the 1750s, the labor system increasingly became one of racial spoils, while indentured (often white) servants were replaced by enslaved Africans. This drastic shift did not go unnoticed; many in the North were increasingly stung by the contradictions of their own struggle against Britain and America's enslavement of others ("the execrable sum of all villainies" in the defiant words of Reverend John Wesley). And for growing numbers, slavery became a matter of deep conscience. But when antislavery advocates opened their mouths to call for freedom for people of all colors, slave owners bitterly fought back. Where abolitionists clamored for eradication of slavery ("Shall not our Lord, in due time, have these Heathens also for his inheritance"), slave owners clung ever more tightly to what they saw as the most treasured creed of republican liberty—property rights, including slaves. Where the abolitionists, heirs of the Puritan notion of collective accountability, viewed slavery as the most heinous of all social sins, slaveholders icily saw only economic ruin, social chaos, and racial war. Where abolitionists regarded slavery as evil against God's children, pure and simple, slave owners increasingly formulated it as nature's positive good: the foundation for peace, prosperity, and racial comity. And where abolitionists preached slavery as a violation against the higher law, slave owners countered with their own version of the deity: the need to maintain Union. And here was the rub: for in this they had the support of most of America's leaders, and virtually all the Founding Fathers. Washington owned slaves; Madison owned slaves; Jefferson owned slaves; John Jay owned slaves; and even Benjamin Franklin had owned slaves. Slavery was accepted with an insouciance and inevitability and even enthusiasm that is almost impossible to fathom today. But that doesn't mean the haunting contradictions and moral sentiments weren't there.

They had long been there.

∞

IT BEGAN AT outset, with the Declaration of Independence, those soaring words evoking "a decent respect to the opinions of mankind." But at the heart of America's claim to liberty was a cancer, eating away from within. What of the slaves? How could the rebels say "all men are created equal" while some 600,000 blacks were scattered throughout the colonies, who by law and custom were treated as chattel and enjoyed no rights at all? Actually, in his draft version of the Declaration, Jefferson—rather disingenuously—sought to blame American slavery on the British and King George, charging that the king had "waged a cruel war against human nature" by attacking a "distant people" and "captivating and carrying them into slavery in another hemisphere." But when the draft went to the full Continental Congress, it wouldn't fly, especially with the delegates from the Deep South. The slavery passage was removed, though the word "equality" remained in the text—a glaring anomaly that cried out to be rectified.

And the chances for rectification would come sooner rather than later. Early on in the Revolutionary War, some daring voices openly questioned the institution, going as far as to urge that slaves be made into soldiers. One of these voices was John Laurens. Laurens was not some misty-eyed Quaker or a dreamy follower of Rousseau, but a vaunted patriot, a lieutenant colonel, and one of America's most valorous fighters—wounded twice and captured once, he would eventually die in battle—he also happened to be Alexander Hamilton's closest friend and an intimate member of Washington's military family. Hamilton and Laurens were both ardent abolitionists who saw emancipation as integral to a free nation ("I think," Laurens intoned, "we Americans, at least in the southern colonies, cannot contend with a good grace for liberty until we shall have enfranchised our slaves."). Laurens also smartly tinged his views with realism, seeing the blacks as a vital source of manpower. While five thousand blacks eventually did play a significant role in keeping the Revolutionary War machine alive, serving proudly alongside the patriots, most were relegated to noncombat roles. But Laurens, the son of a prominent slave owner, was prepared to go further, even if it meant shaking up the edifice of the South. With his tight-set lip and formidable bearing, he told his father—who also happened to be the president of the Continental Congress—that he would submit to his inheritance in the form of a black battalion, freed and outfitted to fight for South Carolina. Would it work? Reason suggested yes, especially after Savannah and Augusta fell to the British in 1779. Now he minced no words. Resigning his commission on Washington's staff, he rushed back to defend his home state, but not before he asked Congress to establish four black battalions for the Continental Army; Hamilton supported this effort, writing, "I have not the least doubt that the negroes will make very excellent soldiers with proper management."

But they would have to be given a chance first, and the creeds of the day were too mesmerizing; so was the need for regional unity between the North and South. South Carolina, which processed more slaves arriving in North America than any other state, viewed the Laurens proposal with unmitigated horror. In a region perpetually fearful of slave insurrections, obsessively searching slave quarters for weapons, the South Carolina legislature's animosity toward this proposal was palpable. It was bluntly rejected. This didn't prevent Hamilton from ruminating philosophically: "An essential part of the plan is to give them their freedom with their muskets. This will secure their fidelity, animate their courage . . . and ope[n] a door to their emancipation." In other words, Laurens's plan was not simply a matter of humanity but of self-interest. Abigail Adams, writing in Massachusetts, agreed. "It always appeared," she noted with pathos, "a most iniquitous scheme to me—to fight for ourselves for what we are robbing and plundering from those who have as good a right to freedom as we have." But these suggestions, like the others, were met with a sneer. And ironically, it was the British who promised—and made good—that runaway slaves defecting to the British side would be given their freedom. Twenty thousand would jump sides by war's end.

Equally ironic, or tragic, was the fact that back in London it was now the patriots who were mocked. "How is it that we hear the loudest yelps for liberty," Samuel Johnson ridiculed, "from the slavedrivers of Negroes?" Horace Walpole nodded in agreement: "I should think," he observed skeptically, "that the souls of the Africans would sit heavily on the swords of Americans."

They did—to some at least. And the issue would *not* wither. The dreaded specter of slavery stalked the Constitutional Convention in 1787 as well. Once more it was not merely Federalist versus anti-Federalist, but slave owner pitted against abolitionist. The typical view is that the convention arrayed large states against small states; this is true, but other issues were at play as well. No less an authority than Madison himself conceded that the states were divided less by their "size," and more by their "having or not having slaves." As it happened, the debate, taking place in the August heat, was complex, it was convoluted, and it was knotty. Charles Coatsworth Pinckney of South Carolina, a staunch and much admired Federalist, thundered that "South Carolina and Georgia cannot do without slaves." At the same time, Virginia's George Mason, one of the largest slaveholders attending, vigorously attacked the institution and especially the slave trade. And though they were meeting in secret, behind closed doors, this didn't prevent members of the New York Manumission Society from delivering a petition to promote the "attainment of the objects of this society." But their efforts went nowhere.

In a sense, it was lunacy. The Founders were about to celebrate the golden age of a new constitution, of Washington, Madison, Hamilton, and Jefferson, to rejoice in the blessings of republicanism and liberty and bask in the dignity of the individual and freedom from government oppression, all the while embracing . . . bondage. It was like the Ptolemaic universe being invented and reinvented. Didn't the colonists put it aptly before the war, when they declared the slave trade "inconsistent to the principles of the Revolution and dishonorable to the American character"? But reason was one thing, practical matters another. To be fair, the task was an immense one, worthy of Solomon. For starters, the revolution was fought as much over the rights of property as the rights of man. Moreover, slavery didn't carry the stigma that it would to later generations—and no model of a multiracial society existed yet in history. Even the British general Lord Cornwallis, in a last-ditch stand at Yorktown, thought nothing of infecting the freed American slaves who had defected to his cause with smallpox and marching them along the enemy lines in the hopes of unleashing an epidemic among the Americans. Most agreed then, and even do today, albeit with a heavy heart, that whatever tragic consequences for this young country struggling with its national identity, the whole enterprise would have been stillborn had the Founders embraced emancipation.

In fact, the Founders sensed this too. The Constitution is rife with tepid evasions about slavery: the very words "slave" or "slavery" themselves appear nowhere in the document, masked instead by the less offensive euphemism, "persons held to service or labor." (Madison acknowledged that it would be wrong to admit in the Constitution the idea that there could be property in men.) In another fateful tradeoff, Northern states agreed that five slaves would be equal to three free whites in apportioning Congress, the infamous "three-fifths" compromise. Still, the door was left open for abolition: The Southern states, recognizing that the value of their own bondsmen would rise, cynically agreed that the importation of slaves might cease in "twenty years"—in 1808.

The tragedy is that a number of the moderate Southern members themselves were no great believers in the institution, even as they stoutly defended it—again, for the sake of Union. It was an equal tragedy that many Southern members who were deep humanitarians and some of the most ardent defenders of individual freedom remained woefully blind, or purposefully blind, or simply turned their backs on the evils of the institution. The result? This delicate balance of America's paradoxes—a republic based on an inherently undemocratic foundation—would haunt the country for nearly eighty years, to be resolved one day in an ocean of tears and blood. But the Founders needed to buy time. As historian Max Farrand reminds us, it is difficult to grasp the fears of the Founders. Yet they were real and they were legitimate. It is hard for us to imagine their world—and it was the only world they knew. They lacked our perspective. They could not hold up a mirror to the future. And they could only be guided by the past. The South consequently demanded concessions and the North was ready to make them. Madison himself was never in doubt about this: He glumly ruminated that without the compromises, "no Union could possibly have been formed." Even the staunchly abolitionist Hamilton, for whom these measures were anathema, saw this too.

Another paradox loomed: Yes, the Americans had somehow unleashed one of those rare tectonic jolts that forever changes the course of history. But in one glaring way they remained stubbornly wedded to the past. Thus the evasions, thus the compromises, thus the circumlocutions, thus "persons held to service" and the "three-fifths" compromise.

Thus the tortuous circle of slavery lived on.

∞

IT HAD BEEN in existence seemingly forever. As early as 8000 B.C. Libyans enslaved a Bushman tribe, and the Egyptians stole slaves from what is now Somaliland. Indeed, as historian Hugh Thomas points out, slave labor constructed the great hydraulic system of ancient China; it erected the majestic pyramids of Egypt; and it built the glorious Parthenon. In ancient Greece and Rome, slaves themselves worked as domestic servants and as managers of brothels, not to mention as doctors and lawyers and even majordomos to noblemen, and as warriors—the Athenians sent slaves to fight for them at the battle of Marathon, though they did free them first. At the end of the Roman republic, there were an astonishing two million slaves in Italy alone.

Remarkably, almost all great societies embraced servitude. Among the earliest slaves were light-skinned Germans—Saxons, with "beautiful faces"—but black slaves existed too, Nubians and Ethiopians captured by the Egyptian pharaohs ("blacker than night," it was said, "blacker than an ant, a jackdaw, a cicada"). What mattered was not the color of the slaves' skin, but the simple belief that the condition was assumed to be for "eternity." There were doubters too, from the beginning: Diogenes, not unlike later abolitionists, counseled that the man who relied on captive labor was the true slave, but such reflections had no effect on practice. In Rome, as in Greece, however, manumission—the freeing of slaves—was not uncommon.

With the end of the Roman Empire, everything from vast aqueducts to legal codes lay in near ruin; so did once-beloved gods and hallowed beliefs. But not slavery. Scythian slaves could still be purchased in Antioch;

Gothic ones were hawked in Rome too. In the Middle Ages, slave markets prospered as never before: Saxons, Angles, Wends, and Avars could all be bought at Verdun, Arles, and Lyons, often at great fairs. Curiously, though, sometime in the Middle Ages slavery vanished in northern Europe, but it continued to thrive everywhere else, prospering especially in the Mediterranean. As always, conquest was the tool of choice; and for Christians and Muslims alike, the institution had sanction in Roman and canon law, in the Bible, and also in the Koran. It was, too, the product of continuing war between Christians and Muslims. Following in the footsteps of the Romans, who had conquered Carthage and promptly enslaved its entire population, the Moors overran Visigoth Spain and quickly subjugated vast masses of Christians; 30,000 alone were sent to Damascus, as booty due to the caliph. Yet Christian representatives also vied with the Muslims for their precious cargo. There was extensive bartering: Middlemen offered gold, ivory, ebony, dyed goatskins, or malaguetta peppers, in return for European luxuries such as glass beads, weapons, woolen goods, or even people; not uncommonly, black slaves from Guinea might be swapped for blond ones from Poland.

And inside every Muslim Mediterranean court, especially those of al-Andalus, there existed a multinational throng of servitude: Greek, Slav, German, Russian, Sudanese, and Ethiopians or Nubians. Borrowing from the caliphs of Baghdad, the Umayyad rulers of Cordoba employed a slave army of sixty thousand "silent ones"—"silent" because these captive legions of Germans, English, and Slavs could speak no Arabic. Meanwhile, Christian white slaves (*esclavos blancos*) crowded the markets, most of them abducted on the open seas or in quick coastal raids on Spanish or Italian ports or villages. All the same, black slaves, particularly girls, were prized by Arab merchants as servants, concubines, or, in the case of young men, as soldiers.

But the enthusiasm for Africans was not simply a private interest among the Muslims; they were also popular as slaves in Java and India; even the Chinese seemed to have been fond of East African slaves, provided by Muslim merchants in Canton. And while the main Arab route across the Sahara to Morocco sliced from Timbuktu to Sijilmasa, some of the traders were Jewish, Berber, and, as it happened, black as well. Increasingly, though, other nations were also lured by the complex traffic in human souls. The Dutch had a thriving slave trade, and so did the Sun King, Louis XIV, not to mention the Portuguese and Spanish. Interestingly, the racial mixture was still unimportant: French royal galleys contained a polyglot mix of Russians,

Poles, and Bulgars, as well as blacks. Meanwhile, Muslim Barbary pirates seized and enslaved an estimated 1.25 million Europeans and Americans between 1500 and 1800, at one point even carrying off virtually every inhabitant of the Irish village of Baltimore in 1631. For that matter, slaves were already widespread in Mexico, Peru, and Brazil, and even the Cueva Indians had slaves, secured as captives in battle. Moreover, in Latin America a new occupation was found for slaves: as human sacrifice.

The early eighteenth century marked the increase of slaves in North America as well: in South Carolina, where they cultivated rice; in North Carolina, where the colonists complained of "distemper'd negroes"; and even in Quaker Pennsylvania, where they cleared fields and put up houses. Actually, for a brief interlude, from 1733 to 1748, the trustees of Georgia forbade the import of slaves, but the prohibition collapsed when it was decreed slaves were "the one thing need[ed]." Soon, all across North America advertisements were found offering "strong hearty stout negroes," or "very likely negroes," or "several lusty negro men." But in an age of increasingly enlightened thought, a tinge of guilt, or outright humanity, also began to govern the trade: Hence regulations were created. The Dutch insisted, "Do not permit any Negroes, slaves, or slave women to be defiled." But despite the efforts of humanists, there were heartrending tales of atrocious treatment, some told by slaves themselves who survived the perilous journey across the Atlantic—the dreaded Middle Passage. Slaves were "tightpacked," like sardines, in wrenching heat and underground dungeons, forced to endure suffocating filth and intolerable smells, often from their own urine and excrement; many were brutally secured by locks ("spoonways"), manacles, chains, or head rings. To sleep, they had to lie on top of each other. To calm pains of hunger—depending upon the circumstance, they were often barely fed—they were provided sticks to gnaw on. As the slave Equiano wrote sadly: "the shrieks of the women, and the groans of the dying, rendered a scene of horror almost inconceivable."

Indeed, so horrible was the Middle Passage, so frightening was the "green sea of darkness," that captured slaves would frequently try to end their lives, throwing themselves into the sea if they had the chance, or banging their heads against the ship, or holding their breath to "smother themselves," or dying of hunger by "not eating." In these cases, it was not uncommon to force slaves to eat by breaking their teeth because "they would not open their mouths." In other instances, a special device was used: a unique pair of scissors, or *speculum oris*, in which blades were forced between the teeth of the rebel slave, and an attached thumbscrew turned in order to "force the jaws apart." The wretched situation was only aggravated by disease—dysentery, the flux, smallpox, and sometimes simple dehydration; the mortality rate was appalling.

Some perspective is perhaps necessary; in general, human life on ships was held in shockingly low regard; white crews were frequently subject to a sadism that bordered on criminality too. But few approached the willful cruelty of a "slaver." On one voyage, a woman died after childbirth; her infant was then callously left in the broiling heat of the sun, and the British captain ordered it "thrown overboard." He was later acquitted of murder, as there had been "no premeditated malice." Sometimes the worst atrocities were inadvertent. Storms were always much feared, but by the slaves most of all. One Portuguese captain reported that upon a rising storm, it was then that the din of the slaves, chained to one another, became "horrible." He writes: "The clanking of the irons, the moans, the weeping, the cries, the waves breaking over one side of the ship and then the other. . . . Many slaves break their legs and their arms, while others die of suffocation." Other times, the atrocities followed slave rebellions: It was typical that slaves would be mutilated. Their hands were cut off, passed around for all to see, and subsequently "hoisted" like flags. Then, after a day of writhing in agony, the men would be beheaded, the most cruel punishment of all, for many Africans believed that if put to death but "not dismembered," they would return to their own country again.

One incident, however, would be so horrific that it would stand out among others, and rouse international attention: It was aboard the Liverpool slave ship *Zong*. In 1781, sailing with 442 slaves from São Tomé in western Africa for Saint-Domingue in the Caribbean, the ship lost its way. The water supply dwindled, and many slaves died or fell ill. The master, Luke Collingwood, gathered his officers and informed them that if the ship's slaves died "naturally," the loss would be borne by the owners of the vessel. But if, he added, they "were thrown alive into the sea"—to protect the safety of the crew—it would be the loss of "the underwriters." The first mate protested, to no avail. One hundred twenty-two slaves were flung into the sea: fifty-four were thrown overboard on the first day, November 29; forty-two the next day; and even after a rain, which improved the water shortage, another twenty-six on December 1; meanwhile, ten more slaves jumped into the deep on their own. The case became a scandal on both sides of the Atlantic, but largely because the insurers refused to pay. When it was brought to jury in England, the lord chief justice shamefully remarked that the killing of the slaves was the same as if "horses had been thrown overboard." And the solicitor general blandly concluded that a master could drown his slaves without a "surmise of impropriety."

Soon, though, there was a price to be paid for this cruelty. By 1783 it would no longer wash. Suddenly opponents of slavery seemed to be raising their voices louder and more furiously, and not just the standard suspects of pacifists and Quakers in Britain and North America, and intellectuals in France. The clamor grew in England, and more broadly in France too, the two movements reinforcing each other. Echoing Abbé Raynel—and more distantly the giants of the Enlightenment, Voltaire, Montesquieu, and Rousseau—Jacques Necker wondered in a widely read essay, "how we preach humanity yet go every year to bind in chains twenty thousand natives of Africa." In 1787 the agitation took the form of the prominent Thomas Clarkson Committee established in London, "For Effecting the Abolition of the Slave Trade." The group included not just the indefatigable Clarkson, but deans and doctors, majors and great businessmen, tutors in colleges, bishops and squires, prebendaries and archdeacons, not to mention members of Parliament, including such dominant spirits as William Wilberforce, one of the most eloquent men of the era, and the young prime minister himself, William Pitt, often considered "the best brain that had ever graced English politics." The symbol of the campaign, which became legendary, was devised by no less than the master potter Josiah Wedgwood. It consisted of a chained Negro on bended knee poignantly asking the question: "Am I not a man and a brother?"

The question was haunting and so was the image: And the once fringe Quaker cause of abolition had now become a national and international movement. Within months, the French quickly followed suit, forming the Sociéte des Amis des Noirs (Society of Friends of the Blacks), which included such enlightened aristocrats and would-be-revolutionaries as the Duc de La Rochefoucauld, the Marquis de Condorcet, the Comte de Mirabeau, and Lafayette himself; by 1790 Mirabeau, in one of his most impassioned speeches in the Assembly, powerfully mocked the brutal slave trade, *"Et ce commerce n'est pas inhumain?"* And back in London the passionate feminist Mary Wollstonecraft taunted Thomas Jefferson's racial views and the "abominable traffick" in slaves, while even Adam Smith, the foremost economist of the day, now insisted that freemen would work "better" than slaves. For his part, the tireless British statesman Edmund Burke, who spoke out often, brilliantly, and effectively, assaulted the very idea of slavery, deriding it as a "a state so improper, so degrading, and so ruinous to the feelings and capacities of human nature that it ought not to be suffered to exist."

One image of cruelty particularly struck the members of Parliament. As General Tottenham testified about slaves in the West Indies, "I saw a young man in the streets in a most deplorable situation—he was entirely naked—he had a collar about his neck, with five long spikes projected from it. His body before and behind his belly and thighs were almost cut to pieces with running ulcers in them, and you might put your finger in some of the weals. He could not sit down, owing to his breech being in a state of mortification, and it was impossible for him to lie down, owing to the projection of the collar round his neck. . . . The field negroes are treated more like brutes than human beings." And beyond the cruelty was the sheer degradation of the institution: surgeons examining slaves' eyes, ears, lifting their arms, probing their necks and squeezing their joints and muscles, pinching their breasts and their groins, and making them jump, like common farm animals. Indeed, it is possible that Alexander Hamilton's grandfather, as a surgeon in the Caribbean, did just that; doctors were often called upon to certify the health and performance prospects of those being dispatched into bondage.

The accounts of the brutalities, the whippings, the torture, all as a matter of routine, without any legal limitation, had a vivid impact, not just in England, but increasingly in the young United States as well. But here, unlike in England, unlike in France several years hence, while the states were interested in the slave trade, the friends of liberty were divided.

Still, the dissenters *were* eloquent. The Quakers declared in 1790 that slavery was "a national sin" and would lead to a "national calamity as sure as there is a righteous God who rules among nations." General Thompson of Massachusetts followed suit, ruminating: "Shall it be said that, after we have established our own independence and freedom, we make slaves of others." In a private letter to Madison, George Lee Turbeville, a friend of Washington's, derided slavery as a "great evil"; Madison himself had written in *The Federalist* that slavery was a "barbarism of modern policy." Vermont agreed, having abolished the institution in 1777; Maine, Massachusetts, and New Hampshire soon followed suit. And too often forgotten is that in 1778 and 1783 even Virginia and Maryland had barred slaves, briefly, from their ports. But no less than states like Quaker Pennsylvania laced their laws with sunset provisions and restrictions; meanwhile, in slave-heavy New York, which did not enact a form of emancipation until 1799—and even then, it was more ruse than reality—slaves could be seen on almost any New York street, and the city had more slaves than any other in the young republic except Charleston.

Yet if the enemies of slavery were gathering, so were the enemies of abolition. The imperatives of union were too strong. As Samuel Brown of Philadelphia wrote in the *Independent Gazetteer*, in the Constitution, "the words, dark and ambiguous . . . are evidently chosen to conceal from Europe that, in this enlightened country, the practice of slavery has its advocates among men in high stations." To many, the pivot on which this debate would turn seemed to rest on the last great act of a dying man: Benjamin Franklin.

∞

HE HAD LESS than two months to live. But outside of George Washington, it would be hard to think of another Founder who by temperament, by reputation, by history, and by sheer force of intellect and presence could hold sway over the debate taking place in Congress as Franklin: Franklin the visionary, Franklin the polymath, Franklin the lover of liberty, and Franklin the homespun jack of all trades whose likeness is to be found in nearly all of America's defining moments.

Known worldwide, Benjamin Franklin was the most recognized of Americans outside of the president, and the best known statesman of his era. Jefferson called him "the greatest man and ornament of the age," and Turgot, the Frenchman, memorably remarked: *"Eripuit Coelo Fulmen, Sceptrumque Tyrannis"* ("He snatched lightning from the sky and the scepter from tyrants"). His extraordinary touch covered the young American republic he helped give birth to: as a celebrated inventor of the Franklin stove, bifocals, and the lightning rod; as a celebrated printer and publisher; as the first postmaster general; as America's leading diplomat (he was its envoy to France and Britain); as one of its great politicians (he was president of Pennsylvania); and, of course, as a Founding Father—he was the only one to sign all four decisive documents of the republic. He helped write the Declaration of Independence; he secured the alliance with France; he assisted in negotiating the peace with England; and he sat in the conference that drafted the Constitution. But if history was beckoning, for that matter, if history seemed to be his—it did—it hadn't always. Little in his family background suggested he would or could amount to much. The hallmark of his life was his beginnings. Born in Boston in 1706, he was the youngest son of a family of seventeen—his father was a candle maker—and virtually everything about him was self-made, a telling counterpoint to the charge that the young America was merely for the propertied and the well-to-do, for great families and the well-connected. Starting from scratch, over the course of a long, distinguished life, his genius spanned the worlds of politics and literature, science and social reform; it would become part of the very fabric of America.

Part of his genius was his timing. He was an indentured servant to his older brother's newspaper, but ran away (illegally), moving to Philadelphia, then the most dynamic seaport city in the colonies. Another part of his genius was his diligence and innate talent: Though he had only two years of formal schooling, he became a lifelong autodidact, teaching himself French and Latin, Italian and Spanish, math and science. And underneath that wry smile and bemused veneer, he had, as biographer Walter Isaacson points out, a fierce entrepreneurial streak. Success came to him quickly, almost effortlessly. Actually, the "quick" part of the story was true, but not the effortless part. It was only through hard work, driving ambition, and formidable abilities, and just the right contacts, that Franklin became the owner of the *Pennsylvania Gazette*, and then developed an almanac, *Poor Richard's*, that mixed wry wit with mottoes, helpful aphorisms, and poems. Though he cribbed from Jonathan Swift, he creatively made the almanac in his own image, and it was uniquely American. In it, he introduced the clever wisecrack, and he embraced the idea of the self-made man, disseminating this new ideal as part the mind-set of the country. His words echo to this day: "in this world nothing can be said to be certain but death and taxes." Or too: "laws too gentle are seldom obeyed; too severe, seldom executed."

Franklin rapidly became one of the leading men of Philadelphia. And once enriched—astoundingly, he was able to retire from business in his early forties—he remade himself, this time as a public-spirited gentleman. He helped establish Philadelphia's first police force and its first fire company, as well as its first library; he played a critical role in paving, cleaning, and lighting the streets, with whale oil lamps; he helped found the American Philosophical Society, the equivalent of England's famous Royal Society, the city's first hospital, and, not least, the Academy for Education of Youth—today's University of Pennsylvania. But even here, due to his restless, searching nature, he was never content; his interests mushroomed, as did his achievements and accolades. He became one of the world's most famous scientists. His experiments—especially with electricity, which were translated into French, German, and Italian, but also his observations on the Gulf Stream, his invention of the damper, his creation of various smokeless chimneys—dazzled European intellectuals and became the stuff of legend. Nothing escaped his attention: He crossed the Atlantic eight times, met leading scientists and engineers, studied geology, farming, and archaeology, as well as eclipses, sunspots, whirlwinds, earthquakes, ants, alphabets, and lightning conductors. And he thought large: He said that America was "favored by nature" and must one day become "a great country, populous and mighty," and when a bystander, glimpsing one of the first gas-filled balloons, asked him what good it was, Franklin flashed a grin and retorted: "What good is a newborn baby?"

Honors quickly followed: a fellowship of the prestigious Royal Society, degrees not only from William and Mary, Harvard, and Yale, but also Oxford and St. Andrews; from then on he became known as "Dr. Franklin." His intellectual life traversed the globe; esteemed thinkers eagerly responded to his letters and, in due course, he was inducted into twenty-eight academies and learned societies—including even the Russian Academy of Sciences. In France his face was everywhere: from painted porcelain to scrolls to printed fabrics. And fashionable society loved him, as did the intellectuals. His *Almanack* was a runaway best seller in Paris.

Then he reinvented himself again, plunging into politics and becoming one of the preeminent politicians in the British colonies. Like the young George Washington, early on he was an ardent empire builder, envisioning a sprawling Anglo-American dominion that stretched to the Pacific by land and sea—under the tutelage of the British crown ("Not merely am I a colonist," he boasted, "but I am a Briton."). He was sent to England as a representative of the Pennsylvania Assembly, and in short order, he represented Georgia, New Jersey, and Massachusetts as well, arguing against the new British program of colonial taxes and stricter regulations. At the same time, he was hardly a saint; though America's grievances were initially about "corruption," a little corruption now and then never bothered him. He lobbied Lord Hillsborough, the secretary of state, to grant him and his partners a new colony of twenty million acres in the Illinois country, then an Indian domain; Hillsborough turned it down.

As the crisis deepened between the colonies and the crown, and pas-

sionate voices of dissent could be heard across America, Franklin, longer
than any of the other Founders, remained a hesitant revolutionary. He fervently sought to maintain the British empire and doggedly remained loyal
to King George—in his words, "the best of kings." Actually, independence
was a heartbreaking crusade for him; he once remarked balefully that "In
England I am too much an American, and in America, too much an Englishman." So year after year, he sought to patch up the empire, hoping to avoid
a civil war, and year after year he sought tirelessly to placate the British and
to undo the mistakes of a heedless London ministry—even when the London papers called him an "old snake," "a living emblem of iniquity," and a
"Judas," and even though he was cruelly humiliated—twice—by the British
authorities, including being denounced publicly by the Privy Council. In
fact, he lived in England so long that he thought of staying there.

But when Franklin eventually turned on the British, he was a devastating opponent. He now helped push the colonies into open rebellion and
became a charismatic leader of independence.

And he never looked back.

∞

HIS 1778 MISSION to Europe to secure the French alliance—he was,
remarkably, seventy-two at the time—was essential to America's military
victory. And while in Paris, he was no mere diplomat; tackling his aspara-
gus with his fingers and cutting his food with a knife rather than a spoon,
he was treated as another Rousseau, and a more tantalizing one at that,
being an American rustic rather than a mere Swiss. The Comte de Ségur,
who spent years in Catherine the Great's court and volunteered to fight in
the American Revolutionary War, said of him: "His bearing [was] simple
but dignified, his language direct, his hair unpowdered. It was as though
the simplicity of the classical world, the figure of a thinker of the time of
Plato, or a republican of the age of Cato or Fabius, have suddenly been
brought by magic into our effeminate and slavish age." Strong words of
praise, but they were widely shared. And Franklin knew how to give as good
as he got: In Paris, at Congress's behest, he awarded a ceremonial sword to
Lafayette for his services, with the engraved image of a rising moon and the
motto *"Crescam ut Prosim"* (Let me wax to benefit mankind). It was a typical
Franklinesque touch.

So extraordinary, so multifaceted, so charmed was Franklin's life that it is tempting to see it as flawless. Not so. For all his simplicity he loved the good life: In Paris alone he had 1,041 bottles of wine in his cellar, nine indoor servants, and he spent freely, on food, on women, on luxuries. While his circle of friends was among the most fascinating in history—his acquaintances included David Hume, Bishop Berkeley, Adam Smith, Edmund Burke, Voltaire, Noah Webster, Sir Isaac Newton, Jeremy Bentham, Diderot, and Immanuel Kant, as well as the Enlightenment princess Ekaterina Dashkova of Russia, an intimate of Catherine the Great herself—his colleagues often couldn't stand him. John Adams thought him "lazy," "insignificant," and devious, not to mention too accommodating to French interests. Arthur Lee, a fellow commissioner to France, despised him and was convinced that their secretary was a British spy, a charge that Franklin abruptly dismissed (Lee was right). And Ralph Izard actively sought to have Congress remove him as U.S. envoy abroad. Franklin also made his share of mistakes. As an envoy for America, he often complicated matters, especially with the British, rather than solving them. He was vitriolic toward his critics, calling them "bug writers" and "dirty stinking insects"; he once called John Adams "a raving madman." His friendships repeatedly fell apart, with his brother James; with his closest friend, John Collins; with his business partner, Hugh Meredith.

And his winning smile, sharp wit, and famous humanitarian spirit wholly deserted him with his family. Vain and self-absorbed, he was bitter toward his estranged, illegitimate son, William (who would become, with Franklin's help, Britain's royalist governor of New Jersey); almost unimaginably, he would even teach his grandson, William Temple Franklin, to despise his father. He was distant with his daughter, and didn't even bother to attend her wedding. And while squiring countless women around Europe on his arm (Madame Brillon once said of him, "*vivre toujours avec vous*"—"it is good to always live with you"), he was indifferent, even callous, to his lonely, common-law wife, Deborah, not even returning home when she began her slow slide into a pitiful death. And for all the simple eloquence of his writings and his brilliance, for all the momentous diplomatic ventures he was engaged in, Franklin was, more often than not, quiet and inscrutable; like Washington and like Jefferson, he knew the power of silence. He once advised, "let all men know thee, but no man know thee thoroughly." To many, friends as well as detractors, he remained a mystery. And to his dying days, he preened about his reputation.

But for all his contradictions, Franklin was special: special because he was worldly and sophisticated, special because he was an ardent devotee of the public good and an unabashed champion of middle-class values, and special because few, in America or in Europe, spanned such a long and remarkable life as he did. He thought large, believing that America should also include the West Indies islands, Quebec, St. John's, Nova Scotia, Bermuda, East and West Florida. And few did so much to create the United States; actually, after years of being abroad, he ruminated, "I wish to die in my own country." While he repeatedly sought to avoid controversy in so much of his career, he also had one other gift: the gift of exquisite instinct—protean and ubiquitous, he knew when to speak and when to hold back; when to weigh in on a problem and when to let it simmer; when to turn on the charm and roll out the famous Franklin smile or when to be cold as ice. He knew how to take a losing situation and turn it to his gain. And he knew how to make himself a citizen of the world. No wonder he once boasted that his face was as well known "as the man in the moon." No wonder William Pitt remarked that Franklin was "an honour" to the English nation and to "human nature," and that Franklin had attained the status of a mythological figure, the nemesis of all evil and tyranny. And no wonder one Frenchman said in amazement: "Who is this old farmer with such a noble air?" Others had asked precisely this for almost forty years.

For these reasons, as much as any, his signature carried stentorian weight when at age eighty-four, wracked by gout, seized by pain, barely able to walk or sit and subsisting on laudanum, a dying Franklin now sought to be uncompromising on one matter facing the young republic: slavery. Here, as in so many other areas, Franklin seemed to be ahead of his time. When a South Carolina delegate argued slaves were property, akin to sheep or dogs, Franklin shot back: "there is some difference. . . . Sheep will never make any insurrections."

So, on that grim winter day in 1790, could Franklin's endorsement of the petition once again move the country and make it whole?

∞

IT WAS NOT to be. The assembled representatives remained deaf to his clarion call. This, despite the fact that Southern moderates, like Madison, initially praised the abolitionist cause in quite favorable terms—terms that paralleled Northern praise for the petitioners. But the matter was too incendiary, and everyone knew it. Keeping the republic whole rendered any early action against slavery, even so much as discussing it, exceedingly difficult. So in the end, the Northern and Southern moderates embraced procedure and defended the petitioners' right to be heard in Congress, but they were woefully silent on principles and the petitions' antislavery calls. Wise or misguided, Madison observed that the federal government was duty-bound to accept all petitions, even if it were to swiftly deny the request, but his involvement stopped there. The House then split the difference; it rebuffed the Senate, which had tabled the petitions, as well as the Deep South, which unanimously voted against even considering them. Instead, the antislavery memorials were referred to a select committee.

But that would not be the end of the matter. If Congress did not end this "slave business," the head of the select committee quietly communicated to his antislavery colleagues, then the entire South might "rise up in a flame." No one was willing to take this chance. While Madison himself thought the proslavery arguments "shamefully indecent," he cherished the union even more. And as Madison went, so went the moderates. As the moderates went, so went the Congress. After several days of debate, on March 23, the majority that had favored Franklin's petition disintegrated, and an exhausted House voted 29–25 to accept a heavily amended report that basically buried the matter. More starkly, it said in effect that slavery in the South would be out of bounds of congressional discussion—forever. The founding generation, always obsessed with its capacity to set precedents, had inadvertently set a strong precedent on the very matter they hoped to avoid. Without intending it, Congress had determined that the new republic could—indeed it would—sanction, even underwrite, bondage. The stage was thus set for a collision one day between the integrity of republican government and the right to hold, sell, and buy men, women, and children as though they were cattle.

Was this a tragic failure of the Founders? Tragic, absolutely yes. But could they have done differently? Here, the what-ifs are enticing, but so are the consequences. Accept abolition, accept disunion—that was the fear, and that was the threat. Moreover, as Bernard Bailyn has pointed out, before the revolution, slavery was rarely seen as a problem; after the revolution, there was never a time when it wasn't a problem. He further notes that in the context of their day, what is so striking is not that they did not get rid of slavery, but that so much was done to discuss it, to debate it, and even to eliminate it, in the North and Northwest at least, despite the fact that it wasn't outlawed. And as the historian Joseph Ellis has noted, no responsible statesmen had ever contemplated, much less endorsed, a biracial American society. Is this persuasive? It is telling that in this debate, Hamilton, a staunch abolitionist, remained stubbornly silent rather than jeopardize his financial program; it is equally telling that John Adams, another ardent opponent of slavery and never one lightly to hold his tongue, also remained quiet. And so too was George Washington, who had taken a personal vow never to purchase another slave and hoped that the institution would be abolished by "slow, sure, and imperceptible degrees."

Here Washington, who so often had the last word, was himself acutely aware of the stakes. He commented that the "slave business" had been at last "put to rest and will scarce awake."

Yet it would not. In three weeks, on April 17, Benjamin Franklin would be dead. But not his ideas. As the Founders would soon find out, the issue would be raised again within a year—and again, and again, and again. But once events heated up abroad, few would take notice. For a great war was coming their way. It would roil a continent and envelop the young America in its wake. John Adams once said, "Our deficiencies . . . are owing wholly . . . to our inexperience in the world." And as Joseph Galloway had earlier warned: "The practice of conquering and dividing territories and kingdoms, is becoming fashionable in Europe," and he feared an independent America would soon be subject to partition by the great powers.

As the globe fell into greater tumult, then terror, that day now seemed to be coming.

CHAPTER 5

RUSSIA

✦

A T THE END OF Catherine's grand expedition to the Crimea, the Frenchman Comte de Ségur noted wryly, "We had to get back to the cold calculation of politics." To this, he could have added the word "war." The victor's prize was now dangled in front of the empress's eyes: a new Jerusalem rising around the Black Sea, with Constantinople, the City of Caesars, at its center, ruled by her grandson Constantine—and governed from the heart of Russia. But for all her audacity, was it worth it?

The question lay heavily on Catherine's mind as the two great foes, Russia and Turkey, lumbered into conflict. Every morning she would rise at six and light her own fire and begin fifteen hours of work. While downing black coffee and crunching on biscuits, she attacked the mountains of paper brought to her each day and dutifully initialed them, made comments in the margins, signed orders, and dispatched missives, and anxiously waited for word from the military campaign unfolding at the mouth of the Black Sea and the Bosporus. She had not yet heard from Potemkin, her commander-in-chief. But she knew her grand plan of victory without bloodshed had failed; unlike in the Crimea, the Ottomans were not about to yield or be intimidated. Catherine had hoped to play for time, but it was too late for that. The final storm was brewing, but which way would it break? Already, enlightened Europe—Prussia, Sweden, Britain, Holland, and even France—was lining up against her. And soon a new set of near-catastrophic events would menace Russia as the continent crept slowly toward its next butchery. But for now, her fleet had, she believed, been deci-mated off Ochakov by a great swale. And in most respects, save for man-power, Russia was woefully unprepared for war.

Yet fired by imagination and history, she remained steadfast. And if her imperialistic actions were at the time provocative and a bit puzzling, she still confronted her Turkish nemesis with a quickening pulse. Fear, she knew, was a frail foundation for military policy. At one point she had blus-tered to Ségur: "You do not want me to drive out your protégés the Turks? Indeed they do you credit! If you had neighbors in Piedmont or Savoy who were murdering or capturing thousands of your countrymen, what would you say if I took it into my head to defend them?" But however persua-sive this logic may have been—and in truth it smacked of a self-serving rationale to defend her imperial ambitions—it was only the cold logic of bloody battle, by land, on sea, and in the hearts and muscle of her men, that would decide Russia's outcome with the Muslims. And the tsarina's success or failure would also depend as much on the vast empire arrayed against her as on her own tactics and temper, her own commanders, or her own fight-ing men.

Which meant the center of the Islamic world: the Ottoman empire. It meant the sultan's domain.

∞

OTTOMAN LEGEND HAD it that one glorious night, the first sultan, Osman, dreamed of a beautiful, enormous tree growing from his navel, a tree whose shade "compassed the world," including far-off mountains and mighty riv-ers. As it happened, the dream would prove to be more than prophetic. From its humble beginnings, the Ottoman sultan's orbit grew to be vaster than that of a Roman emperor.

Every acre in his domain had been subdued by Muslim blood and vanquished by the ruthless scimitar. For some four centuries, his provinces dominated an immense swath of the planet, enveloping much of south-eastern Europe, a quarter of the continent, reaching westward across the lengthy coast of Africa to the Moroccan border and into the Middle East, and straddling the banks of the Caspian Sea, the Mediterranean Sea, and the Dalmatian Coast. Meanwhile the Black Sea remained the sultan's pri-vate lake, his "pure and immaculate virgin." And inside the empire's mam-moth tracts of arid plains and blackened mountains, its extensive reach of rivers and verdant valleys, lived more than twenty-five million subjects, and claimed such great cities as Baghdad and Belgrade, Algiers and Cairo, Athens, Jerusalem, and Damascus, not to mention the lush Balkan territories of Walachia and Moldavia. And fulfilling a dream that had eluded both of the great Persian kings, Darius and Xerxes, who had hoped to absorb the whole of ancient Greece into the Persian Empire, it included this cherished soil as well. Religiously, the empire owed its allegiance and fealty to Islam; socially, its beating pulse was the holy cities of Mecca and Medina; and politically, its center was the Ottoman court. Among the Muslims, the Ottoman Turks were, of course, the ruling overlords, but the empire was also a great multinational state rivaling those of the Hapsburgs and the Romanovs. And the sultan reigned over millions of Christians as well: Greeks, Serbs, Hungarians, Bulgars, Walachians, and Moldavians. Indeed, from Vienna to Persia his armies were feared by Christianity and Islam alike, and the former territories of the empire would one day yield an astonishing twenty-one modern nations.

At the empire's core stood Constantinople, proud and ancient, where the sultan was supreme. But of necessity, this multilingual dominion of peoples and religions was administered not just in Constantinople, by a vast bureaucracy, but locally, by a multitude of "pashas" (literally, the "sultan's foot"), princes, beys, aghas, governors, sheiks, mullahs, and emirs. Year after year, carts groaning with gold and silver and precious jewels shuttled in from the north to the walls of Constantinople to pay tribute. And year after year, Ottoman fleets ruled the waves from Algeria to India, while the Barbary states of Northern Africa raided coastal towns in Europe. Remarkably, these Barbary pirates would reach as far and deep as the Irish and Cornish coasts, making slaves of the English king's subjects, and eventually assaulting the ships of the young United States as well.

When the Ottoman sultan Mehmed II first overran Constantinople in 1453—actually, it was the thirteenth Muslim attempt to take the city and the first to victoriously breach its walls—he galloped straight to the doors of the Roman emperor Justinian's venerated Church of Hagia Sophia, the Church of Holy Wisdom. There, below a mosaic of Jesus Christ, he sprinkled earth on his turban to proclaim his humility before Allah, drew his sword, and growled, "the buildings of this city fall to me." They did. And the message was clear. Henceforth Mehmed, barely twenty, was not just the Turkish khan, or the Arabic sultan, or the Persian padi-shah, but the Kaisr-i-Rum: "Caesar of Rome." To Westerners, however, he was the "Grand

Seigneur" or the "Great Turk," but he was just as often the "Emperor." "He is and remains Emperor of the Romans," George Trapezuntius, the Cretan historian, announced in 1466, "and is also the Emperor of the whole earth." Trapezuntius was not far off. As early as 1529 the Turks attacked Vienna and seemed about to burst into Germany: Then and later, the Christian world shuddered at their Muslim hoofbeats. To them, the Turks were an imponderable mystery as well as a terror.

And that terror only grew. By the sixteenth century, under Sultan Suleiman the Magnificent, the Ottoman Empire, at its height, *was* practically the whole world, a feat all the more remarkable given that Britain's Henry VIII and Russia's Ivan the Terrible both ruled in the same era. This was the golden age for Constantinople, when the city overflowed with immense riches, when lavish imperial pavilions rose up along the golden shores of the Sea of Marmara, and great festivities lasted three whole weeks. Suleiman himself rode the crest of the age: He was a champion of the arts and sciences, a lover of literature; he supported the finest poetry, the grandest music, and the best of modern philosophy. But foremost, he was an emperor—and he meant to subjugate all of Europe into his orbit. In central Europe, then, whenever Tatar horsemen were on the move or the sultan's murderous guardsmen were mobilizing, Christendom felt all the old fears. Popes regularly dreaded that the enemy might break into Italy, and throughout Germany, by special order, "Turk bells" were erected, ever ready to sound the alarm.

But in truth, the picture was more complicated. Actually, the Ottoman empire tolerated other religions more so than most Western kingdoms—certainly better than the Moors or Jews had fared in Catholic Spain (indeed, when Spain expelled its Jews in 1492, the Ottomans opened their arms), and better than the Muslims in all of Christendom, or for that matter, Protestants in France after 1685 or Catholics in Anglican Ireland. Christian provinces were permitted their own systems of government, their own class structure, and their own bureaucracy. More than that, the Ottomans actually sought out Christians to serve in Constantinople's imperial administration, and even to form an elite corps within the sultan's own guard, the Janissaries. Invariably though, for Christians, the keystone to advancement was conversion to the Islamic faith. Countless Christian boys, most ruthlessly taken by force from the Balkans at a young age, were given a painstaking education intended to exterminate every memory of their family, as well as to extinguish every vestige of Christianity. In turn, these young boys, often the best-looking and the best-born, grew to comprise brigades of unflinching and fanatical followers, ready to assume any role; they were thus boatmen, butchers, and slave dealers, but also pages in the palaces and apprentices in the civil service, some even climbing to the uppermost echelons of the imperial administration. In the end, it was the sultan they served, it was Allah they worshipped, and it was the Koran they blindly followed.

The fabled Janissaries were a unique caste, the praetorian guard of Istanbul. As children, and later as "slaves of the gate," they spent their whole lives in barracks. Theirs was an exacting existence: They were barred from marrying or having children, and their complete fealty was to the sultan; they were his personal guard as well as his dedicated infantry, not to mention the principal military force in Constantinople itself. In peacetime, when they garrisoned the Seven Towers or enforced law and order, they brandished only a scimitar, but in battle they wore their trademark blue uniforms with bejeweled headdresses and wielded their weapon of choice, be it a javelin or a sword, an arquebus or a rifle. And without roots, ties, interests, or ambition outside the rarefied military brethren to which they belonged, they became the ideal fighting force. But like other praetorian guards throughout history, in time they became a feared political force as well. In the fourteenth century, there were only 12,000 Janissaries; by 1776, there were almost 45,000. As the years passed, the strictures were loosened: Older Janissaries were permitted to retire, to take wives and have children, and eventually they became a special, hereditary class unto themselves, not unlike the powerful Russian Streltsy. And ultimately they were as feared by their own masters as by their enemies: During one revolt, the Janissaries mutinied and slaughtered an astounding twenty thousand Turks, so many that the emperor "could stand on the dead bodies as though on an island." Grand viziers and even mighty sultans were toppled at their command; not unsurprisingly, the Janissaries held great sway over policies of state as well, even forcing Suleyman himself to raise the siege of Vienna in 1529.

When not at war, the Janissaries lived within the fabled city of Constantinople, today's Istanbul. Not far from here had once stood the ancient city of Troy; and high above loomed Mount Ida, from whose 5,800-foot peak the gods were said to have witnessed the Trojan War. Constantinople itself could be reached from the south only by sea, through the astonishingly slim thirty-eight-mile-long Dardanelles, and from the north by the Black Sea, both of which afforded it unique protection. This, of course, befit the city's fabled status as the capital of the Muslim world, the military, administrative, commercial, and cultural hub of the mighty Ottoman Empire. It was also one of the more glorious diplomatic posts of the world—as the Bailo of Venice wrote in 1583, "*È certo che se nelle altre corti de' principi è necessario splendore, in Constantinopli è necessaríssimo*" ("It is certain that if in the other courts of princes splendor is necessary, in Constantinople it is most necessary"). And it was one of the most beautiful cities on the planet: One Ottoman poet insisted that "heaven itself gasped with envy" at the majesty of the city, while for his part, Hans Christian Andersen felt that in Constantinople he had entered one of his own tales—"Everything looked as if it were outlined in flame," he wrote. "Everything was enveloped in magic light."

Indeed, it was hard not to be awed. Rising from the gleaming emerald waters of the Bosporus, the city was a thriving mix of races and religions, with its domes and mausoleums spread amid sumptuous spice bazaars and flowering fruit trees. During Ramadan, the mosques were packed with believers and the minarets were aglow with candles. During the rest of the year, the warehouses and wharves were laden with goods from across the globe, and the city itself was a hodgepodge of colleges and coffee kiosks, hospices and public baths, its long avenues filled with blooming trees and its gardens awash in wild roses. Off in the distance glistened the snow-capped mountains of Anatolia. And most magnificent of all, watching over the hallowed city from high above on the Byzantine Acropolis, was the *saray*, or as it was better known, the Topkapi Palace.

This was the seraglio of the sultan.

∞

INSIDE HIGH, THICK walls and an increasing progression of rarefied courtyards, stretched dozens of buildings, a state within a state, an entire finely tuned apparatus created to indulge the whims of one man, the sultan. Throughout the year, for his joy, huge wagonloads of booty and tribute arrived from the provinces—sardines, salted "tunny-fish," rice, sugar, raisins, lentils, garlic, coffee (coffee houses thrived in Constantinople a hundred years before London or Paris), macaroons, dates and figs, honey and henna, saffron, plums in lemon juice, mutton, and even cartloads of snow. Beyond the colonnades, kitchens, and long avenues edged with tulips, the sultan's private domain bustled with flocks of warbling birds and herds of animals, as well as some five thousand servants waiting to attend to their master's every need. And after traversing an intricate series of gateways, a traveler would at last reach the visible mark of the Ottoman government, the Bab-i Ali. In the West, it was known as the Sublime Porte.

Sublime was the right word. In a manner that would more than rival the opulence of Versailles or St. Petersburg, let alone London or Vienna, the sultan's universe was an elaborate mosaic of ritual and pleasure. Amid the ornate pools, golden domes, and gilded *baldachins*, water poured from magnificent dragon-headed bronze fountains. And from the moment the sultan had his inaugural cup of sherbet and pronounced it *"Peki"* (very good), his table was waited on with detailed precision by the chief attendant of the napkin, who was aided by the senior of the tray servers, the fruit server and pickle server, the sherbet maker, as well as the chief of coffee makers and the water server—and a doctor standing by the sultan's side as a precaution against poisoning. There were also the footstool carrier and the keeper of the sultan's robes, the chiefs of the laundrymen and bathmen, the message-keeper and the master of the turbans. In addition, there were grooms and gardeners, pipe lighters and door openers, two thousand musicians, fifteen hundred mimics, and a fantastic array of jugglers, acrobats, snake charmers, and fortune tellers, not to mention the retinue of dwarfs and mutes, the mutes holding a special pride of place for the sultan: They were his messengers.

And then within lay an even more guarded world, the secluded, scented environs of the *harim*, or *harem*, Arabic for "sanctuary." The harem was an isolated realm of floating veils, conspiracy, and, at the snap of the sultan's finger, sex. Governed by a rigid protocol that was unalterable even for the sultan himself, the harem was closed to everyone but him or, of course, his guests. Guarded by eunuchs, the harem was reached by a lone passage, leading to a complex labyrinth of richly appointed apartments, pavilions, mosques, secret doors, courtyards, baths, and hundreds of turtles crawling among the flowers. Once inside, the sight was like nothing seen anywhere else on the planet, not in all of Europe, not even in Baghdad or Cairo or Tripoli, certainly not in the New World. Surrounded by erotic frescoes, magnificent Turkish carpets lay draped over the floors, incense burned, and soft sounds of the *cura* (lute) drifted in. Meanwhile, naked women reclined amorously, savoring thick Turkish coffee and nibbling on fresh figs. Throughout, musk-scented fountains gurgled to keep prying ears from overhearing the sultan's most private words.

The sultan's women, several hundred all told, were theoretically slaves, and since under law Muslims could not be enslaved, the harem was thus composed entirely of nonbelievers: Russians, Albanians, Greeks, Venetians, with many more coming from the Cis-caucasus, because the blond-haired, fair-skinned Slavs were legendary for their looks as well as their ardor. (Ironically, until the late seventeenth century, the tongue of the court was not Turkish or Arabic, but Serbo-Croat—the language of the Janissaries.) After being renamed, the slave girl stepped inside the harem's gates, locked by two keys of iron and two of bronze, where she often remained ensconced until death.

The harem itself was a microcosm of the empire. The sultan was revered as a demi-god, and all were slaves of the sultan, who was in turn indistinguishable from the state itself. Girls, and in very discreet cases if the sultan also had homosexual tastes, boys, entered the harem as young as ten; from then on, every moment was spent waiting for the crucial signal: the sultan's silent nod of approval. Once a woman became *ikbal* (bedded), she was rewarded with her own suites, servants, fine diamonds and silk clothes, and even gold ducats.

In the early years, the presiding eunuchs, like the harem women, were white, enslaved from the Caucasus; but by the beginning of the seventeenth century, eunuchs were delivered from one of the many slave caravans that traversed the upper Nile around the Sudan, then castrated at Asyut before being shipped downriver to Alexandria. In another maddening paradox of the sultan's world, since the Koran banned castration and Muslims were ashamed to perform the deed, the act was left to a Christian sect, the Copts. In turn, these captured black children, soon to be given pet names like Tulip and Saffron, were then sent east as tribute—human treasure for the sultan. But the eunuchs, who were themselves slaves, and indeed also servants of slaves, the harem women, frequently commanded substantial power through their access to the sultan. Over time the chief of the black eunuchs, often an African, ruled the whole seraglio, controlling the finances for Mecca and Medina and sometimes even ranking as high as third in the empire, after only the sultan and the grand vizier themselves.

But always, the sultan reigned supreme. When he passed, those nearby had to scatter quickly; no woman, or man for that matter, was allowed to enter his presence without having been summoned, and it was impermissible to speak to him without being spoken to. Sitting on his throne in the Gate of Felicity, he would be surrounded by the grand vizier, the chief black eunuch, and a fantastic collection of his palace guard, actually two of them: the *peyks*, with gilded helmets, standing at attention to his right, and the *solaks*, wearing feathered helmets, to his left. During the Feast of the Mevlud, one of the grandest festivals on the Islamic calendar, the grand vizier sat facing Mecca, while hundreds of the sultan's household gathered to celebrate the birth of the Prophet Mohammed; at the same time, raised high above, was the sultan himself, hidden from the faithful, praying. On festive occasions the sultan sat on his throne in his audience hall while his mother and daughters sat on sofas, where there might be poetry readings, singing dwarfs, or prayers accompanied by the effusive distribution of sweets. And it was in this hall that the sultan infrequently deigned to receive foreign envoys, who in a sign of obedience always donned Ottoman caftans before prostrating themselves three times. They would enter to find the sultan sitting, silent and immobile, on his marble throne, clad in a fine robe of golden cloth edged with suede or, just as likely, precious black fox, and wearing a white turban that housed a giant ruby. And invariably, he was visible only in profile.

In this way, no heathen might glimpse the full visage of the Shadow of God in This World and the Next.

∞

AT ITS HEART, though, the sultan presided over an armed empire. Traditionally sultans were a skilled line of remorseless, cunning, and vigorous leaders; indeed, the early sultans spent most of their time on the frontier fighting, rather than in the capital. But over time, engulfed by the debaucheries of their harem, surrounded by conniving eunuchs, and suffocated by elaborate rituals dating back to the Byzantine emperors, the Ottoman sultanate was slowly weakening. True, when the sultan was sure, visionary, and courageous, the state flourished; but when he was vacillating, decadent, or simply lazy, the state deteriorated. Another complication was the fact that there was no set law governing succession, which meant any accession to the throne was often accompanied by a royal slaughter. Ottoman rites, almost uniquely, held that as soon as the sultan rose to the throne, he—in the Islamic world it was always a he—would immediately have his brothers strangled, usually with a bowstring or silken handkerchief, thus spilling no imperial blood and thereby eradicating any threat or competition. Sultan Mehmet III murdered all nineteen of his brothers, many just small children—incredibly, they kissed the sultan's hand first—and to ensure that no rivals could emerge from anywhere else, he then killed his father's seven pregnant concubines; meanwhile, outside the room, tiny little coffins lay in wait.

But in 1603 the new sultan, Ahmed I, put a stop to this horrible tradition. A potential rival's fate, however, was scarcely any better. The sultan now consigned them to a special pavilion called the "Cage." Here, cut off from all contact with the outside world, save for the companionship of eunuchs and sterile concubines, they were entombed by carnal pleasure, left ignorant by neglect, and lived in constant terror of the ubiquitous bowstring. Should one assume the throne or be released into the world at large, he would emerge dazed and fearing for his very life. And predictably there were dire consequences for the health of the empire, for it frequently found itself without a ruler who had the gravitas or political savvy to manage the great Ottoman state.

But someone had to rule. If not the sultan, then who? When the sultan was weak, or stupid, or decadent, or all of the above—in one case, Ibrahim the Mad encased his beard in a network of diamonds and idled his days tossing gold coins to fish, that is, when he wasn't searching for the fattest women in the empire—the dirty business of administering the empire fell to the chief minister of state, the grand vizier. Appointed by royal seal, he oversaw the armed forces, the bureaucracy, the foreign policy, the trade policy, everything, that is, but the seraglio.

Perched behind a latticed bay, the grand vizier was thus supremely powerful, and on occasion, even engineered a sultan's demise. But in an empire perpetually ruled by fear and force, the position also carried incalculable risks. Failure in war, not to mention peace, meant dismissal or exile, or just as often, death by strangulation. The skill required to accede to this exalted station was matched only by the skills required to survive it; in fact, many grand viziers were more renowned for their deaths than for their rule. The dismissal of a vizier was announced by recall of the imperial signet ring, including the key to the Ka'ba at Mecca; in practice, the vizier knew what this meant, and each man dutifully carried his personal will with him if sent for by the sultan. "Am I to die?" he would often ask. "Yes, it must be so," would come the reply. The sultan's death sentences were then conveyed with great subtlety: by the soft tap of the foot or the opening of a single window, and while he listened to the shrieks of the dying from the next room, the execution was usually carried out by his trusted mutes with string wound around the neck. On other occasions, an ax or dagger was used. In either case, it made for a grisly sight: The heads of top officials, and sometimes their bloodstains as well, were laid out on white marble pillars in the Topkapi Palace, some for as long as one hundred years, although even here there was a rigidly stratified hierarchy—significant heads were filled with cotton; lesser heads with straw; and minor heads displayed in corners and cupboards, while piles of human innards, noses, and tongues also decorated the palace. Female victims had their own special punishment: They were weighted down with stones and drowned in the Bosporus.

No one had clean hands in Constantinople.

∞

In Catherine's time, sixty percent of viziers were executed. Still, when the empire fell into near anarchy in the previous century, it was a series of grand viziers who resuscitated the empire's fortunes. The first, Memmed Korpulu, attempted to stanch rampant corruption by executing an astonishing 60,000 men. In doing so, he revived the morale of state and rebuilt the military. By 1683 an Ottoman army of 200,000 drove back the fleets and armies of Austria, Venice, and Poland, raced up the Danube, seized all of Hungary, lunged farther west, and, for the second time in history, the sultan's triumphant warriors stood at the gates of Vienna. Even Louis XIV, an often implacable enemy of the Hapsburgs and a tacit friend of the Turks, feared for the fate of this hallowed Christian city.

Western Europe trembled. As in 1529, the inner chambers of the continent now lay bare to the Muslim invaders. Of necessity, throughout the anxious summer of 1683, the West rallied. Legions of men from the German states sprang to arms, as did forces of the pope (Innocent XI), as well as of Poland, Russia, and the Republic of Venice; even the French secretly joined in. The Ottoman forces were repulsed and driven down the Danube; the sultan ordered his vizier strangled. From then on, it was quicksand: The great Ottoman Empire's fortunes waned, and just as suddenly, Belgrade, in Ottoman hands since 1526, fell to European armies. Wrath was now the thread binding the Western forces together. As their victories mounted, so did their morale. The Venetian navy laid siege to Athens, destroying the magnificent Parthenon in the process. And the Eternal City of Constantinople itself lay open to attack. By the war's end, the Ottomans had lost Transylvania and Croatia, though they still held Romania and the rest of the Balkan peninsula. Then, five times in the eighteenth century, Russia and the Ottomans waged campaigns against each other.

Finally, an uneasy stalemate ensued.

∞

BY CATHERINE'S DAY, though, the Ottoman sultanate was slowly rotting. As in the past, it was enfeebled by mass strangulations, constricted by fickle palace ritual, and suffocated by Islamic religious fundamentalism and the treacherous ways of the sultan's court. Rich and isolated, the intellectual power of the empire was also in decline: One Turkish official was convinced that Spain was in Africa, while the Reis Effendi, the foreign minister, was convinced that warships could not navigate the Baltic. And remarkably, the sultan himself was often more interested in his harem than in affairs of state ("I am for you a slave in chains," Sultan Abdul Ahmed pleaded with one of his mistresses. "Come for me tonight, I beg of you"). The empire's military power lapsed as well; the once magnificent Islamic fighters could become alternately undisciplined, arrogant, and unpredictable. It was even whispered—whispered because to say it louder was to court death—that Constantinople no longer enjoyed the traditional *al-mahmiyya*, that is, the divine protection from Allah against disorder.

But if the empire lay like a motionless log buffeted by the stream of human advancement, over time, the Ottomans did grasp some elements of needed change; here, they were forced to act like just another European state. The formula was straightforward: balance of power. Perceived enemies of Russia—France, Prussia, Sweden, even Poland—became possible allies of the Sublime Porte. And as Catherine was to learn, these powers would not idly wait while the Russian fist mercilessly pounded the Turks.

So as Catherine prepared for war, there were ominous signals still. Indeed, as late as 1774, Sultan Mustafa III felt comfortable enough to assume the gloried title of *Cihangir* (World Conqueror). He knew, rightly, that the Ottomans continued to possess fearsome military resources, a messianic spirit fueled by Islamic xenophobia, as well as a rich heritage. And inspired not simply by nationalism or patriotism, but by that most poisonous of emotions, racism, they had not lost their sense of divine mission, their belief that the Muslims were still the chosen people on this earth. War with the Ottomans was thus a risk; it was always a risk.

As the conflict escalated, this was more true than ever. Seething with anger, the Ottoman forces planned to make good on the warning of Mustafa III after the annexation of the Crimea: "I will find some means of humbling those infidels!" Then, not unlike today, recruits from across the Islamic world poured in, from the plains of Mesopotamia to the hills of Anatolia, from Barbary ports to the Balkans, all eager to join the cause; even a young Napoleon Bonaparte would soon flirt with serving the sultan. And despite Louis XVI's secret communiqué to the Turks, warning that "war had become a difficult science," the *askari*—the private troops—and the muftis were in a fighting temper, and the mob was thundering down the streets of Constantinople waving its knives and screaming for blood. Meantime, the Reis-ul Kuttab gloated hotly that the Muslims preferred a "glorious death" rather than suffer the "perfidy and painful tortures" the Russians were preparing for them.

As it happened, Catherine's commander-in-chief, Grigory Potemkin, also knew this. From past experience—he had fought against the Turks in 1770—he was all too familiar with the Ottoman army. Exotic and sinister, they were a dazzling sight: Wearing breastplates embedded with jewels and wielding engraved sabers and gem-encrusted pistols, they would work themselves up into a frenzy. Then, fueled by opium, they would charge, emitting "frightful howlings" (Prince de Ligne's words) of *"Allah! Allah!"* And Potemkin also knew they were unpredictable: They might one moment break ranks and flee "like rabbits," at another, savagely fight to the death. And he also knew that to the cry of *"Neboisse!"* ("Be not afraid!"), any Russian captured would be instantly beheaded. Potemkin himself once rued that "dreadful slaughter" or desperate carnage was invariably the result.

Already shivering with spasms of fever, Potemkin now fell into emotional despair when his Sebastopol fleet was seemingly crushed by the great storm of September 9, 1787. "My career is finished. I've almost gone mad," he scrawled nervously to an old friend. And then this: "Assign the command to someone else," he begged Empress Catherine in a communiqué. "Really, I'm almost dead."

In truth, it was a muddle, this Catherine knew. But if she were to succeed—and she realized the killing would soon begin in earnest again—it would depend on two men: first and foremost, her commander and former lover, her secret husband and the icon and glory of the Russian state, Grigory Alexandrovich Potemkin.

And, strangely enough, on an American hero seeking to regain his days of military glory: John Paul Jones.

∞

No one on the world stage was neutral about him. Few characters have been as original or extravagant as he was. And few have been as maligned, or misunderstood, or underrated in history. But of all the men to carry the great campaign against the Turks for Catherine, few were as incorrigibly flawed, yet ironically suited, as Prince Potemkin.

Where does one begin to describe him? In his single remarkable life, he represented not only the two sides of Russia's history, half compassionate and cosmopolitan, half savage and pagan, but the constant struggle in every soul between good and bad, ebullient and cynical. He was a conquering soldier, a shrewd diplomat, a wily politician, a naval architect, a military strategist, a literary advisor and art collector, and a tireless empire builder. His role as the power behind the Russian throne—he was Catherine's former lover and twenty-year soul mate—made him one of the most formidable eminences grises in modern history. In an age filled with titans, past and present, he was larger than life in every sense of the word, and was Russia's legendary answer to some of history's great advisors, whether a Woolsey of Great Britain, a Richelieu of France, or an Olivares of Spain. Had he been born elsewhere, in Western Europe or even the New World, he could have vied with a Hamilton or a Madison; he would likely have dwarfed a Necker or a Lafayette; and he would have stridden comfortably through the courts of a Louis XIV or a Frederick the Great, or complemented a William Pitt or a James Fox. Had he been American, even a George Washington might have found a place of pride for him.

He defied easy categorization. Potemkin was always a tenuous balance of conflicting impulses and opposing forces: inspired yet quixotic, relentlessly ambitious yet recklessly indulgent, at once inspired yet debauched, uncommonly brave yet psychologically frail. He was widely read, highly intelligent, deeply cultured, and a brilliant mimic with an encyclopedic memory. He was also a formidable Greek scholar and an expert in theology. Prince de Ligne, who knew all the masters of the age from Frederick the Great to Napoleon, once described Potemkin as the "most extraordinary man I ever met." An equally admiring Jeremy Bentham, the English creator of utilitarianism, called him a friend and dubbed him the "Prince of Princes," while one of Potemkin's staff members considered him "Czar in all but name." Yet his inconsistencies, which were extreme, and his flamboyance, which was legendary, and his ruthless scheming, which knew no end, led his enemies to alternatively dub him Cyclops, after the one-eyed giant from Homer's *Odyssey*, and the less flattering "Prince of Darkness." History has been equally unkind, choosing to record him as a cross between Louis XV's vain and manipulative mistress Madame de Pompadour, and the last tsarina's wild-eyed monk Rasputin. Lord Byron, for one, would have agreed; he put down Potemkin's successes to little more than "homicide and harlotry." But in truth, Potemkin was an exceptionally alert, incisive, and intricate man, a sui generis being if ever there was one.

He cared little for social decorum: Famed throughout Europe for his palaces, his jewels, his parties, and his women, he wandered around the empress's apartments naked under an open bearskin dressing gown, baring his hairy chest and munching on apples and raw vegetables, usually turnips or radishes, or obsessively chewing his nails. His face would be furrowed, his hair uncombed, his gaze at once piercing and caressing, naïve and cunning. More often than not, he might look as though he had just woken up, or had been sleeping off a hangover, only suddenly then to burst into fits of manic activity. But beneath that facade was sheer genius. And yet he was everything his detractors said he was: egotistical, voracious, gigantically energetic, volatile, and unrelenting. No wonder France's Duc de Richelieu thought him "an astonishing confection of absurdity and genius" while Comte de Ségur said he showed the "genius of an eagle" *and* the "fickleness of a child." All this, of course, was part of his magic, and also part of his allure.

More than six feet tall, massively built, extremely good-looking with a mane of luxuriant hair, he was irresistible to women—"the best *chevelure* in all Russia," it was said—and he repaid the compliment by being an insatiable womanizer. Moreover, Potemkin had a knack for being at the right place at just the right time. An obscure sergeant major in the Imperial Guards in 1762, he played a prominent part in the coup that toppled Catherine's feeble husband, for which Potemkin was massively rewarded with money and serfs. Far more important than that, he caught the new empress's wandering eye. Later he made his real breakthrough by an audacious gamble: To an aghast audience, he mimicked Catherine to her face. The likely sequel was the executioner's ax. But instead Catherine laughed, fell under the spell of his animal magnetism, and Potemkin's fortune was made.

Actually there were few clues he would even ever get this far. Neither born to a great bloodline—the Potemkins were inconsequential nobility—

nor having powerful patrons or ties in the court, Grigory came into the world in 1739 near Smolensk. Brought up by doting females and a mad father—here, perhaps, is the key to his mercurial nature—he hankered for advancement. His problem was not innate talent—he had an abundance of that—but of cracking through the procrustean world of Russian nobility. The only chance of accelerated promotion, it seemed, was through royal favor, or luck, or fate. Potemkin was determined to harness all three.

With his mother, he went to Moscow, where, thanks to one of her admirers, Potemkin enrolled in the university. Here his brilliance paid off when he was taken to St. Petersburg and presented to Empress Elizabeth. The young Potemkin was smitten, both by Petersburg and by the court; he wanted more and he wanted it now. But rash and impetuous, he had his share of failures. Among other things, he was thrown out of Moscow University for dissipation soon after his return there. Yet he borrowed money, returned to St. Petersburg in 1761, and managed to join the elite Horse Guards, not long after Elizabeth's death.

A conspiracy within the guards soon developed to overthrow Elizabeth's Prussophile successor, Peter III, and raise Catherine to the throne. The chief plotters were led by Grigory Orlov, Catherine's new lover, along with his brother. Potemkin made sure he was a part of the coup. The rest is history. Actually, what followed smacks of a classic legend: When Catherine mounted her magnificent gray steed by the Winter Palace to galvanize the Russian troops behind her, she realized she had lost the sword knot to her saber, her *dragonne*. In one exhilarating and unforgettable moment, the young Potemkin instantly galloped over to her, ripped the *dragonne* off his own sword, and gallantly handed it to her—with a bow. From then on, for the rest of their lives, they would, in one manner or another, be by each other's side.

To be sure, Catherine had many admirers, and in the perpetually shifting alliances of the court, Potemkin's rapid promotion was a sure route to isolation, retribution, or even assassination. Indeed, it is believed that he lost his eye in a midnight brawl with the Orlovs, ostensibly over "billiards." But their star was on the wane and his wasn't. His real chance came during the first war against Turkey in 1770, where he was a cavalry commander. He acquitted himself ably at the great victory of Chesme, proved to be a brave soldier in Wallachia (today's Romania), then triumphed in a magnificent battle near the Ol'ta River ("A glorious and famous feat"). Then he seized Bucharest and survived the messy chaos by Silistria, where there was house-to-house fighting against the Turks on the Danube; all the while, Catherine implored him "to keep out of danger."

He had also shown himself to be a sound strategist, and soon Catherine made him a lieutenant general. Potemkin and Catherine gradually became closer, until in early 1774 their mutual attraction was consummated—quite likely in the private *banya* (steambath) of the Winter Palace. He was thirty-four and Catherine forty-four. Once again, Potemkin showed impeccable instincts: Prior to that moment, Catherine was torn between two court factions, the Orlovs and the Panins, and was bored with her current lover. Her image was suffering in Europe; the puritanical Frederick the Great even mocked that Orlov had been recalled from all offices, "except that of fucking." Sensing a vacuum, Potemkin melodramatically withdrew to a monastery, where he became a holy monk, grew a beard, and prayed ostentatiously. Equally ostentatiously, he also let it be known that he had begun to hate the world, "because of his love" for the empress. Flattered, Catherine summoned him, and by mid-February they were not only making love in her apartments, but were inseparable. "My dear," she wrote, "you are so handsome, so clever, so jovial, so witty . . . I've never been so happy."

The relationship was based on laughter, mutual admiration, and appreciation of power—and sex. It was the strangest sight in all of Russia: Where she liked to awake at the crack of dawn, he liked to gamble and drink for days on end, talk late, and lie in all morning; where she disciplined her emotions, he was brooding, mercurial, prone to protean depressions and wild extremes. Where she was majestic and dignified, he was ferocious, untamed, almost bohemian. Both could be cunning and ruthless, in war as well as politics, yet the greater the crisis, the steadier she got. By contrast, Potemkin was often a prisoner of his restless temperament, his vanities and delusions, his rapid plunges from hysterical excitement to black despair. And he was a notorious hypochondriac, frequently sickened and secluded, and suffering from constant fevers, incessant flus, numerous colds, not to mention sensitivity over his useless left eye. But she was always enamored with him and frequently morose without him. And with his gift for humor and mimicry, he, as no one else could, made her laugh. They had pet names for each other: she called him animal names—her "lion of the jungle," "Golden Tiger," "Golden Cockerel," "Wolf," "Cossack," and "Kitten," as well as simply "my heart," and later, "my darling husband." In turn, Potemkin addressed her as "my beauty," or "my darling," and mostly, by the traditional Russian "dear *Matushka*" ("little mother"). And her love for him, however turbulent and complex, was absolute. To Potemkin she wrote: "What a trick you have played to unbalance a mind, previously thought to be one of the best in Europe. What a sin! Catherine II to be victim of this crazy passion." And: "Enough! I have already scribbled such sentimental metaphysics that can only make you laugh. Well, mad letter, go to that happy place where my hero dwells. Goodbye *Giaour* [Turkish for "non-Muslim"], Muscovite, Cossack."

Legend—it is almost certainly true—has it that Potemkin and Catherine secretly married on June 4, 1774. Even King Louis XVI privately began to refer to the Empress as "Madame Potemkin." But for all of Potemkin's many titles, his real position quickly became impossible to categorize. What was sure was this: As his biographer Simon Sebag Montefiore points out, Potemkin became in many ways a coruler of the empire, a fact to which contemporaries paid heed. Sitting atop this notoriously unsteady but powerful throne, which rendered her vulnerable, Catherine looked to Potemkin for his canny advice, his emotional support, and his shrewd bureaucratic eye. He had an instinctive feel for politics and diplomacy, as well as how to assist the empress in playing the various interest groups of her notorious court to her advantage. And he was a survivor, even weathering a plot against him, in which the conspirators planned to do away with him by "break[ing] his neck." However peculiar it was—and it was—this torrid love affair, and political alliance, became as remarkable for its achievements as for its romance, as endearing for its humanity as for its power, and would take on epic proportions, becoming reminiscent of Antony and Cleopatra, Louis XVI and Marie Antoinette, Napoleon and Josephine.

To be sure, their sexual affair would eventually cool, but his influence only continued to grow. Among other things, it would be Potemkin who handpicked new "favorites" for Catherine, who, as she grew older and plumper, required men younger and leaner. Nor did this change as scandal flourished aplenty, especially when Potemkin had affairs with three of his young nieces brought to St. Petersburg to be ladies of the empress's bedchamber. But Catherine didn't mind; if anything, she seemed to take pride in his rampant lovemaking elsewhere. For that matter, she seemed to take pride in everything he did.

It couldn't have always been easy. From the moment he awoke, his every day was unpredictable. While still in bed, he received visitors in his dressing gown, then roused himself for a cool bath and a short morning prayer. But after that, his moods swung from unrestrained highs ("Can any man," he once said, "be more happy than I am?"), to crippling lows ("I'm good for nothing . . . let me rest, a little"). When he was depressed, he retreated into a near paralytic silence. He refused to sign papers, machinations of the state would grind to a halt, and a significant part of the Russian government would simply stop. Sometimes he sat alone, like a catatonic Gulliver, soothing himself with music and pouring emeralds and rubies from hand to hand. But when he was happy, his tastes were gargantuan. He relished fine food, devouring caviar from the Caspian, smoked goose from Hamburg, oysters from the Baltic, suckling pig from Poland, anchovies from Livonia, and figs from Provence. He was an inveterate gambler, forever playing faro, whist, or billiards in his palace, often winning (or losing), staggering sums; he was ostentatious too, binding up banknotes as books and displaying them in his study. Still, everyone who met him was astonished by his vast knowledge: His library housed works on theology, war, agriculture, and economics—from Plutarch, Voltaire, and Adam Smith, to Locke, Newton, and Gibbon's *Decline and Fall of the Roman Empire*; he amassed more than 1,065 foreign tomes, filling eighteen carriages.

And in a voice thick with emotion, he constantly queried those around him on every conceivable subject that seized his brain. Indeed, one observer raptly noted, he was "the biggest questioner in the world." Among the great questions of the day, of course, were revolution and politics. Where did he stand on these matters? The three major revolutions of the era—the American, the French, the Polish—alternately captivated and disgusted him. Thus he grilled the Comte de Ségur at length about the Americans, for whom the Frenchman had enthusiastically fought. Then he proceeded to rip back, echoing Montesquieu, that republican institutions could not have a long life "in a land so vast." Of the mounting French Revolution, he told the Comte de Langeron, "Colonel, your countrymen are a pack of madmen." And to his dying day he clung, as did Catherine, to the view that absolutism was the most effective means for ruling an empire as immense as Russia's. For Potemkin, enlightened despotism was not some primitive aberration, but was sanctioned by some of the best minds of the day, and conflict, not amity, was the way of the world. Intuitively he was an idealist, but in his own checkered way: "Improve events," he once said, "as they arise."

By 1782 he had securely fended off jealous courtiers and outmaneuvered intriguing rivals. But now he saw his chance at destiny, a dream that propelled him with unstoppable zeal. Where Peter the Great had thrust

Russian foreign policy and Russian morals toward Western Europe, Potemkin had a vision of a vast Russian empire stretching not to Vienna or Paris, but southward, to the Crimea, the Black Sea coast, to the Caucasus and all the way to Constantinople. With the empress's blessing, he threw his superhuman energy into the conquest of the Crimea and indeed would eventually give Catherine the Black Sea coastline.

It was Potemkin who in 1783 had boldly led the annexation of the Tatar khanate—without war, though not without bloodshed—and gave her the crown jewel of the Crimea, which established Russia as a force in the Levant for the first time and which set the stage for Russia's spectacular entrance into the center of European politics in the Napoleonic era. It was also Potemkin who favored light troops and developed a close relationship with the Cossacks, whose forces would later bring Napoleon Bonaparte to his knees. In the ensuing years, Potemkin would set his sights on developing this whole region to the south, including today's Ukraine, bringing in settlers and immigrants—an astonishing one million in little more than a decade—and planting trees, vines, fruit orchards, and mulberry plantations; manufacturing Italian silk; establishing factories and shipways; importing British gardeners, as well as building an entire fleet that would have made Peter the Great beam. He also founded a string of towns along the vast, empty steppes, such as Kherson, Sebastopol, Ekaterinoslav, Nikolaev, and finally Odessa. And it was Potemkin who stage managed the triumphant imperial procession that laid out his striking achievements before the admiring and astonished gaze of the tsarina and Emperor Joseph and their assembled retinue during the historic Crimea trip of 1787.

But were these towns real? Or was this all some quixotic fantasy? To this day, the very phrase "Potemkin village" evokes the myth not of docks, towns, and palaces, but phony constructions made of pasteboard; it is further believed that Catherine saw the same peasants and the same flocks over and over again, who were simply moved down the River Dnieper each night ahead of the empress's entourage. But while myth has a significance all its own—the phrase "Potemkin village" is still shorthand for political deception and misinformation—it is, Montefiore notes, more than likely that this story was part propaganda, stoked by malicious European envoys flush with jealousy. The Comte de Ségur himself wrote that every year there were "flourishing new villages in places . . . left before as deserts"; many of these towns are today Russia's great industrial and military centers.

Wayward and dominating and always eccentric, Potemkin then ruled this region "like an emperor," from the River Bug to the Caspian, from the Caucasus almost to Kiev. Alternately stamping his feet and sputtering with rage, or flashing his famous wit, he acted as a blend of merchant, financier, absolute despot, and entrepreneur; one observer even compared him to an old Israelite in his desert Tabernacle. Such was his aura that when he was absent, he alone was the subject of conversation; when present, he engaged every eye.

It was Potemkin who had started the Great Game, with his dream ultimately of recapturing Constantinople from the Muslims, the perfect revenge for the Orthodox Church. But now, as the war heated up, would he be up to the test? He could well recall the barbaric slaughters of the first Turkish war; a victory in the second would be the apogee of his public life. As ever, he remained a whirlwind of contrasts: He dreamed of retiring to the church and immersing himself in mysticism, but never let religion inhibit his carnal pleasure. He was torn by the superstitions of the motherland, even as he tutored himself on the intricate web of European politics. He relished his great riches and thrived in the rough and tumble of Catherine's court, but nothing made him happier than the wild exuberance of frontier development and being surrounded by devoted Cossacks. And he lived on gold yet would one day die on the grass—he didn't know it yet, how could he?—but exhausted by hard living, he would collapse on the Bessarabian steppes and expire in a few short years. His bones would be interred in the Ukraine, his innards would find their way to present-day Romania, but his heart would always rest in Russia. As the poet Derzhavin penned in tribute, "Roar on, roar on, oh waterfall." But that was in the future. For now, though, there was a war to fight.

And Potemkin was paralyzed by gloom. Many years later, Winston Churchill would one day grumble: "War, which was cruel and magnificent, has become cruel and squalid." That would become the case here. For all of Potemkin's humanity, he had a capacity for brutality, an unshakable faith in the superiority of the Russian people, and he could positively relish the prospect of leading a nation of warriors. But at this stage, almost fifty years old, mired in desperation, he knew that the glories he hoped for would not be as quick, or cheap, or simple as he had hoped. Would he see his fate and his future destroyed? He was responsible for a military theater that stretched across all points of the compass, from the Swedish and Chinese frontiers to the Caucasus and the Kuban, and Poland and Persia as well; it was a vast portfolio that mandated not only extensive understanding of land tactics and sea maneuvers, but overarching strategic insight. In the grip of nervous collapse, that vision was now missing. He fell into deep despair.

"Despite all the measures I'm taking, everything's gone topsy-turvy," he now anxiously wrote to Catherine. He hovered at the edge of insanity. "I'm exhausted, Matushka," he wrote on September 19, 1787, his words dripping with emotion, "I'm good for nothing." And also this: "Assign the command to someone else. . . . Really, I'm almost dead." Catherine hastily wrote back to buck him up, "You're ill and I'm well. [But] in these moments, my dear friend . . . you belong to the state, you belong to me." She fervently urged him to realize, "nothing is lost." She prodded him by saying the Sebastopol storm "was equally harmful for the enemy." And she insisted that there be no withdrawal from the Crimea because it was "not in danger."

Not yet. As his hulk of a war machine readied itself, and the Ottoman armies lay in wait at Ochakov, a defender's dream, and then began thumping toward their foe, it would unleash a welter of butchery, not to mention fatally fray the prince's waning energies. For this reason, perhaps, Catherine was taking no chances. Thus, part of her strategy was also giving a high command to a man who spoke no Russian, had not been in battle in eight years, and whose political outlook was anathema to her. It would lead to one of the most curious and ill-fated partnerships perhaps in all time: between one of history's great enlightened despots, and one of its most colorful republicans and naval heroes.

It began with a meeting in Paris brokered by America's representative, Thomas Jefferson, and the Russian ambassador—and John Paul Jones. It would end in ferocious battle against the Ottomans, as well as in tragedy. And not surprisingly, a revived Potemkin would once more play a key role.

∞

THE IDEA THAT the contemporary world, a world of air travel and instantaneous communication, has only been bound together in the twentieth and twenty-first centuries is a latter-day form of chauvinism. It is also untrue. The eighteenth century was stitched together in ways we can scarcely grasp, even by today's standards. The great nations of the day, and their leaders, were all intimately tied together, watching one another, marveling at one another, and reacting to one another—whether from the bustle of French salons in Paris to the young American capital in Philadelphia, from the luxury of St. Petersburg to candlelight dinners in Monticello and Mount

Vernon, from the splendor of Vienna to the mysteries of the seraglio in Constantinople. Political figures of the day, great philosophers, ardent rebels and revolutionaries, all freely crossed and recrossed borders, switched allegiances, spoke in foreign tongues and fought for foreign causes with great relish, and shared dreams beyond their national boundaries with an alacrity that has little parallel in the modern world. So it was that the efforts of three men in Paris—the author of the Declaration of Independence, a Virginia planter who served the Polish king, and a Russian diplomat—would bring the son of a gardener from a Scottish island into the service of no less than the Empress of All the Russias.

He was the most celebrated naval commander of his day, and widely regarded by history as one of the founders of the U.S. Navy. Born Paul Jones, Jr., he was better known as John Paul Jones. Living in Paris since 1783, he had been at loose ends, looking for ways to keep his famous name alive and to regain glory at sea. In 1787 the U.S. Congress gave him a gold medal, but no flag or fleet—only permission to seek a French commission. He didn't get it. His mood plummeted and so did his military fortunes. He felt marooned, depressed, and defeated, seemingly at the end of his career and isolated in the world. It was then that Thomas Jefferson, through secret diplomatic channels, heard that Empress Catherine was interested in hiring Jones to command a fleet in the Black Sea against the Ottomans. The match was a curious one, as was the inevitable question: Why would Jefferson, a devoted republican and sworn enemy of monarchs and despots everywhere, be willing to recruit America's greatest naval man for the tsarina?

The answer is uncertain, but important to probe: Russia and America were both still seen, oddly enough, as native, virgin countries straddling a corrupt Old World and enjoying a New World kinship, sharing common interests in trade, in foreign policy, and in migration; already, for example, Germans in great droves were either emigrating to Pennsylvania—or to the southern steppes of Russia. And both Russia and the United States shared problems with the Muslim world: Russia with the Ottomans, and America with the Barbary states of North Africa—Algiers, Tunis, and Tripoli, which were part of the sultan's empire. For some years now, the Barbary states had mercilessly engaged in piracy, kidnapping, and blackmail against the Americans; routinely, American ships were attacked and their sailors seized and held in Barbary dungeons or sold off as slaves. Moreover, Britain, which stoked the Barbary powers in attacks on the Americans, used Turkey as a counterweight against the Catholic powers on mainland Europe: so why shouldn't the United States reply in kind, by helping Russia make life hard for the Turks? Even then, the idealistic America, feeling its way in the world, was not above a bit of realpolitik—it couldn't afford not to be. Moreover, Jones in Russia would, Jefferson clearly reasoned, help cement the ties between the two nations.

But was Jones himself interested? Not at first, not even with Jefferson's recommendation. "I regarded this as a castle in the air," he later wrote. To him, Russia was a strange and forbidding empire, filled with a bizarre hodgepodge of Oriental scheming, Asiatic luxury, and monarchical despotism. Moreover, a Parisian friend had actually sternly warned him about the intrigues of the Russian court, advising him that it would be better to "go to Constantinople" than enter the service of Russia.

But the courtship of Jones continued. The empress had first heard about Jones from her Parisian ambassador ("Jones is," Simolin briefed her, "one of the greatest sailors of the time.") and found his celebrity irresistible. Nor was she to be denied: Russia needed experienced sailors, and she would get them. So one day, over breakfast, Lewis Littlepage, a friend of George Washington's and aide to the Polish king, and Baron de Simolin, the Russian ambassador, dazzled Jones with flattery. Jones would, Simolin said, have carte blanche; he would receive great glory and great honors; and in less than a year, he would achieve international fame by making the Ottoman Empire "tremble." Still, Jones appeared hesitant. "I can never renounce the glorious title of a citizen of the United States," he told Jefferson. But this was all smoke. His ambition whetted, Jones packed his bags.

Braving the cruel Russian climate—like the American emissary before him, Francis Dana, Jones was exhausted from the lengthy trip and dangerously afflicted with a lung infection from the cold Baltic winds—he met with Catherine in the first week of May at her magnificent palace of Tsarskoye Selo, the elegant provincial town dominated by the life and gossip of the court. Overlooking the Neva, just fifteen miles from St. Petersburg, he was alternately awed by the magnificent baroque palace, infatuated by the English gardens he saw, and overcome by the empress's uncommon magnetism. "*Je me laissai séduire*" ("I let myself be seduced"), he later admitted. By then, this was understatement. Received in splendor by the wealthy Russian aristocracy, Jones now saw Catherine for what she was: the longest reigning monarch in the world, head of the great Romanov dynasty, and one of Europe's mightiest imperialists. She had boldly stood up to the British and helped nurse the Americans to independence, and was the idol of the great philosophes of the age, from Voltaire to Diderot. And now the tsarina, the master of all court life, the impresario of all court ceremonies, the bestower of all stars and ribbons, was wooing him as the Americans never did. And it worked.

"I shall never be able to express," he wrote, "how much greater I find her than fame reports." Surrounded by lavishly costumed ministers and generals who wore rows of medals from shoulder to shoulder, his head was spinning with excitement. In a scene replete with deep historic symbolism, he presented her with a copy of the American Constitution. With no doubt a little sneer, she replied, quite presciently, that the American Revolution would "bring about others" as well as "influence every other government." He called her your "Majesty"; she dubbed him "Pavel Dzhones," and smartly made him a *kontradmiral* (rear admiral). Actually, he made a good impression on the empress; in turn, with satisfaction, she informed her friend Baron Grimm, "He will suit our purposes admirably." And thus a fateful match was made.

Of this startling new alliance, the Comte de Ségur wrote in wonderment to Prince Potemkin: "I did not expect having made war in America with Brave Paul Jones to meet him here so far from home but . . . I can't be surprised to see all those who love glory . . . coming to associate their laurels with yours."

And the empress herself informed the prince: I am sending you "one more bulldog for the Black Sea."

∞

BUT FOR ALL his fame and expectations, how would this bulldog fare? An expert in the surprise assault and a virtuoso at overcoming impossible odds, on the surface Jones seemed perfect for the job. Ornery, with a vile temper, Jones was by turns brilliant, inflammatory, headstrong, and imperious— John Adams once called him "leprous with vanity." Still, he was also America's most innovative sailor. To his admirers in America, he was a vaunted "hero" of liberty. Herman Melville deemed him "an audacious Viking"; James Fenimore Cooper, Rudyard Kipling, and Alexander Dumas would all write about him; and in France he was regarded as a swashbuckling romantic. But to the British he was a "pirate," or a "despicable pirate," or that "Pirate Paul Jones," and lurid prints bearing his likeness were peddled in London while English children sat rapt at stories of his bloody deeds.

And to friends and foes alike, he had become the stuff of living nautical legend.

Yet behind this legend was a man, a tarnished and incorrigible man, and a self-made one at that. Born in Scotland, where his family scraped out a fragile existence (he was possibly the illegitimate son of his father's noble employer), he was, all his life, haunted by inner demons and tormented by a blizzard of insecurities. Still, he taught himself to read, he could eventually quote Pope and Shakespeare, he wrote vividly, and he assiduously studied gentlemen hoping to become one. He never did. He embraced the sea at the age of thirteen. By 1764 Jones had crossed the Atlantic eight times in three years, becoming involved both with the slave trade as well as commercial shipping. Eventually he became a merchant captain. It was a disaster. He bullied his crews, and later killed one of his men in a quarrel. Then his fortunes quickly slid. He hastily fled to Virginia, changing his name to escape the authorities. That was 1774. He was a fugitive, stateless, without a profession or any prospects of one, and, for that matter, he was without a name: He was traveling "incog," he wrote: incognito, under a false identity. Then the great rebellion came.

A year later the American colonies were plunged into revolution, and Jones, a staunch patriot, offered the rebels his services at sea. Six months before the Declaration of Independence, he was commissioned a lieutenant in the Continental Navy. He never looked back. With no battle experience and little formal training, he audaciously assumed he would rise quickly to become America's first admiral. Here, he was rebuffed. As it turned out, his temper was sharp, quick, and merciless. So were his vanities. He was a big mouth, a bully, and a glaring showoff. Though he distinguished himself—he captured eight British ships and sank and burned numerous others along the Irish coast—he foolishly treated his fellow officers as cowards and incompetents, and worse still, his superiors with contempt. Denied promotion and frozen out, he subsequently made his way to France, where Ben Franklin became his patron; Franklin arranged for him to take command of a leaky cargo ship, the *Bonhomme Richard* (named after Franklin's *Poor Richard's Almanack*), which was at best a small, slow, and poorly manned pirateering vessel. Yet at its helm, Jones quickly emerged as the war's most dreaded sailor, known as much for his miraculous escapes as for his daring exploits.

And what exploits they were. Time and again, at times and places of his choosing, his tiny squadron of ships outfought, outmaneuvered, and out-bluffed British ships, routinely terrorizing the English coast. His attacks threw the British people into a panic, and they appealed to King George for protection. That protection came in the form of the HMS *Serapis*, a new and powerful frigate of fifty cannons. Thus the stage was set for immortality. Jones's most famous battle was off the shore of England on September 23, 1779, against the *Serapis*, one of the bloodiest ship-to-ship duels of the entire war. Under a rising harvest moon, the two ships were locked in fierce battle, when inexplicably, at a critical moment, Jones was deserted by the rest of his squadron. Believing that Jones's ship had been pummeled into near oblivion—it was outgunned and outmanned and tottering—the British commander shouted out for Jones to surrender. It was now dusk. Lashing himself to the mast—a masterpiece of military theater—Jones offered a swashbuckling reply that was equally immortal: "I have not yet begun to fight," he growled indignantly—or at least something to that effect. His ship reeling, Jones rammed the *Richard* in front of the *Serapis*, causing the two vessels to collide; they then became intertwined in a death struggle. Hand-to-hand combat ensued over the course of three bloody hours, with swords, cannons, and pistols. During a hiatus in the fighting, the terrible toll became clear: Half the men of both ships were dead or wounded. But Jones fought on.

Both ships caught fire from leaping sparks and flame, and were enveloped in rising wisps of yellow clouds that throbbed with flashes of orange and white light—the continued firing of guns and cannon. Spectators gathered on a bluff several miles away to glimpse the spectacle and the carnage. Defying all military logic, Jones lost his ship but won the battle, seizing control of the *Serapis* while the nearly gutted *Richard* began to sink. It was an astounding victory. Of Jones, the *London Morning Post* rhapsodized: "He is no sooner seen than lost." The English press labeled him a "desperado," "a bad man," "a vile fellow." But with such exploits, Jones helped America win its freedom and became a hero of the revolution.

His fame grew. Ballads were written about him ("And you've heard o' Paul Jones?/Have you not/? A rogue and a vagabond;/Is he not? Is he not?"), and Congress honored him with a gold medal. He was inducted into the Society of the Cincinnati. But if his fame grew, so did his idiosyncrasies. He considered buying a farm in New Jersey and contemplated, briefly, domestic tranquillity. But he craved glory too much to settle down, instead growing restless and frustrated. He proposed that Congress build a fleet of battleships to rival those of France and Great Britain; Congress wasn't interested. So he went to France, where he became a notorious rake.

He delighted in recounting his exploits or his conceits. He preened that "the English nation may hate me; but I will force them to esteem me," which they reluctantly did. He boasted that "my desire for fame is infinite," and he achieved that in spades. He bragged that "I would lay down my life for America, but I cannot trifle with my honor," and he pretty much stuck to that too. While he didn't quite look the part of a daring sailor—he was only five-six with a slight build—he did his best: His voice was soft, his jaw was chiseled, and contravening American orders, he designed his own uniform: a knockoff of the Royal Navy's, a blue coat with epaulets, a buff vest, and breeches. And in a bit of further soldierly vanity, he called himself "Chevalier Paul Jones," at one point hiring a French manservant. His vile temper and cocky ways were partially offset by a winning charisma that he could turn on and off, and by his careful cultivation of powerful patrons: Ben Franklin, Robert Morris, Alexander Hamilton, and Joseph Hews of North Carolina. For his part, Thomas Jefferson kept a Houdon bust of Jones in his Monticello home, side by side with the American heroes of the age: Washington, Franklin, and Lafayette. And in France, Jones even received a personal audience with King Louis XVI.

But for all this, he was, in one fashion or another, stalked by controversy or bad luck, or both, wherever he went. While he taunted and tormented the British with his sensational surprise raids, once even stealing the Earl of Selkirk's family silver, his escapades actually played only a small military role in the revolution itself, and he missed the most crucial naval campaign of the war, with the French fleet at Yorktown. And his problems weren't just a matter of his personality conflicts with colleagues, or his grandstanding and self-defeating ways. Peel back that veneer of arrogance, and he was a man locked in loneliness. He fell in love with Dorothea Dandridge, who spurned him to marry the inimitable Patrick Henry. He failed too with Phillis Wheatley, the former slave and black poet. So he visited prostitutes and slept with other men's wives, which gave him neither the intimacy he craved nor the love he so ardently wanted. And when he visited England after the war, rather than receiving respect, he was almost lynched.

Too often, his fierce opinions, or his mouth, or his ambitions got the better of him. Later in life he became a drifter, from Poland, Austria, Amsterdam, London, and finally Paris. In his last days, in the midst of one of the most profound revolutions in history, he would wind up a sad figure, dying in Paris in 1794, alone, at the age of forty-five, his body contorted over the edge of his bed as if this profane but poetry-loving sea dog might finally have sought solace in prayer. While he was in France, only the American ambassador, Gouverneur Morris, paid much attention to him, and even this was negligible. "He had nothing to say," Jones wrote sadly, "but is so kind as to bestow on me the hours which hang heavy on his hands." The beginning of his life was pitiful, and so was the end. But, and it is a big but, not what took place in between.

Few others were so willing to take on the most powerful maritime power in the world as he was. Few others were so deft at the calculated risk in battle. And few others were so prescient in calling for a world-class, modern navy for the young republic—though it would be some years before this would happen. Part George Patton, part Jesse James, part Davy Crockett, part Captain Kidd, Jones was, for all his irregularities and flaws, a perfect American hero. And now this hero was in St. Petersburg, preparing to fight for the Empress of All the Russias against the heathen Turks. He had lost none of his fight, or his spirit, or, tragically, his errant ways with colleagues as well. As it turned out, once introduced into Potemkin's fold, he would fall under the command of Catherine's wily naval leader Prince Nassau-Seigen as head of the Liman fleet, along with such men as the Virginian Lewis Littlepage and the Englishman Samuel Bentham—but like so many others, Jones too would become fodder in the great swirl of Catherine's court and the turmoil of the coming conflict of civilizations.

But first there was war.

∞

IN MAY 1788 the Russians could see a kaleidoscope of masts in the distance—it was the Ottoman fleet, more than one hundred ships, off Ochakov. We can begin the "dance," Nassau boasted. Though his fleet had not been decimated as he had feared, Potemkin was still less than sanguine. He soon scribbled furiously to Catherine, "The Turks have certainly improved their skills since the last war." Actually, it was the French. Since 1783 a staff of thirty military and naval officers from France had helped the Ottomans modernize their forces in preparation for war with the Russians. Indeed, the two nations had been erstwhile military allies since the age of the Sun King, with France often advising the sultan's forces in part to counter Britain's long-standing alliance with Russia; in the latter half of the eighteenth century, not unlike the idealistic Frenchmen who rushed off to help the American revolutionaries in their struggle, French military men, such as Baron de Tott, made their way eastward to the sultan's domain at the behest of Versailles, where they also sought some measure of glory on his battlefields. Added to this, the sultan had recalled the heartthrob of the Istanbul rabble to service: Ghazi Hassan-Pasha, the Barbary *Capudan*, otherwise known as the "Algerine renegado" or the "Crocodile of Sea Battles."

A mixture of imperiousness and audacity, in appearance alone, Pasha was awesome. With his great white mustache and his sleek, fast, fourteen-galley corsair, Pasha, heir to the fierce ancient tradition of Algerian pirates, was among the greatest of Ottoman warrior chieftains. Known for his willingness to use terror as a weapon of war, including Turkish ships on suicide missions—burning hulks to set enemy ships ablaze—he was always accompanied by a pet lion that meekly lay by his side. Despotic and disdainful, he had survived the nightmare of Chesme, had ruthlessly put down the rebellious Mamluks in the Egyptian provinces, and had fearlessly raced back to make a stand at Constantinople. And now his fleet, which he had personally regenerated, lay in wait, by the massive fortresses of Ochakov, hoping to turn the elements of conflict into an epidemic of destruction. Should Catherine be unable to take Ochakov along the Liman, a long and treacherous bay that was thirty miles long but only eight miles wide, the Turks could be free to retake the Crimea. And that would be just the beginning.

∞

ONCE AGAIN, POTEMKIN was stricken with a bout of timidity. Anything could go wrong: Oars could snap, ropes fray, sails tear, their intelligence could falter, and so could the men. He suggested to the empress that the Crimea might have to be evacuated. Catherine hotly shot back, "When you are sitting on a horse, there is no point in getting off it and holding on by the tail." She was right. The first battle, on May 27, was an uneasy stalemate. But Potemkin was at least encouraged ("it comes from God!" he sang out). The prince, often coarse and unscrupulous, could be as notorious for his torpor as for his mood swings. But suddenly he rose to the occasion: He supervised the flotilla's maneuvering, he oversaw the smallest details, he attended to the morale of the ordinary men (they were to have eau de vie—spirits—every day, and meals served promptly, never cold),

and he arbitrated the tensions among his commanders. By this point, however, Jones and Nassau, unable to check their egos or reconcile their own
visions of warfare, were at each other's throats, for all to see. But the battle
would not wait. In the gathering twilight on June 16, the Russians held a
council of war.

Though Jones voiced concern, he boomed, "I see in your eyes the souls
of heroes. It is time to conquer or die."

It was. During one pause in the action, under a full moon, Jones quietly
rowed into the heart of the Turkish fleet by Ochakov. While cannons were
readied off in the distance along the shore, he daringly taunted the enemy,
scrawling in chalk on one Turkish warship's stern: "TO BE BURNED.
Paul Jones 17/28 June." For the moment, it was a war of nerves. It would
not last. And the second battle became a blind and bloody hunt to the
death. The Turkish fleet, stretching from shore to shore, rang with the chilling sounds of drums banging and calls to Allah. Out for blood, Nassau
successfully destroyed ten Turkish warships and five galleys, and took sixteen hundred prisoners. These were the lucky ones. As the fighting raged
on, Potemkin's men, excluding Jones, who nearly wept at the sheer carnage,
took no quarter against the infidels. One Turkish ship, with the rowing
slaves still chained on board, many of them captured Christians, was cruelly
lit on fire with *brandcougles*—homemade Molotov cocktails. In the depths of
the night, the initial flames began innocuously, but then they quickly began
to scale the sails and the riggings, widening and lengthening, and finally
began to leap from ship to ship. Nassau thought the spectacle "beautiful." But to Jones, the moans of the dead and dying were atrocious. As was
what followed: Soon, the crackle of flames intermingled with the echoes of
wild screams of men—while begging for mercy, they were roasting alive.
The sounds were appalling; when the smoke and flames cleared, so was the
butchery. It was like a scene out of the depths of Dante, as the gamey scent
of charred flesh filled the air and blood freely mingled with the briny water.
Then the cries and moans of the men grew weaker and fewer and more
desperate. Then it grew strangely quiet; the massacre was complete. For two
weeks, the bloated, burned corpses, more than two thousand, were found
floating in the Liman. The river eventually turned green.

"Our victory is complete," Nassau declared. Potemkin himself was
jubilant: "I've gone mad with joy!" So did Catherine, who sent Potemkin a
golden sword garnished with three oversize diamonds. But in the ensuing
days, the shadowboxing continued. Lest panic spread through his Turkish

ranks, Capudan Pasha promptly hanged an entire selection of his officers. Potemkin boasted confidently, "Isn't it amazing? I'm the spoilt child of God." By now his words seemed almost Delphic. While Pasha put to sea the remains of his fleet, Potemkin changed tactics and was sprinting with his army across the Bug River, advancing straight to the Ochakov fortress, where he was accompanied by 13,000 Cossacks and 4,000 Hussars. The prince quickly ordered twenty cannon to be brought below the extensively reinforced garrison and personally oversaw the cannon fire, as though he were a junior artillery officer. Was this madness, or, as one observer noted, "beautiful valor"? Actually, it was both. Turkish artillery fire exploded right beside him, killing a cart driver and two horses. Learning of this back in St. Petersburg, the empress was horrified. "If you kill yourself," she wrote, "you kill me too."

Thus commenced the long-awaited assault on Ochakov.

∞

ON LAND, THE killing and dying continued. Ochakov, a ghostly castle on the far shore, was not just a garrison with the strongest fieldworks, artillery emplacements, and traverses to foil attackers thus far, but it also housed an elaborate town of mosques, palaces, courtyards, and barracks. And it was the strategic prize most eagerly sought by the empress, for it controlled the mouths of the Dnieper and Bug rivers, and therefore the very gateway to the Crimea itself. Win Ochakov and keep the Crimea; lose it, and everything could be lost. Once more Potemkin, eager for glory, rashly exposed himself to the enemy. And once more he fooled death. As he stood beside the governor of Ekaterinoslav beneath the fortress, the governor was smashed in the groin by a cannon ball; his agony was so unbearable that he pleaded with Potemkin to shoot him. Within two days, he was dead. Their blood up, the Russians now anxiously waited for the storming of the fortress to begin. The Turks did too. Inside, flags waved in defiance and the walls echoed with bloodcurdling shouts of "Ya Illa, Ya Illah!" Potemkin probed the fortress but to no avail. By July 27, neither side had gained appreciable ground.

The prince then stretched his forces in an arc around the town and ordered an artillery bombardment. All it did was provoke the enemy. Determined to give as much as he got, the next day the pasha ordered first a probe, then a sortie, then a grand assault. Three thousand screaming Turks set upon the Russians, killing some two hundred of Potemkin's best men, who were then beheaded, their heads sinisterly displayed on stakes. An agitated Potemkin wept ("Oh my God!" he cried out in despair, horrified at the terrible "sacrifice"). Yet neither side collapsed, neither side retreated. On August 18 the Turks tested the Russians again, surging forward in a frenzied contest. This time General Mikhail Golenishev-Kutuzov, who would one day bring Napoleon to grief in 1812, received yet a second wound, a bullet that blinded him in the eye. But Nassau managed to drive back the Turks by having his flotilla unleash its guns from the estuary. By the next day, a pall hung over the battlefield. A terrible stalemate ensued. And as the grim winter set in—one of the coldest in memory—there was talk. Potemkin's own men—including Ligne and Nassau—now began to mutter about the prince's egregious mismanagement. And one of Russia's most respected and audacious generals, the legendary Marshal Suvorov, urged Potemkin to make a direct assault.

But Potemkin would not be cowed. Like the Union general Bill Sherman, he preferred to win without fighting. "I'll do my best," he confided, "to get it at a cheap price." His calculations were as simple as they were numbing: Already beset by his responsibilities of maintaining troops from the Caucasus to the Gulf of Finland, he preferred to bluff, outmaneuver, or starve the enemy into surrender. He desperately tried all three. All three failed. Thus far, the Turks had countered his every move. Even Austria's Emperor Joseph had joined the battle—he had earlier honored his word to Catherine and invaded Bosnia and Serbia, where the kaiser was ambushed in his camp and had to flee for his life, hastily retreating all the way back to Vienna. "The Turks are different," Potemkin now morosely informed Catherine, "and the devil has taught them!"

True, but Catherine would hear none of it. "Courage! Courage!" she exhorted Potemkin. Now she was beside herself. Had Potemkin forgotten the reasons for this war? Did he want to see the Crimea invaded? To see all lost? In communiqué after communiqué, she badgered the prince over and over again. First this: "What about Ochakov?" Then this: "Will you take Ochakov?" And then: "When will Ochakov fall?" In the previous Russo-Ottoman conflicts, the combat usually dissolved quickly, with one side collapsing and retreating. But now both lines held. The siege continued and Russian morale sharply plummeted. Panic began to spread through the ranks. In a grim omen, Potemkin's remarkable assemblage of cosmopolitan warriors and Western agents began to unravel, filled with disgust and disillusionment: the dashing Virginian, Lewis Littlepage, left in a huff, and so did

Prince de Ligne. And despite Catherine's hours of wooing John Paul Jones and his considerable role in shoring up Russia's naval campaign, Potemkin had the American hero relieved of command—Jones, deft and valorous in battle still, had proven himself woefully inept at the backstabbing politics of Potemkin's war court. "I eternally regret having had the misfortune to losing your good graces," Jones wrote to the prince. Unmoved, the empress recalled Jones back to St. Petersburg, where, as a fig leaf, he was given command of the Northern Fleet in the Baltic.

Now it was just the Russians.

∞

THE BITTER WINTER came early. Meanwhile, the men were forced to burn carriages for firewood, and their ranks were blighted by the stench of decay and dysentery, illness and night blindness, not to mention lack of substance. Having almost no water and no meat, they scrambled to eat flies for rations, and largely had only wine to drink. They lived among their own urine and feces, staring up at the sky by day; the camp became, one observer grimly noted, all "snow and shit." Maneuvers halted. The elements—the cold, the wind, the snow and sleet—became merciless. So did the exhaustion and desperation. And as the temperature plummeted to a staggering fifteen degrees below zero, the men were forced to pack up their tents and, like shivering rabbits, burrow into the ground for warmth and shelter. Then an epidemic broke out. It was said that forty to fifty men a day perished from the elements and sickness alone.

Growing increasingly anxious—Potemkin chewed his nails so obsessively that he developed pussy infections on his fingers—he took emergency measures. Unusual for a Russian commander, out of his own purse he distributed medical care, money, clothes, and food, but a grain shortage didn't help any. So pacing and fretting along the ice-lined trenches, he now used "all means" to help his men, giving them fur coats, cloaks, hats, and also *kengi*—felt to wrap over their boots. And the frostbitten troops were freely provided with vodka and "hot punch." Still, the dying continued. So did the ubiquitous signs of war: billowing clouds of dense, roiling smoke, stabbing bursts of gunfire, the constant roar of bursting shells. And so did the negotiations. Then, ominously, Potemkin was informed by a deserter that the Turkish commander—the *seraskier*—had executed the Russian officers sent to conduct the talks. More than that, the *seraskier* had declared that he would never capitulate. "Please write to me about this quickly and in detail," the empress bluntly told the prince. On November 11 the Turks made a lightning raid against the Russian forces, killing General Stepan P. Maximovich. His head was hoisted aloft, added with the others to the Turkish embattlements.

A hard snow began to fall.

∞

WITH POTEMKIN'S CAREFUL plans coming to grief, and the Liman frozen, on November 27, back in the capital, Catherine became desperate. "Seize Ochakov," she implored Potemkin, "and make peace with the Turks." It was then that Potemkin signed off on a plan to storm the fortress. The fateful day would be December 6. The assault would begin at four A.M.

That evening, one of Potemkin's most senior aides prepared to die, writing a tender farewell to his sister. Then three shells gave the signal.

Once the attack began, Potemkin, his holy Serenissimus, began to pray.

∞

IN THE DIM, gray light before dawn, 30,000 Russians, in six columns, swarmed in a crowded elbow-to-elbow mass toward the heavily fortified entrenchments. The Turks were waiting, greeting the dense throng of attackers with coordinated firepower. As the Russians scaled the ramparts, they faced a torrent of bullets. The Turks countered frantically, but the Russians had the initiative. The Turks quickly fell back; it was a fatal mistake. The Russian soldiers, mad with "fury" and enraged by their months of suffering, stormed the Stamboul Gate. Soon they were racing around Ochakov itself. Once they were inside, the fighting turned vicious, producing scenes of frightful combat. For hours the men slugged it out in often frenzied hand-to-hand fighting in the streets. The moving columns swarmed in a jumbled confusion of shouts, rumbling wheels and trampling feet, and dust. In despair, the Turks fought back with whatever they had, their guns, their trademark scimitars and swinging swords, and then, in desolation, with little more than their fists. It was to no avail. This day, the Russian assault was too fierce. They conquered the city street by street and house by house. And when the garrison finally surrendered, the violent mob of Russians, "like a whirlwind," made a vengeful appearance. They continued the insensate killing, murdering every man, woman, and child they could find.

Then fresh Russian troops poured into the city.

For the Turks and Tatars, this became a nightmare of unbearable proportions. In the melee, brains were smashed, injured men crawled feverishly on their bellies, the dead were crushed by the next wave of attacking men, and the wounded were entombed alive by slabs of the newly dead or frenzied Russians, stepping from corpse to corpse, so closely wedged together were the casualties. The air was thick with pungent smells: sweat, gunpowder, smoke, and blood. And the Turks were butchered with such rapidity— a staggering 12,000 all told—that the dead fell into heaps. How was the ghastly massacre justified? For the Russians, as for the Turks, it was a holy war. "Turkish blood flowed like rivers," the Russians would one day triumphantly sing. But even Frederick the Great was appalled, blasting the Russians as *"les oursmanes"*—"half bear, half psychopath." Then, until Potemkin finally intervened, the pillage began: Houses were torched, women raped, and Ochakov's precious possessions seized. The booty was immense, the most exquisite prize being an emerald the size of an egg. Potemkin would later send it to Catherine.

At seven A.M., after three hours of bestial fighting, Ochakov was in Russian hands. Like a Roman emperor, Potemkin entered the town surrounded by his elaborate entourage, including an exotic bevy of "handsome Amazons." The bare-headed Ottoman commander, the *Seraskier* Hussein-Pasha—he had lost his turban in the struggle—surrendered on his knees to the prince. "I've done my duty," he sighed forlornly, "and you yours." He added: "Fate turned against us."

It was hard to disagree with his assessment. The Turkish casualties were so numerous that they were stacked for hours on end into wagons, where they were hauled out on the Liman and deposited on the ice. There, still damp with blood and bile, the corpses froze solid, adhering one to another in thick piles.

∞

POTEMKIN SENT A jaunty letter to the empress the next day, "I congratulate you with the fortress." Until then, Catherine had been hovering between anxiety and exhilaration. And that night she was asleep, unwell, and tense.

But upon hearing the news, she declared triumphantly to her messenger, "I was poorly, but you have cured me." Then she exulted to Potemkin, "I take you by the ears with both my hands and kiss you, my dear friend." She added, "You shut everybody's mouths."

In Petersburg, 101 cannons thundered and *Te Deums* were sung. And soon thereafter, two hundred Ottoman banners from Ochakov were paraded through the capital streets by a squadron of troops and a cadre of bellowing trumpeters. But Russia's long-awaited victory was sorely marred. For one thing, the sultan, Abdul Hamid, soon died, replaced by his eighteen-year-old successor, Selim III. True, Selim lamented the recent losses: "God help the Sublime State," he was heard to cry out. But he was also a combative, shrewd ruler, whose ardor for resuming the fight to the death was steeled by Islamic patriots, not to mention the ambassadors of Prussia, England, and Sweden. More than that, Selim persuaded the faithful, from sixteen to sixty, to take up arms against their foes; ladies of the imperial harem even parted with their jewelry to help bankroll the holy war. And he now commanded a fresh army of 100,000. For another thing, Catherine's woes were multiplying; she suddenly had to worry about conflict against Prussia as well as an increasingly rebellious Poland, galvanized by the success of the American revolutionaries, not to mention growing unrest in her own provinces.

To make matters worse, while the fighting still raged in the south, it had become a two-front war. Taking advantage of the Muslim conflict, King Gustavus III of Sweden had attacked Russia on the northern front; actually, he manufactured his own casus belli by orchestrating an assault on his own border, employing Swedish troops clad in Russian uniforms. Preparing to leave Stockholm for battle, Gustavus announced with unfettered swagger that he would soon be "breakfasting at Peterhof." For Catherine, diplomatically the matter was exacerbated further: England, Holland, and Prussia signed an ardently anti-Russian triple alliance. Catherine was now under the hostile watch of the great European nations. Even the tsarina was forced to grasp the essential facts: Her predicament was both perilous and dangerous.

With Sweden on the move, and her forces tied down in the south, it was now a race to the death.

∞

SINCE PETER THE Great's historic victories over Sweden, Russia's influence had kept the Swedes quiescent until Gustavus came to power, seeking to avenge his kingdom's historic humiliations at the hands of the tsars. With most of the empress's crack troops fighting the Turks, Catherine improvised. It was not an auspicious start. She raised a peasant home guard and sent Admiral Greig to blockade the Swedish fleet, now invading through Russian Finland, a mere hundred miles from St. Petersburg. But the tsarina was as defiant as ever. "Even if [Gustavus] were now master of St. Petersburg and Moscow," she insisted, "I would still show him what a woman of strong character, standing on the debris of a great empire, can do at the head of a brave and devoted people." She even ridiculed the Swedish king, composing a burlesque opera in her own hand, performed at the Hermitage, in which Gustavus was depicted as a hapless midget prince. Still, as the war ground on, eventually approaching its second year, the Swedish navy had repeatedly proved its resilience, as it did when it engaged the Russians off Cronstadt. Handed the news of this battle—the officer who gave her the message was covered with dust and gunpowder and was exhausted— Catherine now personally began handling the defense of Russia. The skirmish at Cronstadt was followed later in the summer by a full-scale battle at Svenskund in which the navy, once more under Nassau-Seigen's command, suffered a crushing defeat. The Russians lost some of their best ships and suffered some 9,500 casualties. Almost in tears, Catherine wrote to Potemkin that Svenskund had practically "broken her heart."

What almost broke her heart as well was the death of her ally Joseph II and his replacement by Leopold II, who promptly began peace talks with Turkey. The empress was now living on coffee and rusks and had lost weight so rapidly that her aides had to take in all her dresses. Once Prince de Ligne had called her "impertubable." Now, when things were at their bleakest, she wrote back to him saying, "*Votre imperturbable.*" In fact, uncharacteristically, she began repeating to herself, often in slow, hushed tones, "*J'ai donc de l'imperturbabilité.*" It would be needed more than ever.

On land, things had been going very badly as well, as they had since the early days and weeks of the war. After the fortified town of Nyslott had fallen into the murderous enemy's hands, the Swedes efficiently swept forward, and Gustavus was suddenly marching on Frederikshamn, even as the road to St. Petersburg lay naked to attack. Certain of victory, the king grandly announced he would give a great ball at Peterhof for the ladies of his court and would rip down the statue of Peter the Great. Catherine knew the stakes: Unless something was done, with a single lunge, the Swedes would soon be at the empress's Winter Palace.

To the empress, this was Russia bleeding and dying. Publicly she may have boasted, "he who laughs last, laughs loudest," but privately she wept. Remarkably, in the quivering, silvery light of 1788, Catherine could even hear Swedish gunfire from her tall palace windows. It was at this point that panic engulfed the capital. Then the talk began, flying through the salons of St. Petersburg: It was said that the city would be abandoned, that the Swedes were planning a slaughter. In the offices of the administration, nervous officials secured files and hastily packed boxes of official papers, while precious objects—a mesmerizing shower of furs, diamonds, pearls, sapphires, rubies, topaz, and court papers—were locked away in chests and loaded onto wagons. Meanwhile, on foot, on horseback, in carriages, frightened columns of titled and untitled alike fled the peril knocking at their door, flocking the roads all points south and east, even as foreign ambassadors asked aloud if they should collect their papers and flee themselves—or stay behind until the city fell.

As the hours ticked by, the alarm spread. Hard questions now arose. Could Catherine run the risk of being taken prisoner? Or risk suffering a fate similar to what would one day befall Louis XVI in France, only at the hands of hated foreigners? Alternatively, how could she, the Empress of All the Russias, keeper of the largest empire on the globe and successor to Tsar Peter the Great, submit to the Swedes whom Peter so magnificently defeated at Poltava in 1709?

The answers seemed to be quickly forthcoming. While rumors swirled, five hundred horses were precipitously assembled at each relay post along the road to Moscow in preparation for Catherine's escape. Deeply concerned, the Comte de Ségur rushed to the palace to meet with Catherine and discuss the matter on every Russian's lips: the threat to St. Petersburg mounting by the hour. Catherine was grim-faced, but erect and calm. Ségur, alert and ambitious, pointedly informed her, "Everywhere people are saying that Your Majesty will surely leave this night or the next for Moscow." Her reply stunned him.

Catherine, sure that she was on the verge of something tremendous, had decided on a sweeping gamble: to stay put. Ségur left, convinced she was willfully ignoring the ever-mounting danger.

It was around this time that the empress was fatefully advised to leave for Moscow.

∞

MEANWHILE, IN THE West, the dam was breaking. While Catherine pondered her options, few of them pleasant, a Parisian mob had already stormed the Bastille on July 14, 1789. Days later on August 26, the French would stun their own country, thrill republicans everywhere, and send shock waves through the monarchies of the world—remember, outside of the young United States, they were the world—by passing the Declaration of the Rights of Man. As Catherine had predicted, the ripples could no longer be contained. A French nobleman was quoted as saying: "This dreadful America. Since it has been discovered, it has produced nothing but evil." Catherine, in a bid for diplomatic supremacy of Europe, had ironically helped produce that evil; now she undoubtedly rued her own actions. Converts to the republican cause were found in the unlikeliest of places. Polish patriots, including Thaddeus Kosciuszko, who fought for the Americans against the British, were bolstered further by the stirring ideas of the French Revolution and were agitating against Russian domination. A springtime of hope gripped Warsaw itself, and the Poles defiantly demanded that Russia withdraw its troops.

And as guns muttered in the background, Turkish maneuvers under the new sultan proved to be unusually shrewd. To prevent disaster, Potemkin was forced to clear the Bug, steer his forces toward Olviopol, and march swiftly toward the looming Ottoman fortress of Bender on the Dniester, and then on to Ismail. Yet the winter had come and gone, the war was still raging, and the noose had tightened further: England was demanding that Russia make peace with the Sublime Porte on the basis of the status quo ante bellum, while Prussia had signed a military alliance intended to amputate Poland from Russia, as well as a second alliance in which it pledged to help the Turks reconquer the Crimea. In desperation, the tsarina told her secretary: "Now we are in a crisis: either peace or a triple war with Prussia."

For Catherine, with Potemkin miles away and the tantalizing glimpse of quick victory fast receding, the ironies lay heavy. Each new dispatch seemed to bear out the Russian maxim, "They say it is a wide road that leads to war and only a narrow path that leads home again."

CHAPTER 6

FRANCE

✧

URING THE TENSE WEEKS after the fall of the Bastille, King Louis XVI was wracked by near paralysis. Utterly weary, craving solitary rest and tranquillity from the great unraveling swirling around him, he embraced seclusion in the rarefied world of Versailles. There, amid the seductive confines of his court, he passed the days quietly, avoiding the decisions that affected his ministers, his family, and tens of millions of his discontented subjects. He hunted, as always, three to four times a week, received ambassadors, and went through the motions of the daily royal ceremonies of the most formal kingdom in Europe. He pursued deer and spent time with his children. And observers ruefully noted his silences and reticence ("like a tame bear"), his distant thoughts, the vague, enigmatic quality of his personality, and the forced, vacant smile upon his lips. That is, when he wasn't escaping into sleep, which was often.

But history would not wait for Louis. Newly arrived in Paris, Gouverneur Morris, one of the central drafters of the American Constitution and among the keenest observers of the age, echoed the dispatches that his Russian counterpart, Baron Simolin, was sending to Catherine the Great. Morris was stunned by what he saw. "We stand on a vast volcano," he observed. "We feel it tremble, we hear it roar, how and when and where it will burst, and who will be destroyed by its eruptions, it is beyond the ken of mortals to discern." On July 31, 1789, he also added: "This country is at present as near to anarchy as society can approach without dissolution." He was right.

This was the time of *la Grande Peur* (the Great Fear), its tentacles reaching throughout the country. In the provinces, with food increasingly scarce, the people suddenly exploded. Rumors spread that landowners were stockpiling grain. Hunger bred anger, anger bred suspicion, suspicion bred crowds, and crowds bred mobs. Convinced that landlords would retaliate by letting loose groups of bandits, mercenary gangs then took matters into their own hands, arming themselves with muskets, wielding pitchforks, brandishing scythes, and roaming menacingly. The discontent that flamed in rural hamlets festered all the way to the capital itself. The night was now ruled by mobs, and so was the day—stones were even thrown through the windows of the archbishop of Paris, and Thomas Jefferson's house was broken into. Meanwhile, barges hauling bread to the towns were assaulted and plundered en route, and over the course of six months, frightened peasants acquired 400,000 guns.

The king's close advisor, Jacques Necker, traveling through the countryside, was shocked to see sedition and destruction everywhere. The violence was without distinction. Countless châteaux were attacked—or burned; owners were often summarily murdered on the spot. In the guise of the "King's orders," rich bourgeois were assassinated, monasteries were invaded, and at the Abbey of Murbach, the library was trashed and then torched, and its newer building ransacked. When town officials intervened, they were removed; a number were brutally decapitated. Across the realm, people now inveighed against the monarchy's tyrannies—the lack of food, the lack of accountability, and above all, the widespread corruption. Against this tide of spontaneous anarchy, aristocrats abandoned their manors and sought security elsewhere. It was impossible. Almost everywhere they went, the mayhem was the same.

France's nobility again began to flee abroad.

∞

OVERWHELMED AND STIFLED by Louis's indecision, an exasperated cabinet was left powerless. Not so the National Assembly. On the night of August 4, 1789, a shaken deputy reported: "Letters from all the provinces indicate that property of all kinds is prey to the most criminal violence; on all sides châteaux are being burned, convents destroyed, and farms abandoned to pillage . . . laws are without force, and the magistrates without authority." Whether in a surge of patriotism, or in sad recognition that the revolution was now unstoppable, or both, the nobles of the Assembly began to debate whether to abolish all aristocratic privileges. Suddenly, inside the chamber, the Duc d'Aiguillon, one of the nation's richest barons, stood. His words electrified the deputies: "The people are at last trying to cast off a yoke which had weighed upon them for many centuries past; and we must confess that—though this insurrection be condemned . . . an excuse can be found for it in the vexations of which the people have been the victims." Legitimacy had thereby been granted to the rioters and rebels. Galvanized, one noble after another quickly surged forward to renounce his privileges. Deep into the night, they wept and gestured, embraced and made speeches; and by two o'clock the next morning the Assembly had boldly declared the emancipation of the peasantry, thus sounding the death knell for the feudal system in France. All that remained was to receive the signature of the king, proclaimed by Article XVI as the "Restorer of French Liberty."

But that would have to wait. For the time being the Assembly, functioning effectively as a constitutional convention, was now the oracle and the engine that made and interpreted the laws of state. However, unlike the men who had gathered in seclusion in Philadelphia, this was law-giving by spectacle, and the political theater would prove to be profoundly intoxicating. And absorbing all this from his rented home off the Champs-Élysées was Thomas Jefferson.

In just four short years, the Virginian had seamlessly replaced Ben Franklin as the foreign hero of a tumultuous Paris. It was little wonder. As early as April he had made the frequent pilgrimage to Versailles to hear sessions of the Estates-General. And as the revolutionary fervor took hold, Jefferson, his ambassadorial assignment nearing its conclusion, had flocked alongside jubilant Parisians to the Bastille to watch the mob dismember it brick by brick, and had even contributed 60 francs to a fund for the widows of its fallen captors. Then, at the end of July, Jefferson received a letter from the committee charged with drafting the new French constitution; it invited him to a conference. There are "no foreigners," the letter rhapsodized, when "the happiness of man is at stake." Jefferson begged off, but in truth he was being coy. In secret, he had already been collaborating with the Marquis de Lafayette, now one of Jefferson's intimates, on the boldest document thus far: "*La Declaration des Droits de l'Homme*," the "Declaration of the Rights of Man and of the Citizen." Proposed by Lafayette, aided and edited by Jefferson, inspired, in part, by the Declaration of Independence, and partially modeled on the bills of rights already proclaimed by several

American states, the Declaration was to be the preamble to the new, as of yet unwritten, French constitution.

By any historical measure, then or today, the Declaration's language, like America's before it, was uncommonly stirring. "Men are born," it stated wondrously, "and remain free and equal in rights." Man's natural rights, it declared, were held to be "liberty, property, security, and resistance to oppression." Freedom of thought was guaranteed as "one of the most precious rights of man" and so was religion; no one might be arrested or punished except by due process of "law" and all persons were declared eligible for any public office for which they met the requirements. The document also decreed that laws must fall equally upon all citizens and were the expression of the "general will," to be made by all citizens or their representatives—Rousseau's influence. The only sovereign was the nation itself, and all officials acted only in its name. And, as in America, the government was to be separated into branches. True, it was an imperfect document. Slavery was maintained in the French Caribbean colonies, eligibility for public office was considerably restricted, and civil rights were denied to three specific categories: actors, Protestants, and Jews.

Yet, once unveiled, the document was a resounding thunderclap across the Western globe. To be sure, the thinkers of the Enlightenment, echoing philosophers in the Middle Ages, had referred to the rights of man, and during the American Revolution even Alexander Hamilton had movingly spoken of the "the sacred rights of man." But now, in one of the oldest kingdoms of Europe, republican ideals were suddenly real: In the weeks that followed, to great cheers, the Declaration was embossed on thousands of leaflets and pamphlets, sung aloud in hundreds of town squares, or proudly displayed in countless homes; it became the touchstone of the revolution, and, upon its translation, it carried a powerful message of human emancipation and hope to the rest of the continent and over to America. Of the ecstasy that swept across Europe, the English poet William Wordsworth penned, "bliss was it in that dawn to be alive." Or as Rabaut de Saint-Etienne put it: thus had been created "the political alphabet of the new world." And with it, the revolution seemed to be complete.

But this burst of euphoria was fleeting, for turmoil had begun to stalk the revolutionaries, not simply from outside, but from within. Ever observant, even the reform-minded Jefferson sounded a grim, cautionary note, privately writing to John Jay that the vast Assembly was slothful, the people too "impatient," the remaining aristocrats "seditious," and even the patriots divided. He acknowledged that civil war was openly whispered and "entirely possible." Back in America, George Washington sensed this too. "I do not like the situation of affairs in France," he urgently wrote to Lafayette. "Little more irritation would be necessary to blow up the spark of discontent into a flame that might not easily be quenched." Meanwhile, Jefferson himself tried to stem the tide and heal the wounds. In late August, just as the Declaration of the Rights of Man was being adopted, and upon plaintive entreaties from Lafayette, he held a small dinner for two factions of Assembly moderates. Foreshadowing his American dinner with Hamilton and Madison the following year, which would broker visionary compromises on the disposition of the young country's debt and location of its capital, Jefferson invited Lafayette and Jean-Joseph Mounier, the leader of the Assembly's moderate royalists—often referred to as "the Americans"—as well as Adrien Duport, leader of the Patriot Party, and his allies, Joseph Barnave and Alexander Lameth. All the guests no doubt hoped as well to gain something from the presence of Jefferson, who, as Mounier put it, knew much of "the maintenance of liberty." The meal ended at four, the tablecloth was removed, wine was laid out, and the men began to talk, continuing straight through until ten that night. Jefferson would write that over those hours, he sat as a silent witness to a dialogue that rivaled those of Plato or Cicero. And he ended the evening believing that a bargain among the factions had been struck.

But this was not America; bargains and compromises could not easily survive in this fast-moving rebellion. And as it happened, one of the most resolute holdouts was the king. Louis, hesitant by nature and cautious in most cases, was now doubly cautious here. Even as the nation still believed he would join the Revolution, the king clung to his monarchical impulses and refused to accept both the August 4 decree and the Declaration of Rights as well. But the people of Paris were equally determined to make him submit.

∞

ALL THROUGHOUT THE early days of fall, among the nobles in Versailles and the émigré princes abroad, there was talk of plots and conspiracies, moves and countermoves. But the ministers, including Jacques Necker, didn't know how to advise a sovereign who refused to face the coming storm. So they took matters in their own hands. Plans were hatched to spirit the royal family to safety—to a city in eastern France, eventually Compiègne. But the king, unwilling to precipitate a civil war or allow the crown to become hopelessly indebted to scheming princes and nobles, said no—the memories of England's Stuart king, James II, who had fled in a fishing boat to France never again to regain his kingdom, were still fresh. Pained by his choices, Louis dithered, and wavered, and slept.

By contrast, Paris was suddenly wracked with riots. Bread was again becoming scarce; housewives came to blows waiting at the bakeries. And Parisians, tasting freedom of the press, were now speaking their minds—in various *libelles* (little books), in the dozens of new pamphlets springing up daily, and in works such as Brissot's *Le Patriot français* and Desmoulins's *Révolutions de France*. As these real grievances were fed into the great furnace of an unchained press and truculent sidewalk orators, old shibboleths and reputations were shattered, and fresh expectations heightened. And among the most coldhearted and cold-blooded of these new scribes, not to mention among the most influential, was Jean-Paul Marat.

Like the pivotal Necker, Marat, who would rise amid blood and fury to become one of the triumvirate of the Revolution, wasn't even French. Born in Neufchatel, Switzerland, in 1743, the son of a Sardinian father and a Swiss mother, he always idolized another expatriate countryman—Rousseau. Marat had studied medicine in Paris and Bordeaux, and for a time worked as a physician in London. He analyzed eye disease, gonorrhea too, and then began to turn his attention to politics, at one point offending no less than Voltaire. In 1774, when Jefferson was first setting his pen to paper—even Tsarina Catherine was still in her progressive phase—Marat wrote *The Chains of Slavery*, a furious condemnation of European governments bent on keeping the people in perpetual servitude, for which he was made a member of the patriotic society of Newcastle. In 1777 he returned to France and was appointed a court physician for Louis XVI's younger brother, Artois. Marat had other pursuits as well: Like Benjamin Franklin, whom he assiduously courted, he incessantly inquired into the world around him, publishing treatises on heat, light, optics, electricity, and fire; a number were even translated into German. But here his radicalism was already apparent; from his darkened library near the Invalides, Marat unleashed a scathing attack on Sir Isaac Newton, whom, by comparison, Jefferson considered one of the three greatest men who ever lived (he kept Newton's portrait in his home). So while Marat waited, expecting to be invited to join the Academie des Sciences, the academicians felt him suspect. Marat's entrance was denied.

It stung him deeply. Marat, dwarfish, dour, humorless, and authoritarian, now was seized with an almost messianic fervor, exacerbated by a constant array of pains that rendered him sullen to the point of violence. He had rampant "psoriasis," which was only ameliorated by sitting and composing missives in cool baths. His cumbersome head was too big for his body, one eye sat higher than the other, and both were strangely yellowed. And doctors bled him repeatedly for his many ailments. Thus he worked not simply with the ferocity of a man whose many ambitions were unquenchable, but of one who was living on borrowed time. "I allot only two of the twenty four hours to sleep," he boasted. "I have not had fifteen minutes play in over three years." To those who saw him, and that was infrequent as he preferred near total solitude, he was often unwashed, twitching, and clad in filthy attire. Many believed him mad, and for good reason.

Increasingly, his character, and his judgment, were affected. For all his courage and ambition, he became known for his delusions of grandeur and his erratic bursts of anger, for his vanities and his paranoid diatribes against such historic moderates as Necker and Lafayette, and for his bloodcurdling cries for mass violence. He saw enemies everywhere, even among friends and allies. Still, the stunning success of his journal, *L'Ami du peuple*, stemmed not merely from his vilifications but from his steadfast support for the voteless working class, the sansculottes. And in his madness was vision. As few did, he never misjudged the intelligence of the people or the temperament of these heady early times. He not only sensed chaos, but enthusiastically stoked it. His ideal polity was akin more to despotism than democracy, where the ruler was held in check by recall or revolt, not unlike in republican Rome. He even intimated that he would make a fitting Caesar, worthy of the people of France.

Most of the National Assembly, and even many Jacobins, despised and feared Marat, and for good reason. But not the sansculottes, who embraced his philosophy and would one day be willing to risk everything for him. And how could they not? He spoke their language, he embodied their fears, he breathed life into their aspirations. In those autumn days, as the people went hungry and the anger of the populace smoldered, and the king and Assembly alike failed to provide much-needed bread, it was Marat and seemingly Marat alone who issued a call for action.

To march against the "perfidious conspirators." To the gates of Versailles itself.

∞

NOT SINCE THE horrid days of the Fronde had such measures been contemplated against the monarchy. But Marat neither sugarcoated his words nor draped himself in hazy generalities. "When public safety is in peril," he pronounced, "the people must take power out of the hands of those to whom it is entrusted." Nor he did not fear bold measures: "Put the Austrian woman [the Queen] and her brother-in-law [Artois] in prison. . . . Seize the ministers and their clerks and put them in irons." Or shy away from big ideas: "The heir to the throne," he insisted, "has no right to a dinner while you want bread." He roused the sansculottes to action: "Organize bodies of armed men. March to the National Assembly and demand food at once," he exhorted. "If you are refused, join the army, take the land, as well as the gold, which the rascals who want to force you come to terms by hunger have buried, and share it among you."

And then finally, he ended with this chilling coda: "Off with the heads of the ministers and their underlings. Now it is time."

∞

LOUIS XVI SENSED this too. Terrified by the riots in Paris as well as the swelling demonstrations near Versailles, the king belatedly heeded the warnings of his ministers—that he needed troops untainted by radical ideas to guard him. Late in September he sent for the Flanders regiment, whose loyalty he saw as unwavering. The regiment quickly came, and on October 1, the king's Garde du Corps welcomed it with the traditional banquet in the Opera House of the palace. The two regiments mingled in an unmistakably monarchical spirit, toasting the health of the king and cheering themselves hoarse. When Louis and Marie Antoinette appeared, the hall then deliriously resounded with cries of "Long live the King!," the orchestra struck up "O Richard, o mon roi," and soon the guards were replacing the national tricolor emblems on their uniforms with cockades of the royal colors, white and black. One report, in the revolutionary paper L'Ami du peuple, later claimed that the discarded colors, now so dear to the revolution, were trodden under dancing feet.

Yet that evening, and the next, the king and queen felt blissfully reassured of their security, confident in the same certitudes that had sustained the monarchy for almost a thousand years. In the mirage of the cloistered world of Versailles, the queen even believed the nation and the army "must be dedicated to the King as we are ourselves." But within hours, as the story of the fête traveled to Paris, enlarging with each mile, it was clear that something had gone terribly wrong. In Paris, the people were outraged. Marat and other journalists were once again insisting that the people must compel both the royal family and the Assembly to move to Paris—where they would be under the iron thumb of the populace. Remarkably, though, the king still believed all was well: On October 5, despite overcast skies, Louis quietly went hunting, and Marie Antoinette made plans to redesign her gardens at Trianon. It was only after lunch that the queen realized something was deeply amiss. Louis's remaining brother and sister rushed into her room in great agitation. Their eyes contained warnings, and so did their words.

In the distance, through thickening fog and a drenching autumn rain, riders were coming with reports of the dull thud of feet, thousands strong. Then Marie Antoinette was informed: The people of Paris were marching on Versailles itself.

∞

IT HAD BEGUN that morning, as the tocsins clanged steadily throughout Paris. While gangs of men, wielding pikes, streamed menacingly into the plaza, the market women of the city (*poissardes*), hungry and unemployed and seething with frustration, took the lead in forming a brigade to march on Versailles, twelve miles away. As they walked under darkened skies, they called out to passersby to join them, and their ranks soon swelled by the thousands. At first the march was neither surly nor violent. But along the way, the mob, now numbering some eight thousand strong, armed themselves with sickles, sticks, and guns, and were waving their fists and howling for the king.

Meanwhile, without the king, Versailles fell into paralysis. The royal ministers pleaded with the queen to seek safety in Rambouillet in the southwest, only an hour away. Instead the queen immediately called on aides to ride off and find the king. As the minutes passed, members of the court gathered by the high windows to watch and to wait; soon, they too felt overwhelmed.

Now events began to spin out of control. The ministers flocked to

Necker and La Tour du Pin, who directed the War Department. But absent Louis, no decisions could be made; the most they could do was to shut doors and seal passageways that had been opened since the time of the Sun King. "We will let ourselves be captured here," La Tour du Pin darkly predicted, and "be perhaps massacred." Only then were hasty precautions taken: the Versailles National Guard and the Flanders regiment lined up for battle, their thin ranks the only thing between the would-be rioters and the gates of the royal courtyard. Finally, at three in the afternoon, the king rode through the gates in a flurry, and immediately huddled in the council room with his ministers and Marie Antoinette. For an hour they heatedly debated the options: Should emergency measures be invoked? Should the queen and the royal children flee under escort to safety? Should the king lead the troops and meet the Parisian mob head on? (Catherine or George Washington almost surely would have done this, and so would have the Sun King.) Or should the king stay put and negotiate, as Necker believed?

By four o'clock, no firm decision had been made. And the first wave of the Parisian women was arriving.

∞

ENTERING VERSAILLES UNDER a pouring rain, they massed in a haphazard array, trembling with excitement and drenched. Before the high gates of the royal palace, they angrily demanded food, shelter, and access to the king. Marchers also overran the Assembly, which was meeting nearby, and demanded bread for the crowd. Jean-Joseph Mounier, Jefferson's celebrated dinner guest and leader of the moderate royalists, was the presiding delegate; he hastily took six women, including a pretty seventeen-year-old flower girl as the spokeswoman, to accompany him to see the king. Allowed entrance to the royal apartment, the girl was so awed at seeing Louis in the flesh that she whimpered, *"Pain,"* and collapsed. When she was revived—Louis had actually brought her smelling salts himself—he vowed to produce food for his tired and hungry subjects. Upon leaving, some accounts have her leaning forward to kiss his hand, an incredible breach of royal etiquette, but it no longer mattered; instead, he spoke as gently to her as if she were a daughter. At the same time, the atmosphere outside had seemingly relaxed enough that many of the Parisians fraternized with the troops; several of the younger women, taking off their skirts and petticoats to wring out the water, disappeared into the barracks where they received succor and warmth. But this comity would not last long. For as Louis's luck would have it, only on this day did the Assembly learn that the king had refused to ratify the Declaration of the Rights of Man. Suddenly too, the delegates were infuriated.

Night fell. And whatever options the king had were few and increasingly fleeting. With Versailles in disarray, Louis himself panicked. He promised Mounier that he would sign the Assembly's Declaration. In turn, Mounier desperately pressed the king to leave for Normandy as soon as possible for his safety; so did the other members of the cabinet. Yet the more he resisted, the more the ministers began to beg Louis. Distraught, the king instead muttered cryptically, "A fugitive King! A fugitive King!" But precious time was wasting. "Sire," one of the ministers desperately implored, "if you are taken to Paris tomorrow, you will lose your crown." At that, the terror-stricken war minister literally flung himself at the king's feet, urging him to flee. Only then did Louis give the orders for the carriages to be prepared.

It was too late.

∞

THE ARMED MOB hurled itself on the king's carriages, severed the harnesses, and took the horses away. By then the sky was black, and the rain washed down in torrents. Horrified, the king withdrew. Suddenly, around midnight, a mud-spattered and exhausted Lafayette himself arrived, with fifteen thousand National Guardsmen. But contrary to the king's fervent hopes, it was not the makings of a rescue. Many of the guardsmen had marched in support of the Parisian mob, and some were even fomenting plans to force the king to return to Paris. The thirty-two-year-old Lafayette, by now a prisoner of his own army, had ridden out with them, hoping to restore order and defuse the situation. He was received by Louis and promptly pledged his protection. "I thought it best to die here at Your Majesty's feet," he said touchingly, "than to perish pointlessly at the Place de Greve." But he also promised the king that the Parisians' intentions were peaceful. Reassured by the hero of the American war, Louis entrusted the security of the château to Lafayette, and, worn out, went to bed.

The château doors were shut and the candles extinguished. By two in the morning, everyone had fallen asleep.

Lafayette quickly took charge. He posted the National Guards outside the palace, the first of several fatal mistakes; as it happened, they remained more loyal to the Revolution than to the king; they also despised the king's bodyguards. Only the positions inside were turned over to the king's trustworthy personal guards—the Black Musketeers—but they were painfully few in number, the bulk of them having already been dispatched to Rambouillet to prepare for an escape. And the king's crack forces—the Swiss Guards—had been sent on to the neighboring city of Rueil. Still, a quiet descended over the palace. At three in the morning, a reassured Lafayette announced: "I will answer for everything."

Soon thereafter, he headed across the road to his wife's family mansion. Exhausted and filthy, he finally collapsed on a couch around five A.M. In barely an hour, he would be frantically awakened.

∞

OUTSIDE, HORDES OF soaked Parisians had sought cover in the stables, the taverns, the mansion courtyards and churches and even the Assembly, although the majority remained camped in front of the palace itself. They lit bonfires and drank and sang into the early morning hours. A horse was butchered, cooked, and eaten. Here and there, random gunshots were fired into the air. By four or five in the morning—accounts vary—there was a lull. Light after light was extinguished.

The great mob too seemed to be at rest.

∞

IT WAS NOT. Suddenly, just upon the first fingers of daylight, the weary, angry crowd became a sinister throng that poured through two unguarded openings in the gates and into the palace, where they surged through the chambers and the hallways. Armed men, screaming and swearing, massed under the king's window, while others forced their way up the stairs to the private apartment of the queen. Lafayette was wrong; the mob had turned violent. Now it was the voices of revenge that would be heard the loudest. Terrified courtiers, ministers, deputies, and servants cringed in every corner, crawled beneath tables, and flattened themselves behind sofas. Then the first blood was drawn. A bodyguard, a mere boy, was ruthlessly murdered and dragged into the courtyard half dead, becoming little more than a bleeding trophy. Here, in a grisly reprise, one of the sansculottes sliced off the guard's head with an ax to cheering and applause. Within minutes the invaders reached the queen's chamber. A second bodyguard was massacred, as the mob now vowed to "cut off" Marie Antoinette's head, "tear out her heart," and "*fricasser* her liver."

"Save the Queen!" the cry ran out. In her petticoat, she leaped out of bed and raced to the king's room; she was deathly pale, but calm, although the king was missing. No one yet knew that, in darkness and wearing his bathrobe, Louis was already fleeing through the secret passageway that connected his apartment to hers. Meanwhile, the desperate and woefully outnumbered palace guards stoutly continued to resist the invasion; defending the last rampart that separated the royal family from the assailants, another was butchered.

Below the stairway by the king's room lay a decapitated corpse from the queen's guards, while inside her room, a howling pack fell upon the queen's empty, unmade bed, slashing it vociferously with axes and sabers. And thirty more royal guards had fallen into the hands of the shrieking, pitiless mob. At last a frantic Lafayette appeared. ("I was completely trusting," he later confided, "the people had promised me they would remain calm.") He knew immediately that the royal family, now huddled just beyond the Salon de l'Oeil de Boeuf, had been spared death only by the lock on a lone anteroom door—the king himself had run through a blackened corridor with the dauphin in his arms and a servant clutching at his bathrobe. But Lafayette, with his own innate genius for political theater, steadied himself and helped quiet the tumult by convincing the king to appear on the balcony. Under duress, and with as much dignity as he could muster, the king did. There was a momentary hush. Lafayette began to promise, in the sovereign's name, an end to the bread shortage. Then the cry began: "To Paris!" Then louder, "To PARIS!" Then louder still, "TO PARIS!" until all other sounds were silenced. Frightened, the king withdrew. Then the brazen crowd demanded the queen, alone, without the shelter of her children. Shaken and still in her dressing gown, she stepped forward, making a deep curtsy, momentarily swaying the crowd. But interlaced with their cries of love was the steely demand that she too come and live in the capital.

Dazed, the king hastily retreated and began to pace. Outside, the fracas continued to grow; the crowd was vehement. Up until now, Louis had felt free to take some of his bravest stands—or at least to equivocate—precisely because he believed that when the choice came, his loyal subjects would rally around him. After all, France had sustained one thousand years of glorious monarchy; it had enjoyed two centuries of grandeur under the Bourbons; and through decades of war and peace, it had cherished his family. Now, in the seat of his own government, in the pounding heart of the nation, the people had apparently not only deserted him, but openly humiliated him— and even threatened his wife's life. As the bellicosity of the mob grew by the minute, the powerful mystique of the monarchy seemed to diminish by the hour. The king's inner struggle was irresolvable. He viewed with grow- ing terror the massive array of forces gathered around the palace, the brutal slayings of his own guards, the clenched fists and screaming taunts. Ashen- faced, Louis took a moment to fix his voice, then fatefully announced to his ministers and then to the crowd below: "I will go to Paris with my wife and children; I entrust what is most precious to me to the love of my good and loyal subjects." But blanketing them all was a feeling of profound dread. For their part, the men and women of the mob roared in ecstasy, "*Vive le Roi,*" promising to hug the king to death.

Outside, applause rang out along with an accompanying chorus of gunfire.

∞

AROUND NOON, AS Thomas Jefferson was waiting at the port city of Le Havre for a packet boat to begin his journey home to Virginia, a remark- able procession formed. In front marched the National Guard and the now disarmed royal Garde du Corps; then a coach with the king, the queen, and their two children, escorted by Lafayette and the Comte d'Estaing on horseback; then an array of carts carrying sacks of flour; then the victo- rious Parisians, singing and chanting and openly tormenting the royals ("We're bringing back the baker, the baker's wife and *le petit mitron*"—"the little baker's boy!"). Women sat astride cannons or hung menacingly on the carriage doors, while in front of the royal carriage, men waved the severed, blood-drenched heads of the palace guards; at Sèvres, they paused so the heads could be dressed and powdered. Meanwhile, back at Versailles, bands of marauders were rampaging through the relics of the palace; left behind were the queen's slippers—she never had time to put them on—as well as a delicate silk stocking awaiting her royal foot. Her gilt panels were smashed and her painstakingly decorated door was splintered.

The queen feared she would not reach Paris alive, although by day's end, the royals were met by Jean Bailly, the mayor of Paris. With no irony to his voice, he gave them the keys to the city on a velvet pillow, proclaiming, "What a beautiful day, sire." That night, while torches glowed outside and a new crowd gathered, the king and the rest of the royal family slept in flimsily prepared beds in the barren apartments of the Château des Tuileries, where French royalty had made their home before the Fronde rebellion that nearly upended the young Sun King; it had since served as the queen's pied-à-terre on her Paris jaunts. Shortly thereafter, the Assembly joined the royals in Paris and was housed in the palace's theater.

In a single day, the populace of Paris had wrested control of the revolution. From this point on, their wrath, and the Assembly's, would amplify and enlarge, as together they stared down the most powerful monarchy in Europe. They were now flouting what had once been a God-anointed king. Meanwhile, inside the Tuileries, the dauphin, heir to the throne, awoke and plaintively asked his mother: "Good God, Maman, would today still be yesterday?" In a profound way, it was, and would forever remain. Now a prisoner in his own capital, now subject to his subjects, now, with a heavy heart, the king unequivocally accepted the Rights of Man as a fait accompli.

And soon the queen was heard to murmur to a member of the court: "Kings who become prisoners are not far from death."

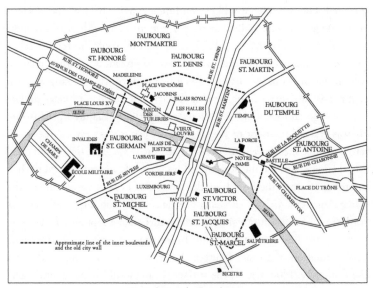

Paris during the Revolution

∞

WAS THE KING, at times pathologically indecisive, at other times unusually resolute, in fact a hostage? He was, for sure, a man now increasingly without hope or means. The city before him was openly hostile, while his aristocratic peers had largely abandoned him. Some 150,000 would flee France, among them many officers in the army. Others, like his cousin, the Duc d'Orleans, had cast their lots with the mob. And for its part, freed from royal resistance, the Assembly zealously took matters into its own hands and began to legalize the achievements of the revolution. Even as the king continued his twice weekly public dinners, and had his *lever* and his *coucher*, the Assembly set about the slow task of writing its constitution.

The first great question: Should it retain the kingship? It did, and it allowed it to remain hereditary as well. Yet still, the Assembly believed that the dazzling power of the throne was crucial to maintaining France's tenuous social order; moreover, many of the revolutionaries, like Lafayette and his rival, the Comte de Mirabeau, like many in the countryside, were still monarchists at heart. They also feared that without the king, wars of succession would ensue—such bloody schemes were already being considered in Orleans' home in the Palais-Royal. But the king's reach was also sharply curtailed: The Assembly would permit him a yearly "civil list" for expenses (a not ungenerous 25 million livres per annum); any additional expenditures would require the Assembly's consent. He would be free to select (or fire) his ministers, but each would have to provide a monthly financial accounting and could be arraigned before a high court. The king was still commander-in-chief of the army and navy, but could no longer declare wars or sign treaties without first obtaining the legislature's permission. He was granted the right to veto legislation, but it was only a "suspensive veto"; if three successive legislatures passed the vetoed bill, they could override the king, and the measure would become law. In truth, in more than a number of ways, the French model tentatively began to approximate the young America's with its checks and balances, the one difference being it bordered on a constitutional monarchy rather than a republic. To be sure, it was a gross humiliation for Louis: In the span of weeks, he had become a pale imitation of former Bourbon kings, let alone enlightened despots like Catherine. However, on the surface, Louis would retain considerable authority—no less than George Washington and not significantly different from King George. And he was king still.

But there was one undercurrent with ominous portents: The Assembly, operating on multiple agendas, also stipulated this—if Louis left the kingdom without its permission, he could be deposed.

But it was not just the problem of the king that plagued the Assembly: It also had to wrestle with the thorny matter of France's staggering fiscal woes. So on November 2 the Assembly, more inclined to worship Rousseau and the Romans than God, brazenly decided to nationalize church property to pay state debts, not unlike what Peter the Great had done in Russia; the vote was 508–346. Then, on December 19 it authorized the sale of negotiable "assignats"—notes entitling the holder to ecclesiastical property, the proceeds of which would be used to further address the economic crisis. Having now, in effect, gone to war with the Church, the Assembly watched in desperation as inflation climbed, the assignats dipped in value, and the treasury sank further into debt. (Within months, these paper notes smacked of the worthless "Continentals" that had been printed a decade earlier by the rebellious American colonies to finance George Washington's armies.) Unbowed, the Assembly only twisted the knife deeper. It suppressed monasteries and subsequently unveiled a "Civil Constitution of the Clergy," which in essence required priests to become employees of the regime.

This was the most far-reaching measure yet. The effect was to create a citizen clergy, bound to the principles of the revolution. There was a risk entailed by this deadly bargain, however. All priests had to swear an oath to the new constitution. Thus a schism was born: Their loyalty would be either to Pius VI, the pope in Rome, or the new French nation; there were two churches in France, one clandestine, maintained by smuggled funds, the other official, financed by the government; one unsworn, or "refractory," the other constitutional, and, at least as considered by themselves, patriotic. Unlike the radical Abbé Grégoire, of the 134 bishops in France, a staggering 127 were unwilling to take the oath; of the 70,000 parish priests, 46,000 declined as well. In retaliation, the assembly extracted a high price: Renegades would be forbidden from exercising their priestly functions. Thus, the seeds of eventual civil war, even holy war, were sown. A great majority of the population sided with the conservative Church, and Louis himself, a devout Catholic, openly balked at the new constitution.

Nor was he alone. Others, for different reasons, increasingly rejected the principles of the draft constitution. The volatile delegate Maximilien Robespierre would spearhead a vocal minority against limiting the right to vote to property owners, arguing that it was a breach of the Declaration of the Rights of Man. But the middle-class deputies were undeterred. With the nobility, or vast portions of it, in flight, they believed that it was at last the bourgeoisie's moment to govern. So, heedless of the king's protests, the Assembly triumphantly moved forward to make France a constitutional monarchy. Never was this more apparent than as the first anniversary of the Bastille dawned. On June 5, 1790, the entire country was invited to send its now federated National Guards to join the newly constituted government on the Champs de Mars, just outside the old city, to celebrate the anniversary of the fall of the Bastille. As word of the invitation crossed the country, soldiers from all of France converged on Paris, and in many cases, the Parisian National Guard, some eighteen thousand strong, gleaming, giddy, and shouting "Long live our brothers!," marched for miles out beyond the city to escort them. Idealistic foreigners were roused as well. Led by a rich German known as "Baron Anacharsis Cloots," thirty-six of them appeared at the bar of the Assembly on June 19 to petition for the honor of French citizenship and an invitation to the feast of the Federation as "an embassy of the human race." The Assembly granted their wish.

The preparations for the national fete were immense; so was the task. The hilly field of Mars had to be sculptured and terraced to create a vast amphitheater for the occasion. It had to be able to hold 400,000 men, women, and children, not to mention the king, clerics, deputies, and soldiers, all of whom would together commemorate their country's rebirth. But, two weeks out, the job was still not done. So, as the Marquis de Ferriers recorded, the people came: rich and poor, adolescent and infirm, "the whole population" with sweat and shovels and wheelbarrows and song, "Ça Ira!" ("It will go!"), to remold the sprawling hills. Even France's first emperor, Charlemagne, would have been moved.

On July 14, 1790, beginning at dawn, despite swift breezes and a driving rain, at last the "great day" arrived. The crowds filled the streets; wine, ham, fruit, and sausages dangled from windows; and then some 18,000 soldiers entered the Field of Mars in proud procession. They came in thickly wedged files of eight men deep, or more, accompanied by bands and banners and sturdy patriotic songs, with another 300,000 Parisians, many wearing the *bonnet rouge*, the replica of the red Roman cap worn by slaves when they gained their liberty, jubilantly cheering them on. Four ancient vases burned with incense, and eighty-three spears had been driven into the ground, each with a new department banner fluttering magnificently in the wind. It was John Paul Jones, back from Russia, and Thomas Paine who carried American flags, thus symbolizing the passing of their liberty to France. It was Abbé Talleyrand, not yet excommunicated, who said Mass. It was Lafayette, dismounting from a glorious white charger, both arms stretched out to heaven, who administered the pledge to the assembled National Guard. It was the queen, resplendent in red, white, and blue plumes, watching the troops file past Louis, who called out: "From what province do you come?" and a loyal roar that came back: "The province over which your ancestors reigned!" It was the king who held his tongue and pledged to obey all the new laws of the state. And it was the mass assemblage, with tears in their eyes and hope in their hearts, which answered, "*Vive le Roi!*"

Ringing bells could be heard against the rain, while light arms crackled with salvos and artillery fire thundered. Off in the distance, trees were festooned with colored lights, and when the cannon sounded a salute, thousands of French citizens who had not been able to attend raised a hand toward the Champs de Mars and made their pledge as well. Across the nation, festivities were held, hordes of young people danced in the Champs Élysées, bonfires were lit and fireworks ignited. For once, food was plentiful, and Catholic and Protestant pastors embraced as common Christians and common citizens. No wonder on this day utopian orators issued a "Declaration of Peace for the World."

And that night, at a public dinner, cries of "Long live the king!" rang outside the windows of the Tuileries.

The voices of doom were silent. As tricolored flags flapped gloriously in the wind, who could question that France had not created the birth of a wondrous new age?

∞

ACTUALLY, FROM THE Tuileries, a Bourbon and a Hapsburg could, and did. To Louis XVI and his queen, their new home was little more than a public cage where their every move was scrutinized by the glaring eyes of a restive Paris. And there was always the haunting visage of the mob; indeed, during the great ceremony on the Champs de Mars, however serene the queen appeared on the outside, inside she was trembling. Memories of the Bastille, she confided to one of her ladies, "brings to us everything that is most cruel and sorrowful." But more than cruel or sorrowful, despite all the passion a grateful nation expressed to the king at the Fête de la Fédération, his fate remained as tenuous as ever.

This precarious position was exposed again and again. In August 1790 three of the king's regiments mutinied in the northeastern city of Nancy; some of the rebels were unsparingly shot down by the Royal German regiment; others were carted off to the galleys; still others were hanged at the scaffold. When the news reached Paris, forty thousand rioters clustered menacingly outside the royal palace. They blamed the king (wrongly so) for the "Nancy Massacre," while screeching "*À bas Lafayette!*"—"Down with Lafayette!"; they also called for Louis's ministers to resign. This time Necker, ever the survivor, vanished—for good—to the quiet of Lake Geneva. Lafayette pleaded with the king to placate a hostile population by endorsing the constitution. He would not. Instead, the royal family was drawn further into delicate and highly secret negotiations with the radical Provençal aristocrat, the Comte de Mirabeau. The goal: to save the throne.

It was the oddest of matches. The queen despised the pox-scarred Mirabeau, but Mirabeau, despite his revolutionary fervor, actually revered the crown. From the beginning, the volcanic comte, with his bloodshot eyes and blustery ways, had been a potent force for rebellion and had risen to dominate the Assembly; but Mirabeau was also a pragmatist: Having helped unleash the revolution, he now sought to ride the tiger. Briefly he flirted with the idea of dynastic change, supporting the Duc d'Orleans as the new king; when that effort foundered, only then did he decide that the king himself had the makings of a modern sovereign. But while Louis and Marie Antoinette each possessed a stubborn sense of right and wrong, Mirabeau was frequently dissolute, devious, deeply in debt, or on the take. Not surprisingly, Madame Campan once called him "all eloquence and audacity." Thus, while he believed that an alliance between the king and the Assembly was the only viable bulwark against chaotic mob rule or counterrevolution, he was also quite willing to be bribed to pursue the very policies that he supported. In his own words, he considered himself "paid but not bought"; he even asked the king to make him ambassador to Constantinople.

Actually, where members of the royal household thought him "a mercenary democrat," and the queen thought him a "monster," Mirabeau quietly despaired of France's fate. As far back as September 1789, when the country was still basking in the glory of the Bastille, he direly wrote to the

Belgian nobleman living in France, Auguste de La Marck, "All is lost. The King and Queen will be swept away, and you will see the populace triumphing over their helpless bodies." And in October he added this: "If you have any influence with the King or the Queen, persuade them that they and France are lost if the royal family does not leave Paris."

Though Louis was still temporizing, he had enough reservoir of doubt that he agreed to bankroll Mirabeau's defense of the crown: The king consented to pay his many debts, to provide a stipend of 6,000 livres a month, and to reward him with a large lump sum if he succeeded in mediating between the monarch and the people. In July, to seal the deal, the queen reluctantly consented to give Mirabeau a private interview at her summer retreat in Saint-Cloud. After trading silences, Mirabeau actually trembled when the queen offered him her hand. And upon taking his leave, he whispered complicitly to her, "The monarchy is saved."

In truth, that remained unclear. But from then on she referred to him as "M," while he spoke of her ecstatically as "a man for courage." Was he playing a double game? He was, and brilliantly. But in all likelihood, his concern was real and so was his passion. He ardently believed that the king must leave Paris, perhaps to Rouen or Fontainebleau or Compiegne, where he could rule in an English-style monarchy—adapted to France—without duress. And just as likely, Mirabeau, having saved a weakened, grateful monarchy, also conceived that he would then assume the grandest role of his own life: as Louis XVI's William Pitt, the king's prime minister.

But if Mirabeau began accepting payment, he was no mere hired gun. Elected to the Assembly as a member of the Third Estate, he had no plans to resurrect royal absolutism ("All my life," he once thundered, "I have fought despotism."). Instead, the proposal he submitted to the king's ministers on December 23, 1790, was an ambitious program for marrying liberal reforms with the sanction of royal authority. And given the mood of the day, one that he had done so much to shape, it had great promise. "To attack the revolution," he insisted, "would be to overshoot the mark, for the movement that makes a great people give itself better laws deserves support." He firmly told the king that the spirit of the revolution had to be accepted, and thus much of the constitution, forthrightly pointing out that the effects of the revolution were "so irrevocable," as to be beyond eradication, short of dismembering the realm. But undergirding this compromise was no small measure of tactical cunning: Mirabeau also told the king that the new regime would need even more funds than the old, and here lay the zealots' Achilles' heel. "The veil," he said, "will be torn asunder . . . the people have been promised more than can be promised." He brazenly suggested bringing the radical Jacobins into the ministry to neutralize them, and then advocated placing the National Guard firmly under the king's control so that the crown would never again be hostage to the whims of a makeshift Parisian army or a revolutionary coup. Mirabeau was also thoroughly protean and willing to shift with the times; at one point, he even advocated civil war as a way of introducing order; ironically, this the king frostily rejected. Yet Mirabeau remained the king's last, best chance to reclaim his independence, if not his throne and his life. Moreover, once drawn into the royal web, Mirabeau himself was caught. He would rather die, he told his friend La Marck, than "fail" to fulfill his promises.

But such actions were not without risk. Mirabeau made enemies in the Assembly; he was accused of wanting to destroy the Jacobins, and a pamphlet pointedly charged him with treason ("*Trahison decouverte du Comte de Mirabeau*"); it was even alleged that he was betraying the revolution itself. He dangerously ran afoul of Marat, and had strained relations with members of the court; Lafayette, for one, hated him, while Mirabeau considered the young marquis a self-important, inarticulate mediocrity. But for all this, Mirabeau, with his impassioned pride and striking individuality, was at the zenith of his powers, and everyone knew it. Yet a hard question lingered: Was he too much a monarchist for the Revolution, and too revolutionary for the monarchy? Still, his oratory was unsurpassed, and his control of debates masterly. In fact, in his dealings with the Assembly, he was more than strong, he was implacable. And now, with all his energy, he labored with devotion and bribes, arm twisting and deal making, to save the remnants of royal authority. The Assembly sensed his perfidy—members couldn't help but notice the handsome new town house he lived in, the first-class chef who prepared his meals, the valet who dutifully laid out his bejeweled suits, or the shiny new carriage sporting the family arms—but it remained spellbound by his many talents. On January 4, 1791, he was chosen as the Assembly's president. Yet—and it was a big yet—he couldn't escape his own habits, no matter how self-defeating.

Buoyed by his new riches, he lived lavishly and recklessly. And prey to despondency and aimlessness, he worked all day, gorged himself on food and drink all night, and caroused endlessly with women, often two at a time. He also took fat bribes from Catherine the Great, who hoped to persuade France to join Russia in opposing the European alliance, especially England, arrayed against her; actually, the empress thought Mirabeau should hang from the gallows and *then* be "broken on the wheel." Yet having already been bribed generously by Louis XVI, Mirabeau simply consumed the tsarina's money as well. "I am a mad dog," he once boasted. It was hard to disagree. Then, on March 25, he consorted with two dancers from the opera; the next morning, he fell violently ill. With great effort, he attended the Assembly on the twenty-seventh, but afterward lay in his room, drained and shaking. Word of his sudden illness cascaded through Paris: A multitude of well-wishers thronged to his house inquiring about his condition, and operas closed in a sign of respect for him. No less than Talleyrand informed Mirabeau, "It is not easy to reach you; half of Paris is permanently outside your door." It was to little avail. As the days wore on, Mirabeau grew fainter and weaker. Pale-faced, he started shivering; his sweats became chronic. Then he began to cough up blood. On April 2, 1791, with brilliant spring sunshine pouring in from his courtyard, he announced to a colleague that he would like to have a shave, since "My friend, I will die today." It was morbid and melodramatic and vintage Mirabeau. But as was so often the case in his assessments of the revolution, Mirabeau proved to be prophetic. He was gone by day's end.

On April 3 grief swept through Paris, and then all of France, commencing eight days of public mourning. The Assembly declared that the refurbished Church of Ste.-Genevieve should be turned into a Pantheon for French heroes, its massive portico to read *"Aux grands hommes la Patrie reconnaissante"* (To great men a grateful fatherland). On April 4 Mirabeau was the first to be buried there. The historian Michelet thought it the most extravagant and excessive funeral procession that had "ever been in the world." He estimated that three to four hundred thousand people paid tribute, in the streets and the trees, on rooftops and on bridges, virtually all the Assembly, all the Jacobin club, and some twenty thousand National Guards as well. "One would have thought," it was remarked, "they were transferring the ashes of Voltaire—of one of those men who never die." But Mirabeau had died, and with it, the seeming fate of a king and queen.

The king and queen, of course, did not pay tribute at his death. But Mirabeau's sudden passing, along with what would be the slow but inexorable eclipse of Lafayette—by now, he could not even control the very national guards whom he ostensibly commanded—sent a terrible shudder through their bones. Mirabeau was the one revolutionary figure with the stature and sway who might have made a genuine constitutional monarchy possible. Just as likely, Mirabeau took with him the vanishing dream of a humane, peaceful, democratic republic. And that was not all. One of Mirabeau's last letters to La Marck added this terse, prophetic comment: The queen will not "be able to save her life without her crown." To this, he could have also added the king.

Now, in Paris, in the Assembly, among the sansculottes, every hope of the king's had died glimmering. He must have felt close to total despair.

∞

SUNK IN GLOOM, the king was clearly alone. But as a keen student of history—whatever his failings, this cannot be denied—he knew kings may command, but they may never whine, even when blundering, or luckless. And somehow they must move on. So it was even for Louis. If fate and luck had betrayed him thus far, that too could change. From Russia, Catherine the Great, bristling at the complacency of a fellow monarch, sternly wrote him, "Kings ought to go their own way without worrying about the cries of the people, as the moon goes on its course without being stopped by the cries of dogs." But by this stage, Louis, a virtual captive, was hard-pressed to ignore the baying. Perhaps it would have been different if he were willing to fire on his own people or precipitate civil war, as his younger brother, Artois, suggested from exile. But he was not. Still, all was not lost. The king was being cagey himself. Mirabeau might have been playing a double game, but so had he.

Half prisoner, half symbol of a nation, for the king, an almost ghostly kind of normality ruled over life at the Tuileries. Royal furniture was transported from Versailles; the queen had her favorite dressing table imported, and new pieces were commissioned to enliven the dreary rooms. The king continued to hunt, the queen worked at her tapestry with her ladies, and the royal family went to Mass in public, just as they had always done. They vacationed at Saint Cloud to avoid the summer heat, and on bright afternoons scampered about with the children. And from valets to coachmen and cooks to liveries, there were still more than 150 people who labored at the Tuileries court; the king even went as far as describing himself as "the head of the Revolution," which of course was pure fiction. But peel back the veneer of the royal facade, and inside, the royal family was disintegrating. As the first gray of morning spread in his chamber, the king would awake, tired and trembling. In her confinement, the queen often looked pale and melancholic; once, as she received the English ambassador, her eyes filled with tears. She later confessed that she had seen "too many horrors and too much blood ever really to be happy again."

So it was that the king finally began to think of taking flight.

∞

IN RETROSPECT, THAT Louis had accepted his fate so passively, with an insouciance and apparent inevitability, is almost impossible to fathom. The Sun King would have taken decisive action; Catherine the Great would have taken decisive action; and so, for that matter, would the queen's mother, Maria Theresa. It is hard not to chalk up Louis's hesitancy to fatal weakness, and in a leader, autocrat or democrat, monarch or republican, such impassiveness was unforgivable. But that day was changing.

On December 26, 1790, he signed the Civil Constitution of the Clergy, the bitterest pill yet that he had to swallow. More than ever, the pious Louis visibly suffered. His paralyzing depressions left him in a kind of a semi-stupor. While projecting a sphinxlike aura to the outside world, to his intimates, he seemed hurt, bewildered, marooned, and defeated. Playing backgammon with his sister, all he could do was grunt out the necessary words. That spring, his normally robust health began to wane: He ran high fevers and spat up blood; for a week, he was laid up in bed. The doctors prescribed debilitating emetics and purges, yet his state was so precarious that the National Assembly debated on March 22 whether a regency should take over. In time, Louis recovered, but not his spirits. He sighed to confidants: "How can I have these enemies when I have only ever desired the good of all." Actually, he had his own answer. "The world," he wrote caustically, "is a strange thing."

In coded messages, often in invisible ink, often in lemon juice, royal feelers were put out to courts of the foreign powers of Europe, to Austria, Russia, Prussia, and Sweden: Would they help save the royal family? As the queen noted in one secret communiqué, this was "the cause of Kings, not simply a matter of politics." But the great powers had their price: the King of Sardinia wanted Geneva in return, Spain insisted upon territorial limits of Navarre, the German feudal princes wanted to get their hands on greater privileges in Alsace. And even Austria, ruled by Marie Antoinette's own flesh and blood, was tied down, hampered by the Russo/Austro-Turkish war; this despite the fact that her elder brother Leopold was now emperor. The new emperor unhelpfully advised his sister and the king to wait. Marie Antoinette angrily retorted to the Spanish ambassador: "It is easy to advise prudence and temporizing, but not when the knife is at one's throat." By April 14 the queen was desperate. She wrote again to Vienna, with a heartfelt plea. Could they count on Austrian help: "*Yes or No?*"

The king himself had still not made a final determination. That soon came. Unwilling to take Communion at the hands of a revolutionary priest at the parish church of the Tuileries, he was determined instead "to perform" his Easter duties with a traditional priest. His hope was thus to leave Paris for Saint Cloud, where a refractory prelate could be slipped in; the departure was scheduled for April 18. But no sooner had he entered his coach at the Grand Carrousel than a hissing mob intercepted the king. The humiliation was completed by the fact that the National Guard—despite entreaties from both Lafayette as well as Mayor Bailly—openly blocked the king's passage. Outwardly calm, the king remarked at how strange it was that he who had "granted the nation liberty" was not allowed it himself. In a hushed tone, the queen gravely told Louis, "You must admit we are no longer free." That was, by then, understatement.

That night a mob gathered outside their windows shouting death threats, as though it were October 1789 all over again. To mollify the people, the king celebrated Easter at Saint Germain by taking the sacrament from a state priest. In his journal he dryly recorded: "They stopped us." But that day, he gave his long-withheld consent to the queen to prepare for a flight across the frontier for freedom. Henry IV had done something quite similar, and so had the young Louis XIV.

Now Louis XVI would do the same.

∞

THE DETAILS WERE left to the queen's devoted friend—history also records him as the queen's lover—Count Axel von Fersen, a Swedish nobleman, as well as Baron de Breteuil, one of the ancien régime's shrewder and more successful diplomats. But flee to where? It was decided that the king did not want to leave France itself, but that he should make his way to a "strong city" where, it was hoped, he could restore order and negotiate with the Assembly, and where the people would flock to him. One logical choice, the ancient Roman town of Metz—it was a thriving city when Paris had been just a fishing village—was ruled out as unreliable. The other possibility was the fortified town of Montmedy, near the border of the Hapsburg Empire. Here, General François Bouille, a cousin of Lafayette's and a loyalist to the crown, commanded the army of the Rhine, all the troops of northeastern France. Yet from the start, the whole venture was fraught with peril. The chosen route was a more than 180-mile trek from Paris, often through unforgiving terrain, thick forests alternating with wheat fields, and wooded hills that perilously sloped down into sunlit valleys, where any passerby's loyalty, guardsman or peasant, would be in question.

Slowly the preparations and details were put into place: Money was raised, loyal troops were secured, and plans were readied to discreetly spirit out other intimates of the court as well. The king's brother, the Comte de Provence, and his wife were both to flee separately, rendezvousing in Belgium, at Longwy. Nonetheless, danger of discovery remained. Count Mercy bewailed the perils of getting caught ("Was it," he asked, "really the time for such a bold venture?"). And from Austria, Emperor Leopold repeatedly advised them to beware; "Calculate well," he warned, "the risks." He was right, they were not inconsiderable: Letters to the king and queen were routinely seized, and there was a bevy of hostile spies among them, chambermaids, cooks, National Guardsmen, valets, merchants; the venture could blow up in their faces at any moment. Moreover, the queen herself, backed by the king, raised a serious obstacle. They insisted that the family not be separated, which increased the logistical problems; an escape involving several women and children could not be improvised. It would also require teams of fresh horses, great stealth, impeccable timing, faithful troops, and a considerable array of luck.

In May Louis took his plans a step further. He commissioned his former cabinet minister, Baron Breteuil, to draw up a detailed memorandum should his escape succeed. One of the more interesting questions is whether the king, in quasi-exile, was then prepared to accept a limited monarchy. Or was it true, as was later famously said, that the Bourbons had forgotten nothing and learned nothing? In other words, that he would have restored the old regime? Evidence suggests that at the moment of truth, the king would have ruled out any significant concessions of authority; moreover, Breteuil himself, from the safety of Switzerland, would also later negotiate with Europe's other monarchies to crush the revolution.

The first date proposed was June 6, but was foiled by the arrival of an antagonistic servant sympathetic to the Revolution. A second date was

rejected because it fell right at the Feast of Pentecost. Finally, the date settled on was Monday night, June 20, the second longest day of the year and
the anniversary of the Tennis Court Oath. Around midnight.

All was in readiness.

∞

THAT DAY, THE king and queen went about their routines as if everything
were normal. The king lounged in his study, while the queen joined her
children for their lessons. At midday the entire family heard Mass in the
château's chapel. In the afternoon the king and queen played their normal
game of billiards. And at five o'clock Marie Antoinette took her children
on a drive through the beautiful Tivoli gardens; upon return to the Tuileries, she instructed the National Guards to take them on a similar expedition the next day. Finally, that night, the momentous stakes of what they
were about to undertake struck them: After the customary family supper at
nine o'clock with the Comte de Provence and his wife, the king stood up,
and the four royals embraced tenderly. No one could guess the future; all
hoped they would soon be reunited by the Austrian border.

The king and queen were in bed by eleven. Complicating matters,
however, was the fact that both Mayor Bailly and Lafayette unexpectedly
attended the king's traditional *coucher*; Louis thus had to be careful not to
appear rushed. A "long hour" ensued. Still, at twenty past eleven, the heavy
curtains of the king's great bed were formally drawn by his two valets. Discreetly packed among Louis's things was a magnificent red uniform; upon
arrival at Montmedy, a grand spectacle was planned. He would inspect the
troops amid great rejoicing.

Sometime that evening, the king left behind a declaration on his desk,
dated June 20. In most respects an explicit repudiation of the revolution,
it outlined his reasons for leaving: that he could have departed earlier but
didn't want to provoke civil war; that he had taken up residence at the Tuileries as requested, but then was sidelined by the Assembly ("What remains
to the King except the empty sham of royalty"); the affronts he had to
endure, including being barred from taking Mass as he pleased and witnessing the terrible spectacle of his own National Guard siding with the mob.
The missive was signed as was customary, with the simple name "Louis,"
and it ended by telling the French people that the king would forever be
"your father, your best friend."

As midnight neared, the king, wearing a dark wig and a round hat, soon slipped out unseen to the side courtyard north of the Tuileries, where a carriage was waiting. The area spookily flickered with light from the torches held aloft by waiting valets; Count Fersen, sitting on the box in coachman's garb, was whistling and smoking tobacco. The dauphin and his sister were smuggled out as well. Only the queen was missing. Where was she? Unknown to the king, she was waiting for Lafayette to leave. Within fifteen minutes it was not the queen but Lafayette who appeared, in his well-lit carriage. Fear rose: Had the plan failed without having begun? Then, just as suddenly, peering through the dim light and a gathering mist, the king could make out a hazy person silhouetted against the courtyard. Marie Antoinette? A National Guardsman? It was impossible to tell. No, it was the queen after all, clad in a simple gray dress and large black hat with a long muslin veil that concealed her face. She was trembling, having almost been spotted by Lafayette. At this point they were already delayed, but only minimally. In an unusual moment of public affection, the king and queen embraced, with the king saying over and over, "How happy I am to see you."

Their dramatic dash for freedom began. Sometime between one-thirty and two in the morning, Fersen, as the coachman, reached a large, yellow-wheeled berlin waiting outside the city walls at Porte Saint-Martin. Louis and Marie Antoinette were disguised in plain clothes, the king playing a common steward, the queen a governess—both supposedly in the employ of a Russian aristocrat—and their children as her charges. Having successfully slipped out of Paris—in truth, this was considered the most difficult part of the mission—they were to rendezvous with new horses and forty well-trained dragoons at an inconspicuous relay *poste* in Pont de Somme-Vesle, between two-thirty and four-thirty in the afternoon. There the Duc de Choiseul would meet them. Choiseul had fervor and loyalty—he would have laid down his life for the king—but he was a questionable man for such a delicate job; arguably he lacked the training or prowess needed for the mission.

Meanwhile, Baron Breteuil climbed into his own carriage and set off for Montmedy to await the liberated king in triumph; at the same time, the king's brother and sister-in-law, with one attendant each, were accomplishing their individual escapes. They would be reunited at Namur in Belgium. It was hoped the king and queen would have comparable success. And indeed, around two o'clock, as the carriage bumped and jostled through the countryside, the situation seemed secure. The king himself was finally able to relax a bit. In fact, he was in such good spirits that he joked, "Lafayette is now in real trouble."

The king was sorely mistaken. A horse stumbled, causing a break in the harness and a precious delay of two hours. Unknown to them, the Duc de Choiseul, after having stayed beyond the time limit, feared that the whole mission had been aborted. Unsure what to do next, and facing angry locals who thought his troops were part of a tax collection scheme, he fled. The king and queen had thus lost their military escort. Eventually arriving at Somme-Vesle, the royal party was aghast. No Choiseul? No dragoons? Staring out his window, even the normally placid king could not mask the dread he felt, as if "the ground had opened up beneath him."

But having no alternative, the berlin rolled on.

∞

Back in Paris, the morning of June 21 had long since dawned. As the sun streamed in, the royal curtains to the king's bedroom were drawn back at seven o'clock. The bed was empty. Anxious valets scurried to the dauphin's room. He too was gone. And the queen as well. Rumors flew. Word of the slow, bleak truth spread throughout the city. By eleven o'clock a massive, infuriated crowd was swarming outside the doors of the Tuileries, screeching insults, waving sticks, and shouting, "They've gone! They've gone!" Violence flared; inside, the king's portrait was torn to shreds. Lafayette breathlessly rushed to see his friend Thomas Paine. He asked, "What to do?" and exclaimed, "The birds are flown!"

"Let them go," Paine blandly replied.

But the Assembly thought differently. That morning it made a fateful decision: to assume all executive power and place the king's ministers under its jurisdiction.

Then an arrest warrant was sent out across the kingdom. A furious manhunt for the king now commenced.

∞

On the surface, the king's delay was hardly fatal. True, it cost precious ticks of the clock. But there was little reason to believe that Louis wouldn't make it to Montmedy. Remarkably, however, even now the convoy deferred entirely to the established habits of the royal family: They had paused for the king to stretch his legs; for breakfast at eight; for the children to take a breath of fresh air; even to chat with locals at Chaintrix along the way. There were other thoughtless mistakes as well; at one point they failed to lower the blinds, meaning they might be noticed. And later the fugitives even dozed off. But for all this, there was an almost odd serenity to the trip: in semi-darkness, they had passed ancient garrisoned towns, with their magnificent Renaissance and baroque architecture; in fresh daylight, they rolled alongside small feudal villages whose cottages huddled behind the protective walls of their overlord's castle; they drew beside the tree-lined banks of the Lorraine, not far from Domremy, the little town where Joan of Arc was born in 1412 and not far from the Church of St. Pierre-aux-Nonnains, believed to be the oldest in all of Europe; and they edged near the Vosges mountains, lined with fir-covered slopes, and where the renowned town of Baccarat made some of the finest crystal in the world. This was the so-called Middle Kingdom, a fascinating amalgam of French and German influences, the region where the Treaty of Verdun in 843 had split Charlemagne's empire, and where a millennium earlier barbarian tribes had swept westward in their mighty conquests.

And here, eighteen hours later, at Sainte-Menehould, it appeared that the king had gotten a break. One hundred forty dragoons were waiting for him. Yet again, however, his luck was tainted. Word of the king's escape from Paris had already reached this sleepy village. National Guards were summoned, and a detachment of troops loyal to the Assembly appeared. The dragoons' colonel rapidly directed most of his men to a neighboring village, but they were soon surrounded by hostile forces. Suddenly the royal peril grew. Once in Sainte-Menehould, the town's postmaster general, Jean-Baptiste Drouet, got a fleeting glance at the passenger, a heavy-set man. His face bore a remarkable likeness to the image emblazoned on one of the government's assignats, that of Louis XVI, the king. Hearing this, the municipal council gave the order to chase down the speeding black and green berlin. Within an hour and a half, Drouet had launched a furious pursuit of the king.

THE KING'S PARTY, meanwhile, had reached the ancient town of Clermont, rising 100 meters from the plains and overlooking the Aire valley. There was no time to waste, this they at last realized. It was, in fact, worse than

Louis believed. The original plan was that the berlin should turn sharply north on a secluded road through wooded hills. Unfortunately, the new quartermaster leading the party missed his turn, ambling on for some time toward Verdun. It was at this point the decision was made: to take the road to Varennes.

The name itself is legendary; it wasn't always. Varennes, near the frontier of the Austrian Netherlands (today's Belgium), was about to be immortalized, actually, twice. This little town would be famed in subsequent history, not for its pottery or churches or gorgeous stained glass, but as the scene of some of the greatest wartime carnage known to man—we think of it today as the Argonne Forest, a tangled path of land bearing silent witness to some of World War I's bitterest savagery, where gains were measured in yards and inches between blood-soaked trenches. But that came much later. Otherwise undistinguished, including by the lights of 1791, Varennes, a small town of about one hundred inhabitants, would within hours be known as a momentous turning point of the revolution as well. It began when the royal party lumbered into the town at around eleven o'clock at night. Varennes had, as it does today, a main street that descended a steep hill to a bridge over the Aire River, and another section of the town, including a castle, on the far side. And it was here the culmination of the king's flight was about to be played out.

The king still had a half hour lead over Drouet, and stood a mere half hour from liberty. By now, however, there was a dire need for fresh horses. Yet incredulously, no one in the royal party had the slightest notion where new horses could be secured. As a result, the king's companions, and also, remarkably, the king himself, were left to grope in the darkness, pounding on doors. It was a terrible blunder. A barrier was soon erected at the top of the town. And Drouet, arriving in a mad dash, informed the procurator of the local commune, Monsieur Sauce, a grocer, "Your passenger is the King!"

The king? The town's fire bells began to sound; local National Guards were summoned. The king's berlin quickly set off to escape, but as it entered a vaulted passage, Louis watched in dumb horror as a dozen soldiers unthinkably appeared with bayonets fixed and rifles crisscrossed. The king was fatefully detained; incredibly, a grocer had helped capture his most Christian majesty, King of All France and of Navarre. For the moment, the pretext was the "irregularity" of their passports; Louis was thus "persuaded" to accept the "hospitality" of Sauce's house until morning. Of course, without troops of his own, he had little choice. A crowd began to gather outside the windows, armed with pitchforks and muskets and clubs, while inside, the royal family settled in the back room of the Sauce grocery to await their fate. The next few hours were by turns terrifying, confused, and helpless.

But once more, there was a chance. At one in the morning, an ear-splitting shout filled the evening air: "The HUSSARS!" The Hussars were Choiseul's men. In a chance act of fortune, Choiseul had been alerted by the glaring lights of Varennes and the ringing tocsin; he made haste for the town, frantically reaching the king himself with his dragoons. It was still possible, even now, if the king ordered it, for the troops to force the liberty of the royal family. Actually, the choice should have been all the easier: For at the very moment, the Marquis de Bouille, who had already assembled a large force to support the king at Montmedy, was also sprinting toward Varennes to rescue the king. He would be able to arrive by seven-thirty the next morning. For all his bad luck, Louis still might make it.

But the king, overwhelmed, declined to command his troops. His face turned white. For one brief terrible moment, no one spoke, and those in his party could barely breathe. Was this incorrigible stupidity? Or weakness? In truth, it was the march on Versailles all over again, with Louis clinging to the familiar role of the father-king and recoiling from any use of force. Instead, he waded into the crowd himself and attempted to pacify them, responding with a curious mix of candor and obfuscation. He flatly announced that he had no intention of leaving France, and, moreover, that he would return to Varennes after establishing himself at Montmedy.

"And what if your foot slipped over the frontier?" a member of the crowd shouted menacingly.

Meanwhile, Choiseul was still hoping to take the king to freedom. As the throng turned wrathful, one of the loyal soldiers, in a tone that ignored the danger around them, boomed out to the king: "Sire, we await your orders."

None was forthcoming. "What should we do?" the king asked pathetically. "I have forty hussars," Choiseul firmly told the king. "We will try to get through, but we must not lose a minute, for in an hour's time my hussars will be won over." Again the king faltered. Seeing himself as a prisoner, he was also unwilling to risk his family's lives should a skirmish break out. ("Can you guarantee," he asked Choiseul, "that in this unequal combat a musket shot might not kill the queen, or my daughter, or my son?") Instead, Louis's singular determination, or rather, his hope, was that the

Marquis de Bouille would arrive in time with his decisive force of men. If the king could hold out until morning, he could still escape.

So it was now a sprint—between General Bouille and the Assembly.

∞

AT SAUCE'S, AN exhausted king and queen nervously waited. By daybreak, emissaries appeared—from the National Assembly. Antoine Barnave and Jerome Petion, not Bouille, arrived in Varennes first, bearing orders for the immediate return of the king, "unharmed," to Paris. Finally the king summoned his wits to fight back. He stalled: The queen was exhausted, he insisted, and had to rest; the children too. One of his waiting women also feigned illness to bide respite. Breakfast was served; the king picked briefly at his food, then pretended to nod off. Barnave, however, was not so easily gulled.

At last the unspeakable had happened—Louis, the fugitive king, was now captured. It was the most incongruous of scenes: As they left Varennes, lost in a human sea of a howling mob and navigating tenuously through thick clouds of dust and trampling feet, the blazing sun rose over the village. When the carriage was only a blur on a distant horizon, Bouille arrived, and the king was gone. And with it, any chance to escape. It would soon cost him his throne and his life.

∞

THE DUC DE Choiseul, who would rapidly be arrested, never forgot his feeling of "inexpressible anguish," as if he were seeing King Charles I handed over to his executioners. He was not far off the mark. Shortly after the party left Varennes, a local landowner, the Comte de Dampierre, approached the carriage to pay his respects to the king; he was hacked to death in plain view of the royal family.

For nearly four days, the captives were leisurely driven back to Paris, led by sixty thousand of the National Guard; the last stages of their return, through suffocating heat, were dust-stained and bedraggled and pensive. On the way, Barnave sat in the royal coach between the king and the queen; twenty-eight years old, Barnave was spirited, had undeniable good looks, and was a prominent member of the Assembly; a brilliant orator, he had been among the guests at Jefferson's illustrious dinner, now almost two long years ago. But Barnave had also been steeped in the waning chivalry of the ancien régime and, despite his revolutionary ardor, felt the tug of royalty-in-distress before him. While arguing politics with the king's sister en route, he silently considered what fate might await the queen and the children she so carefully cradled. And he was struck by how the royal family was in no way like the spurious gossip being peddled by the journals: The queen lovingly bounced the dauphin on her knee, and Louis and Marie Antoinette were far removed from the haughty caricatures they had been made out to be. By the time they reached Paris, Barnave was the queen's servant—a new Mirabeau. In turn, she was taken by his "animating and captivating eloquence."

At the barriers of Paris, there was yet another vast crowd awaiting; they had been lingering there for hours. It was an unsettling sight; the same for the eerie, unnerving stillness in the streets. A chastened Lafayette had ordered that the usual sign of welcome for the King be suspended; at the same time, an order was clearly posted, "Whoever applauds the King will be flogged; whoever insults him will be hanged." So it was that a strange, almost macabre quiet reigned as this creeping procession of despondent and slovenly prisoners passed through a long line of National Guardsmen standing at attention, arms reversed. There were no calls, no cheers, even as tens of thousands of people were ready to explode at the wave of a hand, until the royal party reached the Tuileries. Meanwhile, in most parts of the city there was an unsettling feeling of expectation, and fear. From the windows of his apartment, the American ambassador, Gouverneur Morris, tensely wrote: "Things now change hourly in the city." He was correct. The flight to Varennes was a profound turning point. Until then, the king had still clung to a last shred of power. By the time he was returned to Paris, power had permanently passed to the Assembly—and to the people.

Yes, Louis had kept his throne. But how much longer could it remain his? The realization slowly dawned: With the flick of a finger, that too could be dispatched.

∞

FOR NOW, THE Assembly deflected the heated calls of the sansculottes for the immediate overthrow of the duplicitous "traitor-King." Ironically, many of the deputies still remained constitutional monarchists at heart. So the official story—the fiction really—became that the king had been abducted by counterrevolutionaries. Nevertheless, the king was suspended by the

Assembly, an unprecedented event in the history of the French monarchy. But this was not enough for the radical leaders: They hotly protested that the king be immediately deposed or tried. And in the streets and the coffee houses, the salons and the clubs, for the first time patriots spoke openly of establishing a republic, especially those who had been to the United States.

Nor was it just talk. The radical clubs and journals now called for the people to again assemble on the Field of Mars; fearing chaos, Mayor Bailly hastily declared martial law. But on July 17, 1791, two days after the king's supposed "acquittal," fifty thousand arrived, while six thousand affixed their names to a petition calling for the king's trial. Violence broke out—two innocent people were apparently murdered by the crowd—and Lafayette ordered the National Guard to scatter the rebels. The rebels angrily refused, answering back with a hail of stones. Irate and feeling betrayed, the soldiers returned fire, thus bringing the pledge of "peace for the world" made just one year before to an ignominious end. Something had unalterably changed in Paris: Jacobin leaders suddenly became fugitives on the run, and the moderates—Bailly, Barnave, and Lafayette—had just as suddenly become enemies of the Revolution. Actually, for Lafayette, it was doubly humiliating: The general and hero of the American Revolution, the man who had fervently sought to graft an American utopia onto his kingdom and country, now contemplated his own escape from the mounting chaos of France.

Meanwhile, a thoroughly chastened Louis XVI appeared before the Assembly on September 13, 1791, and formally endorsed the new constitution. He no longer even merited a throne, only a plain armchair covered with fleurs-de-lis. As he read the oath, he realized the deputies were sitting and still wearing their hats; never had such a mortifying spectacle occurred before a French king. Louis XVI's eyes filled; his voice quivered. Upon returning to his queen at their desolate palace, the humiliated king broke down and wept, and asked her forgiveness for the terrible tribulations of their incarceration. And the queen? She had long since pledged, "my duty is to die at the King's side," and continued to preach, "Courage, courage." But her last portrait tells a far more accurate story. Her gaze is sad, she has grown thin, and her hair has turned white from her endless sufferings. By many lights, she was unrecognizable.

Now there was no escape, no freedom, no rescue at hand. Only a helpless king and queen, frightened and utterly alone. Inexplicably, to the king at least, he who had boldly saved America, he who had summoned the

Estates-General, he who had refused to precipitate civil war, both the man and the crown were now despised, forgotten, scorned. The Tuileries had been transformed from a palace into a fortified camp and, more precisely, his prison: People had to be patted down before entering, spies were ubiquitous and National Guardsmen camped in tents next to the château, and in the evenings, all the heavy doors were slammed shut and double-locked.

Wrote Marie Antoinette: "We are under surveillance, day and night."

∞

FOR THE ASSEMBLY, it was a race between holding together its fragile facade of unanimity and finishing its last measures. At this point, the battle was less between nobles and peasants, or even the king and his subjects, than a family feud about the future of the French nation. Yet after two years, the deputies were spent: They had overseen the dismemberment of the feudal system; they had done away with hereditary privileges; and they had elevated the people above the whims of unfeeling despots and pampered aristocrats. More than that, they had given the people laws: equality under the law, freedom of expression before the law, and an end to arbitrary imprisonment. They had streamlined the once-haphazard provincial administration. They had humbled the once-arrogant Church by seizing its inordinate wealth and guaranteeing freedom of worship. And they had codified these changes in a great new constitution, one embraced by the people and belatedly accepted by the king, all as a glorious step toward everlasting prosperity and a renewed, unshakable French harmony.

There remained one last act: Having finished its work, as, in effect, a constitutional convention, on September 30, 1791, "the most memorable of all political assemblies" adjourned itself forever.

∞

IN ST. PETERSBURG, for Catherine the Great, tired and almost sixty, watching these events, it was the same bitter story. Varennes was the Bastille redux—in spades. It was everything she despised: a weak monarch, a rabble out of control, rule not by reign or reason but by untutored barbarians. And the accompanying international madness, ready to stampede across Europe, was far from over.

All her life, she had lived by instincts, hunches, and flashes. She always had a hard, almost medieval capacity for witnessing senseless brutality with a clear, untroubled gaze. While her beloved Potemkin might toss at night in fitful sleep, dreaming of corpses at Ochakov, of Ottoman guns flashing in the night and his men shivering in the cold, of silent troops lying exhausted and not knowing what butchery might come next, she was fascinated or delighted at the prospects of victory. And where other monarchs felt desperate, she felt challenged. Of France especially, she longed for a Caesar to once again subjugate Gaul. "When will this Caesar come?" she rumbled. "Oh! Come he will, make *no* doubt of it!"

But as the winter passed, in the midst of her own troubles—on the Swedish front, the Ottoman front, the mounting unrest in Poland (Polish patriots, now encouraged by the French Revolution as well as the American, were demanding ever more heatedly that Russia withdraw its troops), the madness in France—had come a personal disappointment that cut her deeply. Louis-Philippe, Comte de Ségur, France's emissary, her friend and confidant, had left Russia to become an ally of the Revolution. Ségur was everything she relished in a confidant: charming, clever, and worldly. The son of a French war minister, he had gallantly fought beside Lafayette in the American war and had met George Washington; he was friends with Marie Antoinette, Diderot, and D'Alembert; he had traveled with the empress to the Crimea as one of her treasured "pocket ministers"; and he was an intimate member of Potemkin's circle as well. Once the prince audaciously received the Frenchman only in a fur wrap; Ségur subsequently invited Potemkin to dinner and impishly greeted him in the same garb. Later Ségur announced to Potemkin: "I am devoted to you for life." He had the wit befitting the grandeur of her court ("the first poet in France," Catherine remarked of him, "is unquestionably Ségur"), and she found him enchanting to boot. In the Crimea, they had debated America at length, they had discussed the Puritan colonists, and they had even talked about his ride across the virgin lands of Venezuela.

But soon the French contagion overshadowed the continent and beyond: In Ireland, Britain, America, the Hapsburg Empire, even in the corners of the Ottoman Empire, even in Russia, even around her own court, even with Ségur. As it turned out, Ségur too had been swept up by the events upending France, recording with enthusiasm the spirited passion of the merchants, the tradesmen, the "bourgeoisie," and also some of the young gentry, not to mention the Frenchmen, Danes, Germans, Englishmen, Hollanders, and, yes, Russians, all rushing breathlessly into the streets of St. Petersburg and wildly embracing one another, "as if," he recorded, "they had been delivered from a heavy chain that weighed upon them."

With some sadness, Catherine had granted Ségur a farewell audience; she knew he desperately wanted to return to his homeland and see the blessings of liberty, and that he wanted to see them with his own eyes. She would not stand in his way.

As they parted, Catherine clearly seemed to be pained. Her eyes sad, she mustered a warning and a heartfelt good-bye. "Tell the King how much I wish him happiness," she insisted, adding, "I am sorry to see you go: you would do better to stay with me and not go looking for storms, the full extent of which you may not foresee. Your fondness for the new philosophy and for liberty will probably incline you to support the popular cause."

She continued: "As for me, I shall remain an aristocrat. That is my calling."

But it was the next exchange that was the most remarkable. Catherine, ironically echoing George Washington's admonitions to Lafayette, then proclaimed that Ségur was going to find France "very feverish and ill." Ségur bowed his head and responded, "I do fear it, Madame, but that is why it is my duty to return." They parted with affection and mutual esteem. Yet by 1791, when Ségur had fallen fully in with the Revolution, Catherine, with the wrath of a monarch betrayed, would coldly dismiss him. "Fie!" she sputtered. "He is as false as Judas . . . With some people he poses as a democrat, with others as an aristocrat." And also: "We saw the *Comte de Ségur* arrive here . . . Now *Louis Ségur* has been stricken with the national disease."

So it was at the dawn of a new year, as the French rebellion continued with no foreseeable end. Ségur had been replaced by his young aide, the twenty-six-year-old Edmond Charles Genet, a name destined for notoriety and infamy: notoriety because, as a courtier and diplomat of the old regime, he would have the rare distinction of being part of the great revolutions in three worlds: Russia, France, and the United States; infamy because he would one day arouse the ire of George Washington himself and be a candidate for the Terror's blade as well. By most criteria Genet was impressive, and for the empress, just the right antidote to Ségur's departure. He had blue-blooded manners and high-minded tastes. And his patriotism was unqualified and absolute. Indeed, Genet grew up at Versailles itself; two of his four sisters even became ladies-in-waiting to the queen, and Henriette, in particular, would become Marie Antoinette's loyal friend, Madame de Campan. His résumé was also striking: A career bureaucrat, he was a master of six languages (by the age of fifteen), an acquaintance of James Watt and Joseph Priestley, the discoverers of the steam engine and of oxygen, respectively, and an inventor in his own right. He was also a rising star in Louis's ill-fated monarchy: He had already served in Berlin, Vienna, England, and even knew the early American representatives in France—Benjamin Franklin and John Adams. There was thus little reason for Catherine not to consider Genet as an ally of royalty and a friend of the court of St. Petersburg.

He was. Initially, hearing the news of the turmoil in France, Genet, like the empress, was shocked. Recoiling in horror, he applauded Catherine's security measures ("prudent measures") to wall off, in the empress's words, the "poison" emanating from France. He feared that if infected, the Russian peasants, "who are all slaves," would "throw off their chains" and "massacre" the nobles, and that the flourishing empire would be plunged back into "the most frightful barbarism." And he similarly reeled in disgust at the many French nationals in St. Petersburg already contaminated with, as he put it, the "craze . . . to pull down the government." That was back in 1789.

Yet in time, Genet, a royalist by birth, a monarchist by temperament, a diplomat by profession, increasingly became rhapsodic about the Revolution. By 1791 he talked glowingly of the "nation." More than that, he spoke movingly of its "sovereignty" and of majority rule as a "law of nature." He even praised "democratic ideas" and saw what Catherine did not; that Russia itself was boiling. While Catherine fought her wars abroad and silenced her dissidents at home, he began to see immutable currents of discontent, social as well as political, that were corroding the great empire's structure: the rumblings of millions of downtrodden serfs, the hidden daggers of praetorian guards, the machinations of courtiers, royal princes, and scheming nobility, the barbarous independence of Cossacks and Tatars, and the early stirrings among the tsarina's non-Russian subjects as well. Suddenly, in the empire of the tsars, Genet glimpsed, or so he thought, the "real germ of democracy" and, at the very least, a profound thirst for liberty.

And perhaps most fatefully, fatefully because here his predictions were eerily prescient, he eventually began to reflect on a "popular" rebellion in Russia, perhaps similar to those in America or France.

"All would be annihilated," he predicted of the land of the tsars. "All would perish by fire and sword."

With their transforming ideas on government and liberty, the eminent philosophes Voltaire, Montesquieu, and Rousseau helped pave the way for a new age, inspiring not just the American and French revolutionaries, but even Russia's Catherine the Great.

George Washington's two favorite sons, Thomas Jefferson, the secretary of state, and Alexander Hamilton, the secretary of the treasury, fought bitterly over the direction of the new American nation, particularly about the French Revolution, which deeply divided the young republic.

Washington's first cabinet was the most impressive in America's history, and included such distinguished figures as Edmund Randolph, Alexander Hamilton, Henry Knox, and Thomas Jefferson.

King Louis XVI played an indispensable role in supporting the American rebels, but soon found himself unable to control the revolutionary tide that swept France and stunned the world. It would cost him his throne and his life.

More reactionary than her husband, Marie Antoinette once confessed, "My fate is to bring bad luck." But this onetime queen of pleasure would suffer immeasurable indignity and loss on her own often poignant journey to the scaffold.

When the Bastille fell, it was a sound heard around the world: George Washington was sent the key to the fortress, while upon hearing the news, Russians danced in the streets. Louis XVI asked, "Is this a revolt?" and was told, "No, Sire, it is a revolution."

The conquerors of the Bastille paraded the heads of their victims on pikes, including the fortress's commander, presaging the ghastly Terror.

Moved by the horrors of the slave trade, Benjamin Franklin, in his last dying act, petitioned the new congress to end slavery, but James Madison, "the father of the U.S. Constitution," and his fellow legislators allowed the plea to die from benign neglect. Seen below are the appalling conditions of a tightly packed slave ship.

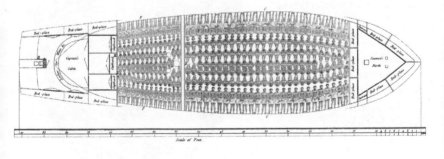

Russian empress Catherine the Great helped midwife the American rebels to independence, but after subjugating the Muslim Crimea, waged a holy war against the Ottoman Empire.

It was the inimitable Prince Grigory Potemkin, the empress's lover and closest adviser, who personally helped mastermind her drive into the south and commanded her bloody war against the Turks.

The vast, sprawling Ottoman Empire was ruled by the sultans from Constantinople, today's Istanbul. Despite Louis XVI's injunction to his Muslim ally that "war has become a difficult science," the Ottomans bitterly fought back against the Russians. Here the grand vizier in Constantinople is seen giving an audience to French emissaries.

Heroes of the American Revolutionary War fanned out across the globe, to spread the ideals of liberty. Ironically, John Paul Jones, America's great naval warrior, joined Empress Catherine's navy in its struggle to subjugate the Ottomans.

France's envoy to St. Petersburg, the Comte de Ségur, presciently warned Catherine that the "winds of freedom" were now blowing everywhere.

Another revolutionary veteran, Thaddeus Kosciuszko, led a fierce revolt against Catherine in his native Poland.

Jean-Paul Marat, the most extreme of France's ruling triumvirate, incited Parisian mobs to mass murder with his inflammatory editorials.

Georges-Jacques Danton rallied a fearful Paris against the foreign coalition seeking to crush the Revolution, and later pronounced, "Terror should be the order of the day."

The fastidious radical Jacobin Maximillien Robespierre came to embody the bloody excesses of the Terror. Still, an ocean away, the participants in America's Whiskey Rebellion would enthusiastically toast him.

On June 20, 1792, a French revolutionary mob stormed the Tuileries, forcing a perilous showdown with the king. Seven months later, he would be sent to the guillotine.

Louis XVI's ghastly beheading sent shudders through the capitals of Europe and Russia, and would intensify the resolve of its monarchs to destroy the Revolution.

Though Thomas Jefferson gloried in the French Revolution, George Washington was appalled at the chilling death of a king who had been his staunchest ally. He icily received revolutionary France's new envoy, Citizen Genet, who audaciously sought to foment insurrection on America's borders.

While the French revolutionary juggernaut was engulfing western Europe, Empress Catherine, seen here flanked by the Continent's monarchs, cynically led the dismemberment of the ancient kingdom of Poland. Below, Russia's dreaded Field Marshal Alexander Suvorov is seen eating the severed head of a French soldier, which referred to the Russian defeat of the French in Italy.

General Swallow all a Feasting on a French Fricassee!!

As the Terror mounted, these three watched the dream of an American-style revolution inexorably fade in France. Thomas Paine joined the revolutionary government, but was then slated for the guillotine.

Washington's beloved protégé, the Marquis de Lafayette, fled France in fear for his life and ended up in an Austrian prison.

America's envoy in Paris, Gouverneur Morris, reported back to George Washington, "At the end of every vista, you see nothing but gallows."

As Napoleon's armies devoured Europe "leaf by leaf," America's second president, John Adams, feared the young republic would be next. No less than Russia, he brazenly stood up to the French Directory in Paris while weathering mounting dissension from Jefferson's Republicans at home.

PART III

TERROR

Twin Guillotines

CHAPTER 7

FRANCE

✧

THE ONCE FUGITIVE KING was now a prisoner in all but name at the Tuileries. Even the servants rustling in the hallways were overwhelmingly spies for the revolution. Louis and Marie Antoinette had to calculate their words, or not speak them at all, while under the strains of their confinement, the king's sister, Princess Elisabeth, icily complained about the queen's meddling and wanton indiscretion. For her part, Marie Antoinette had no use for Elisabeth's advice. In their brief periods alone, Louis XVI's two women, his wife and his sister, bickered and seethed, while the king largely drifted along. Tersely, Elisabeth wrote her brothers who had fled abroad, "It is hell at home."

Years later, Louis XVI's actions, at the brink of disaster, have often been cited as evidence of incorrigible stupidity and weakness. It is hard not to reach a similar conclusion. Messy times often favor messy solutions; this he never understood. So he continued to eat large meals, as though the walls weren't collapsing around him; he continued to nap, as though he had scarcely a care in the world; and he continued to conduct his *lever* and *coucher*, and to attend dinners, concerts, and brilliant displays of fireworks, as though he weren't a captive in his own capital or his powers hadn't been painfully stripped from him, when of course they had. It was the oddest of sights: his bizarre passivity, his seeming naïveté, his reluctance to fight back or to submit outright, and a sphinxlike quality that could lead him to declare war, for example, as though he were blandly ordering his meal's next course. While Marie Antoinette would collapse in a chair, sobbing aloud, Louis could stand there impassively, staring at the ground. And any king with so little will as to sit blithely playing backgammon while his nation revolts would seem to deserve nothing: not his monarchy, nor compassion.

But there was more to it than that. In the king's hermetically sealed world, the only world he could envision, he was still surrounded by those whose hearts burst with affection for him. And there remained the sea of closet royalists: in Paris, in the countryside, in the foreign courts of Europe, who desperately wanted to save the monarchy. And he had powers still: Even the Assembly proclaimed that the nation could go to war only if the king asked it to do so. The king also was considered to have immunity for actions he took as monarch. And he retained the right of veto. While such small vindications did little to exorcise the demons and dangers that rode Louis all throughout the summer and fall, they continued to cast their own spell. And thus, remarkably, with blinkered eyes, he could not see the chasm beneath—not yet at least. Winston Churchill would one day write about Tsar Nicholas in Russia, another deposed emperor: "He had made many mistakes, what ruler has not? He was neither a great captain nor a great prince. He was only a true, simple man of average ability, of merciful disposition, upheld in all his daily life by his faith in God . . . At the summit where all problems are reduced to Yea or Nay, where events transcend the faculties of man and where all is inscrutable, he had to give the answers. His was the function of the compass needle . . . Advance or retreat? Right or left? Democratize or hold firm? Quit or persevere?"—but Churchill could have just as well been speaking of Louis XVI, for those remained the battlefields awaiting Louis as well.

And behind closed doors, the monarchs of Europe plotted. "The foreign powers," Marie Antoinette wrote privately in August, "are the only ones that can save us." Soon the country—and a continent—would be lurching into war precisely for this reason.

∞

BUT BEFORE THERE was war, there was peace—and the crafting of the new nation. It began with the elections for the second revolutionary congress, which were held under the sway of the now seemingly ubiquitous journalists and the equally ubiquitous political clubs. With censorship and restraint a thing of the past, leading journalists rushed their ideas into print, and their words quickly became an engine driving the revolution. In 1790 alone, Paris had 133 journals—nearly all were radical—while there were hundreds more in the provinces. Ironically, it was Mirabeau who had warned the king that the surest way to preserve his throne would be to buy off some influential journalists. Napoleon himself would seize upon the same sentiment: "The advent of cannon killed the feudal system. Ink," he solemnly predicted, "will kill the modern system."

It did, but not alone. For then there were the clubs, the other new fulcrum of command. One of the earliest was the Breton Club, which gathered in the refectory of an old Jacobin monastery by the Tuileries. Later, as its membership grew, along with its fervor, it spilled over into the library and then the chapel. It was here where a former Assembly deputy named Robespierre found another megaphone for his ideas. And before long, the club would be known by another name, one fated for immortality. History records it as the Jacobins.

The Jacobins quickly became a force of nature. In 1791 an Englishman in Paris reported that their "clubs abound in every street." They did. By 1794 there were some 6,800 Jacobin clubs alone, with some 550,000 members, creating a formidable organized underground, zealous, hardwilled, hardhearted, and unbending. When the journals took up their policies, which was often, their power was magnified that much further. And as the prestige of both the club and the press expanded, as they soon did, the influence of the Assembly declined, and they, like the king before them, were forced to bend to the Jacobin will or face an ungovernable mob—or outright insurgency.

But as democracy flowered, the Jacobins were not alone: In this sense, this was not a tale of one France, but of many. As early as 1790, some radical leaders, finding the Jacobin dues too expensive, formed the "Society of the Friends of Man and the Citizen," which Parisians soon dubbed the Cordeliers Club; it met in the former monastery of the Cordelier Friars. Here a host of revolutionary names destined for notoriety and infamy alike found their platform: Marat and Hebert, Desmoulins and Danton. Then there were the moderates. Believing the Jacobins too extreme, men like Lafayette, Bailly, and Talleyrand formed the "Society of 1789" to support the tottering monarchy; they met in the Palais-Royal. Another group with monarchical leanings, led by the queen's new favorite, Antoine Barnave, formed a club fleetingly referred to as the Feuillants, which met in an old Cistercian monks' convent. It was, of course, a fitting metaphor of the times—the tug of war with the Church, the altered mind-set of the day—that failed monasteries were given new life as homes to political provocateurs. And it was perhaps equally fitting that they would lead France in peace, then rebellion, then war, then terror.

The varied complexions of the clubs were amply represented in the election of 1791, which sent to the new Assembly a wide swath of the nation, including a significant minority committed to keeping the king. These 264 Feuillants arranged themselves in the right section of the Assembly (helping bestow an appellation on conservatives to this day). The 136 deputies from the Jacobins or Cordeliers sat on the left in the highest seats of the hall, called the Mountain. In the center sat 355 delegates who were not yet labeled or refused to be labeled; they were called the Plain. Virtually all were middle-class, most were lawyers, and many had already held office under the new regime as mayors or magistrates. Thus the revolution stubbornly remained in the hands of the bourgeois.

But in the new Assembly, the most vocal bloc was the Girondins. Nearly all were members of the Jacobin Club, and virtually all joined the Jacobins in deploring the monarchy and resisting the church. Where they parted ways, the difference was less in philosophy than geography; they sprang from outlying regions of the Gironde known for their bustling commercial or industrial activity. As such—and this was not a new story—they distrusted having all France ruled by Paris and its officious, restless populace. The Girondins met regularly, weaving threads of rebellion, entrapment, and enlightenment, and life and death: Benjamin Franklin's friend and a noted mathematician, the Marquis de Condorcet, was their theorist and philosopher; Pierre Vergniaud, a lawyer from Bordeaux, was their president; and their stentorian voice was Jacques-Pierre Brissot. Brissot was a special case. The son of an innkeeper, he was a modern-day adventurer who had traversed Europe and even visited Philadelphia; he was, too, imprisoned briefly in the Bastille, and in 1788 was a founder of the Société des Amis des Noirs, working for the emancipation of slaves. An accomplished storyteller and witty performer, he was also a natural leader to captain the people of France to a grand utopia. At the same time, his personal appeal was palpable; like a magnet, he drew all matter of filaments into his sphere. Part street-savvy agitator, part resourceful politician, Brissot soon took charge of the nation's foreign policy, fatefully coming to believe that war was essential to saving the revolution.

The Girondins' spark also fascinated such prominent intellectuals as

America's Thomas Paine and the German nobleman Anacharsis Cloots, who had once begged the National Assembly for citizenship as the "representative of the human race." And in their salons, the headquarters of rebellion, the Girondins stoked their ideas and embellished their passions, until the day when many of these leading lights would nobly be carted off to the scaffold.

∞

THE REVOLUTION WAS reaching a critical stage. For nearly three years, Europe's governments had largely sat silent, that is, while they could. But silence was increasingly no longer an option—and not simply because of the queen's secret appeals to the monarchs of the continent. The tenets of the French Revolution, like those in America, were too easily understood and even more easily exported. Trumpeting the rights of man irrespective of status or class, nationality or race, they rang from country to country, crossing borders and overrunning frontiers; as a result, pro-French and pro-revolutionary groups sprang up across Europe. In Poland, those resisting further partition and Russian encroachment embraced the French example (along with that of the Americans) and took it as their own; in Hungary, landlords cited it in their disdain for Joseph II, who was himself a reformer; and even in England, at least at the outset, leading voices in Parliament were genially convinced that the French merely sought to imitate them. But it was the downtrodden in Europe who were the most aroused: In quiet whispers, impoverished Silesian weavers prayed that "the French would come." In loud voices, strikes broke out in Hamburg and peasants readied to rebel elsewhere. In utter surprise, one English diplomat was stunned to find that even in the Prussian army, there was a "strong taint of democracy" among officers and men. And in Belgium, where privileged elements were already struggling to throw off the yoke of Austrian rule, a second rebellion erupted, patterned after the French. Meanwhile, revolutionary elements were stirred in the Dutch and Austrian Netherlands, not to mention in Switzerland too. As Gouverneur Morris wrote to George Washington, "The French disease of Revolt is spreading."

It was. Not for years, perhaps not for centuries, had intellectuals, reformers, and radicals felt the shared sense of such a new era dawning: In England, the seat of ancient liberties, the noted political activist Dr. Richard Price began to correspond with the Assembly in Paris, hoping to remake the British Parliament. Inventors of fame and significance, James Watt and Matthew Boulton among them (the pioneers of the steam engine), clamored for change; so too did the oppressed Irish, who instituted their own rebellion. And everywhere, the impassioned were captivated: an impassioned Friedrich Hegel in Germany, an impassioned William Wordsworth in England, and an impassioned Mary Wollstonecraft, who increasingly saw themselves as citizens of the world.

Yet as the revolution gathered in force and decibel, so did the movement against it. Edmund Burke, the political philosopher and English parliamentarian, now entering the summit of his remarkable life—he had been one of the most dynamic voices supporting the American rebels—was aghast at the tendencies of this new breed of radical, French or English alike. In 1790 he published his immortal work, *Reflections on the Revolution in France*. Actually, for France, he wept. He predicted anarchy, which happened, and then chaos, which also happened, and then dictatorship, which would happen too; for his own England, he soberly warned the English to calibrate carefully any wider expansion of their own liberties. And for all the world, he condemned a revolutionary dogma premised not on experience and tradition but on ephemeral conceptions of right and wrong, proclaiming instead that each nation and all peoples must rely on their collective history, their own collective character, and their own collective destiny.

Once the revolution entered its more vociferous phases, Burke soon began to beat the drums of war as the only means of stanching France's new barbarism. But his *Reflections*, widely translated and ardently read, also drew a stinging and eloquent reply from one of the age's greatest polemicists, Thomas Paine, in his wildly best-selling *Rights of Man*. Having galvanized the American people to the rebel cause in 1776, he now sought to do the same in France; brilliantly written and powerfully argued, this was his most influential work. In the short run, both Burke and Paine fell on willing ears: In Sweden, King Gustavus III dreamed of leading a crusade in defense of monarchists, while in Sweden's mortal enemy, Russia, a wrathful Catherine even forbade Russians from referring to the "revolutions of the heavenly spheres."

Remarkably, the divisions split the Americans too; in the United States, where George Washington sought to hover above the fray, Jefferson, along with James Monroe, James Madison, and a young Andrew Jackson, increasingly leaned toward the Jacobins, Paine, and the French, while Hamilton and his allies tilted toward the aristocrats, Burke, and the British; Virginia's governor, Henry Lee, was so captivated by the rebellion's fervor that he had to be talked out of joining the French Revolution by Washington himself. As far south as the Spanish colonies, American ideas of independence were strengthened, and the famed Venezuelan Francisco de Miranda, back from his Crimean expedition with Prince Potemkin, became a general in the French army. In the days and months to come, in every nation on the European continent, and in America and Russia too, there were radical or at least pro-French pockets that increasingly alarmed their own regimes; in every nation, including France, there were implacable enemies of the revolution; and in every nation, there were movements whose fidelities or sentiments lay abroad.

The world had not witnessed anything like this since the upheavals of the Protestant Reformation, nor would it see anything similar until after the monumental Russian Revolution split the globe in the twentieth century.

∞

AT FIRST, WITHOUT a doubt, official Europe had been of two minds about the revolution. Catherine had no desire to be sucked into the maelstrom of Western Europe, not so long as she had a two-front war to fight and unrest brewing in Poland; she instead preferred to agitate her neighbors into action and to collude with the émigrés. For his part, William Pitt, the British prime minister, also ignored the dire calls of Edmund Burke. A leading founder of the new Tory Party and prime minister since 1784, Pitt was hoping to shepherd through his own reform program and feared that it would founder on the shoals of war. Thus, by default, the leading antirevolutionary became Marie Antoinette's brother, Leopold II, now the Hapsburg emperor. But Leopold too was slow to action: Having inherited a divided nobility and an equally clamorous empire, he saw little choice but to resist the demands of his sister, as well as the unyielding demands of the émigrés. That, however, would soon change.

Indeed, by 1791, France's elite émigrés were making inroads. Coblenz, south of Cologne, became a hotbed of royalist claims. There, under the Comte d'Artois, a mini-Versailles was created, suffused with ostentatious deference to the royals, all-night gambling parties and dances, all protected by royal bodyguards. And there, twenty thousand troops were assembled. By August, Frederick William of Prussia was finally engaged; he cynically saw a chance to enlarge his kingdom along the Rhine. And Leopold too, though not wanting war, was no longer inclined to do nothing. Meeting with William in Saxony on August 27, he issued the infamous Declaration of Pillnitz: It was at once a step across the Rubicon, albeit a cautious one, inviting other rulers to join their efforts to restore in France "a monarchical form of government which would be in harmony with the rights of sovereigns and promote the welfare of the French nation." But it was also premised on a famous caveat: Leopold would take military action against France only so long as all the other powers would join him; and from what he knew of England's Pitt, he believed that alliance would never come to pass. But it would. In fact, the entire structure of monarchical Europe—and with it, the stability of the global peace—was beginning to disintegrate.

Actually, outside of the kings, it seemed everyone favored war. In Coblenz, in Austria, in St. Petersburg, in Belgium, wherever the French émigrés had gathered, these exiles greeted the declaration with joy, announcing they would soon march into France with the troops of civilized Europe to vanquish the treasonous and punish the unrepentant. But in France itself, increasingly royalists and revolutionaries alike, for their own reasons, also saw war as their only recourse. The queen, of course, had sent missive after missive to her brothers imploring them to take up arms on her behalf, and the king had lobbied the rulers of Prussia, Russia, Spain, Sweden, and Austria-Hungary to mount a combined army to return him to the throne. Lafayette also itched for war, mistakenly believing that it would curb the excesses of the Jacobins. In the provinces, the people too considered war necessary to drive out the last vestiges of the émigrés, whose heinous agents still roamed the countryside collecting tribute. Above all, the Girondins favored war. Ironically, at first the revolutionaries misread what Leopold really intended and took the fiery threats of the émigrés at face value. But far from being intimidated, they promptly turned their fury against all the sovereigns of Europe. And fully expecting Austrian and Prussian legions to march on France, they believed counterattack was the only viable strategy. So with the stakes rising, within France the timetable for a fight accelerated.

But more than that, the revolution now adopted a frightening new philosophy—frightening, that is, to the rest of Europe. The Girondins believed that the revolution could never flourish in France until it had spread outward across the continent, and a ferocious struggle against the counterrevolutionary forces was seen as a necessary "cleansing process." As such, they became something quite new in modern history: a messianic party of international revolution. Many Girondins dreamed of a glorious war in which French armies would storm into adjoining countries, unite with joyous, welcoming local rebels, unseat corrupt regimes, and set up a union of republics.

Curiously, the only serious opposition to the war came from a handful of radical democrats. Robespierre was one of the notable dissenters, opposing the war on the grounds that it was the working classes who would end up dying for it and the middle class who would reap the benefits. But this was Brissot's and the Girondins' hour, not his. Brissot thus helped transform the contest into a political morality play—good arrayed against evil in a succession of stark tableaus. If the European monarchs were never quite the stalking monsters they were conjured up to be, the French people were not quite such bloodstained, righteous crusaders either. But once the drama was cast, each side played its appointed role to the hilt. "The time has come," Brissot insisted, "for a new crusade, a crusade for universal freedom." Bold words, but they still weren't enough—yet. Then Emperor Leopold died. He was succeeded by the militaristic Francis II, and the youthful new Austrian emperor was much more willing to yield to the hue and cry of the exiled aristocracy. He quickly resumed negotiations with Prussia, and just as quickly, by April 20, 1792, France's Legislative Assembly resumed its fateful debate. Would it be peace? Or war?

With only seven dissenting votes, the Legislative Assembly, though mired in unrest at home, preemptively declared war—upon Austria only. There was a method to the madness, as they hoped to divide the allies. In this, however, they sorely miscalculated.

It was Louis, with sadness painted all over his face, who made the formal declaration of war. Then, on April 26, Rouget de Lisle composed the stirring anthem, "The Marseillaise." Thus began twenty-three years of revolutionary and Napoleonic struggles, which would encompass all the great powers, leave the continent choking on its own blood—with a toll of more than two million dead or maimed—realign boundaries, topple rulers, kill a pope, weaken empires, divide America, and presage the ghastly bloodshed of the twentieth century's two world wars.

∞

YET EARLY ON, the Girondins erred, not having considered the state of the French army. On the eastern front, it outnumbered the Austrian troops by 100,000 men to 45,000, but more than 3,000 officers, almost all men weaned on the ancien régime, had already quit their regiments rather than swear a new oath of loyalty that omitted the king. Meanwhile, those who remained had little will to fight: Troops mutinied, equipment broke, and ammunition was often unavailable. Three cavalry regiments even defected to the enemy. And back in Paris, a panicked Legislative Assembly demanded the king's signature to create an armed camp of 20,000 National Guard around the city, as well as to deport renegade priests. But at the mention of this, the king's face grew cold. He not only balked at signing, but provocatively fired most of his cabinet.

When news of the king's veto circulated in Paris, it was interpreted as a sign that Louis was expecting an alien army to strike, putting an end to the revolution. Frantic plans were set in motion to evacuate the capital. One eloquent Girondin orator, Pierre Vergniaud, pointed bitingly to the Tuileries and declared, "From here I can see the windows of a palace where counterrevolution is being plotted." Meanwhile, behind the scenes, the Girondin leaders called for a mass demonstration.

So called, they came.

The king's nightmare deepened once more. On a steamy day, June 20, 1792, in the sinuous streets and winding alleyways of Paris, an excited, sweaty crowd of men and women sprang to arms in a trance of ardor, even elation, then forced their way into the courtyards of the Tuileries. Patriots and hooligans, mischief makers and troublemakers, not to mention fervent followers of Robespierre, Marat, or Brissot, they shouted taunts, piercing their fists in the air and demanding to see *"Monsieur et Madame Veto."* All day long they prowled about, brandishing a menacing array of weapons—pikes and sticks and knives and hatchets decorated with tricolored ribbons. Then, finally, they stationed themselves in front of the entrance of the Tuileries itself.

It was at this juncture that the mob broke into the king's apartments.

∽

Louis was standing there in his room, his sister clinging to his coattails. The elderly Duc de Mouchy sat firmly in front, his arms wrapped backward around the king, prepared to defend him with his own life. Some fifty rebels surged forward. One carried a stained doll, with the label dangling, *"Marie Antoinette à la lanterne"*—thus evoking the Parisian mob's ancient means of removing their enemies: hanging them from a lamppost. Another held a bullock's heart; it read, "the heart of Louis XVI." Others simply hoisted placards on which was scrawled—"*Tremblez tyrans!*"—"Tremble, tyrants!"

A sword was pointed directly at the king. The king gave a hard look. This was insolence of the highest order; it was also a sign of how vulnerable he was. Was this a prelude to more butchery? Or perhaps the denouement once and for all, when he and the queen would be finally slaughtered? This much is known: Elsewhere in the Tuileries, the queen was in a separate room, cowering with her children, when she heard the blows of hatchets splintering wood and smashing doors to the dauphin's chamber. Her dilemma was existential. Should she flee to the king, to live or die by his side? The servants said no, reminding her that her presence might pose additional dangers to Louis XVI. In despair, she and her children instead fled through a secret exit where they hoped to find safety in another part of the Tuileries. Now, listening to the scratch of her own breath, her mouth tightened and eyes alert, the queen could hear the stamp of feet and the rattle of axes hacking through more doors. An ever-growing number of invaders was looking for her; meanwhile, the dragnet, cursing, dirty and bellicose, was closing in. For several hours, the mob roamed about, laughing and yelling and dangling a miniature gallows that strung up the queen. At one point Marie Antoinette's composure broke down; fighting back tears, she shrieked, "Save my son!"

But that was precisely the problem. There was no one to save her son. Not the National Guard, not Lafayette, not the Assembly.

Trembling, she waited.

∞

THEY WERE CRAWLING all over the Tuileries now. Unmoved in the face of danger, the king found his monarchical soul. His voice remained firm, he never wavered, and with great equanimity, he adopted the small bonnet rouge offered to him on the end of a butcher's pike, putting it casually on his head. "*Vive la nation!*" the invaders shouted. "*Vive la liberté!*" Louis responded gaily with a toast to the health of the people. For three hours, it then became a tense standoff. He stood behind a table and listened to the arguments and threats of the people to withdraw his vetoes. In turn, he stood his ground, answering with the plummy voice of the sovereign that this was hardly the fit place and circumstance for considering such high matters of state. "I demand the sanction of the decree against the priests," one rebel angrily replied, "either the sanction or you shall die." The king refused to give in. Actually, as the day passed, he personally quieted down the rebels; eventually they drifted off and boasted that they had given the king a "good scare." That was certainly true. By this time, all the doors of the royal apartments were broken, furniture was smashed, and the king's children were whimpering with fear. While a dozen deputies from the Legislative Assembly eventually came to the king's assistance—it was too little, too late, and too halfhearted—order was not restored until twilight. This time, the rebels, insatiable but exhausted, dispersed back into the city.

As the sun set, when it was all over, the king summoned his family. "I still exist, but it is by a miracle," the queen later wrote. Now Marie Antoinette rushed into his arms, the children fell at his knees, and Madame Elisabeth embraced her brother. They had survived this melee. But with yet another fateful anniversary approaching, the storming of the Bastille, hovering over them was the dreaded question: For how much longer?

And in an ominous sign, even as they pledged their continued fealty, the decree against the nonconstitutional clergy was enforced by the Assembly—despite the king's veto.

∽

As July 14 approached, the king and queen were touched by a sense of earthly doom. They again pondered whether they had to attempt another royal flight. Schemes were urgently discussed: One possibility was fleeing to Compiegne with Lafayette guarding them—this was the marquis's idea; another was to set off for the outskirts of Rouen in Normandy, where the Duc de Liancourt retained some dependable troops; still another option, hateful as it might be, was to flee to England. Were any of these plausible? With an air of sadness and resignation, they knew none would be as easy as Varennes. Despondent, the king and queen also feared that they were laboring under an illusion, that they might actually be safer in the capital than in the French provinces, where revolutionary violence was mounting daily. There was another consideration: With war having broken out, the queen—we don't know the king's exact thoughts—was convinced that the royal family would have to be rescued from the capital. Prussia had entered the war in early July, joining the Austrians, under command of the Duke of Brunswick. In secret communications, the duke informed the king that he planned to "deliver" the royal family to freedom from the "very walls" they were in.

But as the summer dragged on and the heat rose, so did the temper of the people. The Parisians were anxious and they were hungry. Outside the bakeries, long lines of women stood for hours waiting for a ration of bread. Workers milled in the streets, worried and grumbling, while soldiers gathered among their comrades and listened from supper to dawn to the speeches and exhortations of revolutionary agitators. And with each passing day, the radicals increasingly doubted the sincerity of the king and railed against the Assembly, believing it would halt the revolution now that the middle class had consolidated its political gains. Thus Paris, in the summer of 1792, ripened for its next phase of revolution.

Marat, Danton, and Robespierre became the angry thread enlarging the reach of the Jacobin Club to wider political sympathies. Soon it became a torrent. Dissatisfied with the Assembly—peasants everywhere felt the bite of soaring food prices and were also grumbling about delays in redistributing land—the workers of the industrial cities were coming to see themselves as one with those of Paris. And their unrest was intensified by the war. With the fate of civilization seeming to hang in the balance, when the Assembly asked the country's departments to send local National Guard detachments to celebrate the fall of the Bastille, the recruits wept and surged into Paris. One rebel detachment, 516 strong, left Marseilles on July 5, promising to depose the king; as they marched through France, they sang: "*Allons enfants de la Patrie/Le jour de gloire est arrivé!/ Contre nous de la tyrannie/L'étendard sanglant est levé...*" ("Arise, children of the fatherland/The day of glory has arrived/Against us tyranny's/Bloody standard is raised...") It was the new anthem, "The Marseillaise"; henceforth it would be known as a resounding cry to rise up against tyranny. Back in Paris, armed men stormed up and down the streets, exhorting, "Long live the sansculottes! We shall hang all the aristocrats." A sense of excitement began to rip throughout the city.

Surprisingly, in recent months, the king had been allowed an astonishing amount of freedom to ride out into the surrounding areas. But even venturing out of the Tuileries had now become dangerous. He ceased riding his horse; on Bastille Day, and many days thereafter, he wore a thickly quilted waistcoat, akin to a metal breastplate, as protection against assassination. With demonstrations increasing in frequency and hostility, the king and queen were unable even to walk in the gardens. In constant fear of assassination herself, the queen adopted a little dog, which slept under her bed to warn her if an intruder entered the room—the dog would remain with her until the day she died.

The Marseilles and other regiments of the guard marched into Paris after July 14, and, at the Paris Commune's behest, stayed; notwithstanding the king's wishes, the city had become an armed camp after all. The Commune—Paris's local government—was now firmly in the hands of radical leaders; more than that, week by week, indeed, sometimes hour by hour, it was slowly but inexorably becoming the new government of the capital. And while Paris brooded, the nation—*la patrie*—was officially proclaimed in danger of invasion.

They were right. On July 28 the city was stunned into alarm and indignation by a manifesto—issued three days earlier—from the Duke of Brunswick at Coblenz, announcing that he had taken command of a united army at the French frontier and that he was charged with putting "an end to the anarchy in the interior of France," as well as to restore the king to "the legitimate authority which properly belongs to him." Anyone daring to oppose Brunswick's troops would be "punished immediately according to the most stringent laws of war." And he further charged that "The City of Paris and all its inhabitants shall be required to submit at once and without delay to the King." But it was the manifesto's final lines that chilled Parisians' blood: If the king's liberty were "not immediately assured," the duke promised that he "will inflict an ever memorable vengeance by delivering over the city of Paris to military execution and complete destruction."

The Duke of Brunswick was no ordinary aristocrat. A veteran campaigner, he had fought brilliantly for Prussia in the Seven Years' War and then later in Holland against the Patriots. To the people of Paris then, to radicals and republicans alike, the Brunswick document was a bitter challenge to forsake the revolution—which of course was no longer thinkable—or to repel the attackers, no matter what the price. "A horrendous anarchy suddenly prevails," Baron Simolin wrote back to St. Petersburg. Yet it was more than anarchy. While the enemy gathered its forces and prepared to invade, it meant that those in France who wanted to depose the king no longer had to say so in a whisper. Now it could be openly proclaimed.

In a burst of patriotic excitement, on July 29 Robespierre addressed the Jacobin Club with a stinging reply, defying Brunswick and calling for an end to the monarchy and the establishment of a republic with universal suffrage. The next day the provincial soldiers still milling about the city stirred up further agitation: The Marseilles troops banded together with other detachments to promise their assistance in deposing the king. Out of such drops do squalls form: On July 31 one of the forty-eight *sections* into which Paris was now divided issued its own proclamation, declaring Louis to be "a despicable tyrant" and calling on all to strike "this colossus of despotism." By August 3 forty-six *sections* had declared that Louis was "no longer King" of France.

Three days later, on August 6, angry demonstrators massed in the Champs de Mars to again demand the king's abdication. However, the helpless deputies of the Assembly did not act: Angry words in the street or formal petitions were one thing, but actually doing away with the monarchy was another matter. Once more, the Assembly proved it was no longer driving events; that clearly belonged to the radicals. On August 9 early risers found that Marat, the returning prophet, had taken matters into his own hands, calling on the people to storm the Tuileries, arrest the king and his family, and render his royalist stooges "traitors" to the nation. Since October 1789 there had been an abatement of massive, popular violence in Paris; but now the coming of the war and dissatisfaction of the lower classes were about to lead to frightful new explosions. Marat was the spark. And predictably, Louis was the first target.

That night the drums of the National Guard could be heard echoing across the city—"All of Paris is under arms," Gouverneur Morris recorded—as a disorderly body of 2,000 troops hastily arrayed themselves around the palace. Whether they were friend or foe of the king was as yet unknown. Meanwhile the tocsin tolled steadily; it was a call to assemble the people around the Tuileries the next morning. Their cry became: "No more King!"

Inside the palace, that night was spent waiting for the attack. Ultimately, of course, the security of the king depended on the quality of the troops available. Outside the palace, an armed force of aristocratic volunteers—300 of them, carrying swords, rifles, even mere shovels or fire tongs—bravely raced to the defense of the king. But the bulk of the defense of Louis would fall not to the National Guard—experience already dictated that they couldn't be relied upon—but to the elite Cent-Suisses du Roi—900 or so fiercely loyal and meticulously trained Swiss Guards. In most ways they were a special breed: They lived segregated lives in their own barracks, spoke not French but their native tongue, and rigorously adhered to their own customs; at ceremonial occasions, they still donned the ancient uniform of the conquerors of Switzerland. But since the sixteenth century they had been reared for one mission, and one mission only: to keep the king alive. Wrote one Swiss Guard on August 1: "We have all decided to surrender our arms only with our lives." Once the mob, first in the hundreds, then in the thousands, began to collect around the Tuileries as early as three A.M., armed with muskets and pikes and swords, some even with cannon, that prophecy seemed more powerful than ever.

As the showdown neared, with breathtaking speed the Swiss Guards, dressed in red, braided with gold, drew themselves together, arms shouldered, to form an unbroken wall. In the gardens, the courtyards, and lining the great staircase, they stood silent for the murderous encounter to come. Eerily, morning broke with brilliant sunshine and stifling heat. Also eerie was the king's demeanor. Throughout these tense summer days, Louis's energy had ebbed, his spirits had flagged, and he had hovered near the edge of melancholy; but in August, in his own way, he came alive. The king, for all his flaws, was not a cauldron of insecurities and resentments, of black rages or Gallic detestments. Nor did he overly personalize everything, as did other monarchs—say a Henry VIII or a Frederick the Great—who were prone to cry: *my* country, *my* people, *my* policies. Convinced he would soon be brought to trial, he destroyed numerous documents and set up a hiding place for other papers. He had also been reading, in his words, "incessant[ly]," the history of Charles I, the beheaded king of England. One day, pausing, he gazed at the queen and announced to her that the events of France were an exact replica of the English Revolution; he hoped his readings would help him better cope with the crisis he now knew was about to break. On Sunday, August 5, the family prayed together in the family chapel. It would be their last time.

The fateful hours began inauspiciously. As evening descended, the king made an extraordinary concession: He omitted the finely tuned ritual of his *coucher*. Even on June 20, after the king's humiliation at the hands of the sansculottes, the ceremony was carried out. Now, symbolically, in the blink of an eye, a way of life was hurtling to a close. Instead, as the minutes dragged on, there was disarray in the king's bedchamber: nobles crowding around, servants sitting on the ground, aides leaning on console tables and chairs. Etiquette no longer mattered: The king was not yet undressed, he still wore his purple coat, and his hair was unpowdered. One loyal National Guardsman observed that the king's face was flushed and his eyes unusually red. When the bells began to ring in the distance, an anguished Louis stood by the windows and listened.

Suddenly, the tocsins stopped. A haunting silence descended over the Tuileries.

∞

ONLY THE CHILDREN were in bed; Marie Antoinette, her face harassed and sunken, lay down on a sofa, side by side with her sister-in-law. At two o'clock, Parisians were informed that the Tuileries was about to be attacked. By four o'clock, the queen hurried to the King's bedchamber. Assassins were coming; would he put on his breastplate? He refused. Louis replied he was prepared to share his fate "on equal terms." At five o'clock in the morning, the sky was streaked with red—*rouge matin chagrin*—as the sun climbed; while Paris woke, the royal family estimated that 10,000 men were swarming toward the courtyards of the Tuileries. Louis, now pale as "a corpse," inspected the defenses. The Swiss cheered him, but to his horror, the unrest among the National Guards only grew. They openly jeered ("Good God!" bellowed the naval minister, "It is the King they are shouting at!"), and the mutiny spread quickly to other regiments. The queen, watching this scene unfold from the window, began to sob.

By this point, the first shot had already been fired. The Marquis de Mandat, put in charge of the National Guard and instructed to repel force "with force," was cut down and killed by a lone bullet from the revolutionaries, even before he took command. The new commander, Louis Roederer, was a creature of the Assembly, and an unreliable one at that. Outside, voices were reaching a fevered pitch. Then came the pounding on the main door. Should they fight? Surrender? Flee? Once more, Louis had no real stomach for a bloody, pitched battle in the streets of the capital. Years of rule, years of turmoil, years of personal strain had left him even more unwilling to face the prospect of plunging his country into civil war; in any case, it wasn't in his nature. So negotiations began. There were few attractive options. Roederer suggested that the king take refuge in the palace theater, where the Assembly was deliberating.

"We can at least die with you," he then glumly informed the queen. The queen showed a flash of defiance. "We have forces here," she insisted haughtily. "It is time we found out who will prevail—the King, the Constitution or the rebels?"

Baffled, Louis paused. In truth, he had little choice but to agree. After what seemed to be an interminable time, he finally raised his hand and announced: "*Marchons!*" ("Let's go!"). The dimensions of the threat were now staggering. But Louis failed to perceive this; he had evidently learned nothing since 1789. With that, they opened the western garden door, stepped into the courtyard, and waded into a bobbing sea of faces to make their way to the Assembly. Their hope was that their departure would mollify the fury of the mob. It didn't. In spite of an escort of six members of the Assembly, as well as Swiss Guards and grenadiers of the National Guard, they were promptly surrounded. Its fury unassuaged, the crowd pressed around them. The king walked straight ahead, his expression incensed. Some of the royal party were weeping. Others had their eyes closed, or arms raised, ready to ward off expected blows, or struggled to keep their legs moving. The queen was trembling uncontrollably. So thick was the crowd that the dauphin had to be carried by a giant soldier above their heads. In these moments, the royals could look straight into the faces of the bloodthirsty mob glaring harshly back at them. What they saw was pure hatred; with one misstep, the king and queen could have been killed right there.

Tears in her eyes, Princess de Lamballe, the Queen's inseparable friend, shrieked, "We shall never come back."

∞

WHILE ONE CONTINGENT of the insurgents followed the royals, another contingent surged into the palace courtyard and engulfed the Tuileries. Standing before the stairs that led to the king's bedroom were the Swiss forces. The crowd pushed forward. The king's orders to the Swiss Guards were uncertain, a shameful invitation to their own slaughter. Suddenly the crack of gunfire split the air; a shot was fired, most likely by the Swiss. Within minutes, bullets were smacking into the bodies of the attackers; in desperation and anger, the Swiss were now opening fire. Some hundred men and women quickly fell in a hail of bullets. The invaders, motley and exuberant and well armed, retaliated with cannon fire. In the middle of the melee, the king sent jolting orders to the Swiss Guards: to cease fire. They did and paid for it with their lives. Thus began a festival of death. The revenge-minded mob quickly overwhelmed the guards with ruthless reprisals and random terror—men were hurled out of windows, while others were hacked to death hiding in cellars and attics, each becoming its own isolated enclave of carnage. Nowhere was safe. Some were even slaughtered while seeking refuge in the chapel, kneeling and genuflecting and pleading that they had not fired their weapons. It was to no avail. The mayhem continued. So did the mindless barbarism and execution.

What followed was horrific, a grim story of pillage and terror. Those who tried to flee, or pray, or beg for mercy, were often among the first victims. It didn't matter if they were courtiers or Swiss Guards, gray-bearded old men or younger nobles, or whether they were huddled or weeping or scared. They were shot down, cut down, stabbed, stoned, clubbed, or literally ripped apart by mounted gendarmerie. Limbs were hacked off, genitals scissored out and stuffed in gaping mouths, or fed to dogs. The bodies of the guards were literally torn to pieces. For those who escaped, like the queen's faithful aide, Madame Campan—Edmond Genet's sister—or who were only injured, there was the incredible agony of recalling what they had witnessed. And for those few who remained, somehow secure, the worst horror was what the rest of the day would bring. All they could do was bide their time. Meanwhile, corpses of the Swiss Guards, what was left of them, were stripped naked and mutilated; gobbets of human flesh were mounted on pikes and carried in triumph through the streets; other corpses were piled on bonfires and then lit. But for the sansculottes, this was not enough either. Next came the humiliation.

An exuberant prostitute lounged on the queen's bed. The king's wardrobe was emptied and destroyed. Then the mob ransacked the Tuileries from top to bottom: bloodstained hands were wiped on mantels that once glittered with gold; gorgeous inlaid furniture was broken and then burned; wine cellars were pillaged and drained. Bottles were smashed everywhere: on the ground, against the walls, over live and dead bodies alike. Then the dead were grossly mangled: heads, limbs, and other body parts were severed from the corpses, all vivid trophies to the revolutionary fervor. And the fortunate ones who managed to stow away safely, or make it as far as the streets? Many were quickly stalked and hunted down, beaten to death. By day's end, most of the Swiss forces lay slain; those who had been arrested were taken to the Hôtel de Ville (City Hall), where they, too, were put to death. The servants and kitchen staff were also slaughtered. And in the courts of the Carrousel nearby, the meeting point for the flight to Varennes, the crowd, drunk with revolutionary success, set fire to nine hundred buildings, and shot at the firemen who came to put out the flames. The butchery was chilling. Entire structures were ablaze. While fires still burned, everywhere the wreckage of hundreds of royal lives was revealed in vivid detail: shattered dishes, slashed paintings, scorched silver, broken tables and chairs, littered heirlooms and miniatures, and clothes strewn about, of the king, the queen, the ladies-in-waiting, the minor nobility, everywhere the clothes. One young woman, Marie Grosholz, a wax sculptor, would never forget the blood seeping into the gravel or the "appalling objects" spread about the ground as a rich crimson sun vaulted into the sky. She would one day be remembered as Madame Tussaud.

And there was no letup from the hideous scenes. Soon ordinary citizens, oblivious to the death all around them, were lining up to view the wreckage of the royal apartments. And the sansculottes, commanding the day and then the night, hoisted banners aloft made from the red uniforms of the dead Swiss Guards and paraded in victory—likely the first recorded example of a red flag wielded as the banner of revolution.

∞

YET THE FINAL moments of this baleful drama remained to be played out. Amid the carnage, Louis had successfully reached the Assembly around ten in the morning. The deputies were meeting in a chaotic session, arguing furiously and endlessly, but at this point, no longer confidently. "I come here," the king told them with great dignity, "to prevent a great crime." Showing no emotion on his face, he added: "I believe that I cannot be safer than in your midst." Intimidated, not by the king but by the rabble, the Assembly itself was unsure what to do with the royal family. Outside, they could hear the incessant rattle of gunfire; the battle was still raging at the Tuileries. So all that day, and the next, the Assembly debated the fate of the king. On August 10 the legislature voted to replace the king ("The Chief of the Executive Power is conditionally suspended ...") with a provisional Executive Council; at the same time, unwilling to put their own necks on the line, they adjourned themselves—forever as it turned out; many walked out of the building never to return, even as they called for a National Convention to define the French regime. Watching this, the queen muttered, "We will die in this horrible revolution," while Marat jauntily rejoiced in L'Ami du peuple that this was a "glorious day." At the same time, he warned Parisians not to fall prey to the voice of "false pity." With each passing hour, that seemed less and less likely.

Paris had fallen irrevocably. Power now belonged not to the Assembly, but to the radicals, who were in no mood for compromise; they chose to lock the king and his family, under constant watch, in the Temple, an old fortified monastery of the Knights Templar. Louis sought to negotiate how many attendants would be allowed to live with the royal party, observing that Charles I had been allowed to keep his friends with him until he was sent to the scaffold. In the end, the now suspended king was allowed to keep seven attendants for the royal party of five. But that too would soon change.

Before his incarceration, a temporary moment of normality returned; the king had his hair dressed, and then attacked a feast that included two soups, four roasts, and eight desserts, while the queen, haggard and scared, picked at her food. But at six o'clock in the evening the next day, a detachment came to seize the king. The royals were hustled into two heavily laden carriages to make their way to the Temple. When all was ready, the drivers flicked their whips, and the carriages lurched into motion. As they rolled on, the last attendants of the royal family, left behind, took off their caps or discreetly made a deep bow, while the sansculottes hissed as the king departed. The queen whispered under her breath, "It ends with us!" Along the way, Louis saw a sight that had to have rattled him to his very bones. It was the equestrian statue of his grandfather, Louis XV: It had been pulled down and smashed by the mob. One of the commissioners riding with Louis commented smugly: "That is how the people treat their Kings." Acidly, the king replied, "It is pleasant that this rage is confined to inanimate objects."

And then Louis, yielding with little resistance to his fate, anguished over his frail wife and ailing son and waited impatiently for the end. They were shunted into a menacing dungeon in the little tower of the Temple. The doors were locked. That night, the first of their imprisonment, a bright moon came up. Outside, the guards were chanting, "Madame goes up into her Tower/When will she come down again?" And the prisoners began a frightful new existence: living day-to-day, uncertain as to who and where were their friends, wondering whether each morning that followed would find them rescued or taken to the scaffold.

In the meantime, in distant provinces, local nobles still stood quietly and bowed in salute to honor their king. And far off in the east, the armies of the Duke of Brunswick were on the move.

∞

ON AUGUST 19 the Prussian forces, moving at a relentless pace, outfought, outmaneuvered, and outbluffed the revolutionaries. Led by the Duke of Brunswick and King Frederick William II, they charged across the borders, accompanied by a small unit of émigrés swearing revenge. Four days later, flushed with exhilaration, they captured the Fortress of Longwy—it was believed that they accomplished this with the collusion of its remaining aristocratic officers. The word went out: France was being challenged not just from without, but by sedition from within. By the next morning, Brunswick was again on the march. In a matter of days, on the morning of September 2, he had crossed around deep blue lakes, dodged farmhouses, sprinted through the timbered slopes and steaming tangle of the Middle Kingdom—the center of what was once Charlemagne's great empire—and reached the invincible fortress at Verdun.

Here the city was protected against northern invaders by ancient drawbridges standing since the foggy mists of the twelfth century. But the town's defenses failed; the French quickly fell back. An early report sped to Paris that Verdun had fallen; it hadn't, but it did that afternoon. At this point, the gates of Paris lay exposed to the enemy; the Duke of Brunswick expected to break bread in the capital by October.

A disconsolate reality settled over the city. That evening Lafayette, an arrest warrant hanging over his head—the radicals had now unmistakably turned on him—hastily defected to Austrian-occupied Belgium, hoping to make his way to America via England. And General Dumouriez was already warning that all that stood between the Prussians and Paris was their "Thermopylae." So with the Prussian-Austrian forces amassing victory after victory, Paris wavered between resolute reply and paralyzing fear. Politically it was hamstrung too. This was the revolutionaries' direst crisis since the start of the rebellion, more nerve-wracking than the king's flight to Varennes, more harrowing than the growing internecine strife in the Assembly—the rightist deputies had almost stopped attending altogether, and only 285 remained of the body's original 745 members—and more confusing than the seizure of the king. In these convulsive weeks, the revolutionaries were not sure which way to turn. But as the days passed, a strange emotion prevailed throughout the insurrectionists themselves. They realized that this was not the first time revolutionaries had been laid siege to, or cut to pieces by assaulting legions, or dispirited or enervated, yet somehow found the fortitude to prevail. Now, faced with the prospect of losing everything, they sought a new man to lead them, a rebel who could rescue France, who could save the revolution, and who could keep the people free. With war and fear the new driving forces, the rump legislature made a fateful decision: It voted to choose Georges Danton to head the new executive council as minister of justice. It was Danton to whom the sansculottes now looked; it was Danton who was to brace the nation and harness the militants; and it was Danton who was to lead the revolutionary movement either to its glorious next stage, or to its tragic demise.

History knows this Georges Danton well, or at least it thinks it does. But did his own people? That is a tougher question. Was he a dedicated revolutionary, a bloodthirsty psychopath, or a bourgeoisie manipulator? In truth, Danton was all these things, and more. He was not only complicated but remarkably young, only thirty-three years old; incredibly, his extraordinary life would end in just two years. But it would seem like another lifetime. In the fall of 1792, he was judged to be the ablest and strongest character of the revolution—and in this time of crisis, he certainly seemed to be. Although his part in the overthrow of the monarchy remains obscure, he was largely credited with its success. Little wonder. As a leader, he commanded the revolutionary stage as did few others: He was at once an ideologue but also a realist; he was cunning but also diplomatic; and he was willing to risk all to keep the revolution safe from either foreign invaders or mounting internal anarchy. He briskly defied his peers: Where Brissot and Robespierre were marked by moral intensity and single-minded determination, not so Danton. To save the cause, he would cooperate with anyone— with Marat, with the king, with the Girondins. Accordingly, he made more than his share of bitter enemies: Robespierre was jealous of him, Marat pilloried him, the king was leery of him, the Marquis de Lafayette was attacked by him, the Girondins were appalled by his very presence, and the moderates trembled when he unleashed his scorn. But when confronted with the aura of his revolutionary past, or his ability to make swift decisions, or his often profane speech crashing like lightning over the Assembly and the proletariat crowds, he inevitably cast a spell on his detractors.

He began modestly. Born one of ten children in the farmlands at Arcis-sur-Aube, in the region of Champagne, he rejected the church and instead followed his father, who held a minor legal office, into law. He studied in Reims, then found success as an attorney in Paris. Danton didn't look the part of a tribune of the revolution: His lips and nose were disfigured from several childhood farm accidents, his skin was badly marked from smallpox, and his voice was gruff and brutal. With his face furrowed and his thick, flattened nose, he looked more like a bruiser in a boxing ring than the streetwise leader of the sansculottes; in truth, he was a bit of both. His character, intricate and knotted, however, was not as arrogant as his face or his words. True, he was everything his detractors insisted he was: anxious and ill-tempered, incendiary and volatile. But between his dramatic gestures and expressive features, he could also be uncommonly charming and charismatic. And he thrived on laughter, carousing, gambling, cursing, all sorts of indulgent delights, especially pretty women.

It would later be rumored—the exact truth is difficult to pin down—that he promised that "whatever happened," he and his friends would "watch over [the Queen] and her children." No wonder the historian Jules Michelet later called him a "serf of nature," and insisted he was more beholden to the "tyranny of blind desire" than the revolution whose spirit he bore. In fact, in his lifestyle he was the antithesis of the fastidious lawyer Maximilien Robespierre, whom he despised. But he could have been the Comte de Mirabeau's revolutionary twin: He borrowed and pilfered money, purchased a comfortable home in Arcis, three houses in or near Paris, a large estate in the Aube, and acquired considerable parcels of church property. Many wondered how he garnered these vast sums and just as many whispered that he had been bribed by the king, which he almost certainly had. But if he took the king's money it was to work for the sansculottes. There is no evidence that he ever betrayed the central tenets of the Revolution or wavered from its more dramatic actions. His great dream, one historian has written, was a "vast table in which all of France, reconciled, would be seated to break the bread of fraternity, without distinction of classes or parties." But at the same time, he was not afraid of fratricide, or hesitant to engage in betrayal, or mired in guilt or the agony of tough decisions demanded by a ruthless proletariat dictatorship.

Curiously, he was also suspicious of the very democracy he was helping to create: In 1792 America's Gouverneur Morris noted that Danton always maintained that "a popular system of government for this country was absurd; that the people were too ignorant, too inconstant, and too corrupt to support a legal administration; that . . . they required a master." And among the revolutionaries of Paris he was a rare animal: His ability to improvise passionately was almost without peer. Yet he remained a natural politician, often telling people what they wanted to hear.

His education was also formidable: He could read both English and Italian fluently, and his library included 571 volumes in French, 72 in

English, and 52 in Italian. But unlike Robespierre or Marat, his words were less often sprinkled with passages from the classics than with the plain-spoken language of the people; a number of Jacobins darkly suspected that he was one of the few revolutionary leaders who did not regard himself as a latter-day Roman, and they were right.

Religiously, he was a renegade, eschewing Catholicism while flirting with deism ("I recognize," he once said, "only the cult of justice and liberty"); he was essentially an atheist. Yet his first wife, Antoinette Charpentier, was not only a royalist but a Catholic; actually, here one sees that behind his iron facade was a sensual, almost romantic side: A week after her death in 1793, raging with sorrow, he reopened the earth to embrace her corpse. And as the Revolution accelerated and the blinding fugue of Robespierre would become all but unstoppable, Danton would retreat almost serenely to his house in Arcis, where he would stand for hours, immobile, dreamily gazing out his window.

But at the moment of the Revolution's gravest turning point—Brunswick on the march—he had courage. On that chilling morning of August 25 when it was learned that Longwy had fallen, the Girondins proposed that the government withdraw—flee would be more accurate—to Blois, behind the Loire. Danton objected vigorously, asserting it would destroy the spirit of the people ("Our enemies had taken Longwy," he roared, "but Longwy is not France"). When the frightful news of September 2 arrived in Paris—the news that Verdun had fallen and Brunswick could be racing to the capital—deputies were applying for passports under false names, and the gates of Paris were slammed shut. But again, Danton was unflinching. He argued that he would rather see Paris burned than fall to the Prussians. And more than that, he believed the time had gone for hesitation and come for action, as well as for building new armies and fortifying them with esprit de corps and confidence. This is precisely what he set about to make happen. As Paris slipped into convulsions, Danton hurried to the Assembly to stir the nation to courage. It was his most passionate and thrilling speech, delivered with the full force of his formidable lungs.

"We ask that anyone refusing to give personal service, or to furnish arms, shall meet the punishment of death," he boomed. "The tocsin we shall sound is not the alarm signal of danger; it orders the charge on the enemies of France." His next phrase would then rouse a country, define a new army, and ring throughout the entire century: "To conquer," he bellowed, "we have to dare, to dare again, always to dare—and France is saved!" ("*De l'audace, encore de l'audace, toujours l'audace—et la France est sauvée!*")

In scarcely more than a week, Danton had emerged at the helm of the new French state. But as he held his colleagues spellbound, the revolutionary wheel was about to take its bloodiest turn to date.

∞

NEVER WERE PERIL and promise wrestling with each other more than in these high-pitched weeks. The emotional fever of the Revolution was already swelling beyond control. This was not surprising. Most revolutions have dark sides lurking within them, and the Duke of Brunswick's manifesto predictably drove revolutionary leaders to behave as if a knife were at their throats. Revolutionary France was also becoming stiffly doctrinal and unbending. Still, a contributing factor was the growing conflict between religion and the state, and the effort to make worship of the state a substitute for religion. The result was a series of acts that relentlessly pitted the support of the church against the achievements and stability of the revolution.

Thus, in early August, literally a few hours after the king was formally dethroned, the measures began, the harshest yet taken against the church. The Commune disseminated a list of priests believed to be harboring antirevolutionary sentiments; many of the clerics were promptly rounded up and sent to prison. Days later, the crusade continued: On the eleventh, the Assembly ended any role for the church in education; the next day, the Commune banned the wearing of religious vestments in public, which was soon enlarged to a nationwide decree. On August 28 the revolutionaries dug the blade deeper, decreeing that all priests failing to raise their hand in allegiance to the civil constitution would be deported; these clerics were given just two weeks to leave France. Within days, in some cases hours, nearly 25,000 priests fled abroad.

In the meantime, the Assembly received a utopian petition calling on it to grant "the title of French citizen" to all foreign philosophers "who have with courage upheld the cause of liberty." It concurred, grandiloquently bestowing honorary French citizenship upon George Washington, Thomas Paine, James Madison, Alexander Hamilton, Joseph Priestley, Jeremy Bentham, William Wilberforce, Thaddeus Kosciuszko, and Friedrich Schiller. Alexander von Humboldt trenchantly summed up this new religion of freedom; he had come to France, he rhapsodized, "to breathe the air of liberty, and to assist at the obloquies of despotism."

It was wartime in Paris and people felt it. Some felt it with grief; others with anticipation. The apprehension came and it went, but mostly it came. And so, in the early morning of September 2, the nation looked for ways to profess its love of the *patrie*. Able-bodied men mobilized to repel the advancing invaders at Chalons, Soissons, and Reims. Workmen rushed out to fortify the city's defenses. And in a comparable burst of patriotism, women defiantly carried muskets for their loved ones, often bearing loaves of bread or large sausages on their bayonets.

But then, as a kind of collective sociopathy began to grip the city, a group of armed assassins marched to the prisons with an altogether more ominous mission in mind: to organize an attack upon the inmates inside. Including priests.

∞

SOMETIME AROUND TWO P.M. on Sunday, September 2, six coaches carrying thirty priests jostled along the road to the Abbaye jail. A menacing crowd began to yell taunts and insults; a priest sought to push them back with his cane. That was enough. The leader of the mob plunged his saber twice through the open window, and soon the crowd, swelling and ranting, fell upon the prisoners; so did some of the *federes*. Seventeen prisoners were hacked to death. These insurrectionary volunteers—self-appointed at that—thereupon boldly declared that they would not fight enemies on the frontiers until they had disposed of enemies in Paris. Now, intoxicated by the euphoria of impending death, a frenzied convoy raced off to the Carmelite Convent and assassinated all its imprisoned priests. Late that afternoon, the mob was reinforced by roughened criminals and common thugs, as well as by troops from Marseilles, Avignon, and Brittany. The idea made the rounds to return to the Abbaye; they did, where they forced all the prisoners to march out, while the crowd stood in ad hoc judgment of them. Whether they were saying prayers for the dying or being told to climb on a pile of corpses to swear an oath of loyalty to the nation, whether they were male or female, whether they were Swiss Guards or clerics, Black Musketeers or ex-servants of the king or queen, most were forced to run a gauntlet of men who bludgeoned them with swords, knives, pikes, axes, sabers, iron bars, and even a carpenter's saw. Screaming, running back and forth like trapped animals, they vainly tried to fend off their executioners with their hands or arms. Blood ran in streams from the bodies in the street.

The first wave of terror had begun.

∞

THE MOB, HAVING anointed itself judge, jury, and executioner, was soon ratified by the governing Paris Commune. Strangely, in the beginning, the work of the executioners was relatively restrained. Deputized by the sectional authorities, each received 6 francs for a day's work and as much drink as they needed, plus the promise of 24 shillings if they did not plunder the bodies of their victims. Outside the Abbaye, benches were brought to the streets, and the crowds seconded their judgments with nods and acclamations. There were moments of compassion: Inexplicably acquitted, one royalist was escorted home by men carrying torches who refused payment for their services; another would-be victim was restored to his children by a weeping crowd. But for the most part, as the bloodletting grew, so did a new ferocity. Torture became their vocation. They took a perverse delight in extending the suffering of the condemned, while spectators cheered enthusiastically. One woman had her breasts cut off and a bonfire was set between her spread-eagled legs—to wild shrieks of acclaim. Afterward the bodies would be stripped and thrown on a heap of corpses, or dismembered, the bloodied flesh paraded on the end of a pike. To take a break, executioners would drink brandy while sitting on the naked remains of their victims, until the butchery again commenced. Meanwhile, the cries of the slaughtered rang out in the streets and alleyways of Paris.

The response of the authorities was a telling coda on this newest phase of the revolution. The Assembly dispatched representatives to the Abbaye; the street-side tribunals would not even deign to listen to them. After some debate, the Assembly finally agreed that the most prudent approach was one of acquiescence. The communal authorities were less conflicted. They spoke of purging the earth of counterrevolution, as if joining in a secular version of the Last Judgment. Then they sent their representatives; in a show of support, one deputy starkly congratulated the killers: "Fellow citizens, you are immolating your enemies; you are performing your duty."

That duty continued. The following day, the judges and executioners rushed off to the other prisons, the Conciergerie and La Force. Now the paroxysm of violence took on the character of a ritualistic, apocalyptic mass

murder. With fresh workers and new victims, this torrent of rage continued, and the executions went on. Treason, actual or real, was to be punished.
In La Force was the most attractive victim thus far: the famous Princess de
Lamballe, the beloved consort of Marie Antoinette. The princess was once
exceedingly wealthy, was still, despite all, extremely lovely, and her loyalty
to the queen was without peer. She was also so fragile that she often found
the indignities of everyday life too much to bear. Tragically, earlier in the
Revolution, she had been among the émigrés fleeing to safety; but upon the
tender pleadings of the royal family, she unwisely—but nobly—cast her lot
with the queen and returned. For a time she was imprisoned in the Temple
with the king and queen, until on August 19 the Commune ominously
decided to remove her to La Force prison. The queen had wept bitterly at
this, and in vain did everything she could to prevent the separation. She
failed.

Lamballe was now forty-three-years-old and heartbroken, and, by
noon, she would be dead. Dragged into the courtyard of La Force, she
was asked to denounce the king and queen. She refused, declaring with
great serenity, "I have nothing to reply, dying a little earlier or a little later
is a matter of indifference to me." Summarily, she was killed in the most
gruesome fashion, with furious blows of a hammer to her head. Stripped
naked, her body was mutilated, then dragged through the streets. Then,
and only then, was she beheaded. As if it were the Middle Ages, her heart
was cut from her body, roasted over a fire, and—accounts vary—either
eaten by a zealous patriot or brandished on the end of a sword. Somewhere
along the way, Marie Grosholz was forced to make a wax model of her face.
From there, Lamballe's head, soaked in blood and wrenched into a silent
scream, was borne on a pike to a hair salon, where her long blond hair was
washed, combed, and curled; then her head was returned to the pike. At the
same time, Lamballe's naked body was impaled on the tip on another pike.
Drunk with wine, the revolutionaries now set off for the Temple. Where
the king and queen were imprisoned.

∞

AFTER HIS ARRIVAL at the Temple on the thirteenth, the king had learned
that the revolutionaries could no longer be taken at their word. He was not
allowed the servants he had been initially granted, except for one valet. It
was on August 19—the day Brunswick's armies crossed the frontiers—that

the king's cherished attendants, most especially the adored Princess de Lamballe, had been removed for interrogation or incarceration at La Force prison, never to return. The whole family was shunted into the Small Tower, a forbidding medieval edifice that stood sixty feet high, with various turrets attached to it; it was cold and damp in the winter, and clammy and uncomfortable the rest of the time. In spite of this, the family adopted an odd, almost tranquil schedule under the watchful eyes of the municipal guards who day and night monitored their royal visitors' every move. Each day the king awoke, promptly prayed, and dressed and read until nine; the queen joined the family for breakfast. In the mornings the king worked with the dauphin on his penmanship, as well as teaching him Latin, history, and the classics, while the queen and her sister-in-law steeped the young princess, Madame Royale, in music lessons. When they were granted permission, they took walks in the garden; these were some of their happiest moments. After lunch the king played backgammon or piquet with his sister, then napped; Marie Antoinette did needlework and the children reviewed their lessons or played games quietly. At seven one of the princesses read aloud, while the rest of the family dutifully listened. An intimate dinner was held at eight.

But the hardships quickly mounted. Many of the guards enjoyed offering little insults to their captives; one jailer named Rocher took particular delight in blowing his pipe smoke into the royals' faces; once, brandishing a saber, he told the king that if the Prussians came to Paris, he would personally kill Louis. Indeed, the king, who was always called "Sire," or "Majesty" (even by the queen), was suddenly "Monsieur" or even "Louis." His sword was seized, and all food was thoroughly examined for contraband. By now it was clear that the four commissioners who watched over the royal family day and night were bullying revolutionaries, seasoned by years of privation and bitterness.

Nevertheless, the family managed to strategize about the future, speaking in coded conversation during board games or exercise outside; once they used the second finger of the hand on the right eye to signify Austrian success on the Belgian front. And despite everything, Louis struggled to remain hopeful and unafraid for the future. There were pleasant surprises: The Commune inserted a valet named Hanet Clery to "manage" the little royal household. Clery proved to be a devoted royalist. And there were simple pleasures too: The turret of the tower was lined with books from the archive of the Knights of Malta; Louis read a book a day.

One of his greatest hardships was lack of timely news. The king often depended on a blend of rumor and fact that drifted into the Temple walls. He did not know, until later, about the invention of the guillotine. He did not know, until after the fact, that Lafayette had fled the revolution to Austria. He did not know, until the dreary reality had settled into the rest of Paris, that Longwy had fallen. But other times, he was kept up to date on events outside; the king's valet faithfully reported everything he heard—he was allowed to shop three times a week for food—and criers, bribed by royalists, read out the news in a loud voice alongside the Temple walls. In braver moments, a woman in a house below the Temple scribbled news on a large placard, which she furtively held up for the king to read; this was how he found out about the fall of Verdun.

Despite their narrow confinement, the king and queen still cherished the hope, however remote, that they might be freed by foreign armies; upon hearing about the series of French defeats by Brunswick, their hearts beat with a quickening pulse. The same upon learning of the royalist movements cropping up in far-off provinces. On September 2, feeling more buoyed than they had in months, the royal couple quietly walked in the garden while their children scampered about, when suddenly the whole family was hustled inside. What was happening? They listened. Drumbeats sounded in the distance. From inside the gray Temple walls, they heard menacing noises, breaking in an ever disconcerting crescendo. They realized Paris was again in chaos, but what did it mean? Brunswick? Or more wanton violence from the mob?

On September 3 the king was barred from walking in the garden; unknown to him the Assembly feared that the royal family would be assassinated. He still had no real hint of what was taking place beyond his prison walls, though news of the terrible truth—the massacres—began to trickle in. That afternoon, as the royal family began to dine, a deafening clamor erupted. The door flew open. A municipal guard and four commoners barged into their room, adamant that the king and queen go to the window. The once august and unapproachable personages were now not just wholly approachable, but vulnerable. One of the delegates demanded, "I advise you to appear if you do not want the people to come up here." The king was calm, but he knew exactly what this meant. He neared the window.

Below, the militants were parading beneath the royal cell at the Temple, laughing and singing in an ecstatic frenzy, while Princess de Lamballe's head was bobbing up and down. Earlier in the day, the royal family had been told that the princess was safe. Sensing this was no longer true, the queen eventually sought to inch her way to the window; her legs trembling, she never made it. Meanwhile, the children were wailing uncontrollably. The king's valet, Clery, peering through the blinds, saw Lamballe's twisted face upon a pike, but said nothing. Marie Antoinette, frozen with horror, lost all composure. To spare her the hideous sight, his eyes reddened, the king—or in a rare moment of compassion it might have been one of the jailers—explained to her what lay below.

Marie Antoinette fainted. Meanwhile the crowd milled outside for hours, cat-calling for the queen to come, look, and kiss the "sweet lips" of Lamballe. They also tried, unsuccessfully, to enter the tower. Eventually the rioting ended. By five o'clock the mob wandered off.

Marie Antoinette wept all through the night.

∞

By September 4 the bloodthirsty delirium was unstoppable. Watching the savagery, Gouverneur Morris gasped, "Gracious God, what a people." The ad hoc tribunals now moved to the prisons of the Chatelet, Tour St-Bernard, and St-Firmin; in Salpêtrière, rape replaced murder. At Bicetre, an insane asylum, all forty-three youths, from seventeen to nineteen, were slain. Some of the killing was especially sinister: For no apparent reason, beggars and prostitutes were slaughtered. So were children, both boys and girls as young as eight. That evening, Morris wrote in his diary: "It is now past twelve at midnight and the bloody work goes on!" For two more days, while the piteous cries rang into the evening air, the massacres raged, not just in Paris but Versailles and Reims as well; all told, there were some 1,400 victims, only thirty of whom were nobles. Many of the condemned begged for mercy; they merely wanted to die quickly. Many were denied. Only after four days did the executioners withdraw, having transformed the glory of revolutionary France into a saturnalia of blood. Meanwhile, the walls and floors of the prisons, and the streets and courtyards beyond, bore the silent scars and gashes of bare-fisted struggles, and the terrible business of mass murder.

Countless corpses lay sprawled, their stinking remains eviscerated by rats or hungry dogs. After the ninety-six hours of unremitting violence, panic, and death, the people were split in their judgment of the carnage: Royalists and Catholics were repulsed and frightened, but revolutionaries maintained that the aggressive actions were a necessary response against "implacable enemies" of the nation. Robespierre sniffed that the will of the people was being "expressed." In other words, organized killing for political ends had become part and parcel of the revolution. Not unsurprisingly, Marat agreed, claiming to have inspired the entire operation. Remarkably, Paris itself remained strangely numbed to the slaughter; even theaters and restaurants opened for business. As for Danton, was he Danton the merciful or Danton the merciless? In this new age of intolerance, when asked to stop the slaughter he simply sighed; then, shrugging his shoulders, he argued, "It would be impossible." Anyway, he added lightly, "Why should I disturb myself about those royalists and priests who were only waiting the approach of foreigners to massacre us?" When a fellow member of the Executive Council protested the bloodshed, Danton ripped back: "It was necessary." And to one dubious young man who wondered if this was so, Danton answered: "I wanted the youth of Paris to reach Champagne covered with the blood that would assure us of its loyalty. A river of blood has to flow between the Parisians and the émigrés." In Danton's mind, the September massacres were a glorious event for the revolution: For the Parisians, there was now no turning back. In turn, those brave men rushing to the battlefields—peasants, proletarians, and bourgeoisie—knew that they could not expect mercy if they failed. Whether at home or on the front, they now had only two choices—victory or death.

On September 6 Danton was rewarded for his steadfastness; he was elected deputy for Paris to the National Convention.

∞

BY THIS POINT, the Legislative Assembly had voted to call a national election for a convention to draw up a fresh constitution suited to the new mood of France and the mounting exigencies of war. And with volunteers being summoned to defend a country now called a republic, it suddenly seemed impossible that anyone, rich or poor, taxpayer or not, should be barred from speaking at the ballot box. Thus, it was not only Danton's star that was on the rise, but Robespierre's as well. He had just won his first significant victory: The convention would be chosen by universal male suffrage, just as he had demanded.

On September 20 the Legislative Assembly concluded its final session—and itself—unaware that in a little village called Valmy, between Verdun and Paris, a historic battle had taken place between the French army under General Dumouriez and Austria under the Duke of Brunswick. It was historic less because of the death or carnage—actually, it was largely a cannonade duel—and more because of the momentous turning point this day ushered in. The battle was a draw, but not the result; Afterward the King of Prussia ordered his struggling regiments—they were also disease-ridden—to retreat, abandoning their strongholds of Verdun and Longwy in the process. French soil was now free again. It was all the more remarkable given that when these French troops had marched off into battle, green, ill equipped, and poorly trained, they could scarcely imagine the noise of cannon and the sounds of whistling shots and shells. They did not know what gunpowder blasting from thousands of rifles smelled like, or what it was to hoist a gun along a battle line, or man a cannon, or taste mortal fear. They did not know what soldiers said when they were wounded or torn apart, or what they looked like when they died. But now they had triumphed, and with them, so did the hopes of all of revolutionary France.

In the battle's wake, the poet Goethe, on the staff of the Duke of Saxe-Weimar, famously commented: "From today and from this place begins a new epoch in the history of the world." He did not overstate.

∞

No LESS THAN in the past, the election to the Third Assembly was in the voracious hands of the Jacobins. As in America, voters chose electors, who met in electoral committees and then chose deputies to represent their districts. But the similarities stopped there. Unlike in America, the process came with a cost: France's elections were by voice vote and took place in public, even as the September massacres were engulfing Paris. In such a climate, who was prepared to speak out against local officials? Very few. The same doom awaited all those who would still betray the revolution. With such a stark choice over a single ballot, the results were predictable. In the cities, conservatives melted into the background, refusing, or fearing, to vote; of the 7 million persons eligible to cast their ballots, a staggering 6.3 million stayed home. Many Catholics, having found sanctuary among a thousand monasteries and hideaways in friendly homes, also remained underground. And in Paris, where the electoral assembly met in the Jacobin Club itself, the result was that all twenty-four deputies chosen were ardent allies of the Commune.

The Convention opened in the Tuileries on September 21, 1792, with 750 new delegates. Once more, virtually all members were of the middle class; prominent foreigners were elected too, like Priestley (the discoverer of oxygen) and Anacharsis Cloots and Thomas Paine. Even the radical Duc d'Orléans, newly rechristened as Philippe Égalité, represented an extremist section of Paris. And the Girondins, who had led France into war, were again well disciplined, erudite, and strikingly spirited. They were also immeasurably buoyed in the coming weeks by stunning French victories on the ground. The troops sent up a cry that would one day intimidate all of Europe, the roaring *"Aux armes! Aux armes!"* of charging French infantry.

On September 25, while a French division rolled into Nice, General Adam-Philippe de Custine began his decisive push into the Rhineland. On October 5 he burst into Worms; on October 19 he seized Mainz; by October 21, Frankfurt-am-Main fell. On other fronts, General Dumouriez now set his sights on Belgium, then a dependency of Austria. He fought a bitter battle at Jemappes on November 6; by day's end, the battered Austrians retreated, leaving a carpet of four thousand dead on the field. Flush with victory, the conquering legions of revolutionary France continued their march, leaving awe in their wake: On November 14 they thrust into Brussels; on the twenty-fourth Liege; and at the end of the month Antwerp. In each of these cities, exuberant reports were sent back from the lines to the convention: The French were welcomed not as occupiers but as liberators.

In Paris the convention beamed. Feeling invincible, it thus announced its most ambitious stroke yet: the December 15, 1792, "Edict of Fraternity," decreeing "war on castles, peace for cottages," and promising fraternal aid to any nation wishing to recover its liberty. It stated wondrously, "the French nation proclaims the sovereignty of the people [in all cooperating regions]. . . . You are, from this moment, brothers and friends, all are citizens, equal in rights, and all alike are called to govern, to serve, and to defend your country."

France was no longer simply a great nation; this decree made it a revolutionary state abroad as well—the first in European history. World revolution was now its stated goal, republicanism its guiding star. With chilling precision, Brissot put it this way: "We cannot be calm until all Europe is in flames." But there were consequences for such ambition: As is so often the case, the newly "liberated" territories began to chafe under French rule, suggesting that one taskmaster had simply taken over for another. And by this stage, the rulers of Europe, angry and bewildered, no longer could stand idly by. Having little choice, they saw the convention's actions as the unmistakable sound of war against all monarchs.

With discussions anxiously taking place in Vienna and London, Amsterdam and St. Petersburg, even in America, the First Coalition against France now took shape. Undaunted, revolutionary France remained in a hurry. Already, on September 22, 1792, the dominant assembly factions had declared the First French Republic.

It was now irrevocable. A thousand years after the first king had accepted the throne, the monarchy, and the Bourbon dynasty, had been swept away.

∞

THUS FAR, THE revolutionaries had set out to change everything, from simple crockery designs and furniture, to law codes, religion, and the map of France itself. And on that same day, the convention took its most audacious step yet: It would now control not only geography and ideas, but time itself. Accordingly, the Christian calendar was outlawed, to be replaced with a revolutionary one, in which the years would be renamed I, II, III, IV, V ... (nearly two centuries later, the genocidal Khmer Rouge would do the same thing in Cambodia) and the months would be identified by their weather. Five days of the calendar would be called *sans-culottides*, and were to be set aside as national festivals. In their euphoria, the convention would boldly blot out the Christian cycle of Sundays, saint's days, and such holidays as Christmas and Easter; they would even abolish the week. They believed that this calendar would summon up visions of the earth, and not religious saints, of the simple workingman's tasks, and not of outworn seasons. The new calendar came into use on November 24, 1793; it stubbornly remained in place until 1805.

It was at this moment that Danton also made a calculation that was destined to cast a profound shadow over the remaining years of the Revolution. Under fire from fellow Girondins for his ties to radical Paris, and questioned over his expenditures as minister of justice, he joined forces with the radical delegate Maximilien Robespierre.

Up to now, Robespierre was still a relatively minor figure at the convention. When his name was submitted for the presidency of the Convention, he garnered a mere six votes. But he would not toil in obscurity for long. To those who knew him, and soon all would know him, Robespierre was dogmatic, indomitable, and intense, and would become the Torquemada of the Revolution. The allure of Robespierre is that he transformed himself into a stoic, but he was also a romantic to the end. Doctrinaire and calculating, he was a master of sowing differences among his rivals, waiting patiently for them to unravel on their own. He would then get his chance, proving to be a stern disciplinarian, a capable organizer, and a fearless leader. He instinctively believed that he would one day rule—he was right—and remarkably, prophesied that he would one day be killed; he was right there as well. As an observer noted, Robespierre understood, just as his colleagues understood, that virtually from day to day "he carried his life in his hand."

For the moment, though, it was neither Robespierre nor Danton but Marat, still filthy, still unwashed, and still ailing, who was the most articulate voice of the people. With great fanfare, he celebrated the new republic by changing the name of his periodical to *Journal de la République française.* And together, out of the blood and fury of the times, these three—Marat, forty-nine years old; Robespierre, thirty-four; Danton, thirty-three—would emerge as a singular triumvirate. In truth, they were anything but singular. Danton rebuffed Marat and detested Robespierre, and Robespierre treated Danton with cold contempt while shunning Marat. As the oldest, Marat had only one year left to live; in the Convention, he often sat in stony seclusion, isolated and forsaken. But in the streets, he was never forsaken: His bloodcurdling invective echoed throughout Paris, rousing the sections to riot even while fostering near universal enmity, especially among the Girondins, for whom he reserved special venom. Controversy stalked him from the beginning; Marat was indifferent.

As early as September 25, Marat, with the September massacres as his hammer, was accused of calling for a dictatorship. But when this enfeebled "little man" with a cadaverous complexion attempted to defend himself before the Convention, he was harshly admonished to "Sit down!"

"It seems," he replied, "that I have a great number of personal enemies in this assembly."

"All of us," the Girondins answered back.

Marat once more incautiously raised his cries for a tyranny, again on the Roman model. A deputy suggested that he be removed and charged with treason. Suddenly the hunter of heretics was himself condemned as one. But the motion was defeated. Melodramatically, Marat ripped a pistol from his belt, pressed the barrel to his ear, and declared, "If my indictment had been decreed, I would have blown my brains out at the foot of the tribune." But it had not. And for the Convention, riding high on a crest of euphoria—victories on the battlefield, a French ideological war engulfing the rest of Europe, and a new government at home—the time had come to address the last remaining question mark of this immense historical drama: what to do with the suspended king.

∞

FROM INSIDE THE thick, forbidding walls of the Temple, the king remained a powerful symbol, a human pawn with enormous potential value.

For Louis, his fate by now seemed almost preordained. Imprisoned since August 10, the Bourbon absolutist had prevailed over the constitutional monarch, Marie Antoinette over Mirabeau, the tug of tradition over Rousseau, and the heart over the mind. But what is too often forgotten is how astonishingly young the royal family was: The king was just thirty-eight; the queen, thirty-seven; his sister Madame Elisabeth, twenty-eight; his daughter Marie Therese ("Madame Royale"), fourteen; and his son, the Dauphin Louis Charles, seven.

Louis's bewilderment was understandable; so was his exasperation. Guards entered the Temple rooms whenever they liked, swearing, telling dirty jokes, or taunting the royal family. The humiliations were constant. Graffiti was scratched on the wall depicting macabre images—like the one of a crowned stick figure hanging from a gibbet, marked: "Louis Taking a Bath in the Air." Where the royal family had at least been allowed unfettered walks in the garden every day in the Tuileries, now they no longer even had that, not unless supervised. Family activity was largely limited to what could be done within the walls of the Temple. Meanwhile, every day their rooms were searched, and every day, Louis's bread was sliced up for illicit notes or contraband; even his macaroons were examined. The isolation was painful in its own right: It was on September 21, from the criers, that he learned the monarchy had definitively come to an end; he thus became "Louis Capet," after the surname of the dynasty that ruled France until 1328. It was with glee that one of the guards removed the royal seal from Louis's breast.

Louis's destiny seemed to be sealed at the beginning of October. Early that month, guards entered with sinister news, informing the royal family that the king was to be separated from them and confined alone in the Great Tower of the Temple. The queen and children, overwrought, protested bitterly. It was to little avail. Still, the queen's charm did soften some of her jailers. As a result, the king was at least allowed to join the family for meals, provided everyone spoke "aloud" and in "clear French." All pens, ink, paper, and pencils were removed, another hardship; forks and knives were taken away too, and so were scissors. One day, staring off into space, the king watched his sister bite off a thread and mused: "You [once] had everything you needed. What a contrast." Madame Elisabeth replied bravely that she bore no regrets, so long as she was with her brother.

Their one rock was the valet Clery, who continued to smuggle in information. Yet he was also abruptly removed and taken away for interrogation. But foiling his questioners, Clery soon returned. Meanwhile, in the rooms of the Great Tower, work was taking place: simple blue and green wallpaper was laid on one of the walls, a green damask bed was moved in, white curtains and valances were hung, and a chest of drawers was brought up. So was a silver food service, from which to serve meals. And a harpsichord was moved in, for the daughter's lessons. By October's end, the rest of the family was moved into the Great Tower to join the King. Joyous upon being reunited, the royal family couldn't help but be struck by one ominous sight: For the first time ever, the windows were heavily barred.

With the change of the seasons, the weather increasingly turned cold, and, even with a stove, the family suffered. The mold and dampness of the tower were stifling. Little light penetrated the thickly splayed windows. Once the full force of winter hit Paris, everyone shivered, ill with colds and rheumatic fever, especially the dauphin, and also the king. It was even feared that Louis would die. Unwilling to risk this, the Commune allowed the old royal doctor, Le Mournier—he was in his seventies—to treat Louis. In the meantime, as the days grew shorter, the prisoners were monitored ever more closely, and the convention increasingly debated whether the king should be brought to trial for treason.

The Girondins labored mightily to prevent this from happening, fearing that Louis's execution would intensify foreign assaults upon France. Danton thought so too, but he was outargued by Louis Saint-Just, a young disciple of Robespierre. Though just twenty-five, Saint-Just captivated the deputies with his almost poetic case for regicide. "Louis has combated the people and has been defeated," he held forth. "He is a barbarian, a foreign prisoner of war ... He is the murderer of the Bastille, of Nancy, of the Champs-de-Mars ... of the Tuileries. What enemy, what foreigner, has done you more harm?" This eloquent attack changed minds, but not enough to tip the balance. For several weeks a delicate equilibrium held—the king would not be tried, nor would he be granted his freedom. But the scales tipped when on November 20 an iron box was found hidden deep in a wall of the royal chambers of the Tuileries. Presented to the Convention, it was filled with an astonishing array of royal documents, 625 in all. And on its face, it provided powerful ammunition for the charge of treason.

Here, in one box—the infamous *armoire de fer*—was proof of the king's double game since the beginning of the Revolution: the king's dealings with the moderates, Lafayette, Mirabeau, Dumouriez, Talleyrand, Barnave; his communications with numerous émigrés; his communications from foreign powers; and, of course, his bribes to reactionary journalists. It became impossible for the Girondins to argue that the king, in spite of his professed fealty to the constitution, had meant anything other than to plot the destruction of the Revolution. Thus, by the first frosts of winter, more delegates were demanding the trial of the king; here, they were quickly joined by the Paris Commune. On December 3 Robespierre lent his voice to the cause. But as was so often the case, it was Marat who outmaneuvered everyone. A feuder, a hater, but a revolutionary genius, Marat abandoned decorum and dignity by turning, in effect, to extortion. He mentioned that all voting in a trial should be in public and by oral vote—the *appel nominal*—thus exposing "traitors" and boxing in the reluctant Girondins. Now they too were hostages of the sansculottes in the darkened alleyways of Paris.

Around this time, a translation of the trial of England's King Charles I—he had been decapitated with an ax—was prominently displayed in Paris's bookstalls; it became a best-seller.

∞

DECEMBER 11. ON a cold, windswept day, it began with the muffled roll of drums off in the distance. At eleven o'clock, two municipal officers entered the tower and wrested the dauphin away from the king; there was no explanation. At one o'clock, Paris's new mayor, accompanied by soldiers, arrived at the tower to inform Louis that he was to be brought to the Convention to stand trial. As the moments ticked by, Louis slowly absorbed the immutable facts of his situation.

The Convention was packed that day; the back of the hall was actually converted into boxes, as in a theater, replete with spectators. Men dined on sweetmeats and ladies wore their finest dresses, sipped dainty wines, and nibbled on ices and oranges, while the ushers escorted the mistresses of the former Duc d'Orléans, now Philippe Égalité. The Convention pugnaciously read thirty-five charges against Louis. Louis, without counsel, stood calmly before his questioners, pleading lapses of memory, lapses in protocol, failures of his ministers; actually, for the most part, he acquitted himself well (Accusation: "You caused an army to march against the citizens of Paris!" Answer: "At that time I could order troops to march where I pleased"; Accusation: "You vetoed [the decree against refractory priests]" Answer: "The constitution allowed me the free exercise of a veto on decrees"). When the incriminating documents from the box were produced, Louis prevaricated, claiming his signature was forged and even denying the existence of the box itself. At times the audience, for that was what it was, cheered and chortled and hissed between pursed lips. Finally, at some of the charges, Louis lost all patience, crying out, "No, no! No, sir, NO!" At another point, he decried an accusation as "absurd." This time Louis requested counsel; after heated debate, the Convention agreed.

Louis also requested, and received, a four-day deferral to let him employ his attorneys. Remarkably—because he was putting his own life on the line—seventy-two-year-old Chrétien de Malesherbes gallantly offered to defend the king. Even before the Convention, Malesherbes was no ordinary lawyer. Regal, aging, and learned, he had a special status in France: A former parliamentarian and a secretary of state of the royal household, he had boldly defended the philosophes under Louis XV, and for most of his distinguished career had been a tireless reformer. An enemy of royal absolutism, he was an inveterate crusader—for prison reform, for economic reform, for civil rights, for the end of censorship and for freedom of the press, and against the dreaded lettres de cachet. His great-grandson would be Alexis de Tocqueville. And yet he was the very antithesis of the proceedings around him. Where so many other lawyers of the Convention were flamboyant, Malesherbes was the soul of propriety; where they were self-aggrandizing, he was self-effacing; where they were a cauldron of prejudices, he was contained, methodical, a man of immense determination. Conscientious to a fault, he was known for his scrupulous attention to detail and, above all, his integrity. "Your sacrifice," Louis told him, "is the greater because you are exposing your own life, though you cannot save mine." But the awesome fires that had fueled his passions even as a young man had not dimmed. When asked by a deputy at the Convention what made him so fearless, he wanly smiled: "Contempt for you and for life!" Along with his daughter and granddaughter, he was guillotined in April 1794.

But today the accusing finger of treason was pointed squarely at Louis:

He had left France as "a fugitive," the Convention maintained, in order to return as "a conqueror." There seemed to be nothing left to say.

Still, Louis told Malesherbes that his defense team should approach the trial as though he could win. But two weeks after it started, he knew better than that. This was a political trial, not a legal one; even Robespierre trenchantly observed that if the Constitution of 1791 were applied, the National Convention would be on trial, rather than the king. On Christmas Day, Louis sat down in near darkness, composed himself, and made out his last will and testament, not as Louis Capet, but as Louis XVI, the King of France. It was an astounding document, astounding in no small measure for the manner in which it preached forgiveness. Free of cant or rancor, it exhorted his son that in case he should become king, he must dedicate his whole life to the people's happiness—and on no account seek vengeance. And to his wife, Louis wrote with special tenderness, begging her to forgive him "all the ills" she suffered for his sake, as well as any grief that he "may have" caused her.

The king knew time was running short. Before his departure from the tower, in an act of gratuitousness, the Convention informed Louis that the dauphin could not be in contact with both his parents during the proceedings. It was an impossible choice: The king could see him, in which case the little boy would be ripped from his mother's arms, or the boy could stay with the queen, in which case he would be barred from his father's company. Marie Antoinette's heart was already broken; Louis felt he could not break it yet again, so the king resigned himself to solitude while Marie Antoinette gave in to despair. From that day on, they were allowed neither to visit nor to have any communication with each other. By now Marie Antoinette scarcely ate and grew steadily frailer and weaker, barely speaking and frequently crying for hours. On Christmas Day, Louis was forbidden even to see her. On January 1, 1793, it was only through their jailers that the royal family exchanged their wishes for a happy new year.

As best they could, friends outside, as well as on the inside, sought to help. One royal supporter ingeniously placed a magic lantern on a third-floor window of a nearby house and projected letters to provide them with news. On Marie Therese's fourteenth birthday, Clery managed to smuggle an almanac for 1793—a small gift—from the king to his daughter, and on a few occasions, he inventively concealed crumpled notes in balls of string that he shuttled back and forth between the king and queen. One of the

Convention's commissioners—stricken by the dignity and malaise of the royal family—took pity on them and replaced their harpsichord, which had deteriorated. The commissioner himself was soon replaced.

Meanwhile the royal family bided their time; at least what was left of it.

∞

LOUIS'S LAWYERS PUT forth a stubborn defense, and it was not without merit. Louis was portrayed as a "victim of circumstances" rather than an unflinching tyrant; anyway, he had been granted inviolability by the National Assembly, they argued (rightly), and the veto had been awarded to him by the Legislative Assembly. Moreover, the constitution had not given the deputies any rights to try the king, who was perfectly justified in struggling for his life. As to treason? He was one of the most compassionate leaders to have ever occupied the Bourbon throne, and one of France's more liberal monarchs ever. And they eloquently reminded the deputies that it was Louis who gave wings to the Revolution by convening the Estates-General—and who had never knowingly shed his subjects' blood. It was Louis, they passionately asserted, their father figure and giver of liberty, who had invited all Frenchmen to confide in him the injustices of the state. Finally, it was Louis who instituted more reforms than any monarch before him. In truth, the accusations themselves ranged from banal or silly (one charge was that the king pretended to be ill so the public would allow him to go to Saint Cloud in 1791) to serious. For their part, the prosecutors put forth an equally stubborn case. The guilty man brought before them had negotiated with foreign powers for the defeat of the Revolution and, they insisted, deserved no special consideration merely because he had inherited a throne. Perhaps their most damning argument was indeed political: As long as Louis remained alive, plots would fester to use him to undo the gains of the revolution. Better, they reasoned, to make an example that all sovereigns might consider before running roughshod over the aspirations of their peoples. In other words, as Simon Schama has deftly put it, "A king had to die so a republic could live."

To the outside, however, the king remained a sympathetic figure. In the north, at Rouen, a riot broke out on his behalf. In Paris, moved by the king's plight, Gouverneur Morris wrote to Jefferson of the poignancy of the trial. "It would seem strange," he commented, "that the mildest monarch who ever filled the French throne . . . should be prosecuted as one of the most nefarious tyrants that ever disgraced the annals of human nature"; indeed, at one point, Louis's answers actually moved some of his accusers. And in the meantime, efforts were still furiously under way to salvage the king's life. Overtures by foreign powers proposed buying votes for the king; Danton was reportedly the purchasing agent, and Pitt of England the purchaser. But the vast sums required—2 million francs—were more than the continent's royals could stomach.

On January 14, 1793, the question was put to the Convention: "Is Louis Capet, former king of the French, guilty of conspiring against liberty and an attempt against the state?" The voting, by voice, commenced, just as Marat had engineered. Until now, the Girondins had hoped to keep the king alive as a hostage against the armies of the rest of Europe. But Robespierre, who would speak eleven times at the king's trial, maintained that Louis Capet had condemned himself by his own actions, while Saint-Just thundered that Louis could not be judged because he is "already judged." Louis was thus to be convicted not for what he had done, but for what he was. Of the 749 members, 693 voted for conviction, including the king's cousin, the former Duc d'Orléans. There were no votes against. A motion was tabled to submit this verdict to ratification or repeal by the people of France. "An appeal to the people," Saint-Just warned, "would not that be the recall of the monarchy?" For his part, Robespierre, who had always advocated democratic suffrage, now turned his own logic on its head. "Virtue," he said, "has always been in a minority on the earth." He added coldly: "I do not recognize a humanity that massacres the people and pardons despots."

On the next day, when the fateful question was put before the Convention: "What sentence has Louis, King of the French, incurred?" chanting erupted in the streets, as did violence. Pleas were again made to keep the king incarcerated until the end of the war, at which point he would be exiled. Condorcet, who was an ardent revolutionary and had contributed to drafting the constitution, passionately called for the abolition of capital punishment, now and forever; it fell on deaf ears. Thomas Paine had the most ingenious solution: that the king be banished to America, where like the exiled Stuarts, he would sink into obscurity; this move was supported by Gouverneur Morris and the Girondins. But the Convention had come too far to resort to half measures. Marat mocked Paine for revealing a Quaker softness, and in the squares and the boulevards the armed crowds demanded Louis's head, threatening anyone determined to vote for anything less. Deputies, smug only yesterday, vowing never to ask for the king's execution, now began to feel queasy. Most feared for their own lives, and in slow and sepulchral tones cast their lots for death. But there were dissenters. Paine held firm though Danton yielded. D'Orléans, hoping to succeed his cousin, voted for execution. Robespierre, who always opposed capital punishment, now insisted that a living monarch would endanger the republic, while Brissot protested the notion that for 25 million people to be free, one man must die. Finally the roll was called. The voting lasted for thirty-six hours and couldn't have been closer. Filled with parliamentary twists, the concluding tally, taken at three in the morning, was essentially a majority of one—361 for unconditional execution.

On the seventeenth, a devastated Malesherbes saw the king for the last time. As he took his leave, he fought back tears. Louis, squeezing his hand, reassured him: "Tenderhearted old man, don't cry: we shall meet again in a happier life."

Within four days, the king would be dead.

∞

UPON RECEIVING THE news, Louis displayed little emotion, except for a smile of contempt when informed that he had conspired "against liberty." For all his many weaknesses, the king's calm was, witnesses attest, evidence of valor and nobility of spirit; watching his reaction, the minister of justice remarked that he was astonished at the king's "superhuman courage." The king asked for three days to prepare himself spiritually. This was turned down. But upon Louis's request, an obscure nonjuror priest of Irish ancestry, Abbé Edgeworth, was permitted to administer last rites to the king—Louis hoped that Edgeworth's obscurity would save the prelate from persecution. Meanwhile, that evening, under a half moon, the wails of criers beneath the tower broke the frightful news to the royal family, while Louis conferred with his confessor. Finally, at eight-thirty, the family was allowed to see the king. It was a wrenching scene.

∞

FOR SIX WEEKS, Louis had not seen his family and they had not seen him. The queen entered first, leading the dauphin by the hand; the two princesses shuffled in behind. The king, broken in spirit, perhaps, certainly in heart, appeared a changed man. Still, it was the king himself who told them that he had been sentenced to be guillotined in public, on the scaffold at the former Place Louis XV. At this, his family threw themselves into the king's arms. While municipal guards watched them through the glass door, for several moments there was a gloomy silence, punctuated by tender whispers and sobs and saddened glances. Gathering very close, they now sat and started speaking in low voices. For nearly two hours they talked. Marie Antoinette pleaded with him to spend the night—their *last* night, together. Louis refused; he wanted to spend the evening in reflection and prayer. Her eyes melancholy and distracted, she huddled against the king while clutching the dauphin, who was wedged between the king's knees. The little boy gripped both his parents' hands, kissing them and weeping. Marie Therese, disconsolate with grief, howled frantically, and Elisabeth clasped her brother, unwilling or unable to let go. Finally Louis began to cry. In earlier days, accepting his fate, he had asked the Convention to remove his family from the grim confines of the tower to another place "it thinks proper." But who would heed a request from a condemned king? With enormous dignity, he urged his son to forgive his enemies (those "scoundrels") who had sanctioned his death. Then he gave his children his final blessing.

At around ten-thirty, the king rose. Louis and Marie Antoinette caressed the dauphin's hand, and the princesses hung on the condemned king's arms. Once again, they all wept. "I assure you," the king said steadily, "that I will see you tomorrow morning at eight o-clock." "Do you promise?" the queen asked. "Yes," said the king. "I promise." "Why not seven?" the queen then asked. "Well fine! At seven then," the king replied. "Farewell . . ." He kissed them at length as if unable to let go. Then he tore himself away and retired to his room. The door closed behind him. The princesses sobbed as they went to their quarters; Marie Antoinette barely had the strength to put the dauphin to bed. As Louis climbed the staircase, he could hear the shrieks of his family through the massive doors. His young daughter was carried out fainting.

And Marie Antoinette, daughter of Empress Maria Theresa, sister of Emperor Joseph of the Hapsburg Empire, child of Austria and deposed queen of France, lay on her bed fully dressed and trembling uncontrollably. Her remaining thought now was of seeing her husband one last time.

In terror and anticipation, she counted the moments.

∞

To MANY, AS the Revolution had intensified, the king was increasingly dazed, dependent, listless, and awash in events beyond his control or, some have argued, his comprehension. But in recent weeks, Louis had become a different man: the compassionate and forgiving king of his imprisonment. Now, back in his room, he collapsed into a chair, prepared to spend his last night on earth. At eleven o'clock, Clery, tears welling in his eyes, suggested supper. Louis hesitated, but on reflection, accepted the offer, no doubt as much for Clery's peace of mind as his own. The king gnawed on some chicken, downed two glasses of wine, then pushed his plate away. After supper, Edgeworth proposed that the two celebrate Mass in the morning, something the king had neither dared to ask for nor thought possible. For once, the municipal officers acceded to this request. At half past twelve, Louis went to bed. Clery undressed the king and was about to curl his hair when Louis decided, "It's not worth it!" Then, as he lay down and Clery drew the curtains, the king interjected: "Clery, you will wake me at five o'clock."

Clery assured him he would.

IN THE OTHER rooms of the tower, the three women and the dauphin lay sleepless, awaiting their last good-byes to Louis. Finally, at six o'clock, a man came to see the queen. But the knock on her door was not to see the king; rather, it was to get a prayer book. Edgeworth, ministering to the king in his last moments, counseled against seeing his family again; he worried that a final visit would make the parting too inhumane. Hearing that she would not see Louis, the queen was unable even to speak.

From her room, she simply just listened intently. As dawn broke, outside an unprecedented hush descended over the city, punctured only by the rhythmic beating of the *generale*.

It was an unusually cold day, winds swirled, and snow still lay from a heavy fall. As the first thin rays of daybreak revealed the majestic buildings of Paris towering against a still wintry sky, Louis, awoken by Clery in dampness and dark, received Communion from Edgeworth at around six. Just before nine, doors opened with a crash and the commander of the National Guard came to take Louis to the scaffold. The king had wanted Clery to trim his hair to speed the execution, but in yet another small cruelty, he was denied. Louis sneered: "The men see daggers ... everywhere ... Alas they know me very badly ... No, since die I must I will die well!" Upon leaving the room, he weathered one final affront: None of the National Guards had removed their hats.

Escorted by the guardsmen, he descended the stairs.

∞

OUTSIDE, THAT MORNING, the city gates remained locked. Accompanied by 1,200 guards, Louis was delivered to his execution not in a tumbrel but in a carriage. In slow time, the procession began. The carriage crept forward through the morning's dull, steely light, as the people of Paris watched, speechless and still. It crept past the silhouetted crowds, past the buildings that receded into the distance, past the landmarks that were once the glory of the Bourbon dynasty. Perhaps most shocking of all was the quiet. And in the nearly two-hour journey—for a mere three miles—the king could see an extraordinary sight: 130,000 men, practically the whole armed force of Paris, lining the streets, stretching from curb to curb. All of revolutionary France was here. But along the way, Louis's numbed gaze was fixed: While the carriage clicked monotonously on the streets of Paris, Edgeworth and the king were reciting "alternate verses" from the psalms for the dying.

At the square of Place de la Revolution, where the triumphant statue of his grandfather had once stood, Louis stepped out of the carriage. Overhead there was the immense dome of sky, and below there was a vast sea of faces in every direction, some twenty thousand clogging the square and the boulevards. Louis looked pensive but not downcast. However, he was incensed when the executioners sought to bind his hands behind his back. He tentatively glanced over at Edgeworth for aid. "Sire," Edgeworth said softly, "in this further outrage I see only a final resemblance between your majesty and the God who will be his recompense ..." Louis consented.

Supported on Edgeworth's arm, the king, with a weakening stride, climbed the scaffold.

∞

THE STEPS WERE steep, the king's hands were tied behind his back, and he was heavy and feeble from lack of exercise. It was unclear if he would even make it. There was a pause. Suddenly Edgeworth exhorted the king: *"Fils de Saint Louis, ascendez au ciel*" ("Son of Saint Louis, ascend to heaven")—or something to that effect; but whatever he said, it filled Louis with a blast of courage. The king hurried forward to the front of the platform, stood erect, and proudly sought to address the multitude. The head of the Paris National Guard signaled a drum roll; Louis commanded quiet. He now embraced the French not as accused and accusers, or counterrevolutionary and revolutionary, but as countryman and countrymen. "Frenchmen, I die innocent; it is from the scaffold and near to appearing before God that I tell you so. I pardon my enemies. I hope that the shedding of my blood will contribute to the happiness of France and you, unfortunate people"—but at that instant, the guard ordered, "*Tambours!*" and the din of fifteen drummers extinguished the rest.

Suddenly there was a grave silence. Louis was strapped to a plank, there was a hiss and then the thud of the heavy blade, and then for the barest of moments, a perfect stillness.

∞

INSIDE THE TOWER, the queen heard the frantic spectators and shouts of joy, informing her that the king was dead. Amid the penetrating screams of her children, Marie Antoinette had no words or protest; she was nearly catatonic. Finally, it was the king's sister who spoke up. "The monsters!" she screamed. "They are satisfied now!"

After Louis's head was held aloft, people lined up to dip handkerchiefs, pikes, envelopes, fingers, anything they could get their hands on, in the blood that collected in the trough under the guillotine. Hats were tossed in the air, and bystanders linked arms, singing and rejoicing. Robespierre exuberantly declared that the execution "imprinted a grand character on the National Convention" and made it "worthy of the confidence of the French." It was now only the sovereign people who would rule the nation. But not all were so delighted. "On that day," one spectator recalled, "everyone walked slowly, and we hardly dared look at one another." Or as Gouverneur Morris wrote to Thomas Jefferson, "I have seen grief for the untimely death of a beloved parent."

And more terror, and more war, were coming.

By now, France had done something quite unprecedented. With its patriotic festivals, its tricolor flag, its hymns and martyrs and armies and wars, the Revolution had united the disparate elements of the kingdom into a nation and set out to conquer the rest of Europe. In doing so, it had unleashed a new force—nationalism—that would mobilize millions and topple governments for the next two hundred years. And in a curious form of blowback, nowhere was the impact of the Revolution felt so keenly as in the young United States. And nowhere was it so reviled as in the ancient court of Catherine's Russia, itself mired down in conflict and rebellion.

CHAPTER 8

RUSSIA

✦

B Y 1790, AMID THE deepening crisis of war and insurrection, Catherine had been empress of Russia for twenty-nine years. To all who knew her, or knew of her, she was one of the most extraordinary and enigmatic rulers on the earth. The very opposite of Louis XVI, she was an overwhelming personality and a superbly convincing leader, and her prodigious determination and unshakable zeal were almost without peer. So was her cunning, which was well hidden, and her ambition and intense nationalism, which were not. Despite her advancing age, her physical presence alone projected magnetism: Emperors, princes, dukes, and foreign ministers all felt her powerful attraction; for that matter, America's anti-monarchical John Paul Jones had as well. Her conversations were invariably an artful blend of earthy wisdom and learned philosophy. But for the moment, in the midst of a bloody two-front war against both Turkey and Sweden, St. Petersburg was trembling, and it seemed as though the tsarina's world was at the point of collapse.

Would she rise to the task? Catherine was nearing sixty, and, more than ever, the burdens of empire sat heavy upon her. Weary from fatigue and fighting, she was stout and gray, with swollen legs; she also had high blood pressure and now had to read with spectacles and a magnifying glass. Typically, she still rose at six A.M. to begin her morning workday in her bedroom; she also still labored for twelve hours. Here, she signed papers, drafted memorandums, scribbled notes in a little pocket book, and received ministers, signaled by ringing a small bell. She ate only breakfast and lunch, and though surrounded by an army of courtiers, aides, servants, and doting blackamoors, she would ladle out the soup herself; she often washed her food down with a glass of Hungarian or Rhine wine. Around midday, she would take a walk or go for a drive; one of her most gratifying moments was when she threw bread from her windows for the thousands of birds that regularly collected outside. To relax, she embroidered when it snowed, or she read, or she played whist with privileged guests. Unless it was a state occasion, she dressed very simply. But for formal events, there would be great balls, fireworks displays, and magnificent fetes; then she wore a diamond crown, carried a scepter, and draped herself with ermine: Her head was held high, her expression was haughty, and her bearing was assured, so much so that one foreign observer noted that she could have been the "Queen of the world." Indeed, her palaces, and her empire, were more like Versailles with a Slavic gloss.

But with Swedish forces approaching—King Gustavus had bragged about breakfasting in St. Petersburg—her own capital in chaos, the French Jacobins increasingly out of control, the war with the hated Muslim Turks stalled and her cherished "Eastern Project" in tatters, the Poles in virtual rebellion, and Prussia, egged on by England, threatening to join the war as well, hers was a calamity of bottomless proportions. And in the murky chill of the Russian twilight, she brooded that opportunities were slipping away. How long ago was it that the empress, the Austrian emperor, and Potemkin had fondled their dreams of driving the Ottomans from Europe? How long since she was convinced that Russia would capture Constantinople, that Joseph would absorb the Balkans, and Potemkin would be made King of Dacia—today's Romania? How long since she celebrated jihad and crusade coming face-to-face in southern Russia? She confided to Potemkin that the strain was so great that she had almost collapsed, suffering from a "strong bout of diarrhoea" and "colic wind." But she was determined to remain resolute; in this, she drew upon Russian history and ancient tradition. In moments of tumult, the empress took to carrying a tiny snuffbox portraying Peter the Great; fingering it, she asked herself: What would Peter do?

He wouldn't retreat or panic, that much she knew; nor would she. But it was tempting. At this point, the Prince of Nassau had been horrifically routed by the Swedes at Svensksund, where his naval squadron had been victorious just a scant year before. If Prussia were to march on St. Petersburg now, the city would fall and all would be lost. In despair, Nassau begged to be released from command; he also returned to the empress the decorations that she had once so triumphantly bestowed upon him. As with Potemkin earlier in the Turkish campaign, Catherine refused both. "Heavens," she wrote him back, "who has not experienced great failures in his life? The greatest captains had had their disappointments." In truth, these words were written as much to bolster herself as her admiral. But for all her talk, she suffered. And for all her public confidence, she nursed unfamiliar doubts and fretted. One morning, an aide arrived in her study only to find her immersed in Plutarch, "to fortify my soul," she declared. She was convinced that the naval defeat at Svensksund did not bode well for the future.

She was wrong. Surprisingly, the debacle worked to her benefit: Basking in his humiliation of Russia's vaunted northern fleet, Sweden's King Gustavus III celebrated this great and grand victory by announcing that he was prepared to entertain a truce; actually, all the political parties of Sweden were clamoring for him to end this two-year, bloody war. Suddenly St. Petersburg was now safe, and just as suddenly, Catherine contemplated peace, so long as it was on honorable terms. She would not cede one verst of Russian soil, but she did recognize Gustavus as the absolute king in his country—a moral victory for him. He, in turn, wrote the empress, imploring her to forget this war, "like a passing storm." On August 14 Catherine duly signed a peace accord at Verela, based on the status quo ante bellum. "We've pulled one shoe out of the mud," exulted the empress to Potemkin. "When we pull the other one out, then we'll sing Hallelujah!" But not yet, for that shoe was still hopelessly mired.

∞

OUTSIDE OF THE capture of the Turkish fortress of Bender—Potemkin and Suvorov had smartly accomplished this without the loss of a single life—and Akkerman, and though their Austrian allies occupied Belgrade, Russia had not a single major victory to celebrate against the Ottoman Empire. As the conflict ground on and the body count mounted, England and Prussia again urged Catherine to make peace on the basis of restoring all terrain won in the war. And Prussia's King Frederick William II had been stirring nationalist agitation in Poland, promising Polish patriots that he would help them shake off Russian hegemony; the two countries even signed a defensive alliance in March 1790. The empress swallowed every insult, but knew she must bide her time if she were to retaliate. First things first—she could only deal with the Europeans after she had laid waste to the Turks.

Whether out of obstinacy or arrogance, pride or patriotism, the combative Catherine was determined not to yield a scintilla of Russian territory, including the ones she had conquered. She once boasted, "If I were to live 200 years, then of course all Europe will be subject to the Russian Scepter!" Presiding over her vast empire, Catherine had long since forgotten that she was born a German; to the tsarina, her legitimacy was not a matter of blood, but of choice; it was not a matter of realpolitik, but of love; it was not a matter of endurance, but of history. Her lineage no longer stretched to the obscure Christian Augustus of Anhalt-Zerbst, but to Peter the Great of the Romanov dynasty. Russia was now indistinguishable from her; to take any piece of it would be like performing an amputation.

As armies marched, galleys rowed, and ships sailed, a "merry" Catherine (her word after being bolstered by the accord with Sweden) now realized that before Prussia and England fomented their next moves, there was a sudden urgency to smash the Turks. For starters, the sultan had undertaken an ambitious diplomatic campaign to isolate Catherine; for another, he had bolstered his troops on the Danube and in the Caucasus. So to inspire Potemkin—her commander-in-chief insisted that he was "tired as a dog"—she sent him a gold coffee set and a diamond ring. The Prince of Taurida was ensconced in a new headquarters on the Dniester River, where he supervised armies and navies on all fronts while keeping contact with Warsaw, Vienna, the Balkans, and Petersburg. Here, in a decayed Tatar town, nestled among steppes, mistresses, and indolence, Potemkin finally realized that diplomacy was futile. "I'm bored by Turkish fairytales," he told an aide. "Pray to God he will help us. Put all hopes in him, cheer up the crews and inspire them for battles." With that, Potemkin rushed to inspect the fleets, and prepared to fight his way up the Danube to rendezvous with an army beneath the most formidable Turkish fortress in Europe: Ismail.

By November he wrote, "I will make an attempt on Ismail . . . but I don't want to lose ten men."

Thus would begin one of the most ghastly battles of the century.

∞

THE GREAT CASTLE of Ismail was built into a natural amphitheater that was defended by 265 cannons and a garrison of 35,000 men, comparable to the entire British army in 1793, as well as troops sent over from Akkerman. It was the pride of Turkish fortified cities and the symbol of Ottoman power. But if Ismail surrendered, the Russians would have control of the Danube and a clear path straight to the gates of Constantinople itself. Without it, however, the Russians were dangerously bottled up along the river, for Ismail housed a Turkish army that could entrap the empress's legions and slice through Potemkin's rear guard.

For months, Potemkin had attempted a siege of the fortress, but as he studied the maps and reports, he realized that Ismail had no weak point. Indeed, the fortress was all but invulnerable: it was a jagged triangle of jutting walls, ubiquitous ditches, imposing towers, with the Danube River providing further protection. Moreover, French and German engineers had jointly strengthened its "brilliantly constructed" defenses. By every measure, the situation was impossible: The moats were thirty-six feet across and twelve feet deep; the four gates were impregnable. For every two Russians there were three Turks. And in the seven months that Russian forces had been there, they had achieved nothing, nothing except the loss of many men from hunger and the ravages of disease. It was Ochakov all over again. So as Potemkin encamped at Bender, he feared that if Ismail did not fall, then the whole prestige of the empress's forces would be at risk. "The slaughterer must never show his knife," he had once written. "Secrecy is the soul of war." But on December 2, while Potemkin played cards with his harem, his secret weapon arrived: Emaciated, riding in on a Cossack pony with a solitary attendant and looking more like a Tatar foot soldier than the commander of a fearsome army, it was General Alexander Suvorov.

Today, Suvorov's very name evinces a mixture of dread and awe, disgust and dismay. Intellectually at home on the fields of war, he likely would have more than held his own with an Alexander the Great or a William Tecumseh Sherman, a George Patton or an Erwin Rommel. Even in Potemkin's time, the idiosyncratic Suvorov was a rarity: known for his frenetic energy and blunt talk, he seldom bathed and always slept in his uniform, he banned mirrors from his quarters and regularly ate on the floor, and he loved to sing. He was also past sixty; by his own admission he confessed, "I am covered with wounds, sixty years old and the juice in the lemon is all dried up." But he was Catherine's greatest general. And where Suvorov went, glory followed. So did victories. Upon arriving at Ismail in his trademark smock, devoid of badges or signs of rank, he promptly set to work. He reconfigured the emplacements and artillery batteries, ordered ladders to be built and ditch coverings to be made, and, in an astonishingly modern turn, drilled his soldiers nightly in mock assaults of the fortress, even having them scale earthen ramparts just like Ismail's. He also shrewdly engaged in psychological warfare. With the Turks watching, the Russians dug and built, sang and swore; and they trained. Meanwhile, Potemkin waited tensely in Bender, knowing that the Turks were convinced, as were the English and the Prussians, that Ismail was unassailable.

Suvorov thought not. Pointing to the bastion, he told his men: "There is the fortress. Its walls are high, its ditches deep, but we must take it. Our Mother, the Empress, has ordered it!" As it was, Suvorov's peculiar habits and simple ways won him the hearts of his men; he electrified them. So on December 7, a trumpeter and an unarmed Cossack bearing a white flag were dispatched to the fortress with a twenty-four-hour ultimatum; it contained two letters from Potemkin and Suvorov demanding that Ismail surrender to prevent what the prince said would be the inevitable "blood" of women and children. Suvorov was blunter: If Ismail resisted, after the "first shot" it would be "too late for humanity" and "nobody" would be spared. Yet the Turks were defiant; they tauntingly massed on embankments hung with banners—already they were anticipating final victory—while the Seraskier, Aydos Mehmet Pasha, asked for a ten-day truce. Suvorov rejected it. Instead he convened a council of war on December 9. Brigadier Platov, who would later distinguish himself in the War of 1812 against Napoleon, spoke first; he murmured one word: "Storm." One by one the rest of the generals agreed. Suvorov thus ordered the taking of Ismail, from all sides—six columns on the land side and four across the Danube. "Tomorrow," the general informed his men, "either the Turks or the Russians will be buried at Ismail."

Back in Ismail, the Seraskier responded with his own deadly sobriquet to his troops: "The Danube will stop its course, the heavens will fall into the earth before Ismail surrenders."

∞

As a thick mist swirled around the embattlements, the Seraskier's prediction seemed about to come true. The thunder and crash of cannonade pounded the garrison until the ground seethed; then at half past five a whistle pierced the air and a single rocket rose, arcing high over the city. It was the order for the Russians—25,000 of them—to advance in simultaneous columns. With great ardor, they did. As the Russians drew closer, swarming in mass, the artillery inside Ismail opened up, and the perimeter of the mist-shrouded city was lit with stabbing red and orange flashes. The steady pounding of Turkish guns inflicted a horrific toll on the advancing men. Noted one observer, the ramparts were crowned with flames, "a spectacle of horror and beauty."

From his scouting reports, a pensive Suvorov felt that the river side was the most vulnerable. Indeed, according to plan, one of his river columns was the first atop the walls. The advance on land told a different story. Major General Kutuzov, the future conqueror of Napoleon, spearheaded his men twice; twice they were repulsed, taking bitter losses. The next time he muttered tartly, "God be with us!" Throwing himself at the wall again, he succeeded; the soldiers crossed over on the dead bodies of their fallen comrades, and a priest holding a crucifix aloft followed with the reserves. By the time the first hint of daylight broke, wave upon wave of Russian columns had surged onto the outer ramparts.

The enemy put up a tough, unyielding defense. Their gaps filled in; their ranks drew together. As one Russian commander laconically explained, "Every step required a fresh victory." He did not overstate. The Turks' opposition stiffened, bitter hand-to-hand combat began, and the men slugged it out for hours, with a savagery that bordered on the demonic. Russian losses were higher, but as the minutes passed, fresh Russians streamed to the lines. Once inside, they rushed into Ismail like a torrent that, as one man put it, "floods the countryside." Sixty thousand men joined one another in a gruesome minuet of death, while fires broke out and thick smoke gathered into clouds that hung over the tortured city; the combat in the southwestern corner was particularly vicious. Soldiers were slashed with bayonets and swinging sabers, arms and faces were cut to pieces, heads were severed, and as hussars clashed with Janissaries and Cossacks with Negroes, the wounded were entombed alive by slabs of newly dead bodies or screaming troops running forward. Toward eleven o'clock, the Russians, in dense clusters, fitfully inched toward the center of Ismail, enclosing their enemy in an iron ring. The Turks, wielding daggers, fell back. But as late as midday, the outcome was still hanging in the balance.

As the streets were gradually cleared, the Turks defended themselves with cunning and heroism. No place was safe. The organized resistance fought from homes and barracks, alleyways and shops; they were everywhere and nowhere. Ismail became a scene of unimaginable horror. While the rhythmic cries of "Allah! Allah!" rang through the streets, Suvorov's men, having pushed the Turks to the rear, were then overtaken by a ferocious bloodlust, a fanatical drive to destroy everything in their way. Bedlam reigned, and the citadel was given over to plunder; as one Russian recalled, "the most horrible carnage followed, the most unequaled butchery." Now the Russians flocked in with venom and confidence against their increasingly disorganized foes. One by one, Turkish guns were silenced. Little by little, as the ground underneath the Russians became littered with bodies, blood flowed through the streets like rainwater after a sudden storm. Then even women and the elderly were casualties of the Russian fury. Nor did the feverish shrieks of children, their little hands hysterically waving, slow the attackers. More tragically still, thousands of Tatar horses broke free from the underground stables and trampled over the fleeing and the wounded, crushing their limbs and splitting their skulls. Wherever one looked there were now corpses, corpses wet with sweat and urine and corpses caked in blood. Yet the Seraskier and thousands of his finest men, watching the carnage, were still standing their ground.

The resistance did not last. But before they could surrender—actually, they soon wanted to—a Russian soldier lunged toward the Turkish commander, hoping to capture him; with a cold stare he fired, but was then set upon by more than a dozen bayonets. At this, the murderous Russians continued their bloodthirsty bedlam, and systematically picked their way through the rubble in search of fresh prey. No one was exempt, not the dying, the scared, the valiant, those who pleaded to surrender, and those who did not. In the end, some 3,000 Turks lay dead.

Suvorov had ordered that on no account should the "lives of Christians and the disarmed" be taken. It was a futile gesture. Now the looting began; so did the havoc. Carts were upturned and ravenous troops raped Turkish women. While the sweet scents of spices mingled with the sobs of the wounded, and riderless horses stumbled over the mangled dead, the Russians tore apart the city, breaking into houses and stores, seizing more than 2 million rubles. In wanton acts of cruelty, and as dogs barked and innocents wailed for help, they forced their victims to strip before cutting them down; in this way they kept their victims' garments free from blood. Then, in a stench of perspiration and spent gunpowder, frenzied Russians covered themselves in Turkish clothing, men's as well as women's, pieces pulled from the living as well as the dying, or wherever they could find them; even the Cossacks, in wigs and dresses, roamed through the streets, their swords and seven-foot pikes dripping with blood, their legs soaked in mud from the prolonged water assault. Soon the entire city was enveloped in flames, and the bodies were so numerous that the Russians could not step over or walk around them. The butchery continued until four in the afternoon; at that point, the remaining Turks finally gave up. The moans of the near dead and dying echoed into the night, until their cries grew fainter and fewer and more heartrending.

It was over. Thousands of pieces of unrecognizable flesh bled into the ground. Only one senior pasha had survived. The epilogue was lurid. Stirring in a storm of dust and sweat that was infused with gold by the slowly descending sun, the pasha spread carpets amid his decimated fortress and lit his pipe as calmly as if he were safely ensconced inside his seraglio. In this manner, a gateway to the Ottoman Empire now belonged to the Russians. It was one of the most horrific massacres of the century, and, really, of any modern century. By day's end, nearly 55,000—a staggering figure—were dead: at least 12,000 Russians, which paled in comparison to the Turkish toll: about 40,000. By contrast, at Yorktown, there were just 80 American casualties and 500 British.

After three days, the task of clearing began: Turkish bodies were dispatched to the waters of the Danube, while the Russians were carted off to mass graves. Ismail, its final agony complete, shuddered and died. On a scrap of yellowed paper, a rejoicing Suvorov informed Potemkin: "Nations and walls fell before the throne of Her Imperial Majesty. The assault was murderous and long. Ismail is taken on which I congratulate Your Highness."

∞

THE PROLONGED SIEGE of Ismail had worn down both Potemkin and Catherine. The victory was a different story. Though pained by the heavy loss of men (for her part, the empress, with her typical singleness of purpose, was not), Potemkin was ecstatic; he enthused, he was happy as "a Sultan." He ordered the guns to be fired in celebration, and dashed off an enthusiastic communiqué to the tsarina: "Since I belong to you, all my wonderful successes belong to you too." Gaily, she answered back: "My health is improving. I think it's gout which had reached my stomach and bowels but I cure it with pepper and a glass of malaga wine, which I drink daily."

But would the Sublime Porte of the Ottoman Empire, now choking on its own blood, make peace, as Catherine fervently hoped? It would not. As it turned out, Britain, increasingly wary of Russia's hegemony over Eastern Europe, especially after the fall of Ismail, created a "federative system": an alliance with Prussia and Poland to force Russia to give up its gains of war and accept a peace; if Catherine refused, the Royal Navy and Prussia's armies were poised to launch an assault. So while Potemkin raced to St. Petersburg to meet with the empress, Frederick William amassed a force of 88,000 men in East Prussia and England's Lord Hood gathered a fleet of fifty-five ships to bombard the Russian capital. And in Turkey itself?

Sultan Selim III executed his latest grand vizier. Then he too promptly gathered another army.

∞

HISTORY AND GEOGRAPHY and, above all, faith, compelled Turkey not to surrender. This should not have been surprising. At the midpoint of the eighteenth century, while the Muslim world had steadily lost military power since the Turks had been repulsed from Vienna's gates in 1683, the Ottoman Empire still reigned over a great swath of the planet: On the Dalmatian Coast, it touched the Adriatic and faced the Papal States; on the Bosporus, it watched over the sole naval outlet from the Black Sea; and it could still block the Russians from the Mediterranean. Theirs was a unique, impenetrable world, a mixture of the sublime and the simple, the exotic and the fantastic. Actually, crossing from Hungarian territory into Muslim lands, one would at first note few differences, but beyond the Bosporus, that all changed.

Here the economic and the cultural landscape was dramatically transformed. By some estimates, less than fifteen percent of Ottoman terrain was cultivated; the rest was arid desert, or lands with fantastically tall grass, or steep mountains suitable only for mining or herding; and farther afield there was, of course, the lowly fellaheen laboring in the Nile's fertile mud or the nomadic Bedouin, inscrutable and blackened by the sun. Other differences abounded. For starters, religion was more powerful and pervasive in Islam than in Christendom: The Koran was the law as well as the religious gospel, and theologians were official interpreters of the law. By contrast, eighteenth-century Christianity was often wedged between Voltaire and Mohammed—that is, between the followers of Enlightenment, who were upending Europe, and the faithful of Islam, who were much of the rest of the world. For another, in the Ottoman Empire there was no forward-looking reformation and certainly no New World as in America. Instead, year after year, and century after century, the annual pilgrimage of the Muslim faithful repeated its moving drama over the desert and along the well-worn roads to Mecca.

Still, there were schisms: the Sunni and Shi'a sects divided Islam and the empire much as Catholicism and Protestantism had divided Western Christianity; and it was in the eighteenth century that a new strain was founded by the pious reformer Muhammad bin-Abd-al-Wahhab, a sheik of the Nejd, in the central plateau of what today is Saudi Arabia. Even in Islam the Wahhabis were unique. The Puritans of the Muslim world, they denounced idolatry, assaulted the shrines of martyrs and assailed unbelievers as worthy of death, and ordered the stoning of adulteresses. Believing that Islam had become detached from its original teachings, they also allied themselves with the desert raider Muhammad bin Saud—this was the classic combination of the "Book and the Sword"—as well as propounded the freedom of any individual to interpret the Koran for himself, both of which would have even vaster implications for the modern world. And finally, across many Ottoman lands, the clergy themselves had triumphed over the great classical scientists, philosophers, and mathematicians who had once defined medieval Islam: Too often astronomy dwindled into astrology, chemistry into alchemy, and the realm of magic was considered as real as the world of sand and sun.

Yet there was no doubt that the Ottoman Empire believed its Islamic civilization superior to that of Christendom—whether France or England or Russia; America, of course, hardly bore mentioning. Outsiders felt this too. As one diplomat wrote, "The Arabs and Turks, whose books are men's faces . . . and whose glosses are the common saws and thousand old sapient proverbs of their oriental world, touch near the truth of human things. They are old men in policy in their youth, and have little later to unlearn." Or as the English traveler Wortley Montague put it: "The men of consideration among the Turks appear in their conversation as civilized as any I have met with in Italy."

It was hard to disagree. Poets prospered in the empire, for the blistering deserts, the endless dome of sky, the immaculate seas, and the blazing galaxy of stars on a cloudless night roused the imagination with an unconquerable sense of mystery. Even Westerners, like Sir William Jones in 1774, acknowledged their elegance and passion. For her part, the Englishwoman

Lady Mary Wortley Montague was taken by the women, reporting that the Turkish ladies were as lovely in face and figure and as refined in manners as "our most celebrated English beauties." Yet seclusion remained the fate for most Ottoman women, whether wives or concubines or servants. After 1754, they might leave the confines of their homes, if veiled over all but their eyes. Meanwhile, only a father, brother, husband, or son was permitted to enter their apartment, and even after death this separation of the sexes was inviolate: There was one heavenly garden of perpetual bliss for virtuous women, but saved men had their own Elysium, where they could romp with houris—"pure maidens of paradise" who were periodically revirginized.

The Turks took special pride in their public baths and looked down upon the unwashed Christian "infidels." And while the practice of slavery was far more prevalent in Islam, which they themselves acknowledged, the Ottomans saw little difference between slaves in Turkey and serfs in Russia or servants in Christendom. They preferred art to science and handicrafts to industry, and raised flowers with the same tender care as Westerners. Indeed, evidently it was the Turks who brought the tulip to Europe, as well as the chestnut and mimosa trees. And of course the Turks considered themselves unrivaled in pottery and textiles, rugs and decorations, and in architecture too; they looked admiringly upon their intricate ceramics, they basked in the softened gleam of their tiles, and they reveled in the sturdy brilliance of their rugs, whose delicate weaves unfailingly pointed the wor-shipper toward Mecca itself. And Constantinople, though a maze of inter-locking streets and noisy slums, may well have been the most captivating as well as the largest of all European capitals; its population of two million was double that of London, three times that of Paris, eight times that of Rome, and a staggering seventy-one times that of Philadelphia. When Lady Montague gazed out upon the bazaars and the festivals from the British ambassador's palace, she thought the city constituted "the most beautiful prospect in the world." The worldly French aristocrat the Comte de Mira-beau agreed, at one point asking King Louis XVI to appoint him France's ambassador to Constantinople.

However, then as now, much of the Ottoman world remained unfath-omable to Western eyes: Music served both love and war, arousing attack and soothing defeat; the printing press was denounced by the guild of writ-ers as "the devil's invention"; and the harlots of the empire specialized in erotic dances, unfamiliar even to the amorous French or Oriental Russians.

Moreover, beyond war and internal sloth, the empire was not without its own myriad of stresses. Take Egypt, governed by Ottoman pashas and viceroys. True, in many ways it was a thriving nation. Alexandria was one of the greatest ports of the world, and Cairo, among its most populous capitals: Its streets, densely packed with snake charmers and fakirs and the indescribable scent of communal Egypt, were nonetheless surprisingly narrow to block out the blazing sun. The avenues teemed with handicrafts, as the wretched made beautiful things for the rich, while mosques and minarets, more than three hundred of them, fortified the poor. And one mosque in particular, Al Azhar, was "the oldest" university in the world; its three thousand students, spanning the globe from Malaysia to Morocco, came to study Koranic grammar, rhetoric, theology, ethics, and law. Yet by 1769 even the Ottoman Empire was not immune from the changes rippling across the world—or imperial overstretch. Egypt, as it turned out, was seething.

At a time when the American colonies were nearing revolt, and when Russia was perilously close to being in rebellion, Egypt declared itself independent. Its leader, Ali Bey, led Mamluk troops to conquer Arabia and capture the holy city of Mecca, where he boldly took the titles of Sultan of Egypt and Khan of the Two Seas—the Red and the Mediterranean. Later, in 1771, Bey, himself a former slave, dispatched abu'l-Ahahab with thirty thousand men to conquer Syria; he did. But then duplicity fostered duplicity: His emissary mutinied, quickly allied himself with Turkey, and led his army back into Egypt. Bey fled to Acre (in today's Israel) to organize another army, and met the forces of abu'l-Ahahab and the Turks, fighting until he was grievously wounded; captured, he died within a week. That was 1773, and Egypt and its domains were again returned to the Ottoman Empire's fold. No wonder the Ottoman sultans felt superior. Even in the farthest reaches of empire, among the Barbary states, there was a feeling of exceptionalism. For his part, the Moroccan sultan, Mulai Ismail, who reigned for fifty-five years and captured tens of thousands of Christian slaves, sired hundreds of children and considered his achievements magnificent enough to request a daughter of Louis XIV's—the Sun King—for his harem.

So as difficult as it may have been for Catherine to fathom, and it was, one Moroccan traveler joyously exclaimed upon returning from a visit to Europe, "What a comfort to be getting back to civilization!" Or as the Ottoman Sultan Murad IV, wearing a turban to symbolize his victory over

Persian armies after reconquering Baghdad, was told as he proceeded to his golden throne: "You are the pole to which the world turns. The world trembles before you like the needle in the compass."

That pole, of course, was Constantinople, that needle faced Mecca, and that world, the only one that mattered, was the vast domain of the sultan. To Catherine's chagrin, even after the disastrous rout at Ismail, and even as the plague prepared to rage again in Constantinople, the Turks were still not resigned to sue for peace. The sultan was ready to fight on.

Then to Catherine's horror came a new consideration: The French Revolution was spiraling out of control, and with it, the whole continent as well.

∞

"WE DON'T YIELD to the very devil himself," Catherine insisted. But this was just talk. In the north as well as the south, the enemy continued to resist, encouraged by England and Prussia; so in private she finally begged Potemkin to "conclude peace" with the Ottoman Empire.

The time was opportune. Suddenly Russia had a spate of spectacular triumphs: General Kutuzov had defeated 20,000 Turks at Bagadag on June 19, 1791; three days later General Gudovick stormed the fortress of Anapa and captured the Chechen hero Sheik Mansour; and on June 29 Russia's Prince Repnin, the conqueror of Khadzhi-Bei, led forty thousand solders across the Danube and annihilated the troops of the grand vizier at Machin. Only then did the Turks begin to moderate their demands. For Catherine, this seemed to be the opening for an honorable peace. It was long overdue. For one thing, the war was increasingly overshadowed by a French Revolution that was forcing monarchs across Europe to recalculate their concerns; Catherine herself called this new phenomenon of republicanism "a sickness of the mind." For another, the empress was now worried about her precarious state finances, the weariness of her armies, the country's internal difficulties, and the brewing Polish revolt, not to mention the British-Prussian alliance arrayed against her. After four years of bloody fighting, she wanted to get it over with as quickly as possible.

So at the Conference of Jassy in Moldavia, the empress tendered conciliatory proposals. In the meantime, through Potemkin, she bribed Mirabeau (whom she despised) in an attempt to persuade France to join Russia against Britain, while she simultaneously dispatched emissaries to woo and flatter the English. She also even received a secret envoy of Britain's in Petersburg—William Fawkener—attempting the neat trick of dividing the British government among its various factions.

On December 29, 1791, the four-year agony was over: a peace treaty was signed at Jassy. It was not what she would have wanted. Yes, the treaty confirmed that Russia kept all the territory between the basins of the Bug and the Dniester, and solemnly recognized that the Crimea and Ochakov were Russian. But while the whole northern coast of the Black Sea was henceforth Catherine's, the sea itself remained closed because the Turks still controlled the straits. She had not taken Constantinople; her grandson would not be crowned in that city. And four years of incredible sacrifices had rewarded her with only meager territorial gains, the threat of internal riot, and devastation. Still, at least a measure of Russia's prestige remained intact.

But for Catherine, that was cold consolation. For now, a darkening shadow was cast over her empire by the events brewing in France.

∞

YEARS OF POWER, the challenges of revolt and war, the fluctuations of victory and defeat, had accustomed the empress to treat the oppression of the weak by the strong as beyond her means to cure. Yet disturbed by a dozen conspiracies to unseat her, and unsettled by the unrest within her own empire, she became terrified of the French Revolution. It was a striking turn of events. Actually, in the beginning, she was surprisingly indifferent, particularly when it was just the shake-up of an idle aristocracy and an incompetent government; then she was merely irritated by the democratic uproar. She declared to Baron von Grimm, now making his home in Paris, "I am not of those who believe that we are on the eve of a great revolution." To her surprise and dismay, she was, of course, wrong. For then came the earthquake—the fall of the Bastille and the bloody massacres in Paris, all vividly relayed to her in depth by her minister in France, Baron Simolin.

This time she made no pretense of restraining her fury: "How can shoemakers meddle in affairs of state?" she hotly exclaimed. "A shoemaker only knows how to make shoes!" She had a deep-seated disgust for the swarms in the street bent on undermining the monarchy. "The ruling tone in your country is the tone of the rabble," she wrote to one confidant, to which she added, "I don't like justice without justice, and these executions are *à la lanterne*." And France, the empress insisted, was in labor, "giving birth to a rotten, stinking freak." As for the National Assembly? They were nothing but a bunch of hagglers: "If they handed a few and took away from all of them their salary of eighteen thousand pounds," she bellowed, "the rest might change their minds."

But they didn't. So she fervently ridiculed the system "of the hydra with twelve hundred heads," and predicted (rightly) that if revolution took hold in Europe, "there will come another Genghis or another Tamerlane to bring her to her senses; That will be her fate!" And when her good friend Grimm naïvely requested her portrait to present to Bailly, the new mayor of Paris—Bailly would, in exchange, send an image of himself to the empress—she stonily shot back, "It would be unseemly!" Her contempt became insatiable: When a Paris mob forced Louis XVI to leave Versailles and live, as Simolin put it, "like a phantom prisoner" in the Tuileries, Catherine now shuddered at the encouragement given to those who sought similar action in Russia; indeed, already young Russian nobles were dressing in French fashions and sprinkling their talk with French revolutionary jargon; some even had the temerity to sing the French revolutionary anthem "Ça Ira" in Catherine's presence, in her own palace no less. Reeling, Catherine acted quickly with indirect censorship, for the first time even allowing the clergy to forbid the publication of her once beloved works of Voltaire.

In truth, this became a momentous turning point for her. The more the empress thought about it, the more it occurred to her that she detested France, and had never liked it: In her aging, reactionary eyes, it was nothing but a nation of chaos and wanton indulgence; it was also a traditional adversary. Yes, she enjoyed French culture, but that was all—and the honeyed tomes of the philosophes, so appealing on the surface, changed none of this. Just as suddenly, she came to believe that the great French writers, a number of whom she had long studied and admired—Montesquieu, Voltaire, Diderot, Rousseau—were the instigators of France's despicable unraveling. Around this time, the Americans also became equally hateful to her: She condemned their revolution, which she had once professed to admire; derisively called Washington "a rebel"; and bluntly insisted that "men of honor" in Russia could no longer wear the Order of Cincinnatus.

Yet in the months ahead, she continued to dwell luridly on the possibility that the French "fit of madness" might spread. To Catherine, at all costs, Russia had to be preserved from the horrid contagion of the Revolution. When word reached her of Louis XVI's unsuccessful flight to

Varennes, she grew despondent, all the more so because her ambassador to France had secretly aided Louis in his preparations to escape (or so it was rumored). As it happened, Simolin had been forced to flee from the wrath of the revolutionary crowd that had first massed at the Palais-Royal, then on the Champs-Élysées. "They wanted to seize me," Simolin reported to the empress, "and exterminate me for having helped organize the King's flight."

At this stage, relations between France and Russia sharply deteriorated. As French nobles continued to flood St. Petersburg, all Russians—including Simolin—were ordered to leave Paris; in turn, the French newspaper *Le Moniteur* ridiculed the empress as the "Messaline of the North" and French diplomats blasted Catherine as a "hothead." She angrily struck back by outlawing Parisian cravats in Russia; then, in a once unthinkable act, she ripped the bust of her beloved Voltaire out of the Hermitage gallery and consigned it to the attic; if he would rest in the Pantheon of Paris, she would see to it that he would lie amid the dust and detritus of the Winter Palace. Having once so admired the Encyclopedists—Hadn't she supported Diderot and d'Alembert? Hadn't she been their friend and patron when they were being persecuted in France? Hadn't she, the Enlightened Legislatrix, been the idol of liberal intellectuals all over Europe, even publishing Helvetius while his works were being condemned by the Parlement of Paris in France?—she now considered them nothing but fiends, utopians awash with blood on their hands. Yes, they sang sweet phrases of liberty, equality, and fraternity, but in doing so, they had made themselves the philosophers of enmity, inhumanity, and destruction. And more than that, she believed, they had legitimized the vile guillotine.

Perhaps some of those who regarded Catherine as a woman of principle were baffled. They shouldn't have been. She always nailed her colors to the mast, but not always to the same mast. Her sole concern was not liberty but the safety of the monarchy; so she began to focus on France almost at the exclusion of all else. In the name of royalty and civilization, she considered proposing to the Protestant powers of Europe that they should embrace the Greek religion; in this way, they could be preserved from a diabolical Jacobin plague, which was "irreligious," "villainous," and "the enemy of God and monarchs." With a genial little smile, she cuttingly told Grimm, "I maintain that one would have only to seize two or three unimportant little towns in France and the rest would fall itself." Louis XVI might not act, but she was convinced that twenty thousand Cossacks would be more than enough "to flatten everything from Strasbourg to Paris!" But her vow was empty, for the tsarina, always a shrewd practitioner of the old system of balance of power, was careful not to unsheathe her sword; her policy was instead to push and prod—and at times, financially support—the other great monarchies of Europe into action: Austria, Prussia, and even Britain. Why? She still needed a free hand to deal with the mounting cancer of a rebellious Poland, which was taking many of its own cues from the example of the United States and now also from France's murderous revolution as well.

The twilight of autumn 1793 was thus a watershed for the empress. When told of the beheading of King Louis XVI, the empress was so stunned that her doctors feared for her health. Where republicans the world over, from Ireland to America, Poland to Holland, saw Louis's death as a triumph for liberty, and statesmen such as Thomas Jefferson saw it as a necessary evil, Catherine, an enlightened despot to her core, couldn't help but tremble at the grotesque end of a fellow monarch. Regicide? It was as if she herself had been placed on the scaffold. In the days that followed she kept to her bed, awash with grief. Visited by aides, she denounced the act as a "barbarity" and an "obvious illegality," and she sputtered that equality was "a monster."

For three weeks, she was in mourning and wore black; when Marie Antoinette was beheaded, the despair cut even deeper to the quick: she wore black for six weeks.

∞

WHILE DANTON AND Robespierre consolidated their power, everyone awaited Catherine's next decision. By now the empress was incapable of detachment. She hastily broke off all relations with the French government—in America, Alexander Hamilton was urging George Washington to do the same—and pressed the European monarchies to form a coalition to bring France to heel. In the meantime, the increasingly pro-revolutionary French envoy, Edmond Genet, felt the mounting sting of her remarks. Moreover, as Catherine's expression hardened, her joviality faded, and with an ever-rising voice she openly dismissed Lafayette as "Booby the Great," castigating Paris as "a den of thieves," and scolding the revolutionaries as "dissolute scoundrels." Then she went a step further. She canceled the venerated trade treaty that she had negotiated with the Comte de Ségur during their historic trip to the Crimea; she also forbade French ships from using Russian ports. And Genet was expelled in disgrace: "They say that he left Petersburg," Catherine chuckled to Grimm, "cramming a red wool cap down over his head."

As the Revolution's military juggernaut struck out across Europe, suddenly, almost deliriously, she came to believe that the very name of the French must be exterminated, so heinous were their crimes. To the empress, the capital of France was no longer Paris, but the now-thriving émigré court in Coblenz. Meanwhile, at home she went to the extreme measure of ordering a ukase that all the French in Russia had to sign—under threat of immediate banishment—a scorchingly rendered oath saying that they never adhered to "the imperious and seditious" principles now professed in France, as well as of swearing that they regarded the king's death as an act of "abominable villainy."

The French émigrés did, happily. For her part, Catherine no longer regarded them as visitors, but as fresh subjects, owing her their allegiance. There was one notable exception. In 1793 she was thrilled at the arrival of the king's younger brother, the Comte d'Artois and future King Charles X. Prior to the debacle at Varennes, she harbored visions of providing refuge to Louis XVI himself. "That," she cooed, "would be the most remarkable act of my reign." Was this overstatement? To be sure. Nonetheless, a minor princess of Zerbst extending asylum to the grandson of her former nemesis, Louis XV, would have been nothing short of delicious retribution. But with that possibility gone, she warmly welcomed Artois. The tsarina made every effort to impress him; her vanity required that the opulence and ritual of the Winter Palace should equal, even outstrip that of Versailles. It did. However, while she treated her royal guest as a cherished son of France, she found him to be politically naïve, even ignorant. He asked for her help, mainly military aid; for the most part, she demurred, but she did provide a million rubles to assist him in his inchoate insurrection as well as providing a line of credit for the émigrés at the Russian embassy in London. Then, to further prod his crusade against the revolutionaries, she handed him a magnificently jeweled sword that she blessed; it bore the inscription, "Given by God for the King."

The Comte d'Artois bowed, thanking the empress; his face, however, revealed his utter disillusionment. But Catherine believed she had no other choice; as she told a deputy, "I am racking my brains to involve the courts of Berlin and Vienna in French affairs ... so that I may have elbow room. I have [too] many unfinished enterprises." One of those enterprises was a surprising rapprochement with England. This was a remarkable turnabout for the woman who had led efforts to give England a black eye in its war with the American rebels, and who only recently had plotted to ally herself with France against the English; but in a long career of embracing power politics, it was hardly her first dramatic pivot. If foolish consistency was the hobgoblin of little minds, she was eminently disqualified.

Having resolved the Turkish War, she now embraced a strenuous campaign to secure an anti-French agreement with Britain. Already she had help: America's friend, Edmund Burke, had attacked William Pitt for protecting the Turks ("a horde of barbaric Asiatics"), while Charles James Fox asked what possible interests England had in a naval expedition against Russia. She dispatched her envoy, Simon Vorontsov, to rally a pro-Russian lobby of merchants from Leeds to London. For once, her timing was impeccable: Parliament, hearing British merchants mourn the loss of Russian trade, was now ready to give up old enmities. The cry went up: "No War with Russia!" and that, so to speak, sealed the deal. A mutual defense pact soon followed. On April 26, 1793, the Comte d'Artois left for England, and a buoyant Potemkin declared the empress a "spoilt [child] of Providence." His declaration, however, was premature.

∞

As the violence of the French Revolution increasingly intensified, lurching from the Assembly to the Committee of Public Safety, from the beheading of the king to the Terror, from the Girondists to the Robespierrists, like edgy monarchs across Europe, the tsarina watched with ever-mounting dismay. Displaying uncommon foresight, she reflected in 1794: "If France survives, she will be stronger than she had ever been.... All she needs is a superior man, greater than his contemporaries, greater perhaps than an entire age. Has he already been born? ... Will he come? Everything depends on that!" Little did she know that this "superior man" had indeed already come—having forged great glory for himself at the siege of Toulon. Or that her own grandson, Alexander, and Emperor Napoleon would one day decide the fate of all Europe, when the two clashed in the epic campaign of 1812.

"Our rules," the tsarina once said, "were written by philanthropy itself." Still, there was no denying that the French Revolution had turned Catherine from an enlightened despot into a dark, cantankerous one. Her transformation became tragically apparent when confronted at home with some of the most historic reformers ever to walk the soil of Russia: the idealistic Radishchev in 1790—and the public-spirited Novikov in 1792. Between the two, they offered the Russian Empire a springtime of openness and a liberality of spirit.

But with their rise and fall, all the last hopes of a democratic impulse and republican mind flourishing in the land of the tsars died glimmering, not just in Catherine's reign, but ultimately for more than an entire century.

∞

COULD THE EPIDEMIC of the revolution be confined to the borders of France? Increasingly, it seemed not. By 1790 reform was in the air, and everywhere intellectuals of significance began scraping away with their pens and hollering for change.

From the Caucasus to Smolensk, from Moscow to the Baltic, Catherine had once dreamed of a forward-looking circle of intellectuals, discoursing on abstract subjects of philosophy and governance, public policy and literature—and becoming the envy of all Europe. After all, under her reign, Diderot was translated, and so were the plays of Shakespeare, the Greek and Latin classics, and the *Gerusalemme liberata* of the great sixteenth-century Italian writer Tasso; the poet Gavril Romanovich Derzhavin flourished too; so did Princess Ekaterina Dashkova, the head of the Academy of Language who had befriended Benjamin Franklin in Paris and was the first woman honored to join the prestigious American Philosophical Society. These were the mandarins of the new Russia. But now the aging empress suddenly saw them as incendiaries, and every fresh idea became frightening to this former disciple of the Encyclopedists. So as 1790 opened, the empress's unstated motto for Russia (and here presaging Nicholas I) had instead become one nation, one faith, one dogma, and one tsar.

Catherine took an especially harsh line toward literary agitators, and her displeasure manifested itself with particular severity against Nikolai Ivanovich Novikov. His was not a name the world should lightly forget. A journalist, bookseller, and publisher, Novikov could have been the Russian Benjamin Franklin. Expelled from the University of Moscow for laziness, he soon became an intellectual of incessant enterprise and entrepreneurship. In 1769—seven years after Catherine was crowned tsarina—Novikov began editing the satirical magazine *The Drone*, playfully named to rebut another

Petersburg periodical, *The Industrious Bee.* He was only twenty-five, his free-thinking style was enticing, and he rapidly became one of the empress's favorites. He campaigned against rampant government corruption, which she agreed with; he condemned the Voltairean-style secularism of the nobility as inimical to morals and character, which she paid homage to; and he lauded the world of Old Muscovy before the days of Peter the Great, to which behind closed doors she whispered yes. "It is," Novikov wrote, "as if the old Russian ruler had foreseen that, through the introduction of arts and sciences, the most precious treasure of the Russians—their morality—would be irretrievably lost." Nevertheless, Catherine publicly looked askance at *The Drone* and it folded in 1770, but privately, she bore him no ill will.

From there, Novikov joined the Freemasons, a movement that started with absolute fealty to the government but which soon flirted with illuminism, mysticism, and pietism, even as its French and American brothers, the most famous being George Washington, were toying with revolution. In 1779 Novikov became head of the Moscow university press, and published more works in two years than had come from that press in nearly twenty-five. With help from a friend, he purchased other presses, translated foreign books, printed several newspapers and opened bookstores across the country, and sprinkled his gospel of religion and reform throughout Russia; he even instituted an extensive project to promote Shakespeare to the public. And not unlike Catherine, he worked to improve the lot of the average Russian: Engaging in large-scale charitable activities, he founded schools, libraries, hospitals, dispensaries, and also model housing for workingmen.

Catherine carefully watched Novikov, and even contributed anonymously to his newspaper, *The Painter.* But once the French Revolution broke out, gone were the fluffy encomiums for him; and gone, too, was his storied career. In the era of revolutionary France it took courage—or stupidity—to breast the tide; Novikov, though temperate, unwittingly did just that. The empress now chided him for his energetic defense of the serfs and blamed him for his affiliation with the Freemasons; worse still, she feared that he was subverting the existing order. She ordered an aide to examine Novikov's ideas, and the case was investigated with great zeal. Yet when the report came back, it was a surprisingly bold defense of Novikov: "I implore the all-merciful God that there may be, not only in the flock entrusted to me by God and you but throughout the world, such Christians as Novikov."

Strong words, but Catherine didn't buy them. As the months rolled on, she remained suspicious of Novikov, and by 1792 the empress decreed his arrest for "defaming" the Orthodox Church. It was a dramatic about-face for her: Once she had been pulled toward ecumenical openness for Russia. Once the pulsing debates of Petersburg's intellectuals had seemed so compelling to Catherine, as did the eclectic characters who made her smile and set her thinking. Now they seemed tinged with menace. Was the empress becoming erratic or capricious? Not as she saw it: In this uncertain age, she was determined that Russian philosophers would not be allowed to theorize in salons or rage in the streets whipping the Russian intelligentsia into a frenzy. Indeed, lest there be any doubt about this, the empress took it upon herself to dictate the questions to be asked of Novikov.

Though Novikov was not actually tried, his sentence was never in doubt. It was also severe: fifteen years imprisonment in the dreaded fortress of Schlusselburg, the very same site where Tsar Ivan VI had been murdered by the tsarina's men years before.

∞

Was this Russia's new future? It seemed so. Some years earlier, the empress had written in the journal *Vsyakaya Vsyachina* (*This and That*): "We want to walk on earth and not soar in the air, much less climb up to heaven." She added, "furthermore, we do not like melancholy writings." Those, of course, would remain in the eye of the beholder. But soon the tsarina's newfound ruthlessness would be tested in other ways, with the rise of Alexander Nikolaevich Radishchev. Who was Radishchev? Pushkin was his first biographer, but he remains one of history's forgotten greats. If Novikov was Russia's would-be Benjamin Franklin, Radishchev was a variant of its Thomas Paine, but he was neither an implacable revolutionary, as some have painted him, nor an ardent democrat, as some romanticists believe. He was, however, a strikingly wise and remarkably prescient humanist.

He was also, of sorts, a protégé of the empress herself. It was she who had originally sent him to the University of Leipzig, where he was particularly taken by the iconoclastic ideas of Helvetius's *De l'esprit* and the impassioned pleas of Abbé Raynal's tract on European brutality in the slave trade. Radishchev returned to St. Petersburg determined to bring the Enlightenment to Russia; nor was he a voice in the wilderness. Impressed by his ideas and enthralled by his energy, the empress herself promoted him for distinguished service, subsequently putting him in charge of the customhouse, and, to convey an additional dash of prestige upon him, warmly decorated him with the signal privilege of service in the Corps de Pays. Along the way, Radishchev learned English to deal with British merchants, took up English literature, and was especially influenced by George Washington and the Founding Fathers, as well as the American rebellion.

And in 1790 he wrote, anonymously, one of the immortal works of Russian literature: *Journey from St. Petersburg to Moscow.*

∞

NATIONALISTIC, INSIGHTFUL, MINDFUL of the human condition, and understanding of the forces of human history, Radishchev envisioned a better world: His book was both a document and a pamphlet, the narrative of a simple pilgrim's fantastic journey and wondrous musings about it. It was also a deeply subtle and learned work, and, at bottom, an ardent tirade against the evils of serfdom and corruption in Russia. It paid homage to religious orthodoxy, yet it assailed the superstitions of the clergy; it professed obedience to the monarchy, yet it justified popular rebellion against rulers who ran roughshod over the law, whether "a tsar, shah, khan, king, bey [or] nabob." It described the dismemberment of families by conscription, and the abuse of serfs by masters; at one point, cried Radishchev, he had been told of a landlord who had violated sixty peasant maidens. He denounced censorship and pleaded for freedom of the press. He did not advocate revolution, but he asked for a merciful understanding of its advocates. Not unlike the abolitionists, he appealed to the state to one day end the cruelties of serfdom, and even envisaged the liberation of the peasants by a generous act of the regime. His language was poetic: "Let yourselves be softened, you hardhearted ones; break the fetters of your brethren, open the dungeons of slavery." So were his insights: "The peasant who gives us health and life had a right to control the land which he tills." So were his warnings: He predicted, rightly, a terrible revolution in a century's time in Russia (by "slaves raging in their despair"). And he quoted, approvingly, some of the giants of the era, the American Franklin and the Frenchman Mirabeau, both moderates, both, in the end, rebels.

Was this iconoclastic? Or rebellious? Actually, the book passed by the censor. But not the tsarina. After reading thirty pages, the empress herself was enraged—and suspected Radishchev as the author. All this, of course, was ironic, for in truth, the similarities between Radishchev and the youthful Catherine were uncanny. Thirty-five years before, Catherine herself would have agreed with most of this; she too, like Radishchev, once dreamed of freeing the *muzhiks*, of nudging them toward the radiant summit of freedom. Decades earlier she too, not unlike Radishchev in *Journey*, had written a forward-looking paean to progress and reform—her much acclaimed *Nakaz*, the body of liberal laws translated into the language of virtually every country in Europe. And she as well, not so different from Radishchev, had once boasted of her "*âme républicaine*" and observed that liberty was "the soul of everything." But in the tremendous wake of the American and French Revolutions, she felt that it would be dangerous to bring people accustomed to slavery to freedom with the stroke of a pen. With blinkered logic she wondered, just as those in the American South did at much the same time, would they not lapse into anarchy if given independence—or lament the security provided by an all-powerful lord?

The tsarina ordered Radishchev's arrest. Held in solitary confinement, he was tried in secret for treason. He was also denied defense counsel; in yet another maddening irony, the trial violated her own *Nakaz*. But to her, there was no other choice. With its mocking, biting tone, the book spun a story of troops confronting a cruel tsar and "reddening his fields with blood"; with its deft wit, there was fulsome praise for England's rebellion against an unjust king, as well as unacceptable references to her expansionist policies ("the murder called war"); and, unthinkably, there was an "Ode to Liberty." She sneered that the journey's author ("our babbler") would "pipe a different tune" if he stood closer to the sovereign. The Senate condemned the author to death; still, in one act of magnanimity—she knew all of Europe would be watching her—the empress cynically commuted it to ten years in Siberia. By 1792, as Louis XVI was virtually under house arrest, she was no longer taking any chances. All subversive books were to be burned. All Masonic lodges in Russia were to be closed. And Radishchev's fate, and that of Novikov, stood out as an embarrassing blight in her often brilliant reign. Had she listened to them, however, Russia might have been spared a bloody revolution in 1917 that would unravel its monarchy and sweep away the Romanov dynasty forever.

How did it come to this? How did the empress, who had so decisively scuttled the old ways of Muscovy, so blithely turn on her own disciples and so dramatically turn her back to her own spirit of liberality? How did the tsarina, who was so used to being called "merciful and just" and the law's "first servant," become a splenetic autocrat? Was it the case that she was driven to desperation by the French contagion, worn down by years of war, fearful of treason and widespread rebellion, and enervated by enemies who wished her downfall if not her death? To be sure, it was all of these things.

But it also sprang from an earlier spark that had become a firestorm, a decade and a half earlier, just before George Washington and the American revolutionaries were preparing to declare themselves free from the British kingdom. It was the gravest crisis of Catherine's reign, politically and militarily, one that would forever color how the empress saw revolutionaries and revolutions. It was ignited by one man: a Cossack named Emelyan Pugachev. And it began in the remote frontier hinterlands of the Yaik Cossack Urals. What he would initiate, and what he would threaten, would say much about what Old Russia was, and what it would be allowed to become.

And for the empress, and hence the empire, and therefore the world, the implications were profound.

∞

ACTUALLY, EARLY ON Catherine had considered reforming serfdom in Russia, and she herself had once written that slavery was "contrary to the Christian faith." Yet the condition of the serfs was deteriorating: Serf owners were increasingly selling serfs apart from their land, heartlessly breaking up families (one might read in the *Moscow Gazette*: "For sale: two plump coachmen; two girls eighteen and fifteen years, quick at manual work"), using them in mines or for labor, or harshly punishing them at will, or exiling them to Siberia. In turn, long before Radishchev and Novikov, the serf population became restless, especially as they reminisced about their mighty hero, Stenka Razin, who a century before had led an uprising against the landlords. Widespread deprivation bred discontent, and so did class antagonisms. This was never more so than when the rough *muzhik*, hardened by the privations of serfdom and the exigencies of sheer survival, heard the lord talking French so as not to be understood by the servants, or saw him wearing European clothes, or watched him reading foreign books and adopting the manners of a foreign way of life.

But then abruptly, shortly before the Americans declared their independence and before the French took the Bastille, the system of serfdom seemed about to come crashing to an end with the incipient Russian revolution against Catherine.

It began on September 17, 1773.

∞

THAT DAY, FAR southeast of Moscow in the remote world of the Urals, a land filled with Cossacks, Kalmyks, and Tatars, a huge crowd assembled at a lonely farmhouse on the glistening steppes near Yaiksk, the headquarters of the Yaik Cossacks. The Cossacks were a unique community: Descendants of migratory tribes that had largely settled around the Don River, they were leagues of freemen and runaway serfs, escaped criminals and pastoral tribesmen, and vagabond refugees who boasted a heady mix of Muslim and Slavic blood. Ungoverned by tsar or sultan, they became a nation of professional soldiers unto themselves, until the time of Peter the Great. Yet even then, the state's reach was tentative. Proud, free, and courageous, they lived by pillage and subsisted by murder, that is, when they weren't fishing or raising cattle. Each host, whether Don or Yaik, Zaporozhian or Siberian, whether armed gangs on horseback or nomads roaming the hills, lived as a quasi-clan and governed itself according to its own raw democratic rules, electing a leader (an ataman or centurion) in times of strife or conflict. This was one such time, but with an unexpected twist. For standing before them was the tsar himself.

Or so it seemed. He was the charismatic Don Cossack Emelyan Pugachev. Dark-skinned and imposing, he had a striking black beard, flashing eyes, and thick brown hair. On this day he triumphantly declared that he was the true tsar, Emperor Peter III—Catherine's deposed husband—and had just returned from long travels in Egypt and the Holy Land. No one around him had ever seen the real tsar, and few among the illiterate, superstitious peasants doubted him, for as it happened, there was a superficial resemblance; moreover, he displayed the "Tsar's marks" on his body—actually, they were scars—as proof that he was their anointed ruler. Contrary to legend, he mused that he had not been murdered, but that God had saved him at the last moment because the Lord "loved" Russia. After eleven years of absence, he exclaimed that he was going to free his people from the yoke of serfdom and rally them against the hated empress—that "German," that "Devil's daughter." The message had enormous resonance: To these assembled Cossacks, Catherine seemed to be an alien graft on the empire's traditions, a loathsome pretender in the ancient Muscovite world. Her sophistication—she spoke French, surrounded herself with foreign emissaries, corresponded with foreign philosophers—contrasted sharply with the austere morality of the Orthodox Church and the poverty and sacrifice of traditional Russia.

Pugachev, launching into a passionate tirade, promised "every freedom" to the gathering crowd. Indeed, there was nothing that this Little Father didn't offer them: They would, he bellowed, receive "lands" and "waters," "woods" and "wheat," "dwellings" and "silver" and "lead," and even the Yaik River. He proclaimed the end of serfdom, a halt to taxes, and the cessation of military conscription as well. In doing this, he brilliantly captured the seething resentments—the careless snubs of the Imperial Court, the repeated indignities by the military, the ethnic indifferences of the state. And to the men listening to him that day, he was everything the emperor of all the Russians was supposed to be: "sweet tongued," "merciful," and "softhearted."

In truth, Pugachev was a onetime soldier and an army deserter. Born in the Don in 1740, or so it is believed, he had been raised on the lore of Cossack heroes, and had come of age in a time of ethnic strife and global conflict, fighting in the Seven Years' War and putting his life on the line at the siege of Bender. After that, he bore grievances against the government, overplayed his hand, was arrested (twice), and then fled. But he was a gifted orator and his seductive words swayed the crowd, as did his eloquent promises.

In the ancient days of Muscovy, the fiercely independent Cossacks had invariably straddled all sides against the middle, alternately aligning themselves with the Polish-Lithuanian Commonwealth or even Sweden against Russia, while in the same breath befriending the tsars against the Ottoman sultans or Crimean khans. In the eighteenth century, they plundered Russians as well as Turks, but here again, they also served as guards on the Russian frontiers and much-feared cavalry units. However, not unlike with the serfs, troubles in recent decades had been festering between Russia and the Cossacks, and as late as June 1772, they had mutinied against the tsarist forces. A scorned minority, they didn't want to be absorbed into the empress's "Moscow Legion"—that meant shaving their beards, which was sacrilegious; they had quarrels over the persecutions of Old Believers, Russia's equivalent to English Puritans; and the first Russo-Turkish War, approaching its sixth year, had levied an awful toll in blood and treasure, not to mention taxes, which fell hardest on the backs on the commoners. For the Cossacks, then, these were only the latest in a series of treacheries perpetrated by the very people who should have been their zealous guardians: the tsarina, the court, the Church. So when this disheveled Peter III appeared, they were desperate to believe.

After swearing himself in as emperor, Pugachev had three hundred loyal supporters; almost overnight they began assaulting forts, and within weeks, the southeastern borderlands lay in flames. Then, on October 5, Pugachev swept into the local capital of Orenburg with an army of three thousand and over twenty cannons and prepared to lay siege. Zealous, headstrong, imperious, he also proved to be ruthless, strewing a path of wanton destruction behind him. Wherever he struck, terror was the watchword: Bleeding officers were hanged upside down in their captured garrisons, and dumbstruck nobles had their hands and legs cut off, and only then their heads, after which they were left dangling outside their burning mansions. Women fared little better: they were raped or stabbed; even children were massacred or mutilated.

Yet his instincts were uncanny. Time and again, he outfought, outbluffed, or outmaneuvered the tsarina's garrisons, and by November 6, he founded a war college at his headquarters outside Orenburg. With mocking insolence, he formed his own court and surrounded himself with a retinue of courtiers, generals, and even a secretary of state; remarkably, some of his followers had been deputies to Catherine's Legislative Commission. And his marauding troops were a terrifying sight: Comprising Tatar and Cossack warriors, they wielded bows and arrows, lances and scythes; axes, too. Bristling with pride and camaraderie, they were also pitiless; their numbers were now 25,000 strong and swelling.

In mid-October, Catherine was in St. Petersburg. Informed of the news of the rebellion, she was at first puzzled. Not surprisingly, she underestimated the peril; taking it for a minor Cossack mutiny like the one she had crushed the year before, Catherine dispatched a general, Vasily Karr, to suppress it. That was her first mistake. Within weeks, Karr was routed by the "ravaging" brigands and withdrew back to Moscow in disgrace; he would be but one of a succession of generals sent east from Moscow to deal with Pugachev. Complicating matters, many of the tsarina's soldiers felt no desire to fight these "brothers" whose revolt they quietly sympathized with—much as would happen among Louis XVI's soldiers during the French Revolution. And with these initial victories, Pugachev began to attract ever greater numbers of followers, first in the thousands, then the tens of thousands, then hundreds of thousands. He surged through eastern Russia, burning and pillaging, while his faithful Cossacks galloped back and forth astride their magnificent horses. Aglow with victory, he began offering rewards: 100 rubles for a landlord killed or a castle plundered,

1,000 rubles and the "rank of general" for ten nobles murdered and ten castles destroyed. In the dismal weeks that followed, a string of Russian forts manned by weak garrisons fell one by one. And the upper classes in Moscow were equally terrified: 100,000 serfs, whose sympathies went out to Pugachev and his hordes, lived in the city as domestic servants or industrial workers.

As Pugachev's armies took cities, a full-blown revolution threatened to take shape. He was invariably welcomed by reception committees of priests clutching icons and townsfolk on bent knee bearing bread and salt, the customary Russian greeting for dignitaries, all the while pledging their allegiance to "Peter III," but never to Catherine. The story was always the same: Wearing his dazzling red Cossack coat and fur cap with gold tassels, Pugachev would storm into a village as the sovereign, accompanied by Cossack elders. Gallows would be quickly prepared and nobles would be hanged, even as Pugachev took dinner in the local governor's house, often seated next to his petrified wife or distraught children, that is, when they weren't hanged too; those who lived usually ended up in a harem. Then Pugachev would enlist

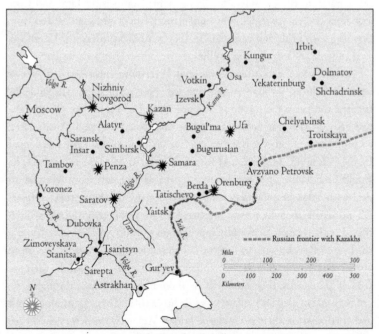

The Pugachev Rebellion

more recruits, commandeer more munitions, ransack the local monies, and gallop off again to the frenzied wail of bells and the burning of incense.

As winter set in, he had encircled the towns of Samara and Orenburg as well as Ufa, then rampaged though the territory north of the Caspian, south of the Urals, and east of Moscow.

Only then, in the heart of Catherine's court, was it clear that he was a real threat to the Russian empire.

∞

BY THIS STAGE, a vast area the size of France now lay under Pugachev's hand, and he had successfully exacerbated all of Russia's ancient hostilities, setting serf against master, pilgrims against the church, peasants against the tsarina herself. Presaging Louis XVI's dilemmas, even Russian troops were increasingly defecting to Pugachev's side; many massacred their officers. And suddenly the rebellion turned a new, more ominous page, becoming a barbarous class war, a *jacquerie*, patterned after the grisly slaughter of French landowners by peasants in the 1358 rebellion. Pugachev wore on his enemies like an affliction: the empire now confronted the possibility of millions of serfs literally stampeding through the nation and slashing their masters' throats. Here then was a danger not solely to the tsarina, but the very underpinnings of the Russian domain itself.

On July 21, 1773, word of Pugachev's latest conquest, the city of Kazan, filtered into Catherine's court in Petersburg. The authorities first fell into dismay, then awe, then outright panic. They had good reason. At Nizhniy Novgorod, the dreaded nightmare seemed to have begun. Serfs rose up and put the entire region to fire and sword. Meanwhile, the grand vizier of the Ottoman empire was inciting the Muslim tribes of the Urals to join Pugachev. And in distant St. Petersburg itself, the common people, "drunk with venom," whispered giddily about Pugachev's "coming." Just as quickly, news of the rebellion had spread all over Europe, and was watched with concern by King George III himself—commentators also linked Pugachev with rebellious confederates in Poland as well. By the spring of 1774, the insurrection was known by another set of soon-to-be rebels: Americans in Massachusetts, New York, Pennsylvania, and Virginia. At this stage, it was the most violent peasant uprising in the history of Russia, and the most formidable mass upheaval in Europe in the hundred years before 1789. The grim spectacle was raised: Would Pugachev march on Moscow? Or on Petersburg? Boasted a triumphant Pugachev: "I have people like sand."

It was then that the empress convened an emergency council at her palace in Peterhof.

∞

ALREADY, EXTRAORDINARY DEFENSE measures were being applied in Moscow. Now Catherine's ministers listened in stunned silence to her proposal. Normally Catherine could enforce her will with a single edict or the tap of a foot; instead the empress, visibly shaken, declared she would travel directly to Moscow to rally the empire. This was not just posturing. Catherine was wild with rage and concern.

But in the end, she didn't have to go to Moscow. Rather than marching on the ancient capital of the empire, Pugachev—after suddenly having lost a series of battles—abruptly fled southward. The Russians pursued, harassing, ambushing, and waging a war of nerves against Pugachev, never allowing him to rest, always denying him open battle. General Suvorov himself raced across the steppes to crush the mutiny. The immediate crisis seemed to pass, although Catherine, while not wanting to give the impression of fear to the outside world, was frantically monitoring the disturbing events across the Volga.

They remained disturbing. Even Pugachev's retreat had all the earmarks of open rebellion. As he prowled his way toward the familiarity of his homeland, his tactical cunning was on full display: Streaming across the steppes, towns surrendered to him, more men flocked to him, great manors were torched, and lords were strung up. As Suvorov found out, Pugachev, bullish and driven, had lost none of his fight: In every direction the road behind him was marked with ashes and ruins and unburied bodies; and at all points of the map, the ebb of revolution had left nothing but anarchy. Even now, he was all hell-for-leather. On August 6 Pugachev attacked Saratov, and amid the rubble, apostate clergy administered oaths of allegiance to him; in turn, twenty-one officials were marched to the gallows, along with twenty-four landowners. But after a series of running skirmishes, Pugachev was scrambling and boxed in. Increasingly cornered by Russian troops, he collected his remaining forces and fled home, to the Don. Despite unbearable weather, Suvorov relentlessly chased him into the farthest reaches of the outback: They trekked under the glare of sun during the day, by the freezing cold at night, and through hordes of mosquitoes. Then his quarry vanished into the desert east of the Volga, a barren landscape scattered with salt lakes, with little vegetation, few animals other than birds of prey, and no population except for an occasional hermit or bands of nomads. Pugachev was four days ahead of Suvorov, who drove his men harder and faster, fearing that to slow his pace would mean more casualties from exposure and starvation. At last there was a break: Suvorov reached the shelter of the steppe forest; nipping at Pugachev's heels, he now planned to hit Pugachev with everything he had.

Collecting his men, Pugachev pushed on to the safety of home. Once nestled there, he hoped to regroup, catch his breath, hone his men, rest his horses, and plot his next moves. Instead, it turned out that his countrymen were beset by famine, and the great warrior was only sneered at. In the wave of a hand, the magic of the impostor's authority began to dissipate, and his men, exhausted by months of hard fighting and nearly nonstop skirmishing, and disappointed by a retreat they didn't understand, began to worry about the consequences of their rebellion. For the tsar impostor, this proved to be fatal. But he never saw it coming. While licking his wounds, Pugachev was betrayed by his own lieutenants, arrested, and turned over to Suvorov and the Russian armies.

For Pugachev, all was lost. It was a striking turn of events: As a revolutionary, he lived like a tsar and had come near to destroying an empire. As a prisoner, he was without any retinue apart from his confused, bedraggled twelve-year-old son. He had once been hailed as a conquering hero; now he was like a wild animal, taken back to Moscow chained in an iron cage, a two-wheeled rolling prison illuminated at night by torches. In the meantime, Catherine moved quickly to crush all remnants of the rebellion. Pugachev's followers were put to death, hung from their ribs by a thick metal hook, an especially macabre way to die; others were beheaded, hanged, flogged, had their ears cut off, or were sent to penal servitude.

With Pugachev now in her hands, Catherine knew that the capitals of Europe were watching her; already she feared that the insurrection was a humiliating stain on her enlightened reputation. Was mercy possible? Pugachev, weak from confinement and constant questioning, hoped so, asking for a pardon because of his courage. But the empress vehemently insisted, "If it were only I who he had offended, his reasoning would be correct and I should pardon him; but this is a case involving the Empire which has its laws." The trial was conducted in secret on December 30–31, in the Great Kremlin Hall. The empress forbade torture, yet privately she was still livid. She caustically wrote to Grimm that the farce of "the Marquis de Pugachev" would soon be finished, and she boasted to Voltaire that no one "since Tamerlane" has done more harm than Pugachev, who "lived like a scoundrel" but will "die like a coward." That happened on January 10, 1775, in Bolotnaia Square near the Kremlin.

Pugachev was condemned to be quartered, then decapitated; but with the eyes of Europe in mind, the empress contrived to grant him imperial clemency, quietly allowing his head to be cut off first. Vengeful crowds gathered while Pugachev was undressed, slowly stretched out, quickly beheaded, and only then quartered. Pugachev's dismembered pieces were draped on a pole in the middle of the scaffold, his head brandished on an iron spike. Suddenly word went out: The Pugachevschina—the dark times of Pugachev—was over. But was it really? The memory of what did happen and what nearly happened would forever haunt the tsarina's reign; never again would she allow the earth to so tremble beneath her feet. Thus by the time the American Revolution had succeeded—indeed, writing to Voltaire, Catherine compared the Cossack region that had embraced Pugachev to the American colonies—the French Revolution was simmering, and her own countrymen, Novikov and Radishchev, were agitating for reform, any assault on the tsarina's institutions, on the tsarina's autocratic powers, or on the tsarina's ideas was construed as sowing social chaos, civil war, and economic ruin. Catherine was taking no chances.

This was also the case in her latest challenge: quelling the mounting menace of Jacobinism in Poland.

∞

It was an irony of Catherine's imperialism that the man who helped inspire some of her greatest difficulties was in fact a puppet she herself had installed on the Polish throne: her former lover of more than a quarter century ago, Stanislas Poniatowski.

Theirs was a curious entanglement: Where Catherine had spoken the language of the Enlightenment, Stanislas had lived it; where Catherine had the sensibility and toughness of a man, Stanislas had the sensitivity of a woman; and where Catherine was surrounded by enemies near and far, so too was Stanislas—but where the empress sought to conquer them, he sought to live with them. "When I love," he once wrote, "I love too passionately.

I am not vindictive." But the empress was. He also wrote: "Though in the first moment of irritation I may long to avenge myself upon my enemies, I am never able to carry out my desire; compassion always comes between." Not so with the tsarina. As a result, they are one of history's more poignant relationships—or perhaps tragic is the word that best captures them.

Actually Stanislas, born in 1732, from a young age seemed destined for a measure of greatness. At the age of sixteen, he undertook extensive travels: He visited Paris and charmed Madame Georffin's salon; he befriended Montesquieu and read Voltaire; and he developed the intellectual nuance of French society then manifesting itself in Rousseau. After five months in France, he picked up English, traveled to England, conscientiously watched sessions of Parliament, and yearned to reshape Poland's government in Britain's image, as defined by Montesquieu. Returning home in 1754, he was appointed high steward of Lithuania, moved on to Russia, and eventually became Poland's ambassador to the Court of St. Petersburg. He struck up a torrid affair with the then Grand Duchess Catherine, and afterward, as empress, she would help install him on the throne of Poland—he was only thirty-two—less because she still cared for him but because (she declared) he would forever be in her debt. He was. All his life he never quite recovered from their exciting tryst, and he always seemed to remember the empress before her lust for power hardened her heart, an infatuation that survived even when she made him her instrument in the enslavement of his own country.

Stanislas was handsome and refined, warm-hearted and humble; he was, too, a passionate Polish partisan. But now he saw the country torn by political chaos and dangerously weakened by military impotence. It also lacked natural boundaries—neither mountainous ranges nor mighty rivers—on either border to protect it from invasion; its very name came from *pole*, a plain. For this, the nation paid dearly: In 1772, as Pugachev was rampaging across the Russian steppes, Russia, Austria, and Prussia were partitioning his country; one-third of its soil was taken, along with two-thirds of the population. Stanislas begged the Western powers to stop the partition, but they were too preoccupied: France was eyeing war with England and was allied with Austria anyway, and England was staring down an incipient revolt in America; cynically, King George III suggested that Stanislas pray to God. He did. It wasn't enough. Western Europe capitulated to the first partition as the only option to the total dismantlement of Poland by Russia; Stanislas accepted it because he had no choice.

Nonetheless, in the ensuing years, Stanislas, while surrendering to Russia as his protector and master, laid the foundations for a glorious Polish enlightenment. A system of national schools was set up, textbooks were written, the universities of Krakow and Wilno were given new life, and teachers' colleges flourished. Meanwhile, when the salons weren't thriving, writers were: The king assembled a striking circle of poets, playwrights, journalists, "men of letters," and philosophers, while, for their part, thousands of Poles (like the Americans before them) discovered Locke, Montesquieu, Voltaire, Diderot, and Rousseau. Under Stanislas, prodded by its poets and publicists, and nurtured by its piety, Poland labored to turn itself into a viable and secure state. The opportunity came not long after Catherine had coldly rebuffed Stanislas during her 1787 Crimean expedition. With Russia tied down in wars against both Turkey and Sweden, it was then that Poland hoped to free itself from its continuing servitude to the tsarina.

In May 1789, just as George Washington was assuming the presidency in New York and Thomas Jefferson was attending the convocations of the Third Estate in Paris, the Polish Diet ordered an army of 100,000 men to be raised and ordered Russian troops to leave Poland at once. Catherine, her forces dangerously stretched across multiple fronts, offered no resistance, but swore this audacity would be avenged. Stanislas was shrewd enough to realize that. So upon his urging, on March 29, 1790, the Polish Diet concluded an alliance with Prussia.

This time, Stanislas himself was inebriated by the winds of liberty. Discarding his affections for Catherine, he spearheaded a new constitution, crafting, in effect, a constitutional monarchy of the republican type. The weighty influence of the recent American constitution, and the spirit of Washington, Hamilton, Jefferson, and Madison was suffused throughout; so was the inspiration of countless Poles who had fought and bled for the colonies in America, not to mention the Virginian Lewis Littlepage, who was one of Stanislas's closest aides.

The constitution was presented to the Diet in the splendor of the royal palace and was ratified on May 3, 1791. At home, Poles greeted it with thunderous applause and memorialized the day in music, art, and prose. Abroad, Edmund Burke called it "the noblest benefit received by any nation at any time" and declared that Stanislas had earned his place among history's great statesmen. It was hard to disagree. While a young, exciting country had been born in the New World, an equally bold new country had arisen in the once implacable depths of the murky Old World.

Every great power except one recognized the new constitution; the one holdout, however, was the empress.

∞

TRY AS SHE might, Catherine could not conceal her hostility to the new Poland and its stinging anti-Russian rhetoric. She coldly declared that the constitution was an emanation of "the revolution spirit," and she shouted that the king's plans for reform were inspired by "the Jacobin Clubs of Paris"—she didn't even deign to recognize the influence of the rabble from the United States. She insisted that France was exporting all its rot to Poland, and maintained that this was contrary to the 1772 partition of the country. Yet Catherine, bogged down in war, was, for the time, helpless to intervene.

That, of course, changed on January 9, 1792, when the Peace of Jassy ended her conflict with the Ottoman Empire. Moreover, by April, Prussia and Austria were embroiled in war against revolutionary France, giving her just the opening into Poland that she craved. So as she energetically whipped the Austro-Prussian coalition against France into a fervor, she vowed to fight the "revolutionary hydra" closer to home.

Standing in her way were two men: her old lover Stanislas, and one of the century's more intriguing revolutionaries: Today, in the United States, traces of him are everywhere—bridges are named for him and so are town squares, his statue rises triumphantly in cities like Washington and Boston, and he even has his own mustard; in turn, he was also one of the earliest to call for a national military academy at West Point. But in the 1790s in the United States, his was literally a cherished household name. He was Thaddeus Kosciuszko, and he straddled the three worlds of tumult and rebellion: American, French, and Russian.

∞

NO MAN IN Poland was more suited to lead an insurrection against the tsarina. He was one of the illustrious foreigners to be commissioned by the Americans in the War of Independence out of an already illustrious list, which included such epic names as Lafayette, Pulaski, DeKalb, and von Steuben. About him, Lord Byron once declared Kosciuszko's very name alone would "scatter fire through ice." It had also been said that he was one of the "most admirable men of the eighteenth century" and a "harbinger of a new era in the human struggle for the highest ideals." This is perhaps overstatement, but if so, not by much. A child of the republican Enlightenment, as much American as European, and a maestro at challenging the odds, Kosciuszko seemed to be Poland's mystical answer to Catherine's rampaging armies and her great fighting general, Marshal Suvorov. That, however, remained to be seen. But not his appeal, which was considerable: Charming in personal relations and adept at professional ones, sometimes gloomy and always idealistic, a soldier of fortune and a self-motivated military man, and much beloved, he was Poland's most innovative and charismatic figure. And there would come a time when the very fate of the country would rest squarely in his hands. As a consequence, in the world of the 1790s, it became a curiosity that two heroes of the American revolution would take such distinctly opposite paths: Where John Paul Jones fought for Catherine, Kosciuszko would lead a rebellion against her. But could such a miracle succeed? For behind the accolades and the encomiums was a man, and a self-made one at that; and behind the national ecstasy was a country that militarily was little more than a shell, and a politically divided one to boot.

But Kosciuszko was a stranger neither to controversy nor to fame. Later in life Napoleon would woo him, and so would Tsar Alexander; in both cases, however, he refused to command Polish legions for causes that would not liberate his country. To his followers and his critics alike, Kosciuszko's reputation became the essence of folklore legend, and he became the subject of living myth. For all his faults, and they could not be discounted, the reasons were not hard to fathom: When, as a younger man, he had aided the patriots during the American Revolution, it would have been hard to find a more romantic figure than Kosciuszko. Disappointed in his military ambitions, threatened with bankruptcy, and heartbroken because of an unfortunate love affair, the Polish patriot journeyed to France in 1776, then hastened to America, to throw himself with abandon into the struggle for independence. His engineering talent proved to be valuable in both North and South, and in one of his last engagements of the war, he barely escaped death. This was not his first brush with dying—or with tragedy. Later in life, fighting Catherine's armies, he would suffer grievous wounds and end up hunched over, suffocated by stinging, chronic pain; nonetheless, he remained dominant in his people's affections and his name would resonate flawlessly on the world stage. In 1815, after the Congress of Vienna reformed the Polish realm but placed it squarely under Russian tutelage, Kosciuszko fired back, angrily growling for sweeping social reforms and defensible boundaries for his nation; when no answer from a callous world was forthcoming, he went into self-exile to Switzerland. But doing good was always never far from his mind: In his lifetime, he freed his Polish serfs and arranged, upon his death, to purchase the freedom of black American slaves, even making provisions to educate them.

He was known to all the leading lights of America. In one congressional debate after his death, he would be lionized as "a friend of man" and "an advocate of freedom," and, in his own day, Thomas Jefferson called him "the purest son of liberty," even as George Washington spoke warmly of his war efforts. Indeed, when Kosciuszko arrived in Philadelphia in 1797, he was greeted with wild enthusiasm—the *Philadelphia Gazette* referred to him as "the illustrious defender of the rights of mankind."

He had fame but little in the way of wealth or meaningful pedigree, and in this sense, was neither fish nor fowl in his home country of Poland. Born to a family of modest noble origin in 1746 in Mereczowszczyna—today's Belarus—he was educated at Piarist college and the military academy of Warsaw, where he later served as an instructor. His outstanding abilities caught the eye of King Stanislas, who sent him to Paris for further military study; there, he also steeped himself in mathematics and civil architecture—which laid the foundation for his military engineering—as well as painting and drawing. Upon returning home, he fell in love with the daughter of a prominent general, tried unsuccessfully to elope with her, and faced the wrath of her father; so he fled Poland to join the American colonial forces fighting for independence. He never looked back.

There were few signs as to how he would perform. His rise, however, was near meteoric: He helped defend the Continental Congress against the British, was soon promoted to engineer colonel, and, in the spring of 1777, joined General Gates at Fort Ticonderoga in New York, helping close the fortifications along the Hudson that led to the capitulation of the British army at Saratoga. Plainspoken and no-nonsense, he largely avoided the politics of promotions, unusual for the status-minded officer corps of the day. Instead he spent the next two years fortifying West Point, for which he was appointed chief of the engineering corps in 1780, and then moved south, to serve under General Nathanael Greene in North Carolina. For the American rebels, these were bleak times; again and again the colonists stumbled, but Kosciuszko persisted. He twice rescued the army from enemy advances, directing the crossing of the Yadkin and Dan rivers; he conducted the elaborate battle of Ninety-six in South Carolina; and he then oversaw a lengthy blockade of Charleston, a masterful act that secured his name. By war's end, he was given American citizenship, promoted to rank of brigadier general in the Continental Army, received a grant of land by George Washington himself, and was even made a member of the Society of the Cincinnati; when Washington tearfully said good-bye to his closest officers at Fraunces Tavern after Evacuation Day, Kosciuszko was among the honored few gathered there.

But upon his return to Poland in 1784, his fortunes waned. He fell on the wrong side of the king and was denied an appointment in the Polish army. Rebuffed, he retired to a small country estate, soon slid into poverty, and quickly slipped into debt. For five years he passed his days in obscurity, but he never lost his dignity or lost sight of his America-honed persona. In truth, he had high-minded tastes and lofty values: Tall, regal, spectacularly handsome, he braided his hair, donned dramatic hats, and dressed himself in furs. More than just well groomed, he was incorruptible. And in 1789, he returned to military service.

Then, in 1792, Catherine's army invaded Poland.

SIXTY-FOUR THOUSAND Russian soldiers stormed into the country and thirty thousand overran Lithuania, intending to stamp out all vestiges of the king's reforms. The Poles resisted bravely, and Kosciuszko, now forty-six, rapidly made his mark: He distinguished himself with a valiant rear guard action at Polonne, fought magnificently at the stunning victory of Zielence, and rocketed to fame at the bloody battle of Dubienka, where a Russian army twice the Poles' size was held to a draw. And for the rest of the war, he became the hope of freedom-loving Poles and revolutionaries everywhere: Thus, he was raised to the rank of general; thus, the French revolutionary government made him an honorary citizen; thus, his presence alone helped kindle the waning resolve of his countrymen.

But not of his king. In the days ahead, Stanislas, now fearing defeat, had given up. He had his reasons. In a base act of treachery, Prussia had refused to honor its defense pact with him, and as a result, the western front lay naked to attack, even as the eastern front anxiously awaited the heavy footsteps of more Russian troops; moreover, supplies were so low that some regiments had to fast for twenty-four hours. The king, near physical and spiritual collapse, defected from the liberal cause, while Catherine prepared to send wave after wave of disciplined, well-armed men against the scattered remnants of the Polish defenses. Kosciuszko went into hiding and prepared to resume fighting. To most observers, it seemed just a matter of time before the nation became a virtual Russian protectorate.

Actually, the Polish affair was far from finished. If one man could answer the prayers of his people it was Kosciuszko—and he fully intended to.

∞

IN JANUARY 1793 Kosciuszko raced to Paris to seek support for his countrymen, first from the Girondins, then from the Jacobins, pledging radical reforms in Poland in return: more precisely, the end of serfdom and a revolt against the nobility, promises that Kosciuszko was only too happy to make. But the outbreak of war with England and the invasion of France by the allies ended all chance of French aid. This was Poland's bleakest hour. Now it stood irrevocably on its own.

But a remarkable emotion passed throughout much of the nation. In underground meetings and by living-room fireplaces, in Masonic lodges and at armed camps, the spirit of independence flickered still. When Kosciuszko returned in August to the exiles in Saxony, he found himself facing new demands for starting an uprising; unwilling to be left out, he quickly hurried from Dresden to a rendezvous point near Krakow to join it. Kosciuszko was promptly appointed commander-in-chief of the national forces, with dictatorial powers no less, but he still found the preparations inadequate. He could fight—or wait. He chose to wait. Delaying the insurrection, he again went abroad, to Italy, hoping to generate further international support. The idea was a sound one, but in practice, this proved to be a dire mistake. The Russians used his absence to ferret out the seeds of the insurrection, breaking some of its supporters and seizing others by arrest.

Remarkably, the enthusiasm of Kosciuszko's followers remained undimmed. Riots broke out in Warsaw, and those left in the underground forced the issue, starting an uprising without Kosciuszko. The day was March 12, 1794. At their urgent request, the general rushed back home, arriving in Krakow amid an enormous assembly of people; it was here, to great fanfare and song, that he solemnly swore to lead a national revolt against the occupying powers. Kosciuszko asked every five houses in Poland to send him a foot soldier, every fifty houses a cavalryman, and begged these recruits to bring whatever weapons they could find. Taking on a force that was vastly superior in quality, he improvised, not only rapidly introducing conscription, but enlarging existing units and reorganizing new formations. Because Poland lacked a war industry, he had few firearms to equip his men. He hastily ad-libbed, arming peasant recruits with pikes and soldiers with scythes, and pioneering new battle tactics of men attacking on the run while backed by artillery fire. Along the way, the old totems of Polish life collapsed. Needing all the resources he could muster, in a blasphemy of the highest order, he took on the nobility and suspended serfdom. And quite shrewdly, he issued a manifesto talking of the "indisputable right of resistance against tyranny and armed violence"—signed by thousands of men, nobles and peasants alike, this manifesto was heavily influenced by the American Declaration of Independence. More men flocked to join his colors.

On April 4 his efforts paid off. With four thousand regulars and two thousand peasant recruits, he won a smashing victory at Raclawice. The effect was electric. Clandestine organizations multiplied under his leadership. Burghers, Freemasons, and army officers hailed him. Even the nobles rallied to his cause. And overnight, he was a national hero. On April 17 he boldly attacked Catherine's occupying forces in Warsaw, as well as a Prussian contingent. Taken by surprise, the stunned Russian garrison, after losing a staggering 3,000 men, abandoned the city; so did the Prussians. For the Poles, the image was unforgettable, and the insurgents were delirious; echoing France's republicans, they triumphantly mocked the empress as "Catherine Tyranne!" while a mob irreverently defiled her portrait and tore it to pieces. Then another uprising liberated Wilno a week later. From across the country, guerrillas now streamed to Kosciuszko's side; by June 1794, he commanded 150,000 men. But only 80,000 of them were suitably equipped, and Kosciuszko's good fortunes proved to be only temporary. Indeed Catherine, awaiting victory, could no doubt content herself with memories of the demise of Pugachev.

A string of sharp defeats inflicted heavy losses upon Kosciuszko's men, with near cataclysmic results: On June 6 an allied army thrashed the Poles near Szczekociny; on June 15 the Prussians took the ancient capital of Krakow; on July 13 the Poles could actually glimpse the terrifying sight of the King of Prussia himself. And on August 11 the Russians recaptured Wilno.

With the remnants of his army, Kosciuszko limped back to his last stronghold. To Warsaw.

∞

BUT KOSCIUSZKO REFUSED to be beaten. Instead, it was his most heroic moment. Acting as both strategist and engineer, warrior and statesman, he kept his resolve. Dressed in a peasant's garb and known as "Father Thaddeus," he was tireless in holding the enemy at bay: He used the city's population to build earthworks, and when the enemy attacked, he personally led the charge with fixed bayonets. At the suggestion of local Jews, he made plans to enroll a Jewish battalion. He was also a strict commander, brooking no disorder. Back in St. Petersburg, the empress was bluntly warned by her generals that there was "enough skill in the rebel [Kosciuszko]" to make her life a festering hell. He almost did. Days slipped into weeks, and weeks into months. Kosciuszko was stoic and so were his rebels. And while the city smoldered, the war settled down to a deadly siege.

For some three months, the rebels held both the Russians and Prussians at bay. The effect was often exhilarating; the morale of his men, gleaming with excitement, soared, and that of his enemies sank. Then Kosciuszko unleashed his most brilliant move to date, stirring up an uprising in the occupied province of Wielkopolska at the rear of the besieging armies. To Catherine's disgust, the Prussian king, Frederick William II, retreated.

The war was still not over; the upstart Kosciuszko had stalemated one of the most fearsome allied forces in all of Europe. And suddenly it seemed as if a near jubilant Poland, once thick with blood, once exhausted by despair, would at last defend itself successfully. But an outraged Catherine was determined to snuff out this lethal flame and to stamp on the ashes until they were cold.

It was then that she ordered Suvorov to reduce Warsaw to submission.

CHAPTER 9

FRANCE

✦

S O IT WAS DONE.

In France, with the execution of the king, there was now no going back. Louis's death was a victory for the radicals, but what kind of victory? Nor was this the only question being asked: Would the blood of the king be enough to satisfy the most ardent followers of the revolution? With their nemesis gone, could peace among the rival factions survive, enabling the *patrie* to finally consolidate itself as a republic, a glorious example for the world over? The signs were not good, and, when the blinders of idealism were removed, perhaps had never seemed good. While in America federalists and republicans were increasingly locking horns, the Jacobins and Girondins were engaged in a similar life-or-death struggle—not just for the direction of the nation, but against each other. For suddenly, as Jacobin-aligned mobs took to the streets and smashed opposition printing presses, now it was the ruling Girondins, like the monarchists and the king and queen before them, who became increasingly divided and desperate and afraid for their lives.

Abroad there were vast shifts as well. Having watched the revolutionaries furiously preaching sedition across the continent, not to mention the beheading of one of their brotherhood, the rage of European monarchs exploded. The greatest change of heart was in England. William Pitt, the prime minister, after straddling the fence for nearly a year, suddenly faced a Parliament and a public aghast at the events roiling England's ancient foe.

To Pitt, France was becoming the stuff of nightmares, an outlaw state; the execution of Louis XVI was, in his words, "the foulest and most atrocious act the world has ever seen." The more Pitt pondered the nature of the new French regime, the more convinced he became that Britain's next generation could lie at France's mercy. There were other considerations too: France was controlling more and more of the European continent, even annexing Belgium, while across the bay in a restive Ireland, there was an explosion of clubs and societies openly supportive of the Revolution.

The New World was convulsed by the events in France as well: Many Americans recoiled in horror at the beheading of a man who had midwifed them to independence, while George Washington and his cabinet furiously met to formulate a response. As it turned out, their concerns were prescient. On January 24, 1793, three days after Louis's death, Europe was rushing full tilt to war—Pitt promptly dismissed the French ambassador and called on Catherine of Russia to demand that France "abandon" its conquests and "withdraw" back to its own territory. But on February 1 the Convention, glowing in anticipation of a fight, moved to declare war upon *both* England and Holland. With scarcely a second thought, in the name of patriotic messianism, the revolutionaries were prepared to sacrifice Paris block by block, to be destroyed rather than dishonored by impudent monarchists.

The response of monarchical Europe was equally decisive: By March 7, the First Coalition—a formidable array including England and Holland, but also Prussia, Austria, Sardinia, and Spain—began its efforts to dismantle the revolution.

A cataclysmic global struggle had begun. Ironically, it was much as Maximilien Robespierre had predicted.

∞

FRANCE, INTOXICATED BY its own success, had overreached. After their initial victories, the lines of the revolutionary armies were overextended, the men lost their fervor, and thousands of volunteer soldiers quit and returned home. Meanwhile, the total number of troops on the eastern front, already ill clothed and badly fed, fell by almost fifty percent. In February the Convention mandated conscription to bolster the army, though they blundered here as well, allowing the rich to employ substitutes. From Lyon to Bordeaux, revolts broke out, and in the royalist Vendée region, anger over conscription prompted a rebellion so fervent that an army had to be hastily summoned from the front to contain it. For the Jacobins, then, suddenly everything seemed to be going wrong and nothing seemed to be going right.

The most spectacular of the French generals, Dumouriez, led twenty thousand troops to invade Holland, but after a setback at Neerwinden in March, on April 5 he defected to the Austrians with a thousand men. While representatives of England, Prussia, and Austria met to formulate plans for vanquishing the revolution, the nation suddenly lay in a panic. And as allied armies took back Belgium and once again threatened to invade France, revolutionaries in Paris cried out, "We are betrayed!"

They were. Actually, it was worse than they feared. This bitter blow was compounded by a growing domestic crisis: Prices continued to rise, the currency continued to fall, trade was languishing, and food was harder to obtain. It was as if the old regime were still in power. And in the Convention, the angry divide between the Jacobins and the Girondins intensified. Day after day Jean-Paul Marat, with the cockiness of a parent lecturing a stupid child, mercilessly assailed the Girondins as protectors of the rich, and day after day, Danton and Robespierre increasingly used the Jacobin Club as a megaphone for their radical ideas—which now spoke of regenerating the *whole* world. Alternatively harassed and fearful, the Convention thus took an extraordinary step: On March 10, 1793, to shore up the crumbling edifice of executive government, it established a Committee of General Security to direct the political police, but this was sheer misnomer. It had total authority to arrest anyone so much as suspected of disloyalty, let alone potential crime, and the committee was soon accompanied by a Revolutionary Tribunal to try any and all suspects sent to it. Gone were concerns about civil liberties; gone were the niceties of the constitution—the judgment of the jurors was absolute; as Pierre Vergniaud warned, France was "laying the foundations of an Inquisition a thousand times more fearful than that of Spain."

Was he right? True, there was no appeal or review process. In theory, this in itself was not necessarily egregious. The revolutionary nation was struggling for survival. But reality told a different story. The tribunal sat in the Palais de Justice, and as the war deepened and its docket overflowed, rather than conduct marathon trials, the tribunal simply truncated its legal procedure and more often than not rendered a quick verdict of guilty.

Or just as likely, it deferred to the *Comité de Salut Public*—the Committee of Public Safety.

Pandora's box was thereupon pried open. This was a body with the muscle of a tyrant and the heart of a rebel. Established on April 6, the Committee of Public Safety (CPS) replaced the executive council. Technically it answered to the Convention, but in truth it became the overseer of France. "We despots," boasted Jean Drury. "Ah, no doubt we are, if despotism is to serve the triumph of freedom." In truth, some said this and some said that. But only in time would the true import of the CPS become clear. The CPS held its sessions in secret, it nominated juries and controlled foreign policy, it supervised the armies and oversaw the bureaucracy, and, of course, it ruled the secret service. Its powers were near absolute. It could confiscate public property and read all private letters. It controlled secret funds—2.5 million pounds—and could suspend any public official. It could summon witnesses at will and just as freely dispense with due process. And with a wink of an eye or a scrawled signature, it determined life or death across the country. Under its rule, thousands upon thousands would be sent to increasingly horrific deaths. Even Russia's Catherine blushed at such extraordinary, capricious powers; the Americans, including many sympathetic to the Revolution, often winced at the committee's bloody excesses as well. The CPS gathered under the large sign of ÉGALITÉ at the Pavillon de Flore, and was first headed by Georges Danton, again called upon to lead a fragile nation in despair. Until the moment of its overthrow, it was seemingly omnipotent.

Danton once more sought to be the peacemaker of France. Having tried to make inroads with France's enemies abroad, he then lowered his giant head and attempted to make peace at home, between the Jacobins and the Girondins: "Citizens," he roared, "now that the tyrant is dead let us turn our energies to valiant prosecution of the war . . . if we must drink blood let it be blood of the enemies of humanity." But this was dwelling in fantasy; the differences were too pronounced, and the progenitors of hate, particularly Marat, too persistent.

Factions and cabals were now everywhere, in the clubs, the convention, the coffee houses, and private apartments. Where in the United States, Thomas Jefferson was discreetly mustering allies to criticize George Washington, Marat, keeping a sharp eye on his enemies, escalated his attacks against the Girondins, but there the similarities ended; almost unthinkably, he now called for his rivals' deaths. "Caesar was assassinated in the public Senate," he cried. "Let us treat the traitorous representatives of the country in the same manner." The violence he inspired mounted with such force and fury that by April 14, the Girondin-controlled Convention decreed that Marat should be tried by the Revolutionary Tribunal for promoting murder and dictatorship; even Thomas Paine supported his impeachment. At his trial, every inch of free space was taken; all of France awaited the outcome. The indictment ran nineteen pages, but Marat, as always, was fluent in his own defense. While the trial unfolded, a jeering crowd gathered in the streets, threatening that "if Marat's head must fall, our heads will fall first." As a result, the Convention was too timid to arrest him, and his judges were too sympathetic to convict him. Marat was promptly freed.

His followers placed a crown of laurels on his head and bore him to the convention, where he angrily promised revenge upon his enemies. From there he was carried to the Jacobin Club, where his ailing frame was invested in the president's chair. Having won this precious victory, he was now all but untouchable. Bristling with rage, he resumed his campaign demanding that the Girondins ("our criminal representatives") be excluded from the Convention altogether; in reality, this meant the guillotine. One can imagine that in total silence, the Girondins who remained suddenly whispered to one another: *"C'est une exécution."*

If so, they were right.

The real blows were yet to come.

∞

BITTER AND RESENTFUL, the Girondins fought back. The honey-tongued Vergniaud, with great rhetorical power and considerable courage, asked of the revolutionary committee: "What revolution does it want to make now that despotism is no longer?" And he asked this too: "Let us prove that we know how to be happy with a republic." Echoing Mirabeau, he then proceeded to make a terrible prophecy: "Citizens, it must be feared that the revolution, like Saturn, successively devouring its children, will engender, finally, only despotism with the calamities that accompany it." But his eloquence was wasted; Vergniaud spoke to empty seats, dozing deputies, and disapproving frowns. Later he appealed to his middle-class electors to rescue the moderates from the tyranny of the mob: "I summon you to the tribune," he wrote, "to defend us, if there is still time, to avenge liberty by exterminating tyrants." Of course, in the new formulation of revolutionary France, what had changed was that the tyrants were no longer the king and the monarchists, but the radical revolutionaries themselves. In the end, it was combat between eloquence and force, and the result was a foregone conclusion.

On May 18, having failed to convict Marat, the desperate Girondins now sought to muzzle their other most vocal adversary: the Paris Commune, the city's radical local government, which had become a violent entity unto itself. A *commission extraordinaire* of twelve was created to investigate these rogue sansculottes, and six days later, the Convention followed suit, ordering the arrest of the so-called *enragés*—radical organizers, militants, semi-military bands, and neighborhood thugs. Among those rounded up were Commune ringleaders, Jean-René Herbert and Jean Varlet, but the Commune insisted upon their release; otherwise, said its spokesman, "Paris will be annihilated." Steadfast, the Convention said no. Remarkably, it was as if the people of Philadelphia were at war with the U.S. Congress. Nevertheless, an outraged Robespierre, at the Jacobin Club, then rallied the citizens to revolt. "It is when despotism is at its height, it is when good faith and decency are being trampled under foot, that the people ought to rise." The next day Marat joined him, denouncing the investigative commission as "hostile to liberty," while a mob, marching under the banner of *Insurrection et Vigueur*, burst through the Assembly doors and called for the prisoners to be released. This time they were. Yet as the Girondins attempted to strike back, on May 31 the Commune raised the stakes: It again readied to invade the Assembly and now boldly called for Vergniaud and his cohorts to be indicted. But a standoff ensued, and the dueling sides recessed for the night.

It did not last for long. Two days later, tens of thousands of armed men chanting *"Purgez la convention!"* ("Bleed the Convention!") surrounded the national parliament and demanded the arrest of the leading Girondins; no trap is so deadly as the one you set for yourself, and suddenly the men who had humbled a king and boldly waged ideological war against the monarchs of Europe were reduced to frightened little pawns. Running for their lives, some managed to escape, with the faint hope of inciting civil war in the provinces, while others were placed under house arrest. Jerome Petion, the former mayor of Paris, successfully fled, and so did the once fervent revolutionary and political philosopher Condorcet. Jean Roland, the former president of the Convention, retreated to the south, but his wife, the famous giver of salons and lover of philosophy, stayed behind, hoping to plead her husband's case before the Convention. She was arrested and lodged in the Abbaye jail; her husband never saw her again. The political shift was now indisputable—the Convention had become the dutiful servant of the new octopus of France, the radicals and the Committee of Public Safety. And under the watchful eye of the people of Paris, violence and intimidation thus commanded the Revolution.

What form was the new revolution to take? A new constitution was drawn up, one remarkable for the degree to which it presaged demands of twentieth-century revolutionaries and idealists; having done much to help the Americans attain liberty, Paine himself contributed to one of the drafts. It codified adult male suffrage and took the pronounced step of adding the universal rights of not simply education and freedom of worship, but of revolt itself. Moreover, with further socialist overtones—though the concept scarcely existed—private property rights were circumscribed by the larger national good. British historian Thomas Carlyle would later enthuse that this was "the most democratic constitution ever committed to paper." In this, he was perhaps correct. Embraced by the Convention on June 4, it was ratified by one-fourth of the electorate, far more than had voted for George Washington as president. But there was a hitch: Suspended indefinitely, the constitution was buried almost as soon as it was born.

For on July 10 the Convention then sanctified the Committee of Public Safety as the emergency government, superseding all constitutional and legal bodies, until the time that "the peace" should arrive.

∞

PEACE—THE MOST elusive concept in France. Ironically, the radicals now fully ruled the Convention, but the Convention itself ruled very little. Abroad, at home, in its own ruling councils, France was a country literally at war with itself.

Upon setting up the Revolutionary Tribunals, Danton had told the Convention, "Let us be terrible so that the people will not have to be." That was the concept at least. So in Paris there was the unrestrained lawlessness of the mob. So every local government in the country was equipped with committees of surveillance, and all citizens were encouraged to denounce anyone suspected of doubtful loyalties. So every day, revolutionary zealots thundered up and down streets all over the country, breaking down doors and dragging away their critics to be beaten or tortured or locked away. So bonfires blazed on hilltops, and the faithful sang martial music at torchlight processions. Meantime, while National Guards ransacked homes, revolutionary propaganda plastered walls and was widely distributed. Priest hunts, too, were authorized, churches defiled, altarpieces slashed, and stained glass windows mutilated. And there remained the omnipresent semi-military gangs, fingering new suspects who displayed insufficient revolutionary ardor. America's envoy, Gouverneur Morris, once averred to George Washington, "Our American example has done them good." But now he chillingly reported, "In the groves [of the Revolution], at every end of every vista, you see nothing but gallows."

Nonetheless, all across the country, in Bordeaux, Marseilles, Toulon, Caen, and Lyon, Girondists mobilized to resist the Paris dictatorship. Their plan was to raise federalist forces and march on Paris, even as Marseilles and Lyon negotiated with the British fleet for supplies. In ensuing weeks, they wrote articles and made speeches, condemned the tyranny of the Jacobins, particularly Marat, established parades of protest, and bided their time to strike.

But one foot soldier among them was not willing to wait. Not yet twenty-five, born to a titled but penurious royalist family, she glowed with beauty and intelligence: We know her as Charlotte Corday. Though she was appalled at hearing that the king was guillotined, Charlotte was no monarchical stooge—she had read Plutarch, Rousseau, and Voltaire, later fell under the spell of the Girondins, and wanted nothing more than for a decent republic to flourish in France; she also became enraged by Marat's diatribes and the purge against the Convention. A descendant of the dramatist Pierre Corneille, she soon conceived of a great and grand role for herself: as a martyr to the *patrie*, a simple woman prepared to do titanic things. To cleanse the nation of a despicable fiend.

She planned to assassinate Marat.

∞

FOR CHARLOTTE, AS the early summer days passed, hate became preoccupation, preoccupation became fixation, and fixation became fanaticism. On July 9, armed with nothing more than her courage, she boarded a stagecoach for Paris. Arriving two days later, under a brilliant sky, she bought a sharp dinner knife with a five-inch blade and planned to enter the Convention to slay the beast in his chair, but was told that Marat was at home. All was not lost: She had heard that Marat, the friend of the people, was willing to see anyone who needed his help. On July 13, a day of stifling heat, she planned to strike. Early that morning she purchased a newspaper—the news was of the latest demand in the Convention for the death sentence against the Girondins—and slipped on a dazzling black hat decorated with green ribbons. She found Marat's address and at eleven-thirty went there, but got no farther than the foot of the stairs. She was told Marat was in his bath, too sick to see anyone.

For Marat, when he wasn't roaming the street with pistols tucked into his belt, the bath was now the cockpit of the Revolution. His disease was a type of scrofula (ironically, it was considered the "king's disease" and long believed to be "cured" by the royal touch), which had worsened, and he found respite from his open sores only by soaking for hours in tepid water while a red bandanna drenched in vinegar was tied around his head. Here the oracle of the revolution labored tirelessly: With a worktable straddling his high-walled copper tub, he kept a pen, an inkpot, and books, and each day composed fresh invectives for his journal. Lovingly cared for by Simone Evrard, his former laundress and present fiancée, Marat steered a country.

From her temporary lodging, Charlotte dispatched a note to Marat asking to see him. "I come here from Caen," she wrote. "Your love for the nation ought to make you anxious to know the plots that are being laid there. I await your reply." But in her haste, she forgot to add her address, so that evening she returned. Once more she was denied entry; actually, this time, Simone was suspicious. But Charlotte brusquely raised her voice, insisting Marat needed to know about "traitors" in Normandy and Caen. Marat, hearing her shout, demanded she be let in. From his tub, he courteously received her. While Simone watched, Marat offered Charlotte a seat. "What is going on at Caen?" he asked. "Eighteen deputies from the Convention rule there," she replied, "in collusion with the *département* officials." "What are their names?" She gave them. Simone, reassured, left briefly to get more minerals for the bath.

Charlotte edged her chair closer.

Meanwhile, Marat wrote down the traitorous names, adding, "Good, in a few days I will have them all guillotined." At that, Charlotte raised her knife and buried it in his breast. Blood gushed from the wound; Marat called out his last words, to Simone, "*À moi, ma chère amie, à moi!*"—"To me, my dear friend, to me!" The screams echoed into the night and the nearby streets. An army surgeon and a dentist rushed in to save him, but there was nothing to save. Outside the courtyard, as the news spread, anguished crowds gathered. Simone held Marat; it was too late. He died in her arms.

Charlotte fled the room, but only halfheartedly. She was stopped by a man who beat her back, first with a chair and then with his hands. In the streets, a growing mob wanted to tear her to pieces. The police hustled her away, to the Abbaye jail. The response of Charlotte, now a prisoner of the state, is immortal: "I have done my duty," she said coolly, "let them do theirs."

∞

AT HIS DEATH, Marat left little more than a trickling of scientific papers, his unpublished *Lettres polonaises* and unfinished death lists, as well as a mere 25 sous. He also left revolutionary France brokenhearted. Derided by his enemies, mocked by his colleagues, overnight Marat became an enduring symbol of the revolution; orators, hymns, and poems even compared him to Jesus. The Cordeliers Club suspended his heart in an urn as a venerated relic, and for two days, thousands came to view it, filing past to get one last glimpse, with "breathless adoration." On July 16, under a great blue sky, his body was carried in slow time through the streets: Clasping hands, silent sansculottes teared up while young girls in white dresses showered flowers upon his coffin. Behind him was his bath, carried by four women marching in careful, measured steps; behind them was Marat's bloodstained shirt, carried on the end of a pike. All of the revolution's leading lights were here: Robespierre, Danton, and columns of grief-stricken radicals who accompanied his corpse to its resting place in the gardens of the Cordeliers. One man insisted Marat's soul, released from its earthly casing, now "glides around all parts of the republic all the more capable of introducing itself into the councils of federalists and tyrants." On September 21, 1794, his remains were transferred to the Pantheon.

Indifferent, Charlotte sat in prison on a straw mattress, caressing a black cat and pondering the afterlife. Her trial was brief. She admitted to her deed, but neither to any nefariousness nor to any guilt. Asked if she were part of a larger conspiracy, she hotly denied that too. She said simply that she was avenging the victims of the September massacres and other numerous objects of Marat's vehemence. "I have killed one man," she insisted, "to save a hundred thousand." Questioned whether she thought she had killed "all the Marats?" she replied with quiet wrath, "With this one dead, the others, perhaps, will be afraid."

Sentenced to prompt execution, she was taken to the Conciergerie. In a letter to her father, she frankly stated, "The cause is good." Then, in one last extraordinary gesture, she asked the court whether she might have her portrait painted before her execution; it took two hours, punctuated by her thoughts about revising it this way or that. At early evening, the executioner arrived. Under a darkening sky, she jumped into the tumbrel, stood upright, calmly looking into the faces of a cold and unsympathetic crowd and disdainfully receiving their curses. It began to rain, hard. In the distance, Danton and Robespierre stood together in the recesses of a window watching the procession. Her face was calm when the blade fell, and she died not recognizing the fatal consequences of her deed: She had transformed Marat from a monster into a martyr—the news was even discussed in the empress's imperial palace in St. Petersburg—and opened the way to the scaffold for the Girondins whom she had nobly sought to save.

The final destruction of the Girondins was now made possible. Vergniaud, their nimble orator, amply described their predicament, intoning, "She had killed us, but she has taught us how to die."

∞

THE DYING WOULD soon begin. During the summer of 1793 the Committee of Public Safety—the "Great Committee," as it came to be called—became the indisputable master of France. It comprised twelve men—today we know them as the twelve who ruled. Upon assuming power, they inherited a France riven by unrest and outright revolt at home, and under siege from Europe's armies massing at its borders. The country was reeling from defeats on the frontiers as well as in the oceans, and every port was blockaded shut. The Prussians were in Alsace, the English flag flew in Toulon, while the British and Austrians were overrunning the northern corridor—and all contemplated marching on Paris. But in under a year, the Twelve would wield dictatorship and terror to forge the country into a single nation; they would establish a paid army of *patriotes pour fortunes*, sansculottes who served as spies and censors for the Jacobin regime; and they would oversee a new crop of generals—most of the old ones were guillotined—that would repel their hated foes with stirring triumphs and emerge victorious.

Victorious, that is, over all except France itself.

On July 27, in a fateful step, Robespierre was made a member of the committee. "That man," Danton fumed, "has not wits enough to cook an egg." But he understood the rhythms of revolution as did few others, and by September the committee sat at the pinnacle of power. One evening, watching the sun set over the Seine, a now exhausted Danton sighed, "the river is running blood." Hardly were more prophetic words spoken.

As with so much else about the revolutionary leadership, the Great Committee were all of the middle class, and all were children of Rousseau and the philosophes—in Robespierre's case, he always kept *The Social Contract* by his bed. But the committee itself was far from monolithic; nor was it simply a Jacobin clique. There were the war bureaucrats, managers of the logistics: Lazare Carnot (the "organizer of victory"), Prieur de la Cote d'Or, and Robert Lindet. There were the prophets and ideologues: Robespierre, Saint-Just, and the crippled Georges Couthon. There were the organizers: Bertrand Barere and Herault de Sechelles. And there were those who led the anti-Catholic crusade, like Jean Nicolas Billaud-Varenne.

Were these men mass murderers? Subject to reelection every month, they were, by any definition, tireless public servants: Working almost informally—remarkably, no chairman was appointed, and no minutes or records were kept—they prepared and guided legislation through the Convention. And with cold efficiency they struggled day and night on their endless challenges, from morning's first light until noon in their sparsely furnished offices and subcommittees; from one to four (or longer) in afternoon debate in the Convention; and from seven until midnight again conferring in their inner sanctum—that is, when they weren't making speeches at one of the clubs, or politicking in private, as was frequent. To be sure, they inherited a culture of death that seemed rooted deep in France's very genetic code, from the Wars of Religion to St. Bartholomew's massacre, from France's many civil wars—six in the last century alone—to the worst of its monarchical despotism. Even before the revolution, it was not uncommon for criminals to be dismembered at public executions, or for corpses to be fished out of the Seine; memories still lingered of the wheel, the rack, the whip, the stake, and the hangman. And also to be sure, they themselves were hardly secure: At any time they could be defied, removed, and carted off to the scaffold; a number were. They too were prey to the caprice of the Paris mob, or hostage to the whims of the National Guard, and at any instant they might be overthrown by an avaricious general, or a battlefield rout, or a rebellious *département*—or one another. They had their reasons, all, perhaps, compelling ones.

But within the span of twelve months, these twelve men would make execution their most fundamental instrument of politics, destroying their adversaries without hesitation, their countrymen without remorse, their own religious figures without pause. No one would be exempt: not the elderly, not the young, not women, and far too often, not even the innocent; eighty-five percent of those executed were commoners, a telling riposte for those, then and now, who insisted the Revolution was simply a class war directed by the bourgeois against aristocrats.

The Twelve would claim the customs of war or the exigencies of revolution; but what is clear is that with the Terror, they would create a death machine nearly unparalleled in history, that is, until the genocidal violence of the twentieth century was unleashed. The Terror *was* terrible: It was ritualistic, apocalyptic mass murder, and it pointed the way toward the worst of modern totalitarianism. It was, one historian has noted, the trauma that scarred modern history at its birth—as the glorious Declaration of the Rights of Man lay stillborn.

With Marat now gone, over the next nine months, one man would stand above all others in France. He felt blessed by God and was on a mission; in practice, he controlled both the Convention and the committee. His name, of course, is Maximilien François Isidore de Robespierre.

∞

ROBERSPIERRE. THE NAME alone fosters the strongest of emotions.

It is at once synonymous with the worst of the Revolution, the Terror, while at the same time with its best, the effort to strike down age-old inequalities bedeviling French society. Yet he remains the most elusive of men. No less than with Danton, for decades historians have debated the true nature of his mysterious and intently private soul: Was he a bloodthirsty tyrant erecting a dictatorship he would head, or a bourgeoisie manipulator wedded to a springtime of republican hope? A great leader of the people and unerring interpreter of the times, or a dreamy madman caught up in a tide of history sweeping the globe? Also to be settled was whether he was a ghastly forerunner to a Pol Pot or a Stalin, a Führer or an Il Duce. What is clear is that in a revolution that spawned a rich array of striking characters, from Mirabeau to Marat, Danton to Talleyrand, Marie Antoinette to Lafayette to Napoleon, Robespierre stands out as the most ambiguous and enigmatic. What is equally clear is this: He was a man who knew himself to be right and who, like Brutus, would sacrifice his own children so that the right principles would prevail. To grasp who Robespierre was is not simply to grasp the revolution itself—its early days, its headiest days, its bleakest days—but it is to see how a nascent democratic republic, filled with such flowering promise, inexorably slid into a prototype of a modern-day killing field.

Some things about Robespierre are not elusive. He was idealistic and strongly motivated, this much we know. In conversation, he could muster a charming smile, but he almost never laughed. He was a lonely man: Rare in France, he was unmarried, chaste, and puritanical; he was also absent-minded. As a child of the Enlightenment, he believed with all his heart in the Rights of Man. As a magistrate in Arras, his hometown, he recoiled from delivering a death sentence. And even as head of the Twelve, he shrank from blood. It has been duly noted that the only execution he ever attended was his own.

His younger years were painful. Robespierre was born in 1758 into a modest and declining family. Orphaned at the age of five when his father, a struggling alcoholic lawyer, vanished, and his mother, giving birth to a still-born child, died, Maximilien, his sister Charlotte, and his brother Augustin were raised by their grandfather and then their two maiden aunts. As a boy, Robespierre was misanthropic and moody; impressionable too. The aunts who watched over him were fussy, fastidious, and keenly aware of the small-est slight, so he became fussy, fastidious, and keenly aware of slights as well. They were prudish, so he became prudish. They shrank from foul language, and so did he. But they always nourished his passion to succeed. And suc-ceed he did, from early on.

A graduate of the famous college of Louis-le-Grand in Paris, he dis-tinguished himself in philosophy and law. From there, he was admitted to the Arras Academy for the Advancement of Arts and Sciences in 1783, then quickly became its chancellor, then its president; he entered competi-tions and with his report on degrading punishments (*"Mémoire sur les peines infamantes"*), won second prize at the academy of Metz. By 1787 he was already recognized for his altruism and for sounding the alarm against arbi-trary justice; he also burned with ambition, at one point sending a copy of his legal pleadings to Benjamin Franklin, whom he addressed as "the most brilliant scholar in the universe." And when the king summoned the three Estates to Versailles, the thirty-year-old Robespierre was among them.

Once the mighty drama began, his fate soon became intermingled with the course of the revolution and the very history of France itself.

The royalist press attacked him from the start, calling him "Demosthenes" and mocking him as a "monkey of Mirabeau's." In fact, he was no one's monkey, not Mirabeau's, not the king's, not Marat's. Frugal, impeccably dressed, with precise manners and enigmatic cat's eyes, he rapidly attracted widespread attention, and just as rapidly became a leader of the radical Jacobins.

He was a sickly man and asthmatic too, often wheezing, frail and feverish. Yet in spite of his ill health and his small frame—he was only five-foot-three—he loomed over most of his rivals. With his hair neatly brushed and powdered, he looked like a ghost of the ancien régime, though in Assembly debates he evinced an original style that reflected his categorical conviction of truth and error. Speaking in the didactic, precise prose of a lawyer, he was impossible to miss: At pauses, he would look up dramatically from his written speech, his greenish eyes staring through tinted spectacles (also green). Of his zeal, Mirabeau sniffed: "That young man will go far; he believes what he says." Still, there were times that his speeches would go on for so long—often hours—that he was the subject of secret ridicule.

No matter. He was eloquent in his appeals for a better society. He was also a man ahead of his time. He called the Bastille a "hateful monument of tyranny," and the people applauded. He fought for universal suffrage, an end to the royal veto, and a cessation of religious and racial discrimination, and they rallied behind him. Where few others cared, he passionately spoke out for the liberty of all Frenchmen, campaigning for the rights of actors and Jews, and castigating his colleagues for tolerating black slavery. For such stands, the people of Paris would one day organize a triumphal procession for him. However, his stands—and his personality—also won him numerous enemies. Marat despised him, and Danton hated him even more; so did Billaud and Collet, with whom he quarreled violently. Lafayette tried to have him arrested, and the king and queen had nothing to do with him. Few were spared his wrath or his tongue: He opposed Brissot's Girondist goal of spreading revolution, he threatened Lafayette's head, he denounced the secret intrigues of the court, and he was a primary force behind the beheading of the king—rallying the hesitant, he spoke eleven times at Louis's trial, and advocated death each time.

His patriotism was unqualified and absolute, there is no doubt of this. And contrary to his bloodthirsty reputation, Robespierre could be quite tender. After his old friend Camille Desmoulins married, he was made godfather to the child and spent hours sitting by a fire, bouncing the baby on his knee while talking amiably with Desmoulins's wife. True, he was unable to relax with friends. Nevertheless his austerity charmed women, devoted ladies banded together as a fan club, and an Englishwoman even wrote him a check to further his crusades. Robespierre returned it, and, in turn, Desmoulins deemed him the "Incorruptible." The name stuck. Yet there is no denying that the Incorruptible could be ruthless: Invariably absent whenever the blood flowed, he thought nothing of whipping the people into a frenzy, passionately orating, "The people are enslaved as soon as they relax; they are conquered as soon as they forgive," thereby sending the mob into the streets in a murderous rage. And when the time came, he was willing to kill anyone in the way, the king, the queen, the royalists, the Girondins, the *enragés*, the rich but also the poor, the strong but also the weak, husbands as well as wives, and even his closest friends: Desmoulins and his wife.

But did he see himself as ruthless? Or as a ghoulish fiend? This is unlikely. For one thing, the Terror was an old Roman idea: If a Roman legion did not do its duty, then indiscriminate punishment would be meted out. For another, Robespierre preferred to preach virtue as Rousseau, his model, had done. Virtue, of course, mandated dedication to family, to work, to civic zeal and the dreams of the revolution. But it also meant getting rid of the enemies of virtue, whoever they were, wherever they were. From as early as 1789, he trafficked in a world of conspiracy theories, warning about newly minted "patriots," old traitors and hypocrites ("who flatter you today to betray you tomorrow"), and the need for constant purges. As the revolution wore on, as the gore deepened, as the daily tumbrels bore victims to the guillotine, Robespierre prophesied that for those who refused to adopt virtue, "the Razor of the republic" awaited. The Terror, he also declared famously, is nothing, nothing "save justice, prompt, severe, inflexible. It is an emanation of Virtue."

∞

IRONICALLY, HE HIMSELF passed through a checkered odyssey. Originally he was a monarchist, then an idealist, then a constitutionalist, then a revolutionary, then the propagator of dictatorship, then the voice of terror. Yet to the daughters of his landlord, he was "*bon-ami*," helping them with their studies and taking them on picnics; and when Louis's carriage passed his

home on its way to the scaffold, Robespierre closed the shutters and told
a young girl, "Something is happening today, child, which you should not
see." He was also God-fearing in an increasingly godless republic; he always
said grace before meals. And in him lurked a reluctant, almost ambiva-
lent streak: Interestingly, he would never quite be the absolute dictator his
detractors insisted he was.

Still, for Robespierre, not the theoretician or revolutionary, but the
politician, there was always an inherent conflict he could never easily reconcile. Austere, chilly, and proud, he had an instinctive distaste for the open
give-and-take of politics. Unlike Mirabeau, a consummate wheeler-dealer
with the common touch, or Danton, who charmed even his detractors,
Robespierre regarded any disagreement as the latest betrayal, any deviation
as questionably motivated. It was once said that he was a lover of mankind
who could not enter into sympathy with the minds of his own neighbors or
colleagues. This was all too true.

Yet as did few others, he felt the pulse of the Revolution, maintaining popularity with the sansculottes by living simply with plain folks, and
extolling the masses and defending their interests. And beyond the elegant
enclaves of the Parisian elite, in his own way he served up ample quantities
of patronage. Though he was a man of refinement, he was at home with
radical street orators and grimy demagogues, alleyway thugs and populist
scandal-sheet mongers; to him these were not coarse vulgarians, but the
real people of France. He seemed intuitively to know that for centuries
the lower classes had wallowed in crowded cities where malnutrition bred
disease, where lack of work bred hunger, where cleanliness was a mirage and
poverty a way of life, and where servitude was constant. "The people," he
once declared, "is sublime." For this, the sansculottes never abandoned him,
even as life in the revolution often became hell and liberty became terror.
So too when their own kind disappeared into Paris's many jails, they still
believed in him. No wonder that even in 1794, after he had outlived his
time and himself, many broke down and wept upon his death.

Perhaps, ultimately, he was ill fated from the start: He was a prophet
of death as much as life, a progenitor of hard medieval terror as much as
of progress, and too often his policies were predicated on his personal rule.
Why? It is a good question. It could be said that the best a man can do is
enlarge his ego to the betterment of his kin, his kind, and his nation, and so
find satisfaction in their widening beneficence. This, after all, is what Robespierre thought he was doing, or at least hoped he was doing, on a grandiose

scale. But in the end, as happens so often to those who claim to know the truth, or believe themselves to be the messenger from God, the politics of the revolution clouded his wisdom, and the Terror darkened his judgment. There was no escaping this butcher's bill. As the revolution intensified, as death became a daily phantom, and as Robespierre felt himself surrounded by potential assassins, he talked increasingly of his approaching end, of martyrdom, and of the future life. He worried constantly about plots and coups, of counterrevolutions and treacheries from his closest allies and his oldest friends. And he was right.

In the waning hours of 1793 and the early, anxious days of 1794, when the Convention was in panic, when foreign spies roamed the streets of Paris and the French army was in disarray, when the morale of the troops, already low, plummeted, and when even French generals were unwilling to follow orders of the government, Robespierre was decidedly a man of his times. Swept up by a sense of high purpose, he would cleanse the lingering sins of the past and institute one bold measure after another to lead France to the Promised Land.

The most notable of these, of course, was the Terror.

∞

IN THE FIFTH summer of the revolution, the air was thick with division and fear, corruption and shifting allegiances, even as the genesis of the Terror was at once evident but also subtle: foreign dangers and domestic disorder, which intensified public paranoia and xenophobia, thus provoking martial law. The First Coalition, under England's leadership and Russia's goading, reentered Mainz on July 23, stormed into Alsace, and had taken Valenciennes, a hundred miles from Paris. The Spanish also made inroads, seizing Perpignan and Bayonne. By August 29 France was like a chessboard, with pieces falling this way and that: French insurrectionists openly surrendered a French fleet to the British, as well as a crucial naval base and arsenal at Toulon; actually, to the dismay of the Convention, they opened up their harbor to the British fleet commanded by General Hood. On the outer reaches, Britain controlled the seas and at its whim was poised to capture French colonies on three continents. On the inner reaches, the allies, smelling victory, resurrected feudal rights as they marched forward and discussed the impending partition of France.

At this stage, France seemed not to have a friend in all of Europe. As if this weren't enough, internally the revolution was cracking at the seams. Lyon, Bourges, Nîmes, Marseilles, Bordeaux, Nantes, and Brest had all drifted into the camp of the revolting Gironde. The Vendée was aboil with rebellion, while aristocrats abroad—the émigrés—and now, even at home, arrogantly contemplated restoration. And the provocative extremes of poverty and wealth again deepened.

In Paris itself, as well as the provinces, the economy was an enemy every bit as much as the insurrectionary armies. Commerce was dead, price controls were a disaster, and fears of famine raced through the nation. Across Paris, Senlis, Amiens, and Rouen, hungry citizens nearly brought down the government. On June 25 a band of *enragés* took to the Convention and insisted that all profiteers be stripped of their ill-gotten wealth and tried before the people. Said their spokesmen to the deputies: "Yours is no democracy, for you permit riches. It is the merchant aristocracy, more terrible than the nobility, that oppresses us.... It is time that the death struggle between the profiteers and the workers should come to an end!" A slave to fear, the Convention acted quickly on this; the next day it adopted a proposal to institute the death penalty for hoarding, which included not just meat and vegetables, but salt and soap, sugar and bread, hemp and wool and oil and vinegar. But far from appeasing the turmoil, this measure only deepened the ferment. The Paris Commune began to call for an "open war" between the workers and the middle class. In other words, a revolution against the Revolution was now taking shape.

On August 30 Danton espoused the chilling phrase: Let Terror be the order of the day. By then, this was stating the obvious. Still, it was against this backdrop—foreign threat, insurrection, strife—that the Committee of Public Safety would spread the tricolor from Lisbon to Cairo, and an industry of terror would hammer a disconsolate nation into an undivided *patrie.* However barbaric, this actually became a time of enormous creativity: With the prospect of being overrun by foreign armies, a desperate Carnot and Barere would announce on August 23 the most innovative plans of the revolution—the *levée en masse*, the creation of a national conscript army, to fall, literally, en masse "like the Gauls on the brigand hordes." All unmarried men from eighteen to twenty-five years of age were conscripted under the slogan, "*Le peuple français debout contre les tyrans!*" ("The French people standing up against tyrants!") No one was exempt from the war effort: While the young were to fight, the married were to forge weapons and transport food, the women were to make tents and serve in the hospitals, and the old men were to "rouse the courage" of the warriors. It was to be an immediate, national outpouring of patriotic fervor, but in reality, it was nothing less than an all-out mobilization of resources amounting to total war: the complete merging of the state with the army. With this bold stroke, human history was thus rewritten. Not only had France revolutionized politics, but thousands of years of warfare as well.

And if (as the CPS decreed) France must overnight become one vast armed camp, Paris would become its teeming arsenal; actually, it seemed nothing less than a republican miracle. Weapons, metal, surplus clothing, all were exacted by the committee. So were church bells, melted down to make ball and shot. Advisory committees of engineers and mathematicians were formed to assist the war effort, and squares and boulevards across the city were transformed into open-air forges producing seven hundred muskets a day. Thousands of mills were nationalized, while contractors were told what to produce and what to charge. Even convents were transformed into factories. And France, hemmed in on each border and at every port, sought to squeeze precious minerals from its own soil; independence was now the watchword.

The Committee of Public Safety called for half a million men; by the end of September France had at least 300,000 men under arms—an extraordinary achievement. They were ill trained, poorly equipped, and still woefully undisciplined; many of them didn't even have shoes or boots. Thus, "inspirational" propaganda became a machinery of the state. The minister of war used journals like *La soirée de Camp* to marshal the republic's cause, and much the way George Washington had Thomas Paine's *Common Sense* read at Valley Forge, these papers were distributed in the field, where troops had nothing else to read. Leaving little to chance, representatives of the committee then went to the front themselves to ensure that their squabbling generals performed; ensure they did—at the battle of Hondschoote, delinquent officers were shot in front of their own troops, and twenty-three generals were imprisoned or beheaded for mistakes, or hesitation, or failure to carry out the committee's dictates. Later, at the village of Wattignies, on October 16, 50,000 green recruits squared off against 65,000 Austrians; this time, the forty-year-old committee member Lazare Carnot, a military engineer by training, fielded a rifle and joined Jourdan's forces as they raced into battle: The village changed hands eight times, and through neither side could claim absolute victory, the outcome was good enough that it buoyed the morale of the revolutionary armies and enhanced the aura of the committee. So did the stunning defeat of the rebellious Vendean army at Cholet on October 17. The northeast frontier was saved.

Thus was Danton's *"L'audace, toujours l'audace"* adopted by France's men of war. Yet while Carnot continued mapping campaigns and disciplining generals, and while the Terror seemingly rescued the republic from disaster, on September 17 a pliant Convention passed the Law of Suspects to ensure that there would be no backsliding. The committee was empowered to summarily arrest anyone remotely harboring counterrevolutionary designs, any returning émigrés, any of their relatives, and even anyone suspected of opposing the Revolution or the war. A simple failure to obtain a certificate of good citizenship was enough to merit imprisonment, and the simple sin of being "indifferent" to the state was enough to merit the guillotine. It was a law worthy of the Nazis and the Soviet Communists in the twentieth century, ensuring that all but the most ardent revolutionists—and therefore virtually all Catholics and *"aristos"*—would live in the recurring shadow of intimidation, indictment, or death. The committee unapologetically defended the measure: "We must govern by iron those who cannot be governed by justice," it declared.

But it was Saint-Just, however, who probably summed up the law's intent best, asserting, "Those who would make revolutions in this world, those who want to do good in this world, must sleep only in the tomb."

∞

THE TWELVE NOW faced a choice as to what to do with the queen. While Mary Queen of Scots had been killed in the late sixteenth century, this was an aberration. After the reign of England's Henry VIII, queens were rarely killed; imprisoned, yes; banished, yes. But executed? Almost never.

Marie Antoinette had aged a lifetime, but she was only thirty-eight. Clad in mourning for her husband, the queen, now known as "Widow Capet," was sustained only by the loving memory of the dead king and the presence of her family. Her daughter, Marie Therese, was now fifteen, Louis-Charles, now eight; both had lived much of their lives in captivity, in the dauphin's case, half his life. Aided by her sister-in-law, Madame Elisabeth, the queen watched in fear, then anguish, as the prolonged incarceration destroyed the health and vibrancy of her boy; he would soon have tuberculosis. And she herself suffered from "convulsions" and "fainting fits." In February and March 1793 she was presented with several plans for escape, but one was aborted because it was too risky, and she refused the other because it entailed leaving her children behind. The committee learned of the abandoned plots; it also heard disturbing reports, which were arousing public sympathy, of the queen's tenderness in caring for her children. Their solution was to remove the boy from Marie. The queen put up a fierce struggle, giving up in the end only because the jailers threatened to kill her daughter. To her despair, the dauphin was kept in isolation, living in filth and darkness, watched over by an abusive alcoholic; for days the queen cringed as she heard her son weeping. Upon her son's removal, the queen haunted her apartments like a spectral ghost with just one hope: to gaze through the narrow slits in her prison walls and catch sight of her boy being taken out for a walk in the Temple gardens.

Awakened in the middle of the night on August 2, after a year of confinement in the Temple, the queen was removed to the dungeon, the Conciergerie; known as "death's antechamber," this was the spot where the September massacres had been wrought. It was a day of stifling heat. Now crammed with political prisoners—she was prisoner 280—the queen had reached the last extreme of loneliness. Here, in darkness, most prisoners lay on a little bed of straw, sleeping by their own urine and excrement, as rats scurried back and forth. Exhausted and ill, she seemed numb to everything: Strangers stared into her cell as though she were a wild beast. Indifferent, she spent her days in a kind of stupor, moving her two diamond rings between her fingers, or lifting her head toward the sound of a harp, a lone echo of her former life.

There were efforts on her behalf. Catherine the Great's ambassador, Baron Simolin, met (in vain) with the Comte de La Marck to explore options for freeing the queen. It is believed that Danton, also in vain, sought to bargain for her freedom. Later that month she consented to another attempt to escape—signaled by two carnations tossed at her feet—but it failed. She was transferred to another room and put under stricter guard. Still, the jailer's maid, Rosalie Lamorliere, jeopardized her own life to become the queen's faithful servant. Finally, on September 2, the committee convened to decide her fate; but the revolutionary calculus was well known. She was guilty of wanton indulgence while the people starved; she was guilty of meddling in the affairs of state to the detriment of the nation; she was guilty of disliking the Parisian masses; and she was guilty of contacting both émigrés and foreign governments, especially the hated Austrians, to strangle the revolution and restore the French monarchy. Was this worthy of the guillotine? To her supporters, she was merely engaging in self-protection. But to her detractors, she had wantonly broken the laws enshrined by the elected delegates of the state, had conspired to wage war on the French nation, and had committed treason.

Still, as with the king, some members of the committee favored keeping her alive, as a pawn to be traded with Austria in exchange for a just peace. But others among the Twelve saw it differently; they called for her death as a way of indelibly binding the executioners together with her blood—as if they weren't regicides already. Robespierre agreed, insisting that the time had come for patriots "to rekindle their vigorous and immortal hatred for those who are called Kings." And there were pressures from the street as well. From the Commune, Hebert, filling the vacuum left by Marat's demise, whipped up public opinion and warned the Twelve: "It is the duty of every free man to kill a king or those destined to be kings. . . . I have in your name promised the head of Antoinette to the sans culottes, who are clamoring for it, and without whose support you yourselves would cease to exist." He added menacingly, "I will . . . cut it off myself if I have to wait much longer for it." Hebert would get his way.

On October 12 the queen was submitted to a sustained and harsh preliminary examination; the inevitable had arrived. The next day she requested three days' respite to prepare her defense. Her letter was ignored. By this stage, she had ceased to hope and perhaps even to fear. On the fourteenth and fifteenth she was summoned before the Revolutionary Tribunal for her trial—Fouquier-Tinville was the chief prosecutor. She had but two things left to do—defend herself vigorously and die with dignity. On the first day, she was questioned for fourteen hours, from eight A.M. to four, then from five to eleven P.M., having barely time to take a sip of bouillon in between. On the second day, the proceedings lasted for a mere six hours, from nine A.M. to three P.M. All told, she faced forty-one witnesses arrayed against her; much of the evidence was hearsay, or gossip, or just pure spite. With the king, there had been a pretense of a fair trial; not so now. Still, there was no doubt that the bill of indictment had a powerful ring of truth to it. Marie Antoinette was accused of urging Louis to issue the veto, which was true; of organizing the flight to Varennes, which she certainly supported; of inviting foreign armies to invade France, which she endorsed; and of transferring millions of francs from the depleted French treasury to France's enemies ("passing the gold of patriots out of the country"), which also was true.

But the Twelve weren't content just to indict her; her very name had to be obliterated. Accordingly, she was accused of corrupting her son morally (raising him as a "royalist"), and sexually (engaging in "incest"). When an answer was demanded for this vilest of charges, she replied with great force: "I appeal to all mothers who may be present." Watching the former queen elevate herself with uncommon dignity and resolve, the audience was suddenly moved. Once they knew Marie Antoinette as *l'Autrichienne*, or simply that *chienne* (or bitch), or as the "she-wolf"; now they were struck by the sight of her, white-haired, wrinkled, and emaciated. Once she had been a fabled beauty whose whims were the talk of Europe; standing before them, clad in black in mourning for her husband, they now saw a sickly woman, her face ravaged by suffering. Once they saw an arrogant woman believed to have uttered, "Let them eat cake"; now, they saw a mother fighting for her life with valor and aplomb against men who were determined to mangle her spirit. Suddenly she was no longer queen, but a woman in distress. A cry of sympathy rose through the courtroom; some of the market women even called for an end to the proceedings, but their protestations counted for little. This trial could end only one way, and that was with a verdict of guilty.

Upon hearing the verdict, Marie Antoinette showed no sign of fear, indignation, or weakness. Only upon reaching the darkened stairs did she falter. Overcome with fatigue, she had to be helped to her cell.

That was when she took pen to paper. To write a letter of farewell. It was nearly four-thirty in the morning and bitingly cold.

∞

FROM HER CELL in solitary confinement, she composed herself to write her final letter to the king's sister, asking her to transmit to her children the injunctions the king had left for them. "My son," she instructed, "must never seek to avenge my death." She finished, "Think of me always. I embrace you with all my heart as well as those poor, dear children. My God! How heartbreaking it is to part with them forever. Farewell! Farewell!" The letter was never delivered; intercepted, it was given instead to Robespierre. Seven months later, her sister-in-law, in solitary confinement and, remarkably, still ignorant of the queen's fate, was guillotined.

The following day, at dawn on October 16, 1793, Rosalie found a tearful queen, collapsed and sobbing on her bed in her mourning clothes. She pleaded with her to drink some soup; reluctantly Marie obliged, but could choke back only a few spoonfuls. "My child, I need nothing," she told Rosalie. "Everything is over for me." Informed she could not wear her mourning dress, the queen donned a plain white pique dress and a small lawn bonnet. Meanwhile, in the distance the drums were beating. At ten A.M. the executioner, Henri Sanson, came to her cell, bound her hands behind her back—despite her protestations—and cut off her hair at her neck. Tied at the end of a long rope by the executioner, like a common farm animal, she was shuffled into a dingy cart. Unlike the king, who rode by closed carriage, to her surprise she was led in a tumbrel to the scaffold, held up only by the fact that she was forced to relieve herself in the corner of the courtyard. Accompanied by soldiers, some thirty thousand, she was taken past cannons in the streets, past the painter David, who from his window sketched the last image of the queen of France for all eternity, past the Tuileries, where her eyes welled with tears at the memory of her children, past jeering, hostile crowds, to the magnificent square at the Place de la Révolution. What must she have been thinking? Her tragic flaws were nearly endless; this is without question. Too often she was totally ignorant of France's appalling social inequities. It had been a fatal blow to her husband when she opposed any curbs on the monarchy and declared the constitution of 1790 to be "monstrous" and a "tissue of absurdities." And arguably, it was the king's blunders, at her urgings, that dethroned Louis. But looking back, there is another equally compelling side to the narrative. Her suffering was enormous; so was the inner voyage she herself traversed, taking her from the pinnacle of royalty at Versailles to a grimy little cell in Paris. Deprived of her children, her boy placed in the care of a brutal master, she felt the simple agony of a parent. In the end, she went from a heedless, frivolous princess to a devoted wife and mother.

All her life she was a pawn: of her mother, the empress; of her brother, the emperor; of the old regime, which she believed in; of the Convention, which she didn't; of those loyalists who wanted to free her; of those detractors who wanted to kill her. By any measure, she was unprepared for her fate from the start. True, she was the daughter of an empress, the wife of a king, and was strikingly attractive. To many in the outside world, she seemed to have a charmed life: Edmund Burke once said that Marie Antoinette "glittered like a morning star, full of life and splendor and joy"; for his part, John Adams was bewitched by her. But the young queen, from the age of fourteen, when she could scarcely even write her own name, was an object always on view, never given privacy, always eavesdropped and spied on. For all her adult years, if she could scroll back that far, she was isolated and lonely and craved friends. She survived the heartbreaking deaths of first her daughter, and then her nine-year-old son just four weeks before the fall of the Bastille. She survived the pamphleteers who lampooned her as a perverted lesbian, a "fucking tart," and later as an incestuous and abusive mother. Even in the heady days of the ancien régime, she was like a bejeweled fly trapped in a vast web of etiquette, rigid practices, and amoral favors. When all of Paris wasn't discussing her menstrual cycles and the king's erections, she was a plaything, handed about to be undressed and washed, her cheeks daubed with bright circles of rouge by one courtier and her caps and cloaks unpinned by another. It was a callous sort of indifference: One time, two duchesses heatedly quarreled over whose right it was to hand her the royal chemise, as the queen stood impassive, naked, and shivering. Later she provoked a scandal when she refused to give birth in public and had the audience of noisy onlookers banished to an adjoining room. Now she faced death with equal courage.

Her mother, the legendary empress of Austria, died at the age of sixty-three, surrounded by a brood of servants and family; her husband, Louis XVI, had his faithful servant Clery until his final hours, and was provided a priest to accompany him to the bitter end. But Marie Antoinette of Lorraine and Austria, widow of the King of France, now Antoinette Capet and *Ci-Devant* Queen of France, was utterly alone.

Ascending the scaffold with "bravado," she let her bonnet fall and gave herself over to her executioners. The board tipped; there was a dull thump. The crowd yelled: "Long live the Republic!" At twelve-fifteen her head fell into the basket, and Sanson held up her severed head to the taunting, ecstatic throng.

∞

THE REVOLUTIONARY TRIBUNAL now shifted into high gear: Executioners could cut off twenty-two heads in thirty-six minutes; that number soon increased to thirty-two heads in twenty-five minutes; a week later, twelve heads in just five minutes.

Since June 2 twenty-one Girondins, who had been seized in Paris, had also been held under guard at the Conciergerie, the waystation to the guillotine. Cried one: "The 72 departments which had declared for us turned around and abandoned us in twenty-four hours." Another ruefully noted,

"We deemed ourselves in Rome and we were in Paris." They were provincials who had loved every inch of French soil, but now they were marked men. Just after the queen's death, they were put on trial; it lasted a week. The irascible Brissot was there, and so was the eloquent Vergniaud. Brissot, of course, was one of the Revolution's earliest and most ardent supporters, and Vergniaud was the former president of the Assembly. But despite refuting their indictments point by point, both were sent to their deaths, singing "The Marseillaise" in the tumbrel as a final act of fraternity; the night before his execution Vergniaud refused to take the poison slipped into his cell, instead predicting that France will "return to its kings as babes return to their toys." As it happened, Vergniaud's other bleak prophecy had already come true: The Revolution was devouring its own children.

Meanwhile, in the Conciergerie, Manon Roland, wife of the former Convention president, pondered her fate with the daunting realization that the tribunal treated men and women with equal disdain. On August 28 she wrote: "The miseries of my country torment me; an involuntary gloom penetrates my soul, chilling my imagination. France has become a vast Golgotha of carnage, an arena of horrors, where her children tear and destroy one another." To her twelve-year-old daughter, she bade a tearful adieu, and to her husband and colleagues, eluding the "assassins" (her words) seeking to bring them before the tribunal, she counseled, "O my friends, may propitious fate conduct you to the United States, the sole asylum of freedom." But Jean Roland and his compatriots would not be so lucky. As for Manon Roland, she was brought before the Revolutionary Tribunal on November 8 and guillotined on the same day. Staring up at the statue of Liberty that David had erected in the grisly Place de la Revolution, she reportedly called out, "O Liberty, what crimes are committed in your name."

To the remaining Girondins on the run, they were endless. France was bleeding at every pore: In what seemed to be almost unthinkable, a mind-numbing progression of the founding fathers of the Revolution followed Madame Roland to the scaffold. On November 10 in a sleety drizzle came the famous astronomer and mayor Jean Bailly, who had once handed the revolutionary tricolor cockade to the king, as well as the keys to the city. Two days later after eating a hearty breakfast of oysters and steak, it was the former Duc d'Orléans's turn. On November 29 it was Antoine Barnave who mounted the scaffold; he, like Mirabeau, had pledged aid to the queen after Varennes. Then came Madame du Barry, the former lover of

Louis XV, who shrieked and begged before her execution. Then Rouget de Lisle, who had composed "The Marseillaise"; and Anacharsis Cloots, the "voice of humanity"; and Generals Custine, Houchard, and Biron; and also the band of brothers who had put their lives on the line in the American Revolutionary War, Admiral d'Estaing, who had led the fleet at Newport and sealed his fate by speaking out on behalf of Marie Antoinette at her trial, and Admiral de Grasse, who had buoyed the Americans at Yorktown in 1783. And just days before Christmas, one of the world's most famous political figures, Thomas Paine, despairing that the Revolution was now acting without "prudence or morality," was carted off to the cold, damp cells of Luxembourg prison, to be marked for the guillotine on July 25, 1794. Also marked for the executioner's blade was the sixty-nine-year-old General Rochambeau, who had commanded France's expeditionary forces in America.

In Paris, for many prominent intellectuals now posing as American citizens, the United States increasingly became the last hopeful alternative to the carnage of revolutionary France; Paine's good friend, the English-woman and feminist social reformer Mary Wollstonecraft—whose work John and Abigail Adams also ardently read—dreamed of leaving France to plant a utopian colony in America. "I have a plan in my head," she wrote. But that plan was aborted because sailing for America had been rendered unsafe, and travel was now restricted by the committee.

At the same time, in the provinces, other Girondins were determined to defy not death—that no longer seemed possible—but the fury of the Parisian radicals. Jean Roland, a fugitive in Rouen all throughout the sum-mer and fall, one day thanked the comrades who had jeopardized their lives to shelter him and wandered off alone. Heartbroken at the execution of his wife, he lay against a tree and penned a note of farewell, which he attached to his chest: "I do not wish to remain longer on an earth polluted with such crimes." He then forced a sword into his body. The famed philosopher Condorcet, who once bore witness to the historic hug between Voltaire and Benjamin Franklin and had counted himself among Thomas Jefferson's friends, composed his last and perhaps most magnificent philosophical tract, then took poison. Etienne Claviere, the former finance minister, com-mitted suicide in prison. His counterrevolution in tatters, Barbaroux shot himself, but survived; the Revolution sent him to the guillotine. And Petion and Buzot, after almost a year of being hunted by the Convention, concluded their last mutual pact, killing themselves together in a cornfield near Bordeaux. On June 18, 1794, their bodies were discovered—half-eaten by wolves.

∞

FOR ALL THIS, the messianic designs of the Jacobins were still held in check by resistance in Bordeaux, Lyon, and the Vendée. Like the Romans long before them, and the totalitarians long after them, the Revolution decreed that if the state were to be made whole, it was necessary to stamp out such intolerable provincial autonomy. To do this, representatives of the Twelve (*représentants-en-mission*) roamed across the country with absolute authority to restore order. Under their leadership, the Terror moved into full swing. Reinvigorating the Revolution, they arraigned suspects and extracted taxes, removed elected officials or appointed their own, conscripted men for the war or confiscated gold and silver, and authorized local committees to act in their stead. And they dispensed with their opponents without mercy.

Chief among their targets would be Lyon, the great capitalist metropolis and the second city of France—it had been the seat of Roman rule at a time when Paris barely existed. Here in this city of 130,000 were the industries that provided one-quarter of all France's exports, plus the military stores for the republic's southern armies. On May 29, a day never forgotten by Lyonnais, savage fighting erupted in the streets, and by early July, an ad hoc alliance of royalists, businessmen, and Girondins ousted the radical government and executed the Convention's leading representative, Marie-Joseph Chalier. The Committee of Public Safety ruthlessly struck back—with an army to punish the renegades. The city was laid siege to; two months later, when Lyon finally fell, the committee decided it would be made an example of.

On behalf of his assassinated brethren—in one remarkable moment, Chalier's severed head was ceremoniously presented to the Convention—Robespierre himself composed the directive: "The city of Lyon shall be destroyed. The name of Lyon shall be effaced from the list of the Republic's cities. The collection of houses left standing shall henceforth bear the name of *Ville Affranchie* [the Liberated town]." On the ruins of Lyon, he added, shall be raised a column attesting to posterity the crimes and punishment of the royalists. It would read: "Lyon made war on liberty/ Lyon is no more."

The city was then subjected to what could most charitably be deemed a colonial occupation. The citizens were disarmed, thousands of men and women were shunted into the prisons, and the homes of the rich were destroyed. One merchant bravely wondered aloud, "Is it republican to tear down houses?" The commissioners gruffly replied: "Shut yourselves up in the privacy of your household." Meanwhile, brutal house searches, for traitors and insurrectionists, were made at night. Presaging the treatment of European Jews in World War II was the requirement that all citizens affix a notice on their front doors indicating each resident who lived inside; it became a serious crime to entertain anyone not on the list. The abuses that ensued were predictable. Denunciations were made of those who had harbored priests, or given sanctuary to émigrés, or verbally defamed the republic. And the guillotine operated with ruthless speed. But shockingly, to many revolutionaries in the town, it was deemed too inefficient. To prove that the Revolution was anything but, on December 4, sixty prisoners were tied by a line of ropes, carted off to an open field by the Rhone, stationed between two trenches, and cut down by a stream of grapeshot from a row of cannon (*mitraillades*); those who somehow survived were finished off manually, with bayonets or sabers, or pounded with rifle butts. The next day 209 prisoners were massacred by a similar *mitraillade*, and then two days later, two hundred more. By March 1794, the executions numbered an astonishing 1,667.

On December 20, 1793, a group of citizens traveled to Paris and begged for the retribution to stop. They failed. Meanwhile, in Lyon, the shooting and the guillotining went on. A barber surgeon enthused to his brother: "What pleasure you would have experienced if, the day before yesterday, you had seen national justice meted out to two hundred and nine villains. What majesty! What imposing tone!" Ironically, among the last of those marched to the guillotine were the executioner, Jean Ripet, and his assistant. When it was all over, an entire slice of Lyon had been annihilated. Back in Paris, Camille Desmoulins observed, "liberty is a bitch who likes to be bedded on a mattress of cadavers."

That was to be the case in another town in rebellion, Nantes. Jean Baptiste Carrier was deputized to pacify Nantes; he did so with an enthusiasm that is almost hard to fathom. He declared trials a waste of time and ordered his aides to fill barges (*noyads*) with anywhere from 2,000 to 4,800 people and sink them in the Loire River. Priests were given priority, but this soon expanded to women and children and the elderly; prisoners had their hands and feet bound and died a slow and horrific death. These mass drownings—jokingly called "republican baptisms"—were unconscionable by almost any standard, but Carrier justified himself by the laws of war. "We shall make France a graveyard," he vowed, "rather than not regenerate it in our own war." For four months, while vultures hovered over the water to feast on human flesh, the brutal executions continued with frightening banality.

Other towns, like Bordeaux and Toulon, escaped mass executions, but were similarly scarred. In the ancient port city of Marseilles, the Twelve even stripped it of its identity: the city was renamed Ville-Sans-Nom—Town Without Name.

The Vendée, however, would not be treated so leniently.

∞

HERE IN THE rocky coastal area of France, between the Loire and La Rochelle, rioting had broken out as early as March 1793. As in Nantes, the source of the discontent was not simply support for the old regime, but the war of the Convention against Christianity. By May 1 there were some 30,000 insurgents poised for rebellion. Their efforts met with stunning success; marching in crowds but attacking in small packs, ambushing, scattering, and escaping, the rebels took Vihiers, Thouars, Fontenay, Saumur, and Angers. There had even been a moment in the middle of June when they might have marched on Paris itself. In August a distraught Committee of Public Safety dispatched General Kleber to crush the Catholic rebels and lay waste to the towns and farms supporting them. On October 17 Kleber won a great victory against the peasant forces at Cholet, and an even greater one at Savenay on December 23.

Military commissioners from Paris were set up with orders to put to death any Vendean bearing arms. Having defeated the rebels in the field of battle, the Convention unleashed its wrath: Whole villages, believed to have harbored rebel armies, were wiped out, while Convention forces were let loose in a blind, wanton slaughter. And slaughter it was, as the old and young, men and women, grandmothers and wives and mothers and daughters were all targeted. Every brutality imaginable was carried out, every sadistic measure employed: Women were routinely raped; boys and girls were routinely murdered; both were routinely dismembered. Two hundred elderly were forced to dig a large pit and drop to their knees; they were then summarily shot, after which they toppled into their own grave. When thirty children and two women sought to flee, they were buried alive. In World War II, the Russians and the Germans would rediscover this very same method for disposing of large numbers of people in a short time.

This had become the bitterest and cruelest of wars, a war of religion. More so than even Nantes, the executioners became increasingly sinister. At Pont de Ce and Avrille, four thousand people were shot in one remorseless butchery; Carrier then suggested, with a straight face, lacing drinking water with arsenic. One general proposed that a cask of poisoned brandy might do the job. And the distinguished chemist Fourcroy was even asked to assess whether "mines" or "gassings" might be used to facilitate the obliteration of the enemy. At the same time, Saint-Just reiterated that the republic "consists in the extermination of everything that opposes it." Looking backward in history, there can be little doubt, as Simon Schama points out, that if the French scientists had come up with an efficient gas chamber, the revolutionary authorities would have been only too pleased to use it. And it is hard not to conjure images of the atrocities wrought by twentieth-century regimes: Hitler's Holocaust, Pol Pot's auto-genocide in Cambodia, Stalin's gulag, and the genocides in Africa.

By mid-April 1794 the subjugation of the Vendée was concluded. Two rebel commanders managed to survive, and took to the hills to become guerillas fighting a *petite guerre*. But their efforts did little to make up for the carnage already wrought: When the apocalypse was complete, nearly a quarter of a million, one third of the population, lay dead, and, like other areas of rebellion, the Vendée too had been stripped of its name, from then on to be known simply as Vengé (Avenged). It was, historians have noted, a human catastrophe of colossal proportions. But not for the tribunal. Its general gaily wrote back to the committee: "There is no more a Vendée citizen, it has perished under our free sword along with its women and children. I have just buried it in the marshes and mud of Savenay. Following the orders that you gave me I have crushed children under the feet of horses, massacred women who . . . will engender no more brigands." He finished with this bloodcurdling flourish: "I have no more prisoners with which to reproach myself."

When it was all over, Robespierre's jubilant comment rang exceedingly true. "A river of blood," he had promised, "would now divide France from its enemies."

∞

BY MAY 1794, as the provincial Terror was coming to a close, it was being intensified in Paris. "My God," Mary Wollstonecraft cried out, "how many fall beneath the sword and the guillotine?" Day after day, rich and poor, high and low, sage and fool, priest and layman, moved lockstep in the dark line leading to the guillotine. So how many would perish? Estimates range from 18,000 to 40,000 (the higher figure being the more probable one); those jailed as suspects, some 300,000. In each case, whatever property belonged to the executed was appropriated by the state. In short, the Terror was not just ceaseless, but actually lucrative for the Revolution.

In November 1793 the Committee of Public Safety was reaching the apex of its popularity, in good measure because of the stunning speed of the *levée en masse* in repulsing the foreign enemy across several fronts. By the following spring, France had some 800,000 to a million men under arms, dwarfing the other powers in Europe, all commanded by a new breed of general—men like Kellerman and Kleber, Hoche and Pichegru. These were the devoted offspring of the Revolution, unencumbered by outworn sentiments, outdated strategies, or the tired ways of the old regime. Their troops were still poorly trained and equipped, but what they lacked in experience, they made up for in revolutionary ardor; outside of the young United States, this was perhaps the first true national army, representing a people in arms and composed of men who were fighting not as mercenaries for a king, but as citizens for a republican cause. And it paid off. Though the French troops were stymied at Kaiserslautern, they bounced back and seized Landau and then Speyer. Then they pushed the Spaniards over the Pyrenees and soon thrust in Catalonia. In July they again entered Brussels and threw the Austrians back to the Rhine. And thanks to the aid of a promising young soldier named Napoleon Bonaparte, they reclaimed the strategically located Mediterranean port of Toulon.

All of France was ecstatic. With these triumphs, as well as the pacification of the Vendée, the committee could now wrestle with France's mounting domestic questions. It was needed, for once again, internal matters were spinning out of control. As Madame Roland had foretold, "The time has come when people would ask for bread and be given corpses"; more than ever, this grim prediction seemed to have come to pass. Unemployment soared while prices skyrocketed. So did the long lines. Rampant shortages forced housewives to wait hours for such basics as bread, then milk, then meat, then butter, then oil, and then wood. Crowds of men and women slept along the curbs to secure their place when the stores opened, while far too often, armed brigands only ended up looting the shops. In the meantime, corruption was rampant; mismanagement threatened the flow of vital army supplies; and rumors spread of alleged "foreign plots" to assassinate the revolutionary leaders.

The iron hand of the committee began to slip. Increasingly the Twelve harangued the people rather than consulted them, compelled them rather than persuaded them, told them tales rather than the truth. Suddenly the people could tell the difference, and they in turn scribbled insulting remarks on the walls of public buildings as well as wrote anonymous letters demanding the dissolution of the Assembly; on one CPS notice, someone had even audaciously scrawled "*anthropophage*" (cannibal) under Robespierre's name. They also argued that under the committee's mishandling of the nation's finances, the bourgeois and *beaux esprits* were fattening themselves on the "blood of the poor." Emboldened, the radical Commune began ever more vociferously to preach that *all* property should be nationalized, or at least *all* wealth. Equally vociferous was one section leader's blueprint for solving the nation's economic ills: sentence the rich to death.

∞

THE INFIGHTING NOW stretched to the committee itself.

At the end of 1793, some of the most powerful challenges came neither from the streets nor from foreign armies, but from their own. Danton, back from his hometown of Arcis—he had rotated off the committee—believed that the French triumphs against foreign armies meant that the Twelve could renounce Terror and even begin a gradual return to constitutional government. How to explain this? For all his ferocity, Danton began to suffer a change of heart. He was once the most powerful man in France; brawny, brutal, cruel, and vindictive, he, as did few others, typified the carnage of the age, and the spirit of the Revolution as well. Yet now, with the monarchy destroyed, not to mention the king, the queen, the hierarchy and the aristocracy, the Girondins, and the church itself, it was time, as he saw it, to restore order, establish peace, and strengthen the republic. "I am not a drinker of blood," he exclaimed. "I am tired of this slaughter." But the national appetite for blood had not yet been sated, and the Revolution, under Robespierre's guidance, began to gnaw at its own vitals even further. And slowly, Danton's enemies began to weave a net around him, and the more he struggled to be free, the more he became entangled in its mesh.

When Britain offered peace—its Prussian allies, like Russia itself, were increasingly consumed by Kosciuszko's revolt in Poland—Danton counseled acceptance, but he spoke in a hubbub and could scarcely be heard. Robespierre thought differently, and instead set about redoubling the Terror, claiming that the nation was still assaulted by the Achilles' heel of the Revolution: deceit, traitors, and foreign conspiracy. But Camille Desmoulins, once Danton's loyal secretary and arguably France's most brilliant journalist, was suffering a similar change of heart. He too called for an end to the Terror, writing, "The Tribunals which were once the protector of life and property have become mere slaughterhouses.... Was there ever greater madness? Can you destroy one enemy on the scaffold without making two others among his family and friends?"

Dramatic words, but in the current climate they were foolish. To Robespierre, ending the Terror and opening up the prisons would be catastrophic. Indeed, he now began to suspect not just the avowed enemies of the Revolution, not just the ultra-radicals like Hebert and the *enragés*, but Desmoulins and Danton as well. The more he considered it, the more he came to believe that Danton's sudden reversal—the calls for mercy, the calls to halt the war—was nothing more than a complex scheme to overthrow the committee and depose Robespierre himself. And in the paranoid politics of the day, he concluded that he and the Revolution could survive only if his old friends Danton and Desmoulins were dead.

Dissension and intrigue now honeycombed the state. Did Danton realize that he was now in Robespierre's sights? That is unclear. But he continued to bait the committee by asserting that the day for violence had passed and the moment of reconciliation had come. "Let us," he insisted, "leave something to the guillotine of opinion." And he grew increasingly insolent. Informed that Robespierre was contemplating his arrest, Danton roared, "If I thought he had even the idea of it, I would eat his bowels out." Yet as it happened, this was just bluster. Instead, strangely dallying with danger, he refused his friends' advice to denounce the committee before the real stewards of the people—the Convention. After four years of Revolution, after the heart-sickening spectacle of seeing his former friends being guillotined, he was exhausted and dispirited and, it seemed, already preparing himself for the martyr's crown. The man who once preached that Terror was to be the order of the day now sighed, "Let them kill me if they will. I would rather be guillotined than guillotine others, and besides, I am sick of the human race."

The committee soon agreed that to allow the Dantonists to continue was fatal to the Revolution. It was, however, a pivotal moment for the Twelve. Executing the king, to be sure, was once unthinkable. But hadn't they done it? Executing the Girondins was once unthinkable. But hadn't they done that too? Still, Robespierre remained reluctant to take this step; he agreed with other members of the committee that Danton had, like Mirabeau, corruptly siphoned off state monies. But Danton was no bumbling royalist or errant Girondist. Robespierre had to acknowledge that Danton had provided enormous leadership to the Revolution, and sentencing one of its most eminent members to death could have woefully unpredictable consequences. In this, he was right.

For his part, Danton visited Robespierre several times, not only to defend himself, but to call for an end to the Terror and an inauguration of peace. Publicly the two continued speaking; in fact, Robespierre evinced no open hostility toward Danton, and reportedly the two even embraced in their last meeting. But privately Robespierre shook with rage at his greatest rival. And unknown to Danton, Robespierre, now seeing enemies everywhere, began plotting his downfall. Cautioned against Robespierre, Danton scoffed, "I shall take him in my hand and spin him like a top!" But he egregiously miscalculated. Remarkably, the man who had melodramatically saved France from Brunswick weakened at the wrong time. While Danton was ignoring the dangers swirling about him, Robespierre was helping prepare the case against him; wheezing and feverish, he then took to his bed. On March 30 the battle between the Dantonists and the Jacobins was over: In a united front, the committee and Robespierre agreed to obtain a sentence of death for Danton, Desmoulins, and twelve others convicted of "corruption" and "royalism." Hearing the news, Marat's sister rushed to plead with Danton to flee the capital and find safe haven in the provinces, or hurry to the Convention to denounce the committee. Sounding strangely like Louis XVI, Danton declined ("If France drives me out, where shall I go?" he protested. "One does not carry his country on the sole of his boot"). That evening the police arrested him and Desmoulins too. Imprisoned in the Conciergerie, Danton fumed: "In revolutions, authority remains with the greatest scoundrels."

By now this seemed almost axiomatic. On April 1 a deputy suggested that Danton be allowed to defend himself. Robespierre flashed a hostile stare: "Danton," he cried, "is not privileged . . . We shall see this day whether the Convention will be able to crush to atoms this mock idol, long since decayed." Then Saint-Just read the bill of charges, a long list of accusations but an extremely flimsy document. It didn't matter. The rest of the deputies, fearing for their own necks, ordered that Danton and Desmoulins be tried immediately.

They were—the very next day. On April 2 Danton and Desmoulins appeared before the tribunal. But as the committee was reminded, Danton was still a popular hero. It has been said that Danton could coo like a dove or bellow like a stentor: Today, with his keen appreciation of public theater and his sonorous voice blaring, he could be heard not simply by the packed courtroom but by the vast multitude of people who had also gathered outside. His lethargy gone, he defended himself with characteristic vigor and thrilling wit: "People . . . my voice will not only be heard by you but throughout all of France. . . . Will the cowards who are slandering me dare to attack me to my face?" He began winning spectators over to his side, so much so that one of his old comrades, François Joseph Westermann, the hero and subduer of the Vendée, insisted on being indicted with his friend, and so much so that prosecutors appealed to the committee to silence this "fearful storm." Once more, the committee relented. The Twelve sent the Convention a charge that the followers of Danton and Desmoulins were plotting to murder members of the committee. On this ludicrous basis the Convention decreed the two men to be outlaws—which meant that they were "outside the protection of the law," and thus could be executed without due process. Upon hearing this decree, the jury declared that it was sufficiently "enlightened."

The accused were taken back to their cells. Murmuring, the spectators filed out.

∞

THE TRIAL WAS a formality. On April 5 the tribunal that had so long been dominated by Danton issued its unanimous verdict of death to all of the prisoners. While Desmoulins's wife, heartbroken, sobbed at his door, Robespierre closed the shutters. But upon hearing it, Danton predicted, "Before these months are out the people will tear my enemies to pieces." And too: "Vile Robespierre! . . . You will follow me." He voiced regret that he was leaving the republic and its government in such a miserable condition ("If only I could leave my balls to Robespierre and my legs to Couthon," he barked, "the Committee might live a little longer."). And upon seeing the jailed Thomas Paine, who after seven months was suffering from nervous collapse and sweats, he held his hand and exclaimed, "That which you did for the happiness and liberty of your country, I tried in vain for mine." In the meantime, from his cell a dejected Desmoulins was less brazen. He wrote to his wife: "My beloved Lucile! I see my arms locked about you, my tied hands embracing you, my severed head resting on you."

It was a cloudless spring day when the condemned men were carted off in five tumbrels to the familiar Place de la Révolution—like the monarchists before them, like the king and queen, like the nonrefractory priests, like their friends and colleagues from the Convention; even the radical Commune leader Hebert had been cut down less than two weeks before. Danton was thirty-four and so was Desmoulins; it seemed like an eternity, and perhaps it was, since that fateful moment when Camille had risen up and exhorted Parisians to head for the Bastille. Passing through a huge and silent crowd, Danton bore up well; not so Desmoulins, who was near the cracking point. Leaning over the red-painted tumbrel he meekly appealed to the people: "I was the first apostle of Liberty; it was I that called the people to arms at the beginning." Pausing at the house where Robespierre lived, Danton, defiant as ever, rose up to his feet and shrieked once more, "I'm leaving everything in a frightful mess. Not a man of them had an idea of government. Robespierre will follow me. Ah, better be a poor fisherman than to muck about with this politics." By now, night was falling. Reaching the scaffold, Desmoulins was third in line, Danton last—thus he could hear the whistle and thud as the blade fell on all the heads before him. For a fleeting moment he faltered, then roused himself, muttering: "Courage Danton, no weakness." As he approached the blood-splattered plank, he altered the ghastly ritual, exhorting Sanson, the executioner, "Don't forget to show my head to the people. It's worth the trouble." A hush fell over the crowd. Eight days later, it would be Lucile Desmoulins, along with Hebert's widow and his Commune compatriot Pierre-Gaspard Chaumette, taking their turn at the guillotine.

The ledger was now wiped clean.

NEVERTHELESS, THERE STILL remained a faint dream of liberty in France, no less than the dream that had once been America's. The problem, as George Washington back in Philadelphia observed, was that "the contrast between the United States and Europe is too striking not to be noticed by even the most superficial observer"; the other problem, of course, was that it lay pale and emaciated in a filthy Austrian dungeon, in the person of the Marquis de Lafayette.

How was it that Lafayette, having blazed trails of glory in America, was so spectacularly unsuccessful in guiding the French Revolution? It is an intriguing puzzle. He was arguably one of the world's last great knights, seemingly ripped from the pages of medieval history, prepared to battle evil; six-foot-one, unusually handsome, and marvelously self-deprecating, he possessed an exquisite bloodline that stretched back to service for Joan of Arc. His family had spilled blood in the name of country, served kings in the name of honor, and amassed extraordinary wealth; he was perhaps the richest aristocrat in France. And he, like his wife's grandfather, was a legendary Black Musketeer. But at the age of nineteen, Lafayette, brimming with passions and compassion and idealism, left it all behind, and defying orders of the king, volunteered to fight and bleed for American independence. Fight and bleed he did: He groped his way through South Carolina's swamps, endured the hideous cold at Valley Forge, weathered the tireless crack of enemy fire in Virginia, and was wounded at the battle of Brandywine, where a young captain, James Monroe, tended to him. Lafayette was a courageous and canny soldier, revered and loved by his troops. And when America's Revolution tottered on the brink of failure, he harbored no illusions save for idealism, and promptly set sail for France to help push Europe's oldest, most entrenched monarchy to ally itself with the upstart rebels against a fellow king. Succeeding, he arrived on American shores with a sizable French armada in tow, then played a critical role in the campaign that ultimately led to Britain's defeat at Yorktown, earning him the accolade of "the conqueror of Cornwallis."

Reared in the Old World, he found his calling in the New, in America's distant battlefields, in American republicanism, in his reverence for the Founders, most of all Washington. Where Robespierre called for his death, and even a fellow moderate like Mirabeau reviled him, Washington welcomed him, "as if he were my own son"; in turn, Lafayette loved Washington back as "my adoptive father." Not unsurprisingly, he named his son after Washington, who was also the godfather, and upon retirement, settled down to the quiet life of a country gentleman, modeling himself as the George Washington of Chavaniac. Yet while he mixed flawlessly with the king and queen, as well as with Frederick the Great and William Pitt (even Catherine the Great had invited this "friend" of Washington's to Russia), one senses that he was always most at home with his trusted and admired American comrades: Thomas Jefferson and Alexander Hamilton, James Madison and James Monroe, Henry Knox and John Jay and Benjamin Franklin, not to mention America's rag-tag continentals, who honored him with the sacred title, "the soldier's friend."

"The United States is the most marvelous land on earth," he once wrote to his wife; the Americans felt the same way about him. It speaks volumes that in the mid-1780s, Maryland made him a citizen and so did New York City, and that much later in his life, President John Quincy Adams would declare, "We shall look upon you always as belonging to us, during the whole of our life, as belonging to our children after us," to which Lafayette tearfully replied, "God bless you, and God bless the American people." Upon his death at the age of seventy-seven, he would be buried beneath soil that he himself had carried home from Bunker Hill.

Yet whatever Lafayette's successes in America, they quickly turned to farce and then calamity when he labored to transplant American-style liberty and constitutionalism in his native land; more often than not, as the revolution intensified, and the bloodshed mounted, he was fatally naïve, or, at times, downright incompetent. As commander of the Paris National Guard, he failed the king, lost control of his troops, never quite understood the passions of a French people precipitously released from the chains of despotism, was no match for the Jacobins, became a prisoner of the mob, and in the end, rather than help usher in a golden age of democracy, was left a hunted man in fear of his life. Was he too American for the French, too moderate for the Revolution? One suspects so. Was the American Revolution that different from the French Revolution, the American political style that different from the French? One suspects so again.

Early on, as tumult engulfed revolutionary France, Washington openly feared for his protégé's life, speaking often of the "deep anxiety" that he felt. And as the Revolution's furies were unleashed, Lafayette felt it too. Then, once the royal family was incarcerated in the Temple and as an arrest warrant now hung over *his* head, he hoped to flee to Britain and settle in his adopted land—America. "What safety is there in a country," he wondered, "where Robespierre is a sage, Danton an honest man, and Marat a God?"

He told his wife: "Let us resettle in America, where we will find the liberty that no longer exists in France." But it was not meant to be, for in France, he was a marked man, and across Europe, monarchs damned Lafayette for having carried this dreaded disease over from America and releasing it on their continent. "His circle is completed," Gouverneur Morris reported to Jefferson. "He has spent his fortune on a Revolution and is now crush'd by the wheel he put in motion. He lasted longer than I expected."

Seized by the Austrians, Lafayette protested that he was an American citizen; unmoved, their response was to lock him up in a fortress prison just across the German border, north of Düsseldorf. American diplomats in Paris, London, and Holland desperately fired off letters in hopes of aiding Lafayette, but it only exposed their powerlessness in dealing with crowned powers like Prussia and Austria, neither of which had bothered to establish formal diplomatic relations with the United States. Meanwhile, the American envoy William Short observed that Lafayette was the man the Austrians hated "the most," while Morris sought to console a worried President Washington ("poor Lafayette . . .").

For her part, Lafayette's wife begged Washington to use his influence to obtain Lafayette's release. But Washington was as helpless as she was—the United States was still allied with Jacobin France, which wanted Lafayette's head, and lacked sway with the monarchies of Europe. The most Washington could accomplish was to publicly offer sanctuary to Lafayette's son in the American capital, as well as to transfer American funds to buy Lafayette books and better food. In prison, however, Lafayette soon became almost unrecognizable: Once days turned to weeks, then weeks to months and months to years, he was covered in rags, his hair fell out, and oozing sores covered his skin.

But unlike so many of his compatriots, at least he was alive.

∞

THE SAME COULD not be said for most others. After Danton's death, a terrified Thomas Paine, shivering with jail fever and reduced to a skeleton unable to speak, awaited his own impending execution; he wrote: "No man can count on life more than 24 hours." Indeed, all the opponents to the Committee of Public Safety had been eliminated or scattered. The Girondins, like the moderates and reformers before them, were dead, or in jail, or on the run in fear of their lives. The once-mighty sansculottes had been marginalized. The clubs—barring only the Jacobins—had been forced to disband, including all women's clubs. The press was under tight watch, and so were the theaters. And the Convention, groveling, deferred all significant actions to the Twelve.

Astride the most influential country in all of Europe, Robespierre now stood alone. At this stage, he saw it as his mission to rehabilitate God in an increasingly godless France. The pliant Convention agreed. Their desire, or rather their hope, was that they could placate the faithful and also assuage the Terror. Having made this Faustian bargain, they sealed it on June 4, elevating Robespierre to the presidency.

De facto had now become de jure. In his new capacity, on June 8 he unveiled a "Feast of the Supreme Being," to be staged by the artist David—the foremost French artist of the era—in which Robespierre, the Incorruptible himself, applied a torch to a statue of Atheism carved for the occasion, and warned, "Today we give ourselves to transports of joy; tomorrow we will be fighting crime and vice again." Many thought the great feast had signaled an end to the Terror; indeed, that day the guillotine had been draped in velvet. But not Robespierre. Two days later he persuaded the Convention to accept the Law of 22 Prairial (June 10, 1794), which amounted to an astonishing tightening of the Terror. It provided only two sentences: death or acquittal. While it encouraged further spying and denunciation (Article 9), it also asserted that an accused could legally be denied all means of defense (Articles 12–18). It further stipulated the death penalty for simply working against "republican government," for spreading "false rumors"—whatever that meant—and even for "defeatism." In the meantime, the prisons, which each night shook with the shrieks and sobs of those who were beaten and dragged to the guillotine, still held more than eight thousand suspects who might rise up or flee; the only way to guarantee their acquiescence was through raw fear.

Fear it was, as from June 10 to July 27 the "Great Terror" began. In under seven weeks nearly 1,500 men and women were guillotined, more than in all the sixty-one weeks before. One Parisian print stated: "*C'est affreux mais nécessaire*" ("It is dreadful but necessary"), and one of the Twelve commented that heads were falling "like slates from a roof." But suddenly something was different. In the summer of 1794 the people now stayed away from the executions; few dared speak freely, spies for the committee roamed throughout

Paris, and life had become a kind of blindman's bluff. Instead, Parisians kept to their dining rooms, drew their curtains, and worried over "the very creak" of a door. And as if a great hush had fallen over the city, social life was desolate. Reporting back to the United States, America's Gouverneur Morris ruefully observed that French festivities had come and gone, "like the shadows of a magic lantern." The Convention itself was a mere vestige of what it had once been: Out of its original 750 deputies barely more than 100 now attended, and the majority of them preferred to sit in silence rather than vote and potentially implicate themselves. Even the Twelve hesitated to run afoul of the new triumvirs—Robespierre, Couthon, and Saint-Just.

Would it ever end? Once more, the French were proving to be decisive on the field of battle. The British blockade that sought to keep American provisions from France failed. Then a French army hurled back the dreaded Austrians and again triumphantly surged into the Low Countries. In six months their cavalry would ride into Amsterdam on ice; the Revolution seemed unstoppable.

But unlike in the past, far from strengthening Robespierre's hand, military success actually made the French people less willing to put up with dictatorial rule and continued Terror. Moreover, Robespierre had unwittingly antagonized all the significant parties on whose patronage he had so long depended: Duplicity now begat duplicity, and betrayal would beget betrayal. And all throughout the grim months of June and July, everywhere his enemies were increasing. The working-class radicals no longer supported him. After the death of Danton, the Convention was afraid of him. And within the committee itself the Jacobin elite was cracking: Billaud-Varenne, Collot d'Herbois, and Carnot were quietly scheming against him.

Embittered by the slanders against his name, he became increasingly irritable and distant. Robespierre also sensed the rising hostility against him, and from July 1 to July 21 avoided committee meetings, hoping this would cool the passions of his opponents. It didn't. His strategy faltered, instead giving his detractors more time to envision his decline.

Thus, on July 26, excitement was in the hall when Robespierre reappeared before the National Convention, hoping to regain his hold on public opinion. This was to be his last speech to the delegates. So often Robespierre's oratory had held the Convention in its thrall—or had outright browbeaten them. Now many of these deputies feared for their own heads, and others were openly hostile. Yet, his voice steady, even dignified, Robespierre answered his critics. He spoke for two hours.

"Citizens . . . Is it tyrants and rascals who fear us, or men of good will and patriots? Do we strike terror into the National Convention?" But inexplicably, Robespierre's speech began to veer in a different direction. He warned that a conspiracy was enveloping the republic. But who was behind it? Staring right at the seats of the radicals, he unleashed a sound of impending ruination. "I have vowed to leave a redoubtable testament to the oppressors of the Republic. I shall leave them the terrible truth—and death!" And for whom did he plan this death? The deputies leaned forward, listening for the names. He declared that slanderous falsehoods had kept him from the Committee of Public Safety, and as a result, it had fallen under the sway of the true despots; in particular, he charged that the public finance was overseen by a devious crook: Pierre Cambon.

Finally, Robespierre made his most damning accusation. "We must say then that there is a conspiracy against liberty, a criminal coalition in the very bosom of the Convention . . . What is the remedy? Punish the traitors, purge the Committee of Public Safety; [and] establish the unity of the government under the supreme authority of the National Convention."

As Robespierre finished, the deputies sat motionless, then as if by force of habit, they burst into applause. The diminutive Incorruptible had, it appeared, bested his foes once more. But then almost inexplicably the Convention erupted. Cambon, the only man specifically charged, leaped to his feet. "It is time to tell the whole truth," he cried out. "One man alone paralyzes the will of the Convention. That man is Robespierre!" He was joined by Billard-Varenne, who had earlier fought with Robespierre and presumed that he would also be fatally named. He challenged: "Let Robespierre's discourse be referred to the Committee before sending it out!"

Shaken, Robespierre stammered: "What? You would submit my speech to the examination of those whom I accuse?"

"Name those who you accuse!" a voice called out. "The list! The list!"

Gathering his papers, Robespierre said simply, "I have flattered no one. I fear no one. I have slandered no one."

But a cry rose up throughout the hall: "Give us the names!"

Coolly, Robespierre rebuffed the assembled deputies. "I am too busy to discuss the matter."

With that, he departed.

∞

ROBESPIERRE'S MISTAKE WAS colossal. There had been a series of unchar-acteristic slips in this fateful speech: the tone, the willful arrogance of a man who held out an olive branch on the point of a dagger, and most of all, the fact that he not simply had incited his enemies, but had left even his supporters uncertain of his wrath. So roused, the Convention even defeated the customary motion to print the speech.

But when he went home, Robespierre was satisfied, unaware of his tactical mistakes. After supper he took a walk along the Champs-Élysées with Eleanor Duplay, a committed friend and his landlady. As the couple chatted, Robespierre's dog Brount played around them. From there, Robes-pierre stopped at the Jacobin Club and delivered the same discourse, draw-ing cries of great approval. "It is my mortal testament," he enthused. "I leave you my memory, you will defend it. I shall drink the hemlock." The artist David, who was there, answered back: "And I shall drink it with you!" At this point, Billaud-Varenne and Collot d'Herbois, who had just been singled out by Robespierre, tried to speak but were shouted down by the cries, "Conspirators must perish!" and "*à la guillotine!*" From there, both men entered the rooms of the committee, where they were stunned to find Robespierre's ally, Saint-Just, at work on their impending indictment.

Deep into the late hours that night, Varenne, d'Herbois, and Fouche, fully aware of the own nearing doom, began a desperate attempt to save themselves. In Fouche's case—he had already had the temerity to cross Robespierre—he had been living like a fugitive, moving from house to house in the dark of night. His daughter was dying, yet he feared even sit-ting at her bedside. So on this evening he was unusually resolute: "Tomor-row, we strike!"

∞

CONFIDENT THAT HE had the Jacobins' support, Robespierre passed a peaceful night. At breakfast he was well combed and powdered, and ate quietly alone, helping himself to fine fruit, fresh butter, pure milk, and fra-grant coffee. When the hour came for him to leave, the Duplay family bade him farewell at the door. Eleanor gently straightened Robespierre's cravat, while her husband took his hand and cautioned him to be "careful today." Robespierre confidently reassured his devoted friend: "The Convention is in the main honest, *all* large masses of men are honest." Why he let his guard down is puzzling. Ordinarily he was the most vigilant of leaders, and the omens of trouble had been unmistakable. But the slow, bleak truth of his crushing errors came only when it was too late.

At the Convention, Saint-Just prepared to deliver the indictment before deputies trembling with anticipation; his mentor, Robespierre, sat directly before the rostrum. When Saint-Just began to reel off his accusations, Tallien, fearful that his own name was coming, rushed to the dais, forced Saint-Just aside, and declared: "I ask that the curtain be torn away! . . . At the Jacobin Club I trembled for my country. I saw the army of a new Cromwell!" Joseph Lebas, an ally of Saint-Just's, tried to intervene, but was shouted down by screaming lawmakers. Now it was Robespierre who insisted upon being heard, but he was also swept aside and silenced.

At this, the Convention grew belligerent, while Robespierre sought again to speak, frantically pushing his way toward the platform. An old friend of Danton's now took the chair, ringing the bell and calling for order. No one had yet dared denounce Robespierre by name, until a minor deputy called out, "I demand the arrest of Robespierre!" For a fleeting instance, the hall fell silent, except for the high-pitched voice of Robespierre, nearly shrieking: "For the last time, president of murderers, will you let me speak?" Tallien shouted back: "The monster has insulted the Convention!" "Arrest him!" the deputies cried. "Arrest him!"

Fleeing the rostrum, Robespierre headed over to his erstwhile Jacobin allies. But far from being welcomed, he was mocked. Speechless, he stared at the room until a deputy announced, "The blood of Danton is choking you!" Facing the moderates, Robespierre pleaded, "Men of Virtue! I appeal to you." Then he sank into a seat.

"Monster!" one yelled. "You are sitting where Condorcet once sat!"

Suddenly it was Robespierre who was dethroned. After a year of holding the deputies in his thrall, his whole reign now appeared a farrago of distortions and calumnies. He and his brother, Augustin, as well as Saint-Just, Lebas, and Couthon, were abruptly dragged from the Convention by the police, to be scattered among five separate jails.

But for Robespierre, all was not lost. His base, the Jacobin Club, declared its intention to conquer or die, and the mayor of Paris commanded that the prisoners be transferred to the Hôtel de Ville (City Hall), where he guaranteed them safe haven; for his part, Robespierre began issuing orders to impose a "new dictatorship." As it turned out, the Commune appeared equally determined; declaring itself in revolt, it called on the National Guard to lay siege to the Convention until it rescinded its decree of arrest. However, though the commander, François Hanriot, had superior forces, he was tentative and his men irresolute, and almost unthinkably, a number of *sections* suddenly threw their lot in with the Convention. In the meantime, knowing that this night meant life or death, the deputies called upon Paul Barras to raise a counterforce, go to the Hôtel de Ville, and rearrest the prisoners, even as they declared Robespierre and his cohorts *hors la loi* ("outlaws"). Paris, and the Revolution, were now literally at war with themselves.

Once again the mayor appealed to the National Guard, but as the hours wore on, the guards were no longer willing to die for the man who had cut down Danton and assassinated Desmoulins. And as rain began to drum on the rooftops of Paris, what force had been assembled simply drifted away. At two in the morning, the Convention's men, converging in two columns, seized control of the Hôtel de Ville. Spotting them, Robespierre, in an apparent effort to spare himself the guillotine, attempted suicide, but the bullet missed his temple and only shattered his jaw. A steelier Lebas blew his brains out. Augustin Robespierre broke a leg in a futile escape attempt from the window. Couthon, already crippled, plunged down a staircase and lay there helpless. Robespierre was then dragged to the Committee of Public Safety, where he was forced to lie on a large table, writhing in agony, only partially mitigated by a blood-soaked bandage concealing his gaping wound.

The following afternoon, at five o'clock, four tumbrels conveyed the two Robespierres, Couthon, and Saint-Just, along with the mayor, Hanriot, and sixteen others to the guillotine. Along the way they heard onlookers cry out, "Down with the Maximum!" while blood was splashed on the door of Robespierre's residence. This was not the same throng that thrilled to the king's execution, or the queen's, or even to the Girondins'. It was an unusually fashionable audience this day. People paid exorbitant sums to stand in windows; ladies showed up in their finest dress, as if for a fete or a festival. And once the crowd began screaming death to the conspirators, they were also screaming life to the republic.

In the end, the man so reviled was his own victim: He had glorified the republic, but could find no compassion for the men and women whose patriotic fervor did not match his own; he had preached universal brotherhood and equality, but then coldly executed the innocent as well as the guilty; he had sanctified the vast, ennobling goals of the Revolution, but had destroyed the very men who tirelessly labored to make them a reality.

As Robespierre feebly wrenched himself up the scaffold, the executioner tore off his bandage, letting loose a howl of agony. While the mob jeered, he was placed half-conscious before the blade, and when his head was lifted aloft, shouts of delirium rose. The National Razor had now claimed its most prominent victim. To an inured nation, one more death out of thousands would hardly have been noticed. But once the executioner let go of the blade, it was a cry from the heart of France: Indeed, all of Paris felt that this one had a special meaning—that its national nightmare might, at long last, have come to a close.

∞

THE CONVENTION ITSELF died a slow death. But with Robespierre's beheading, something profound had changed in France. The Committee of Public Safety was defanged, Jacobin clubs were banned throughout the country, religion was revived, a peace treaty was eventually signed with the Vendean rebels providing them the right to worship, and the radicals fell from power. In a sign of the new times, the dreaded prosecutor Fouquier-Tinville was eventually guillotined too. After this "Thermidorian reaction," the middle class reclaimed the Revolution.

But even the Thermidorians had not lost faith in the Revolution: While they associated democracy with red terror and mob rule, they still believed in individual legal rights and a written constitution. They set aside the Constitution of 1793, which had never been practiced, and drew up a new one altogether. It went into effect in 1795 and established what we know as the first formally constituted French republic: the Directory.

As the Convention prepared to retire, however, it could claim some considerable achievements. Even amid the whirlwind of the Terror, it had liberated slaves and emancipated Jews, it had appointed talented commanders and beaten back a mighty coalition arrayed against it, and it had extended France's frontiers to the Rhine, the Alps, and the Pyrenees. It had also devised metric and decimal systems, established a welfare state, proclaimed universal primary education, and founded or restored the School of Medicine and the Ecole Polytechnique. And it had ruled a revolutionary nation that bequeathed magnificent and inspiring words to the world, as well as to future generations. With a mixture of satisfaction and trepidation, then, after a mere three years—or what seemed like a vast lifetime—it prepared itself for a peaceful demise.

Hoping to learn from the past, this time around, France, not unlike in America, would adopt a bicameral legislature, that is, two chambers. The lower chamber was called the Council of Five Hundred, which could propose and discuss measures but not make laws; the upper chamber, composed of 250 members, was the Council of Ancients—"ancients" being men over forty; this assembly could not initiate legislation, but was empowered to reject or ratify resolutions. In turn, the chambers chose the executive, which was called the Directoire—or the Directory. Housed in the Luxembourg Palace and composed of five men, the Directory soon became the dominant branch of government, a body nearly as despotic as the Committee of Public Safety had been.

But once more France was stricken with another revolt, this time at the hands of royalists, who objected to new restrictions on the freedom of elections. They cobbled together an impressive force of 25,000 men and seized an area of Paris around the stock exchange. Then they marched to positions that commanded the Tuileries—and were poised to threaten the Convention itself. In one of their last acts, the stricken deputies made a fateful decision: They instructed a young general, who happened to be in Paris, to gather men, supplies, and above all, artillery, to come to their defense. He was the hero of Toulon and his name was Napoleon Bonaparte. With a "whiff of grapeshot," Bonaparte killed three hundred of the besiegers and sent the rest of the royalist mob fleeing. The Convention had triumphed over its final test, and a twenty-four-year-old Napoleon, pitiless and determined, commenced his storied career.

∞

So it was done, once again.

Out of the twisted rubble of violent dictatorship, on November 2 the final phase of the Revolution began. Suddenly, almost inexplicably, a country that had been crying out for bread, pleading for work, longing for peace, and praying for respite, would be galvanized in ways almost unimaginable to the modern mind. The distasteful task of governing was again taken up. This time, the mighty sweep of France—led by the Directory, animated once more by revolutionary fervor, and spearheaded by General Napoleon—would be swollen with victory and temptation. It would extend its ideological wings to all of Europe, from the Atlantic to the Elbe, making monarchies everywhere tremble. It would swallow those in its path and establish a ring of pliant satellite states, and even plunge into the heart of the Muslim Middle East, from Egypt to Syria to Palestine. And it would threaten a young American republic and rouse a once reluctant Russian giant to military action. Not since the days of Charlemagne, who gave "laws to the West," had the globe seemed so contained, or so tightly interconnected, or so volatile.

At first the people would greet these events with little more than a faint reaction. But as the mapmaking would continue, as the sphere of military glory would be extended, as entire nations would be sharply divided—are you for the Revolution or against it?—everywhere, it seemed, thoughtful souls and proud statesmen wondered: How long can this Gaulist fervor last? And with ever-heightened stakes, the question would also be asked—though not yet fully answered—in the other young republic: the still-struggling United States of America.

But first, in distant St. Petersburg, the Autocrat of All the Russias, a brooding Catherine the Great, was struggling to stamp out her own revolutionary nightmare, in Poland—inflamed by the two viruses of the eighteenth century—the French Revolution and the Americans.

CHAPTER 10

RUSSIA

✦

As the Polish rebellion continued to rage on and Thaddeus Kosciuszko bedeviled her commanders in Warsaw, the Russian empress Catherine worked from morning to night. She confided to her friend Baron von Grimm: "Four posts, detained by contrary winds, arrive at the same time, three or four courtiers from every corner of the world, so that nine large tables hardly suffice to hold all this jumble of papers." Hold them yes, but what did they say? Little she read provided immediate solace. At sixty-four, and after the unremitting strain of four years of war, she was elderly; she now suffered from repeated bouts of lethargy, an irregular heartbeat, and a chronic shortness of breath. And she could barely walk the stairs on her own. Yet even then, her energy, or rather her determination, did not flag. But she could not shake her concerns that somehow victory would slip through her fingers.

Surrounded by aides and a mass of documents, she thus immersed herself in the smallest details of the war. There were the military matters: Here was the great Suvorov, at the battle of Krupshchitse, then on the move, marching twenty-two miles a day. There were the matters of strategy: Here was her fierce fighting man, General Repnin, filling the gaps on the Polish front. There were the matters of intelligence: Here were her loyal Cossacks, thundering and swearing, serving as her advance guard. And there were the status reports, reports of the vast metropolis of Warsaw, where bloody street battles left a trail of corpses everywhere, now an armed camp.

Reports of her men, bloodstained and blasted and taken prisoner, and the sights, sounds, and scars of siege warfare. And then there were reports of Thaddeus Kosciuszko and his despised rabble. Taking page after page from the American Revolutionary War, there was nothing they wouldn't resort to. They stood at windows of burned-out buildings, scrawny and unwashed, eyes gleaming, holding up guns, some without ammunition. Or they crouched in hastily dug ditches, scythes in hand, poised to kill valiant Russians. They perched themselves in woods and marshes, wielding axes and pikes, hoping to ambush the enemy and roaming the countryside as marauding bands. Exhausted and ghostly, they scratched scabby arms and dressed their wounds. Exhilarated and determined, they wiped their runny noses and shed tears for those already dead. Defiant and dreamy, they chanted verses and sang martial Polish songs. And they fought Catherine's men with whatever they had, day after day.

While General Suvorov was camped out at Brest, Kosciuszko moved his troops from the frontier and Lithuania, as well as Grodno, hoping to cut off the Russians' advance. At the same time another Russian force, under General Ferson, had crossed the Pilica, feinted toward Pulawy, and dashed to the Vistula. Then Kosciuszko himself set off with the cream of the Polish army, marching to engage Suvorov at Brest, until he decided that Ferson could be cut to pieces. Hastily changing course, he raced to engage him at Maciejowice. They would clash on October 10, 1794.

It would be the decisive battle of the war. Meanwhile, as dust was kicked up and the respective armies marched, the tsarina waited.

∞

How LONG HAD it been her lot to wait? Among the rulers of her day—kings, sultans, queens, prime ministers, and presidents—Catherine was nearly indomitable. It is impossible to talk about one of the great behemoths on the world stage—Russia—without invoking her name. It is equally impossible to talk about the eighteenth century—a century she helped shape for a staggering thirty-four years—or the 1790s without summoning up her many achievements. Almost uniquely, she weathered the rise of tumult and the specter of rebellion, the onset of republicanism and the arrival of repeated war, the challenge of democracy and the challenge to authoritarianism, and a clash of ideas in the international arena virtually unparalleled in human history. Almost uniquely, she outlived or outfoxed the exceptional monarchs of the day: Prussia's Frederick II comes to mind, as well as the domineering kings and queens such as Louis XV or Maria Theresa, and such idealistic rulers as Joseph II. But for all this, it is difficult to assess just where to place her in the kaleidoscope of the age. Without question, she lived in a time of some of history's greats, whose notoriety would later seem to eclipse hers: Napoleon, for one, Robespierre for another; the same could be said, almost curiously, of figures like Marie Antoinette, known principally for their celebrity or weakness—or one of history's most impeccable figures: George Washington, the father of the United States. And without a doubt, she reigned in a period marked by the most lively intellectual ferment that the world has ever witnessed: the age of Voltaire and Paine; of Hamilton and Burke and Diderot; of Rousseau, Wilberforce, and Montesquieu; and of Jefferson, Madison, Mirabeau, and Franklin.

Yet there was little Catherine didn't dominate and few setbacks she didn't overcome. She tamed the Muslim world and charmed—or awed—the Christian one. She was a stunning conqueror and a glittering ruler. Yet why in the public imagination of our time has she often been overshadowed by those whom she plainly surpassed in her own? Some might ascribe this to the fact that she was a woman, but this explanation is, for the most part, unpersuasive—or incomplete. Partly the answer is cultural chauvinism: She ruled in the East rather than the West. Partly the answer is political: She presided over the end of an interstice of the age, the era of enlightened despotism, rather than the beginning—the flowering of democracy. We view her accordingly.

Should we though? Yes and no. For there is so much more to her than that. Prince de Ligne called her *"Catherine la Grande,"* and so did Rostopchin, a verdict that has stuck, placing her in the pantheon of a small but distinguished group of some of history's most elite figures: Alexander and Julius Caesar, Charlemagne and Pope Gregory VII, Henry IV and Louis XIV, Tsar Peter and Catherine's own inimitable rival, Frederick II. Upon her death it was written that "the most brilliant star in our hemisphere has disappeared," and *Gentleman's Magazine* rhapsodized that she effected "the most important of all revolutions in the history of human kind, the civilization of so large a portion of the human race, and the cultivation of the wildest and untrodden deserts." Of course, she had already heard the most extravagant praises in her lifetime—as well as the worst insults. She was "Semiramis" but also "Messalina." She was the "Mediatrix of two continents" but also a "cunt." Like her contemporary George Washington (he was a mere three years younger), whom she scarcely deigned to recognize, she unceasingly worked at building her own legend; unlike the illustrious Washington, who chafed at the repeated press attacks against him, her skin was much thicker.

All her life she radiated a majesty that was at once spectacular and unaffected. All else said, in her gut she was a romantic. Who other than Catherine would have ordered a lone sentry to stand in the middle of the public garden of the Winter Palace watching over a stretch of lawn, day and night—the empress had placed him there to protect the first snowdrop of spring. But in truth, her talents, which were unquestionable, and her temperament, which was a rock, were riddled with incongruity. While she loved Russia and its people with unprecedented ardor and passion, she reveled first and foremost in her own personal glory, as treaties and wars exalted her name every bit as much as her empire's. But her triumphs were undeniable: the crushing of the Ottomans, annexation of the Crimea, seizure of the Black Sea ports, partition of Poland, isolation at sea of Britain, titan of two continents—all cloaked in the mantle of justice, or alliance, or empire, or God. The reasons always varied, but the fact was that with each victory, it was Catherine herself whose stature grew. Indeed, nothing with her was straight or simple. She had the heart of a republican and gave her grandsons a renowned liberal tutor, but she always behaved as an autocrat and grew wary of anything that smacked of democratic reform; she had lofty philosophical ideas but she egregiously perpetuated serfdom; she insisted on governing alone, but she constantly enlisted the nobility for support—or her male lovers. She loved life and all its wonders: books, men, animals (dogs especially), trees, games, laughter, and children, but her attention never wandered from the affairs of state. If she had been an American, she would have been part Jefferson, part Washington, part Hamilton, hard as that combination is to fathom. But the description fits. She was a tireless worker, and, at the same time, to paraphrase Voltaire, a charmer who united the wiles of her sex with virile authority. Invariably, however, she combined persistence and intelligence, tenacity and bravery, with a capacity to take formidable risks, though she never feared changing course or extending her hand in compromise either.

And perhaps what is most striking was that she was born a German princess, and a minor one at that, and had scarcely a drop of Russian blood in her veins. How easy it would have been for her to have faded into obscurity, dwindled into mediocrity, or lapsed solely into Asiatic barbarity. How easy it would have been for her to have been overthrown, or imprisoned, or fallen victim to insurrection, or assassination, or never to have obtained power in the first place. How easy it would have been for her to have been the laughingstock of the world's rulers, or the puppet of the empire's court, or the wife of a hapless tsar. And how easily she could have ended up like the Sun King, who after countless defeats and a terrible famine admitted on his deathbed that he loved war too much; or like Napoleon, who from his exile in St. Helena glimpsed that there was more to life than the destiny of France. Operating from her royal cocoon, the empress never had such an epiphany. During her reign, often called "the Age of Catherine," she weathered the Pugachev Rebellion, beat back pretenders to the throne, foiled various plots, ruled for more than three decades smiling, and died in her bed.

Yet from the earliest years, her life was always set for towering theater— or, like the rest of her life, she willed it to become so.

∞

HER NAME WAS not even Catherine. And from her youngest days she was forced to compensate: for her birth, for her sex, for her looks, for her parents, and for her heritage. She was born in Stettin, Pomerania, on April 21, 1729, to a negligible princely house of the Holy Roman Empire, and was named Sophia Augusta Frederika after three of her aunts. Her mother was a minor princess, and through her, young Sophia was a distant cousin to Peter; but her family was so impecunious that she was raised in a small, uncomfortable home and her father, Christian August, was forced to join Frederick's army. Neither parent wanted a girl, and her mother wept at Catherine's birth—an eavesdropper would have thought her stillborn; a year later, to Catherine's lifelong chagrin, her mother rejoiced at the birth of a son. But Catherine would surmount her sex by acquiring the aura of a monarch and the iron will of a commander, all the while being able to woo men with her womanly charms. Even this was not easy: her childhood was plagued by illnesses, and one left her woefully misshapen ("the backbone running zigzag," she later wrote). As a remedy she encased herself in a painful corset that she "never removed day or night" for eighteen months. She was also ugly, or so her mother repeatedly told her, so she became determined to beautify her mind. She did, and by adolescence, she had both looks and brains, to which was added an uncommon ebullience.

As a girl, she had dancing lessons and music teachers and was taught at the knees of tutors. She peppered one so incessantly with questions ("his replies never seemed to satisfy me") and not infrequently, he lost his temper with her—or she with him; once, he grew so irritated that he even threatened her with a caning. Yet she loved her French governess and eagerly devoured the French masters, Corneille, Racine, and Molière, and was soon ready for Voltaire. She developed into one of the finest minds of her day. Still, her family situation was dismal—but then this changed almost by magic. When Russia's Empress Elizabeth invited her to court in 1744, en route Sophia met Frederick the Great; she asked him a "thousand questions," and they talked about comedy and opera, poetry and dancing. She was only fourteen. The journey was long and arduous, made worse by the fact that her parents considered Russia a backward and dangerous place. But she would not. With little more than some lingerie and three or four dresses (in her words, she was "a poor and wandering princess"), she traveled though Berlin, Stettin, East Prussia, Riga, and then reached Moscow after a sleigh ride of fifty-two hours from St. Petersburg.

She was a foreigner in an alien court that despised outsiders; Empress Elizabeth, as it turned out, was suspicious of her background; and her religion was wrong too: Lutheran, not Orthodox. Furthermore, her German accent and manners were painfully anathema to the ways of old Muscovy. But before marrying the Grand Duke Peter, the future tsar, she was determined to learn Russian thoroughly, which she did, and to accept the Orthodox faith in toto, which she also did. On June 28, with a great show of piety, she converted to the Russian faith: Added to her existing names were Ekaterina Alexeevna; from then on, she was Catherine. On August 21, 1745, she married; by October all ties to her homeland were severed when her mother left for Germany. Catherine was sixteen, alone, and in a strange land. Moreover, her marriage, ill fated from the beginning, proved to be tortured and unloving: where she had a keen quick mind, Peter was mentally backward (like "a grown up child"); where she was pious, he was often drunk—or belligerent; where she was determined to become pure Russian, he actually disliked Russia (which he called an "accursed land"), scorning the church, scorning the clergy, and loving all things Prussian. He even surrounded himself with a "Holsteiner Guard" of soldiers, mostly all German. Thus began seventeen years of unremitting unhappiness for the princess.

The marriage was not just a conflict of personalities, but of cultures.

She plunged into the great literature of France, and the classics as well: Plato, Plutarch, Tacitus, Locke, Voltaire, Diderot, and Montesquieu, whose *Spirit of Laws* she insisted should be "the breviary of every sovereign of common sense." These books, rooted in history, steeped in law, and suffused with a progressive spirit, solidified her belief in the benefits of "enlightened despotism." She even read—and enjoyed thoroughly—Benjamin Franklin's *Poor Richard's Almanack*. But if this were the inner world of her intellectual life, she could not escape the outer world. Her days, already miserable, became even more complicated. Peter took a series of mistresses; upon Empress Elizabeth's suggestion, Catherine took lovers as well. In 1751 she became pregnant, but a miscarriage followed. Two years later she miscarried again. Five years later she gave birth to a daughter, but the baby died after four months. Only in 1754 did a baby survive: the future emperor, Paul I. And in 1756 she met a handsome Pole, the twenty-four-year-old Stanislas Poniatowski. But even this too was a mixed blessing. They courted, fell in love, had (almost certainly) a daughter together, then were discovered by Peter, who threatened to hang Stanislas. Terrified, Catherine interceded with Peter's mistress. In a burst of good nature, Peter forgave Stanislas, and even entered into a sexual encounter with his estranged wife and lover.

By then Catherine was becoming the hardened woman we know today. Of one of her affairs she wrote: "If I may venture to be frank, I combined, with the mind and temperament of a man, the attractions of a lovable woman." Was she attractive? Stanislas described her in 1755: "She had black hair, a dazzling white skin, black eyelashes, a Grecian nose, a mouth that seemed made for kisses, perfect hands and arms, a slim figure rather tall than short, an extremely active bearing yet full of nobility..." She finally was coming into her own, and just in time. On January 5, 1762, the Empress Elizabeth died, and the maneuvering had already begun. Attempts were made to enlist Catherine in a scheme to thwart Peter's ascension to the throne; the plotters whispered in her ear that he would cast her aside and make his mistress his wife and queen.

For the moment Catherine demurred. She wouldn't for long.

∞

PETER'S REIGN WAS troubled from day one. He implemented laudable reforms, some brave and quite remarkable, yet he found himself increasingly unpopular. The army despised him as a turncoat, the clergy hated him as a Lutheran, the unfreed serfs agitated for emancipation, and the court mocked him as a buffoon. And the rumors continued that Catherine would be cast aside, that Catherine would be locked up, that Peter would marry his mistress. In truth, they soon seemed to be borne out by events: As the days passed, Catherine was treated harshly, not just in private, which in the court was forgivable, but in public, which was not. As Baron de Breteuil, who would later serve Louis XVI, wrote: "The Empress Catherine is in the cruelest state, and is treated with the utmost contempt . . . I should not be surprised, knowing her courage and violence, if this were to drive her to some extremity."

But St. Petersburg was awash in Catherine's followers. She was esteemed by the army, admired at the court, and often loved by the Petersburg populace. And when one of her closest intimates, Princess Ekaterina Dashkova, was told by Peter that he planned to depose Catherine, she knew she had to act at once. She did—in 1762, after just six months. Her coup d'etat was stunning, and the subsequent death of Peter—history is still unclear if she directly ordered it—was worrying.

Moscow was suspicious of her. And the events created, as well as solved, problems for the young tsarina. Would she be equal to the throne awaiting her? Almost immediately the court was rife with intrigue and sedition. Some wanted Ivan VI, then languishing in jail, to be restored to the throne; others wanted Catherine's own young son, Paul. Around this time the Prussian ambassador reported back home that it was "certain that the reign of the Empress Catherine is not to be more than a brief episode in the history of the world." For his part Voltaire despaired, "I am afraid that our dear Empress will be killed."

And as conspiracies multiplied around her—assassination, deposition, imprisonment—these prophecies appeared about to come to pass.

∞

FOR ALL HER education, and it was superb, and her breeding, which was impressive, and her ambition, which was matchless, nothing about her background recommended Catherine to the overwhelming task before her. She came from Zerbst to govern a country whose language she had not known and whose mores were totally alien. The real Russia, dark, byzantine, ancient, ritualistic, and impenetrable—often called "Holy Russia"—was still foreign to her. Would she be a Marie Antoinette, ill fated from the start? Or a Napoleon, born to reign? The great question now was not would she thrive as empress—nobody really believed that to be possible—but could she escape with her own throat uncut. But in seeking to survive, Catherine turned out to be second to none. Reaching deep down into her soul, no ruler, man or woman, no Russian, noble or peasant, was more fervent in her belief in her nation; no ruler, alien or homegrown, was more attached to aggrandizing not only her own name, but the name of the state. And no ruler was more concerned not simply about staying alive, but about making Russia one of the grandest kingdoms the world had ever witnessed.

Ultimately, it was these beliefs that would give her the ability to go on. In the tense days and months and years ahead, she would need it. Somehow, at the toll of the bell, this foreigner, this regicide, this empress who all her life spoke Russian with a bad accent, would discover herself—and with that, an empire.

∞

DEFYING HER ADVISORS, it became a duel less about her and more over the fiber of Russia.

Actually, most of the diplomats gave her six months; they were wrong. Since the nobility held sway over the Russian Senate, and the Senate made the laws, the choice was between a single autocratic monarch and the end-less jockeying of feudal lords—exactly the same sort of dilemma that the future Sun King had faced in seventeenth-century France. Wary of the Senate, Catherine boldly decided to center all rule in herself. She would, like most of the monarchs of Europe, rule as an autocrat, but hers, she intended, would be autocracy intermixed with philosophy. With courage, ferocity, and vision, Catherine surrounded herself with talented men, secur-ing both their fealty and often their affection as well. She expected much but she paid them generously, in truth, too generously: The state's coffers frequently groaned under her court's opulence and largesse. But it was a remarkable, polyglot court, one richly gilded with French culture—French was the tongue of her palace and guiding animus of its splendor—anchored in Russian mysticism, and ruled by a minor German princess who routinely bested her aides in intellect and statesmanship.

It was also corrupt, as Catherine knew. This she was willing to tolerate. Other things she would not. Plagued by a palace guard whom she could not trust and nobles whom she couldn't buy off, Catherine created a new form of government: She made her lovers the executives of the regime; all told, she would have at least twelve. Each was, during his ascendancy, in effect her prime minister or chief of staff. Was she debauched? Legend says yes, but it would be a mistake to think so. If she were a man—say the Sun King, or Louis XV, or Henry VIII—this wouldn't even be a question. They had their wives, whom they ignored or put to death; or their mistresses, whom they bedded and then forgot. More often than not, their courts were bastions of sexual excess: rampant promiscuity, a retinue of bastards, and quick, fevered liaisons in palace antechambers. In contrast, Catherine was a faithful lover; her affair with Grigory Orlov lasted thirteen years. To Potemkin, whom she almost certainly married, she forever remained like a schoolgirl with a crush, and the sudden death of Lanskoi—her youngest favorite and "dearest friend"—reduced her to inconsolable despair. Reading her letters to them, or her own accounts, we hear the woman behind the ruler—and the heart behind history. Still, her system was a remarkable spectacle: Not unlike a professional football coach recruiting star athletes, Catherine approached possible new favorites with both art and science. She—or Potemkin, who often chose her lovers—sought out men who combined beauty as well as power, political capabilities as well as physical stamina; the gentleman was then brought to dinner, where his ideas were judged along with his demeanor. If he weathered this test, he was sent to a court physician. Only then would she appoint him as her aide-de-camp, install him in a luxury suite, and welcome him into her life. Her infamous explanation was: "I am serving the Empire in educating competent youths." Most were indeed youthful (even when she no longer was), many were competent (or were given tutors), and the treasury spent richly for these favorites—rubles, diamonds, emeralds, titles, estates, lands, admiralships, generalships, ministries, serfs, and, when it was time to let them go, permission to marry.

The last on her list was Plato Zubov, a lieutenant in the Horse Guards and thirty-six years her junior; he became her foreign minister, and she took it upon herself to oversee his education, tenderly treating him like a son. He remained by her side until her death.

∞

BUT IN THE end, there was no doubting that she was the sovereign. Leaving her mark on statecraft and politics every bit as indelibly as England's Elizabeth I, she was the wizard of all court ceremonies and the master of court life, the dispenser of medals and the judge of all court disagreements, the initiator of every war and the architect of all diplomacy. And while Catherine took lovers and shuffled aides around like pawns, she boldly moved to reshape her nation. Thus, though she typically found people more intriguing than ideas, she could, as her court and her subjects were to learn, hold her own in almost any intellectual enterprise: She was equally at home with philology or philosophy, iron works or intaglios, China or chinoiserie, art or architecture; all fascinated her. Yet she also took it upon herself to master the subtlest details of empire, from diplomacy and naval warfare to town planning and criminal law, from the challenges of agriculture to the minutiae of banking. Knowledge to her always had a purpose, and when she chose a bee as the emblem of her regime, it was entirely appropriate. Also appropriate—if she could pull it off—was her desire to reshape Russia.

It was her greatest challenge. Frequently court protocol—and the country—seemed to conduct itself, taking its own course, making its own decisions, running on its own internal energies of tradition—or inertia—exclusive of all human management. This protocol, handed down from a long-forgotten era, remained as obstinately rigid as a block of granite. In the palace, courtiers backed away from Catherine's presence—unless she bade them otherwise. No one ever contradicted her unless told he might. It was improper to speak to the tsarina without being spoken to, and when walking with Catherine, friends did not greet each other or even notice each other's existence unless Catherine did so first. Security—to the extent that she could trust it—was exceedingly tight. A permanent garrison of carefully chosen men from the Imperial Guard provided guard detachments at the palace gates; sentries were stationed inside the palace—loyal Cossacks, elite guards, and even giant blackamoors—in corridors, staircases, kitchens, vestibules, and even in the cellars. While an army of servants moved through the marble halls and silken chambers, these guardsmen were supplemented by the ubiquitous plainclothes secret police, who searched the servants, tradesmen, and workmen, and recorded the names of all who came and went. And at any given moment the tsarina could look from her windows to see the reassuring sight of tall soldiers in greatcoats patrolling the grounds.

Yet amid this phalanx, early in her reign, Catherine, with her bold eyes and wry smile, envisaged a different Russia: more open, more Western, more compassionate, more liberal-minded and egalitarian. Vaguely reminiscent of George Washington, in her copy of Fenelon's *Télémaque* she wrote these resolutions: "Study mankind, learn to use men without surrendering to them unreservedly"; "Search for true merit, be it at the other end of the world"; "Be polite, humane, accessible, compassionate, and liberal-minded"; "Do not let your grandeur prevent you from condescending with kindness toward the small, and putting yourself in their place"; "Do not allow the world to contaminate you to the point of making you lose the ancient principles of honor and virtue." To all this she added the final flourish: "I swear by Providence to stamp these words into my heart."

The young sovereign, driven by romantic urges to accomplish the impossible, took her own words seriously. Yet what she saw as her greatest test was this: the task of ruling a sprawling Russian state that was saddled with some ten thousand often contradictory and equally unfathomable laws. So it was that on December 14, 1766—a decade before the Americans declared their independence and two decades before the U.S. Constitutional Convention—she invited the nation's finest administrative agents and legal experts to Moscow. Her goal? To convene a consultative assembly and begin a process of enlightened reforms. Their task? To completely revise and codify Russian law.

Before they began, Catherine dictated a *Nakaz*, or *Instructions*, laying out the principles that should undergird the new code. Interestingly enough, she based them upon many of the same philosophers and political scientists as Madison and the American Founders—although often drawing radically different conclusions for governance. Thus she embraced Montesquieu's stipulation that republics were only suitable to small states, not large land masses, while the Founders boldly did not; she embraced Voltaire, while they followed Locke; both drew from Blackstone and Becarria. Montesquieu held a special place for Catherine: Writing to the French philosopher and mathematician Jean d'Alembert, she quipped, "You can see from the Instruction how I have robbed President Montesquieu for the benefit of my empire without naming him; I hope that if from the other world he saw me working he would forgive this plagiarism." And as she told the head of the Senate, "The Russian Empire is so large that any other form of government than that of an absolute Emperor would be harmful to it." And too: "any other government is slower in execution and leaves the field open to passions which disperse the power and strength of the state." And finally: "the internal institutions of a country should always develop in accord with the character of that country."

She boldly announced in the first sentence of her *Nakaz* that Russia must be seen as a European state, and she went on to add that it should devise a constitution based upon "European principles." While this did not mean a constitutional government where the sovereign would be held in check by a parliament or popular will, it did mean a system where the tsar reigned within the confines of the law. Could it have been otherwise, as Hamilton and Madison in America would believe? Catherine, studying the map and reading Montesquieu, thought not. For her, space made history. If she ran her eye across a map of the world from Konigsberg on the Baltic Sea to Kamchatka on the Pacific, then from the Arctic Ocean to the Caspian Sea, the Himalayas, Mongolia, China, Persia, Japan, in between lay Russia, with its snaking, anomalous, and vulnerable boundaries. Transportation was difficult, communication painfully slow, and a whole land of communities was as isolated from one another as from the center of government; for up to half the year, inhabitants huddled in scattered villages amid great forests, fierce snows, and icy winds. Thus in 1766 Catherine decreed that the feudal system—the system of mutual loyalty and services between peasant and vassal, vassal and lord, lord and sovereign—remained essential to the vitality of the state.

But if she were a champion of monarchy, she was also a foe of tyranny, and some of her proposals were quite daring. That same year she offered a prize of a thousand ducats for the best essay on the emancipation of the serfs. She further urged that the relationship of master to serf be enshrined and strictly limited by law, and that serfs be allowed to own property; indeed, she was the first Russian ruler ever to even conceive of codifying basic rights. Other proposals were equally forward-looking. She espoused a freer press and more humane laws. She insisted that all trials should be fair, torture should be prohibited, and outside of political crimes, capital punishment should be repealed. And in one astonishingly modern decree she stated that religious worship should be free, since "amongst so many different creeds the most injurious error would be intolerance."

Before being printed, her *Nakaz* was submitted to her advisors. They warned her that any sudden change from existing customs would plunge Russia into disorder. Catherine, a pragmatist, allowed them to substantially modify her proposals, especially those for the gradual emancipation of the serfs. Still, when her *Instructions*, even in their whittled-down form, were published in Holland in 1767, they stirred an admiring chorus among the intelligentsia of Europe who saw in Catherine the standard-bearer of civilization among a backward people. When Voltaire received his copy, he gushed, "Madame, last night I received one of the guarantees of your immortality—your code in a German translation. Today I have begun to translate it into French. It will appear in Chinese, in every tongue; it will be a gospel for all mankind." He added later: "I regard the Instructions as the finest monument of the century," even as the French government forbade its sale in France.

The empress moved quickly to make them a reality. She presented them to a "Commission for Drafting a New Code," which convened on August 10, 1767. With 672 members drawn from various groups, most cross-sections of society were represented: the nobility, the free peasantry—state peasants were actually allowed to elect representatives—the Cossacks, and even the non-Russian tribes (Christian or not), as well as the government. Only the serfs were left out, and there was no allotment for the clergy. In some ways it corresponded to the Estates-General, which met in 1789 in France and was, curiously, more representative than the Constitutional Convention in America—indeed, 13 percent of the deputies were actually illiterate. And like their more renowned French counterparts, the delegates carried with them piles of petitions enumerating every type of grievance, as well as ideas for reform. Once presented to Catherine, they constituted a powerful picture of the state of her empire; they also indicated that while she was a usurper and a foreigner, she possessed a strong hold on the country.

The task was an enormous one: how to find laws suitable for both Christians and Muslims, for the inhabitants of the Tatar steppes and those of the rich lands of the Ukraine, for Muscovites and Siberians. While the commission could not enact legislation, it had free rein to propose reforms on the issues; as the delegates were guaranteed freedom of speech and personal protection, nothing was taboo. Some proposed the emancipation of serfs, where others asked that the right to own serfs be more widely extended—and so it went. The commission held 203 sessions but to the empress's disgust, seemed never to reach full agreement on anything, and was then forced to adjourn *sine die* in 1768, ostensibly because of the onset of the first war with Turkey. In truth, the tsarina was exasperated. No code of laws had been formulated, and this only confirmed to the empress that in a country as outsize and diverse as Russia, she alone had to be the indispensable maker of laws. "The Commission," she retorted, "has given me light and knowledge for all the empire. I know now what is necessary, and with what I should occupy myself." Before adjourning, the commission proposed to confer upon her the appellation "Great"; she refused, instead quipping, "I brought them together in order to study laws and they are busy discussing . . . my virtues." But she did consent to be called "Mother of the Country."

Was the enterprise a failure? Actually, some of Catherine's recommendations did become law: torture was abolished and religious toleration was sanctioned; The Roman Catholic Church was freed to coexist with the Greek Orthodox; Catherine opened the door to Jews in Russia, though they were forced to pay special taxes and required to live in special areas ("beyond the pale"); and religious dissenters were allowed to worship unimpeded—Volga Tatars were even allowed to rebuild their mosques. Then, despite an outcry from the clergy, she transformed all the domains of the church into state property; monasteries and nunneries were closed, and clergy were henceforth paid by the state. Ironically, this was perhaps the most controversial measure instituted in revolutionary France and proved to be central to its undoing. Not so with Catherine. Rather bravely, she further pressed ahead—again over the clergy's objections—with a plan to expand Russia's educational system and, in so doing, to lift up the illiterate masses. Appealing to her confidant Grimm, she asked, "Listen a moment, my philosophical friends: you would be charming, adorable, if you would have the charity to map out a plan for young people, from ABC to university . . . I, who have not studied and lived in Paris, have neither knowledge nor insight in the matter."

But this was false modesty. Bubbling over with plans, she founded a host of new schools for boys and girls—the empress was the first Russian tsar to openly support the education of women. She was also the first leader in recent memory to take such an active interest in her subjects, often visiting the Smolny Institute for girls in St. Petersburg (her "proudest achievement"), where grateful students flung themselves into her arms. And distressed by the empire's lack of teachers, she encouraged young Russians to study education in England, Germany, Austria, and Italy—as it happened, one of those students was a young Alexander Radishchev, the energetic reformer whom Catherine would later turn on.

As her zeal for education became an obsession, the obsession became a policy, and the policy became a tentative fact on the ground—tentative because they were modest steps in a nation that was at once so vast and backward. By 1786, she had founded the first teachers' college. Admiring Emperor Joseph II's reforms—Austria had one of the best school systems in Europe—she wrote to ask him for advice, as well as to "lend her" someone who could help her do the same. He sent her Jankovich de Mirjevo; together they laid the foundations for "popular schools" to be established in districts across the country—elementary schools in all the chief towns, high schools in all the principal cities. They were available to all children regardless of class, and the state provided teachers as well as textbooks. At the same time, she focused on "the dark void," improving the quality of health care in Russia. Appalled by the nation's rampant mortality rate, the empress opened new hospitals in Moscow and elsewhere, staffing them with foreign physicians, plus a foundling asylum, and even a "Secret Hospital" for those with venereal diseases. She further decreed that every county have at least one doctor and one surgeon. Remarkably, at the end of her reign, while infant mortality was 32 percent in London, it was just 18 percent in St. Petersburg.

And in a time when smallpox was a deadly killer—in 1774 Louis XV had died of smallpox in France—Catherine introduced an inoculation for smallpox into Russia; that was in 1768. To calm fears about the inoculation, at potentially great risk to herself, the forty-year-old tsarina served as the second Russian subject for the treatment. The empress was soon reporting that more people were inoculated in one month in Petersburg than in Vienna in a year.

Between statesmanship and governance, impatience and imperiousness, this astonishing woman was equally at home in the realm of ideas and the world of philosophy. Yet if Catherine's intellectual powers were formidable and if the legends that grew about her were equally impressive, in many respects, it was Catherine who authored her own tale—and one that she acted out with great fervor.

∞

ALL HER LIFE she was a shameless self-promoter. But she had help. Catherine the autocrat ardently befriended the philosophes as did no other ruler in Europe. She protected Voltaire when he took refuge in Fernay, she was a friend of d'Alembert's when he was deprived of his pension, and she supported Diderot when he was refused admission to the Academy or was bankrupt. She translated Marmontel's *Belisarius* when it was condemned in the Sorbonne, and subscribed to the *Encyclopedie* when its publication was banned in France. Catherine shrewdly recognized that while these were among the most forward-looking thinkers of the day, they were also the makers of public opinion—and first among them was Voltaire. Long before her ascension, Catherine had relished the genius and irreverence of Voltaire and fantasized about becoming the "enlightened despot" of his ideals. Upon becoming empress, she wasted no time opening a correspondence with him.

Theirs was an extraordinary relationship: He was seventy, and she was thirty-five; he reigned as one of the globe's great philosophers, and she was a novice monarch; he was a galvanizing voice of the Enlightenment, and she came to power in a coup d'etat, by the murder of her husband. But he was sarcastic and witty, and so could she be. He was a passionate idealist and an enemy of radical reform, and so was she. He demanded freedom of conscience and speech, and, in the beginning, she did as well. And he was a recognized genius and a freethinker, while she enthused that his works alone would produce "citizens," "geniuses," "heroes," and "authors"; she also called him "her master," "her teacher," and built the Hermitage gallery in case he ever came to visit her. He never did, but that was mere detail: In turn, Voltaire, the preacher of tolerance, elevated her above Solon and Lycurgus. He exalted her soul and her mind ("my Catherine"), admired her wise laws and brilliant ceremonies ("she reigns with wisdom and glory"), and even praised her little hands and feet. As he once wrote to Diderot: "What do you say of the Russian Empress?...What a time we live in! France persecutes philosophers while the Scythians protect them."

This was an old story. Throughout the centuries statesmen have sought to exploit intellectuals to their benefit: In the twentieth century John F. Kennedy did this brilliantly and so did Franklin Delano Roosevelt; Louis XVI did not, and, as Napoleon pointed out, he paid the price for it with his head. But this was a lesson that the empress had grasped intuitively; even as Catherine opposed French influence in Poland and was defying it in Turkey, Voltaire quickly became her press agent in France. In 1767 he labeled her the "Goddess of the North," and when she went to war with Turkey, he applauded her crusade to save Christians from the horrible Mohammedans. But if Voltaire was taken by Catherine, he wasn't the only one: Denis Diderot, the soon-to-be author of the widely read, groundbreaking *Encyclopedie*, was equally fascinated by her—and for good reason. When Catherine heard that he was planning to sell his library—he wanted to raise a dowry for his daughter—she instructed her Paris agent to buy it for whatever Diderot was asking; actually, she paid him more. Then she requested that Diderot keep his books as long as he was alive ("it would be cruel," she mused, "to separate a scholar from his books"), and to be their custodian at a salary of a thousand livres per year; moreover, she prepaid his salary for a stunning fifty years in advance. With that single gesture, the philosopher became a wealthy man and an instant devotee of Catherine's ("Great Princess, I prostrate myself at your feet; I reach out my arms to you . . . I would speak to you but my soul faints, my mind grows cloudy, my thoughts have become confused"). And when she invited him to visit her, he felt compelled to go. "Once in a lifetime," he enthused, "one must see such a woman."

With great excitement but also reluctance—he was sixty, in frail health, and the journey was long and difficult—in 1773 he set out for St. Petersburg. Twice on the trip he became violently ill, but coughing and spitting, he reached the capital on October 9; the tsarina gave him an audience the very next day—in her study. "Nobody knows better than she," he reported, "the art of putting everyone at ease." She told him to speak candidly, "as man to man." He did, and with such vigor that he slapped her imperial thighs for emphasis ("Your Diderot," the empress laughingly wrote to another French friend, "is an extraordinary man. I emerge from interviews with him with my thighs bruised"). He sought to play the peacemaker and outlined how Russia could be transformed into a utopia, asking her countless questions about the national economy, the fate of "slaves" and the fate of the peasants, the situation of the Jews and the happiness of her "people," the matter of land ownership, land cultivation, and the Russian alliance with Austria and Prussia. He also made suggestions for remaking the nation's laws. Of this, her comment was immortal: "You work only on paper, which endures all things; I, poor empress, work on human skin, which is incomparably more irritable and ticklish." In truth, she later dismissed his ideas as "unadulterated nonsense" and maintained that if she had listened to him, "everything would have been turned upside down in my empire." Nonetheless, she sought out his company and his wit nearly every day, and soon, their "little hour" became a chat that spanned a full five months.

When Diderot was ready to return home Catherine ordered a special carriage for him, more than paid for the expense of his return trip, gave him a costly fur, and detailed an officer to accompany the aging lion to The Hague. On his return to Paris, he lauded her worshipfully. So was this merely an exercise in royal cynicism? Or did she genuinely cherish her relations with the philosophes? Clearly, it was a bit of both. Equally clearly, her money—and her time—had been invested well. The arbiters of opinion, who not so long ago looked upon Russia as a wayward country buried under snow and roamed by wolves, now saw the progressive light in that distant land and the tsarina as an enlightened thinker. Voltaire beat this drum constantly, and so did Diderot. For her part, Catherine knew that she could count on her worshippers, their trumpets at their lips, to support her; thus, when she went to war with Turkey, Voltaire called the sultan "the big pig of the crescent." And when she began her rape of Poland, the philosophes dismissed the Polish revolt as "an Italian farce." Almost overnight the tsarina became the protectress of letters and arts, the "benefactress of Europe" (Voltaire's phrase) and the "hope of Europe" (Grimm's phrase). She was a lay Madonna who dispensed rubles for the sum of knowledge, and the empress who knew no frontiers and recognized only talent.

That was on the surface at least; in truth, she was more discriminating than that—or perhaps calculating is more precise. Catherine never approached Rousseau, viewing him as antithetical to her ideals, and had long since abandoned Locke. But she did cultivate Frederick Melchior Grimm, who became her agent in Paris and whose *Correspondence littéraire* also reached most influential salons throughout Europe. He was younger than Voltaire, and she found him more sensible than Diderot. After meeting with her, Grimm, like the others, was bewitched. "The nourishment of my soul," he enthused, "the consolation of my heart, the pride of my mind . . ." There was one crucial difference. Whereas she became disillusioned with Diderot, and later turned on Voltaire, her friendship with Grimm lasted untroubled until her end.

What were the fruits of this love affair between autocracy and philosophy? Intellectuals and power invariably make for a strange and volatile blend. Catherine was vain, not to mention visibly conscious of her accomplishments and her influence. "Vanity is her idol," Joseph II once scoffed. "Luck and exaggerated compliments have spoiled her." For his part, Frederick the Great thought that if Catherine were corresponding with God she would "claim at least equal rank." But the philosophes were a different matter. In hindsight, we see that mentally Catherine rivaled her favorites. Her musings were every bit as rich and subtle as Voltaire's and Diderot's, while for his part, Grimm wrote, "Had it been in my power to take down [Catherine's] conversations literally, the whole world would have possessed a precious and unique fragment in the history of the human mind." She indulged their appetites but rarely allowed them to determine her policy.

Still, her flirtation with French thought drew Russia's elites ever closer to Europe's orbit and diminished Russia's image as a hulking brute. Like the empress herself, countless Russians embraced French etiquette, culture, and art, and many even flocked to Paris—including her son, Paul. Unwittingly, they would carry back sentiments that eventually paved the way for the revolution in the twentieth century.

But above all, it enriched her in the councils of Europe. Rarely did a monarch offer two such different faces at home and abroad. Catherine was an absolute despot in Russia but a budding democrat in France. Her courtiers were at once nobles who worshipped autocracy and at the same time reformers enchanted with liberty. And she played this impostor's game shamelessly, in one breath, displaying her generosity, in another her mistrust, in one moment, her zeal for Western order, in another, her love of the brooding Russian soul.

∞

BETWEEN COURT INSURRECTIONS, critical diplomacy, and map-remaking wars, Catherine continued to cut her own path; she was a woman who thought hard and worked even harder. It almost defies the imagination that this "arbiter of two continents" also managed to write poems, chronicles, memoirs, plays, opera librettos, magazine articles, fairy tales, a scientific treatise on Siberia, a history of the Roman emperors, and extensive *Notes on Russian History*. For a "universal" dictionary of comparative languages she was compiling—including Finnish, Ancient Slav, Mari, and Votic— she personally corresponded with Thomas Jefferson in 1786, asking for a translation of 300 words in the Shawnee and Delaware Indian vocabularies; with assistance from George Washington, Jefferson added a list from the Cherokee and Chocktaw languages as well. She explored all of them with a defiant gusto. At the same time, with Diderot and Prince Galitzin as her purchasing agents, she amassed an art collection in the Hermitage that would become one of the principal museums in the world: eleven hundred pieces by such masters as Raphael, Van Dyck, Rembrandt, Watteau, and Chardin; an exquisite *Diana* bust from Houdon; a commission to Vernet; 60,000 crowns for the Braancamp collection, not to mention a small fortune for the gems of the Duc de Choiseul and the Duc d'Orléans; and an avalanche of treasures from the Italian, Dutch, and Flemish schools. "I am a glutton," she boasted, once outbuying even Frederick the Great. She also enticed leading foreign artists to come to Russia—touchingly among them was Marie Antoinette's celebrated court painter, Madame Vigée le Brun.

Under the tsarina, architecture also flourished. Resolving to leave her imprint upon St. Petersburg, she flatly avowed, "Great buildings declare the greatness of a reign no less eloquently than great actions." On this score, she left no doubt. Introducing neoclassicism into Russia, her mania became stone and marble ("the more one builds," she jested, "the more one wants to build; it is like drinking"). The Russian treasury may have buckled under the cost of her many palaces, but Catherine always achieved her objective; no wonder one well-traveled visitor wrote, "A Frenchman, after winding along the inhospitable shores of Prussia and traversing the wild and uncultivated plains of Livonia, is struck with astonishment and rapture at finding again, in the midst of a vast desert, a large and magnificent city in which society, amusements, arts and luxuries abound which he had supposed to exist nowhere but in Paris." And Prince de Ligne, after seeing nearly all of Europe's splendor, announced that St. Petersburg was "the finest city" in the world.

Under the tsarina, the sweat and tears of ten million peasants had been transformed into gleaming marble and polished stone.

∞

CATHERINE WAS FAR from flawless. Her achievements should not obscure her failures, or her demons. Yes, it could be said that there was a hurried confusion and instability in the torrent of her ideas; yes, she sometimes plunged too quickly into projects that she had perhaps not thought through, and she was sometimes defeated by the urgency of events or the multiplicity of her tasks. Yes, she could be pugnacious, even stubborn. And rightly so, she was frequently lampooned for her headiness in the capitals of Europe or criticized for her aggressiveness in America. Even so, the results of her efforts were often immense. In this, she differed profoundly from, say, James Madison, who was perhaps more nearly one-dimensional but far more focused; by the same token, the breadth of her ability and range of her interests are reminiscent of Thomas Jefferson, or just as likely, Winston Churchill. Still, there is no denying that after the experience of writing her *Nakaz* and the failed experience of her commission, and beyond the scare of the Pugachev Rebellion and the continuing war with Turkey, Russia settled back into the sleep of centuries. Tragically, while in the 1760s she dreamed of a "new race of men," by the 1780s her grandiose designs had largely been abandoned. Under her rule, 800,000 free peasants became serfs, and their status remained one of heartless, stupefying slavery. And tellingly, though Sir Robert Gunning, the English ambassador, once referred to her court as "a paradise of peace" and where she had begun her reign with pacific aims ("peace" she mused, "is better than the finest war in the world"), she was soon the most bellicose leader on the world stage.

Within a year of her accession, she had sought to make Russia the pivotal power on the continent, openly admitting to her ambassador in Warsaw, "My aim is . . . to become the arbiter of Europe." She manipulated the Seven Years' War in Frederick's behalf; she maneuvered the American War of Independence in the rebels' behalf; and she seized the Crimea and waged war against the Ottomans—twice. In later years she agitated against revolutionary France, and she partitioned Poland—again and again. Despite exhausted armies, untold deaths, and a depleted treasury, she arguably made herself the most powerful ruler in Europe. This was her creed, and in hindsight it was also her more sinister side. And so, by the time the Polish rebellion was in full swing against her in 1794, and Thaddeus Kosciuszko was offering a springtime of hope to a nation, if not the world, a gloomy tsarina was determined to obliterate Poland, piece by mangled piece.

∞

WITH THE TWO armies sprinting toward each other, the end for Poland came quickly. Catherine's murderous generals, fighting at a grueling pace, engaged Kosciuszko at Maciejowice. This time the Russians smothered the legendary rebel general—and it was their most successful attack of the war. The Poles fought and faltered and fought and faltered. Kosciuszko, after shrieking that Poland "was immortal," was himself seriously wounded and then taken prisoner. The heroic revolt was all but over. Suvorov rapidly united the various Russian armies and stormed the entrenched Polish camp at Praga on the outskirts of Warsaw, exhorting, "Work quickly! Glory, glory, glory!" His words were remarkably prophetic. It was Ismail redux: As the Russians poured into Praga, their cries of "Remember Warsaw!" filled the air, recalling the soldiers who were butchered when the city had first risen up against Catherine's army. Now, in reply, his battle-crazed troops brutally slaughtered some 20,000 men, women, and children. Leaving behind this carpet of dead, Suvorov sensed opportunity.

He surrounded Warsaw.

∞

HALFWAY ACROSS THE globe in the United States, the hearts of Americans belatedly went out to their Polish republican brethren. The dispatch in the *Philadelphia Aurora* denounced Catherine for her "plunder," "tyranny," and "caprices," and her "barbarous and implacable vengeance." It spoke movingly of the "wretched fate" and the "abyss of misery" to which Poland was being subjected by the tsarina's armies. And it talked darkly of Catherine's sworn goal of forcing the Poles into a fate "of slavery," and "extirpat[ing] even the name" of Poland.

But the Americans were powerless to do anything. And by this stage, it now seemed, so were the Poles themselves.

∞

THE HEARTRENDING REALITY settled over the rebellion.

The bloody reverse was swift and incalculable, and without their leader Kosciuszko, the Poles were close to total despair. In Warsaw itself, great crowds stood on the banks of the river and watched in dead silence as their countrymen were massacred. Then after Suvorov took Praga, Warsaw plunged into tumult. With the Russians closing in on the Vistula, the city was consumed by fear and rumor. Now panic reached the courtyard of the British ministry as it was flooded with thousands of refugees, hauling their money, their furniture, and their bundled clothes. From a distance, it was unclear who was controlling the city—the king? the council? the rebels?—but Suvorov took numerous steps to induce the surrender of the capital. Stanislas blustered that Warsaw would fight to the end if the safety of the people were in doubt. Suvorov smartly sought to reassure him: Hoping to induce a bloodless, peaceful surrender, on October 25 he presented his terms for capitulation—arms to be piled outside the confines of the town. All soldiers to be allowed to return to their homes. The king to be accorded all honors due him and the possessions of inhabitants to be protected. The terms were extremely generous. At the same time, Suvorov told St. Petersburg, "My business is to destroy ... [the] rebels."

In disarray, the rebels got the message: The uprising collapsed and King Stanislas surrendered Warsaw to avoid an even greater massacre. In a plain officer's hat, Suvorov rode in to receive the keys of the city; the piles of corpses from Praga now only embellished his reputation for savagery: "Europe says that I am a monster," Suvorov bluntly announced. "But . . . is it not better to finish a war with the death of 7,000 [civilians] rather than to drag it on and kill 100,000?" For her part, Catherine was untroubled. Her melancholy spirits soared; jubilant, she wrote back to her heroic general: "Field Marshal [Suvorov]. I congratulate you for all your victories," and she showered him with precious gems. Meanwhile, Suvorov dispatched Kosciuszko and other rebel leaders to St. Petersburg, and sent the king to Grodno to await his uncertain fate. Thousands of Poles fled from their country as the tsarina pondered what to do next. She didn't ponder long. On November 25, 1794, Stanislas signed his abdication. He cried out for Catherine to let some part of Poland survive. It was a stunning reversal of history: For two centuries a strong Poland and weak Russia had warred— Polish generals had once penetrated deep into Russia; Polish troops had once occupied the Kremlin; a Polish tsar had once even been placed on the Russian throne. To all this Catherine muttered acidly: Never again. She decreed that Warsaw was to be punished, "for its evil."

Then she coolly wiped Poland off the face of the map.

As a young girl, Catherine was known for having a burning gaze in her eyes, much like that of a wild animal. Decades later that was certainly still evident. When a grim-faced Kosciuszko was brought as a captive to St. Petersburg, the empress refused even to see the hero of the American Revolution and champion of the Polish cause. She declared with cruel arrogance that he had been "recognized as a fool in the full sense of the word," and she made a great show of deliberately mocking his name: "my poor beast of a Kotushka." By now, she was at least telling the truth. First displayed like a zoo creature in a cage, he was then sent to languish in confinement in the Peter-Paul fortress, where he stayed until after Catherine's death, at which time he was given permission to emigrate to America. Ex-king Stanislas was treated no better: Past fifty, a broken, bitter man, like Louis XVI, he was held under constant guard in Grodno, where he was left to dodge Russian sentinels, play at botany, and write his memoirs.

Then the taking began. After another year of dispute, Russia, Prussia, and Austria signed the third Polish Treaty of Partition: Catherine seized Kurland, Lithuania, and western Podolia and Volhynia—181,000 square miles; Austria and Prussia took the rest. After three partitions, Russia had absorbed some six million of Poland's twelve million souls. In the chronicle of modern times it was a chilling sight.

Across the ocean in a fledgling America, the *Philadelphia Aurora* referred to this as "a new language unknown to the laws of nations." It was right. For with this final act by Catherine, the ancient Kingdom of Poland ceased to exist.

❦

BACK IN ST. Petersburg, there was still more work to be done. As the tsarina aged, she strangely seemed to be at war with herself. But how far she had come in her lifetime. True, she was prudish: No off-color or risqué jokes were allowed in her presence. And her self-deprecation was real. ("I have acted like the jackdaw in the fable," she jested, "who dressed himself in peacock's feathers.") So was her simplicity: She still played parlor games with intimates and teased her friends; she greeted visitors in her nightgown and held audiences outdoors under a tree; and she wore her hair unpowdered and playfully made up words. Humor always meant a great deal to her; so did laughter: She even insisted that in her court "one must be cheerful." And as she once laughed, "given the demands of my trade, I am obliged to let my hand be kissed; but I think it would hardly do for me to perfume everyone around me with my tobacco"—for she liked snuff. To many who watched her, she was as amiable as Charles II of England and Henry IV of France. And yet.

In the closing years of her reign she insisted that she was "merry" and "light as a lark," but this was often belied by the facts. She increasingly indulged in fits of rage unbefitting the omnipotence of a tsarina; growing ashamed of these outbursts, she sought to tame her volcanic moods. At times she became mawkish—or even a bit self-pitying. Other times there seemed to be a barren quality to her—or an uncharacteristically cold streak. Some of this was due to the strain of continuing crises and the unceasing wars; and some was the loss of her sheet anchor: Prince Potemkin. This blow was the one most unexpected—and devastating. In April 1791 Potemkin had thrown the empress one of the grandest fetes that Europe had ever seen: Three thousand guests bowed at Catherine's arrival, thousands of candles reflected in the gold and table service, three hundred musicians played glorious melodies, ballets were performed and poems were read, all the while fountains murmured and precious stones sparkled in the foliage. When it was over, the nobles and guests raised their glasses to toast the glory of the empress, and foreign ministers shook their heads at this incomparable Russian event, an Oriental spectacle of inexpressible riches and folly verging on monstrosity. Potemkin knelt before Catherine, kissed her hand, and burst into sobs, then took his leave to the South, to conclude the Peace of Jassy with the Turks.

But five months later, in October, a courier in black rode at breakneck speed from the Crimean steppes to St. Petersburg bearing the news: Potemkin had died, expiring by the side of the road. The empress immediately felt the enormity of the loss of the man who was her lover, husband, friend, advisor, pupil as well as teacher, military leader, and "almost my idol." She fainted—then was bled three times. Sobbing, she shut herself up in her room and refused to see anyone. "Nothing will ever be the same again," she wept. "A terrible crushing blow has fallen upon my head." Arguably, Potemkin was a pillar for her leadership; and also arguably, she was never the same again.

In the days and years that followed, some vestiges of her early liberalism survived, but only barely: At first she retained the Swiss-born tutor for her grandsons, Frédéric-César de Laharpe, a dedicated republican. In the early days of the French Revolution, informed of his lessons—the talk of social justice, the progressive pleas for equal rights, the love of liberty—she summoned him. "Monsieur," she gently said, "be a Jacobin, a republican, or what you please; stay with my grandchildren, retain my compete confidence, and instruct them with your wonted zeal." With an almost mystical faith, she was confident that young Alexander would one day demonstrate the will of an enlightened monarch in leading a great empire.

But in 1793, after Louis XVI's execution, that changed too. She was shocked to find that her grandson detested despotism, hated violence, and was delighted with the French Revolution; rather than a future emperor, she discovered, to her horror, a sniveling little rebel ("everything I am," Alexander wrote, "I owe to a Swiss."). This time she dismissed Laharpe; with tears in his eyes Laharpe informed his pupil, who in turn began to sob. Yet as was always the case when Catherine acted, she acted decisively; the deed was irreversible. But Laharpe's firing was more than a matter of principle or pique. The more she approached the end of her life, the more she thought about the imperial succession—who would inhabit the throne when she was gone. The heir apparent was her oldest son Paul, whom she despised. The reasons were many: In good measure because the conspiracies to unseat her sometimes proposed to make him emperor with a regency; in equal measure because she thought him weak (his head, observed one aide, "is filled with phantoms"), insufficiently Russian—he loved all things Prussian like his father, Peter III—and finally, because he was mad, not to mention prone to needless cruelties. But Paul's charming sons, Alexander and Constantine, were a different matter. To Alexander especially would lie the destiny of Russia.

So Catherine instead personally attended to their education; she worked day and night to alienate them from their father; and she schemed to have Alexander, and not Paul, inherit her crown. Over time, she isolated her son and assiduously exacerbated his inner demons, even as the grandchildren came to worship her. And while she trained Alexander as a kind of ideal ruler, she also manipulated him, telling Alexander that if he, and not his father, inherited the crown, he would be able to reign employing the republican precepts of Laharpe.

Then secretly—and this would be her undoing—the tsarina prepared a manifesto instituting a new order of succession, to be made public on St. Catherine's Day, November 24, 1796. She locked the documents among her private papers until the time when she would proclaim fifteen-year-old Alexander emperor of all the Russias.

∞

CATHERINE, NOW IN her sixty-seventh year, was ailing more than ever; Rostopchin wrote: "Her health is poor . . . she no longer goes out." Still, she remained as involved in the affairs of state as ever: She continued to follow European events very closely; she rejoiced when she heard news of a retreat by French troops across the Rhine; and she dreamed of seeing revolutionary France, the regicide, crushed. And to expand her influence further, she sought to marry off her young granddaughter Alexandra to the new teenaged king of Sweden. But here, history had curiously come full circle: The betrothal fell apart literally at the altar because he would not convert from Lutheranism to Russian Orthodoxy. On the wedding day, the young king brazenly failed to appear.

In all her years as tsarina nothing like this had happened—and the effect was swift and deafening. The circumstances betrayed even Catherine.

Hearing the news—she was seated on her throne and the orchestra had been playing—the empress was crushed and humiliated. For the first time that anyone could remember, she was a thoroughly shaken woman. She abruptly left the hall, leaning on her grandson Alexander. Her legs were barely able to move, her expression was drawn, and her breath wheezed in her chest. Behind her, the young fiancée fainted and was carried away.

The mortified crowd shuffled off, murmuring.

∞

THAT NIGHT, CATHERINE had a dizzy spell, which was likely a stroke; she recovered the next day. But early one evening she strolled outside Tsarskoye Selo, where she glimpsed a shooting star; in the scented dusk she quietly sighed, "That is a portent of my death." A friend recalled that her majesty always refused to believe in signs. "Yes," Catherine answered poignantly, "in the old days." As they moved on she confided that she felt herself "sinking visibly." Yet in the weeks that followed, she struggled to rally. She told aides that she planned to live to eighty. Exuding the swagger and dash of old, she seemed intent on resuscitating the maxim that she had once written to Potemkin: "I have returned to life like a fly that the cold had benumbed."

Indeed, on the morning of November 6, 1796, Catherine seemed especially buoyant; she was looking ahead to two weeks' time, when she planned to announce the new tsar, her grandson Alexander. The empress talked warmly with her personal maid, vigorously asserting that she had spent an excellent night. She rubbed her face with a piece of ice, finished a cup of hot black coffee, and summoned her latest love, Plato Zubov, to update her on current affairs. Then she rang her bell to call her secretaries, poring over her work as if she were twenty years younger. Later that morning, she retired to her water closet. As time passed and she did not reappear, her female attendants knocked at the door. Receiving no answer, they entered. Catherine was neither in her bedroom nor in her dressing room, but the subsequent sight chilled them to the core: Moments later they found her half collapsed on the carpet against the commode. Her eyes were closed and there was foam dribbling from her mouth, along with a slight rasp in her throat. She had suffered an apoplectic seizure—a rupture in the brain.

They laid the tsarina on her great canopied bed. Zubov rushed to her side. The doctors did everything they could: They bled her, twice. They placed mustard plasters on her feet. They resorted to prayer, anything to keep her body alive. The court physician, an Englishman, considered the empress lost, but insisted on trying "all means" to save her.

Beyond the castle doors, wind howled and snow was falling. For a moment the tsarina seemed to recover consciousness, but couldn't speak. Plato Zubov began to sob. Catherine was breathing, but only with great effort and noisily. The ghastly rattle in her throat increased. The vice chancellor was now there, and outside the royal chamber, a grieving throng of courtiers gathered. The doctors, suddenly powerless to do anything, receded into the background, and the Orthodox priests entered to say prayers for the dying; soon, the archbishop of Russia would be there as well. A trembling fifteen-year-old Alexander was also brought in; his expression was weathered with despair.

Word of the slow, dreary truth spread—the tsarina was dying. Meanwhile, not far from the palace, Catherine's son, Paul, and his wife were dining with friends. They were recounting an eerie dream that they had both had the night before: A powerful hand was drawing them, ineluctably, up to the heavens. Paul had barely finished his tale when a messenger from Petersburg arrived, carrying the news of the empress's condition. Paul immediately ordered a sleigh and raced off toward the capital. He had no way of knowing what lay at the end of the ride. A knife in the back? Poison? The Fortress? Difficult questions arose. But this much he knew: It would be his son—or him. Another aide met Paul to reassure him; the palace was rallying behind the grand duke. After making speed, Paul suddenly ordered the sleigh to halt at the top of a hill. St. Petersburg was visible in the distance, and the snowy landscape was shimmering in the unreal light of the moon. Paul got out and stared toward the sky; tears were running down his cheeks. An aide exclaimed, "What a moment this is for you my lord!" Paul, in ecstasy, answered back: "I have lived forty-two years. God has sustained me. Perhaps he will give me strength to . . . bear my appointed destiny."

And inside the palace itself a familiar scene in Russian history was dramatically playing out: The Winter Palace was frenetic with plots—courtiers pondered fresh alliances, nobles dreaded old adversaries, cliques of army officers weighed where to put their allegiance, and everyone calculated favor or banishment. The air of foreboding was felt by all. Catherine had come to power by a coup and her husband was dead shortly thereafter, while the feeble Ivan VI, a legitimate heir to the throne and a rival to Catherine, was murdered. And Peter the Great had his own son fatally tortured.

As Catherine struggled against death, a spectral figure still clinging tenaciously to the empire, aides anxiously waited for the end with a combination of fear and mourning. Unknown to most of them, in the empress's desk were packets of papers tied up with string, emblazoned with the imperial seal. One envelope bore a black ribbon and was inscribed, "To be opened after my death, in the Council."

This was the empress's manifesto of succession.

∞

IT WAS AT half-past eight that Paul made his appearance in the palace; as he strode in, all the assembled dignitaries knelt to the ground.

Dawn broke and the morning hours arrived, and the empress was still hanging on. Her chest was rising and falling, but her ashen face lay contorted, and with each breath her half-open mouth trembled. A fire burned, but the room was bathed in shadows and semidarkness. Then, with a sickening feeling of emptiness, her retinue watched as her breath became fainter and weaker. Just as quickly, at ten-fifteen the room fell silent. For decades, with a half smile or a shared glance, Catherine could move an empire. Now, one aide spontaneously recorded, "Catherine the Great, having uttered a last sigh, like all mortals, appeared before the judgment seat of God."

By then, Paul had moved quickly, preparing to come to power as much by treachery as by design. He had already emptied the empress's desk and thrown Catherine's will into the flames.

Abroad, the tsarina's enemies would feel that she had earned a far more painful death. To them she was a ruthless, violent despot. In a world of change she had professed liberality and tolerance, yet clung tenaciously to absolutist rule; she had muzzled all opposition, yet after much fanfare, utterly failed to reform Russian law; she had condemned the Terror, yet sanctioned one of the bloodiest days of the decade: Suvorov's butchery in Praga. And she had cynically expanded serfdom. For that matter, nor could her detractors at home forget the families bankrupted by high taxes or brokenhearted at the loss of tens of thousands of sons in her many merciless wars. And then there was the matter of her changes of mind on some of the most profound matters of the day, changes made with a blitheness concealed by her conspicuous gravitas: changes on republicanism (good, bad), changes on reform (good, bad), changes on an open society (good, bad). But across the vastness of Russia, the people as a whole applauded her:

Rivaling Emperor Augustus, she had expanded Russia to wider and safer boundaries; she had added 200,000 square miles to Russia's area, built new ports for trade, and raised the population from 19 million to 36 million people. And by the courage of her convictions and the force of her personality, she had drawn the educated classes of the nation out of the Middle Ages and into the path of the modern world, in literature and philosophy, in science and art. In watching Catherine evolve over the years, one leaves with the sense that she felt she had somehow wandered onto this earth, as the tsarina, in the midst of three decades of war and tumult, for the greatness of Russia. As time passed, Catherine gradually calcified into an icon: It was hard to imagine anyone who could ever surpass her role as Catherine the Great.

Perhaps she didn't deserve all the laurels that came with her high station, but she was, in fact, an epitome of her age. An English historian would one day write that Catherine was "the only woman ruler who has surpassed England's Elizabeth in ability, and equaled her in the enduring significance of her world." And a French historian would compare her to Louis XIV—the Sun King. "The French formed the glory of Louis," he would write, "Catherine formed that of the Russians. She had not, like him, the advantage of reigning over a polished people; nor was she surrounded from infancy by great and accomplished characters."

But in the inner chambers of the Russian court, this suddenly seemed moot. Was it the fading of an era? In the throes of uncertainty and simmering intrigue, there was little mourning; instead, the procurator general boldly announced, "[Catherine's] son, the Emperor Paul, has ascended the throne." And a grim ritual so common to Russia was hastily set into motion. Already columns of Paul's troops were entering the capital in their Prussian-style uniforms. Within the hour, Catherine's body was dressed in white. And by noontime in the chapel, the tsarina's son was to be crowned. In the palace's lengthening shadows, there was a new empress, who kissed the cross and the Bible, then kissed Paul on the mouth and the eyes. There were the imperial children, who bowed. There were the high dignitaries of the court, the church metropolitan, and the clergy, who brushed Paul's right hand with their lips and swore allegiance. And of course, there was the new tsar.

When it was over, Paul tapped his foot and withdrew to his study. The lord of Russia who came to power against Catherine's design had but one overarching goal: to cleanse a past that had been forged by his mother and to pick up a future that had lain dormant since 1762—when his father had been assassinated.

∞

IT WAS, NO doubt, an appropriate symbol for a contentious, bloodstained era. But fresh questions loomed: Was history hurtling forward or backward? Would republicanism last, or would it be strangled by its enemies, abroad and at home? Or paradoxically, by its well-meaning friends as well? In 1796 Tsar Paul would be no less able to escape the convulsions of the globe than the great monarchies of Europe or even the new republic in the United States. The world was, as it had been for the decade, rocked as much by the dread fruits of division as by peace, by war as by prosperity, by fear as by hope. In this immense verge of global history, a great chain of events had been unalterably triggered: France had promised a republic but bequeathed the world an aggressive quasi-dictatorial Directory, which came to power by prodigious bloodshed and would make its neighbors quiver. A pope would be toppled and killed; old alliances would be reassessed and old enmities put aside; and even the Muslim and the Orthodox Christian empires would soon unite. And astride these bloody crossroads of history—the phrase is not too strong—the infant United States would, despite its best wishes, be sucked into the rest of the world's mighty orbit.

Though largely ignored by the great European empires, this rebel nation was perhaps the most interesting case. The still-struggling America was the one country seeking to evade the cruel edicts roiling the globe, from the Caribbean to Cairo, from Paris to Constantinople. But what would be preserved of the Founders' experiment in Philadelphia and what would be lost now seemed to depend upon a president under constant assault: George Washington, and on his closest advisors. And paradoxically, by those at home and abroad who would happily see his government fall.

PART IV

A WORLD TRANSFORMED

Napoleon crossing the Alps

CHAPTER 11

AMERICA

✦

IN 1792, THE MOST striking feature of the Western world, old as well as new, was the hideous death rattle of the war engulfing Europe. The armies of revolutionary France squatted along opposite fronts of the coalition arrayed against them, marching through sleet or mist, or crouching behind trees in the wilds of the Argonne Forest, or encamping amid the red clay of Verdun. In their ardor, or more accurately their zeal, the Parisian revolutionaries swelled with the belief that their forces would soon be teaching French—or at least the vocabulary of republicanism—to Prussian and English children, even as the monarchical coalition furiously counterattacked. Newspapers in Paris glibly talked of great victories to come, of "*citoyen*" and "*aux armes!*," but along the lines, the men knew better—or at least thought they did. The allied armies were better trained, better equipped, and every battle only found the Duke of Brunswick's men stronger. For a thousand years, republics had come and gone while the armies of kings had fought with chivalry and heroism and victory; why should this be any different?

It was, to be sure, a harsh life for these renegade republicans: The officers of the ancien régime, long accustomed to a sheltered upbringing, lived comfortably, but not so the rest of the men. Staggering toward the enemy, they slogged through mud and thin, cold rain, endured constant filth, lived in stinking clothing, and, bellowing the "Marseillaise," raced over jagged soil toward a bestial foe that had thrashed them time and time again. Rations were tight and growing tighter. So were weapons. Morale was low, and the men were exhausted—or scared. They had done everything they could: All able-bodied men were called up, and notices were sent far and wide to the newly created provinces, not to mention to the crippled, the blind, the mad, and in some cases, even the dead—but it wasn't enough. So as the French Assembly repeatedly congratulated itself on the glories to come, it seemed to be little more than a fantastic chimera. Diplomatically the republicans were isolated, and when they were honest with themselves, they knew that militarily the year had been a freak of disaster. The gray tide of the Prussians and Austrians had swarmed across what is now Belgium, recoiled astride Reims and Arnas on the way to the very gates of Paris, while English ships lapped at the breakwaters of Marseilles and Toulon. From afar, Catherine the Great further incited the continent's monarchs, urging them to bring the Assembly to its knees. So if there was a ghost of hope for the French, few among them knew what guise it would come in.

But then, before winter sealed up the passes and the roadways froze, came a stunning turning point: the December victory of the French forces at Valmy.

∞

In America, the country was almost universally gleeful at the triumph of their sister republic. Parades were held, fireworks were lit, glasses were raised in toasts, and thousands of candles burned in windows. This was ironic: ironic because the infant United States was stalked by the phantoms of encroaching predators abroad—and it was not clear that the French were any different. Ironic too because, as John Adams once roared in 1776, "Huzzah for the new world and farewell to the old one!"—yet Americans from Charleston to Boston to Philadelphia saw the two revolutions as one piece. And ironic because George Washington's own government would soon succumb to blows over the great contest three thousand miles away: Are you with the rebels? Or against them? Suddenly there arose the specter of a fragile America haplessly stumbling into catastrophe—or into rampant dissension at home.

For George Washington, it would be among his greatest challenges.

∞

DURING THE COURSE of the last fifteen years, Washington had hammered together an army in his own likeness, persevered in a titanic struggle against Britain, and led a disparate confederacy to victory; now he had to help hold together a fledgling nation. Already he had weathered daunting challenges: the capital (where to place it), the debt (what to do about it), the need for an economic plan (how should it look), the establishment of a national bank (despite its many detractors), and the tragic necessity to delay the debate over slavery. He had created a government where none had previously existed, recruiting the best talent the nation had to offer—Hamilton, Jefferson, Knox, and Jay. And by his careful management, he had skillfully laid the footings for one genuinely national government, rather than thirteen. And he learned from his mistakes. On the surface then, he seemed incapable of failure.

But he was now, as he had always been, a massive riddle of a man. Of course, he was endowed with enormous personal gravitas, an iron disposition, and impeccable judgment; this was never in doubt. And unquestionably he was among the most gifted men of arms that the nation had produced; he was also exceedingly valorous, seeming to court death on battlefields and repeatedly exposing himself to enemy fire; his men marveled at how little concern he had shown for his own person. And he was uncommonly brave not just on the battlefield, but personally: The old yarn is often told about his uncomfortable dentures, but this is actually misleading. In truth, the real story is that he began losing his teeth while in his twenties, and for decades thereafter was in tremendous oral pain; little commented on is the awful procedure by which he had his teeth removed, a grueling process where the teeth were hit with a hammer and banged out, loosening chunks of gum and jawbone in the process. The pain made him irritable and may even have had something to do with his legendary temper, not to mention his frequent silences. As the years progressed, it was often uncomfortable for him to speak and he didn't want to open his mouth unless necessary; when he took his dentures out, he usually stuffed his mouth with cotton or wool. Yet Washington overcame this as he overcame every other obstacle in his life. And here was the secret to his mystery as well as his allure—everything about him seemed perfect: Tall, disciplined, virtuous but firm, composed in crises, with an extraordinary wellspring of physical and mental energy, he was the very picture of leadership.

But if he was regarded as the best of men—he was—to his detractors, he could also be the most disappointing: Behind his forbidding Roman front and exacting bravura, he was impatient and imperious. One moment he was tolerant, as a good republican should be, but he had a cold streak in him and, if pressed, could actually be spiteful. While he subscribed to the notion of a higher power, to some of his critics he seemed to worship only at the altar of himself and the nation—the two being inextricably bound.

One of his eventual difficulties, of course, was that his enormous strengths could also be his failings. David Meade noted that he was "better endowed by nature and habit for an eastern monarch than a republican." Marie Antoinette's lover, Count Axel Fersen, observed, "He looks the hero; speaks little, but . . . a shade of sadness overshadows his countenance, which is not becoming." To some, he appeared almost too remote, so far above everyone else, even his intimates, that he was unapproachable. His good friend Gouverneur Morris once slapped him on the back, only to receive a cold stare. This was the Washington most people saw: He was correct in manner, correct in speech, correct in dress. Cool and brusque, he was also austere, saturnine, grave, proud, ruthlessly ambitious, and (albeit quietly), surely egotistical; these traits in themselves were hardly problems, not in a man of his immense gifts. But if there was a flaw, it was that he saw himself as beyond reproach—suitable for a general or an autocrat, but scarcely becoming in a republican or democrat. He was also more than a touch suspicious of popular opinion, leading Jefferson to complain, "His mind has been so long used to unlimited applause that it could not brook contradiction . . ." Revealingly, Washington would sometimes speak of himself in the third person.

Though they tried, his critics could never quite get a handle on him, no better perhaps than his admirers could. Behind his facade was an inner identity that informed it. He had a thespian's gift for reading the pulse of an audience: Foreign dignitaries felt it; foreign leaders sensed it; his men lived by it; and his nation responded to it. His ornate dress (with a perfectly tied cravat), his extraordinary white stallion, Prescott (with perfectly polished hooves), his favorite greyhound (Cornwallis), and his magnificent chef who made his meals (Black Sam Frances) were all props; so were his slight bow, his stately reserve, his daily walks about the capital, even his silence—he knew exactly how to use his features, his hands, his commanding voice, and especially his imposing frame. From the time he was a robust young man to his waning years when he aged, all his portraits are strikingly similar; it is a fact that Houdon made many busts and many statues but none seem so set in stone for the ages as Washington's. Like Alexander the Great, Genghis Khan, and Henry II—or Napoleon, FDR, and Churchill—like nearly all of history's storied men of destiny, he was always performing.

Was there something unsettling about so much calculated stagecraft? This is a more intriguing question. To be sure, it was an essential ingredient for Washington's leadership. But for all his professed humility ("I did not seek the office with which you have honored me"), for all his modesty ("[my] errors, however numerous"), for all his lack of conceit ("all the prime of his life—in serving his country . . . [who only wanted to] be suffered to pass quietly to the grave"), and for all his talk of retreating to the bucolic splendor of Mount Vernon, perhaps no other president relished the spotlight so much or enjoyed applause more. He would be horrified to hear this judgment, but, put simply, he was more than a touch vain. Like every other creature of vanity, he convinced himself—in this he was hardly alone among his countrymen—that his motives were indeed selfless. But as with so much else he did, here he was almost surely right, for that performance was always in the service of his country—and it was *his* country.

Very few people could ever say they really knew him, but they knew many things about him. He had a love of braid or ceremony—Senator William Maclay actually compared his affairs to "royalty"—yet like all great captains he also had an indefinable presence. He was the only man in America who could walk into a room full of banter and all would turn stone-sober in a matter of minutes; at his weekly levees, such was his aura that people were routinely left speechless. Hamilton could flash his infectious grin or wield his wit, Jefferson his unparalleled ability as a conversationalist; in this sense, both asked to be liked, and were. But by contrast, Washington demanded that he be revered, and was. One searches in vain to see if he had a diminutive; invariably, it appears that he was "General" or "President." He would have made an equally impressive king or tsar or emperor; when Benjamin Franklin died, he bequeathed his Crabtree walking stick to Washington, explaining, "If it were a scepter, he had merited it."

So was Washington a man tied to tradition or a modern man? This is where the mystical divide appears. For all his stilted manner, he possessed first-class brains and an even more impressive wisdom, peering ahead to the future, not to the past. Washington not only acted in cosmic terms, but thought in them as well. This is why he never once retreated from his goal of making America an independent republic, or refused to turn the presidency into a monarchy. This is why he sought so avidly to blend republican practices with royal formality to confer dignity upon his office. And this is why he was a trusted companion to the American people—patriarchal, even fatherly. In a profession—leadership—in which timing is everything, the Virginia-born Washington mastered perfectly the self-deprecating pauses and Southern modesty that went hand-in-hand with an instinct for when to act, when to move forward, and when to take a chance.

In the early years of his presidency he elevated himself above the fray as if he were a sovereign, though of course he wasn't. The effect was quite dramatic. He had the guile and fortitude, he meant to rule others (or more precisely, govern them), and he expected—and received—homage from them. In March 1790 the first medal minted in the United States bore his image; a month before, his birthday became a national holiday. If he failed to receive proper respect, he sulked. Interpreting the "advise and consent" power of the Senate as requiring his personal appearance, much as the prime minister often does in England today, he once personally brought to the body his proposals for a treaty with several Indian tribes. The senators bickered and dithered so long that an incredulous Washington stormed out, and never again returned ("this defeats every purpose," he snapped, "of my coming here"); as a result, the powers of the Senate were dramatically diminished for all time. Some could call this hubris, and did, which was the classic defect of doomed characters in ancient Greek drama; yet at the same time Aristotle prized loftiness as crucial to his vision of the ideal man. The Greeks would have admired Washington, and so did his countrymen; as it turned out, his hauteur was a tremendous asset when he made a great show of touring the New England states in the fall of 1789. Wherever he went, he was saluted, toasted, and cheered as if he were one of their own—and again in the spring of 1791 when he made a grueling tour of the Southern states, covering nearly two thousand miles, a trip that equaled Catherine the Great's spectacular expedition from St. Petersburg to the Crimea. With each stop, from Charleston to Camden to Guilford Court House, he forged link upon link of the nation's history with its future, thus serving as the embodiment of the nation itself. The people were as faithful to his own star as Washington was; a French officer once put it best: Washington had discovered "the art of making himself beloved." Or as one honorific toast concluded, he was the man "who united all hearts."

He was. Still, in a world where egalitarianism was fast becoming the triumphant passion of the time—the French Revolution was in full swing—Washington's Olympian pomp and ceremony began, in dribs and drabs, to evoke malicious laughter or outraged questions. "Alas," as Carlyle would one day observe, "the hero of old has had to cramp himself into strange places; the world knows not well at any time what to do with him." This overstates. Nonetheless, where Washington was once favorably compared with a "Roman hero" or a "Grecian God," one Southern editorial now blasted him as a demigod "perfumed by the incense of addresses," while another sourly wondered if he favored "too much of Monarchy to be used by Republicans." Even the secretary of the British legation characterized the tenor of Washington's administration as akin to that of a "very kingly style." This was not entirely Washington's fault: As biographer Joseph Ellis rightly reminds us, the very idea of a republican chief executive was a novelty; no vocabulary yet existed to characterize a chief magistrate of a republic. And the task of creating one was daunting. For even as Washington was seeking to consolidate Americans as common residents of one nation, under the staggering boomerang of the French Revolution, peoples the world over were simultaneously becoming more cosmopolitan: Some French were becoming less Gallic, some English less British, even some Ottoman Muslims under Selim III less Turkish. Nonetheless, the face of monarchy—the barb that he wanted to turn America into a version of Britain—would haunt his entire presidency.

And there remained the matter of what lay behind his phenomenal facade. Once in office, Washington still yearned for greatness and public adulation. No president received it as much as he did, yet, arguably, few presidents seemed to need it more. He was a careful reader of the newspapers and was acutely sensitive to what they said about him. Thin-skinned, he interpreted—often wrongly—opposition to his policies as simply opposition to him. In time, this shade of prickliness would become a self-fulfilling prophecy. And the time would also come, especially as the French Revolution galloped ever faster, when he would make his share of mistakes and become like a cat's paw for some of the country's greatest Americans—Jefferson, Madison, and Monroe for starters. They would one day doubt his sincerity, his motives, his ideals, even his mental capacity. Nor would they any longer be willing to bend to his imperious ways or genuflect to his whims.

But Washington recognized what few others did. Goethe thought Napoleon's mind to be the greatest that the world had ever produced; were he to speak of judgment, which is actually more important in a leader, this appellation would surely have to go to Washington. He recognized the great sea changes of the day and changed with them; he also saw the complexities of nation building as did few others, and applied the brakes when necessary. Julius Caesar himself would have marveled at Washington. Of Caesar, whose bust Washington once ordered for Mount Vernon, it had once been said that he led a captive people in unparalleled growth, loved history and made it, was fiercely grandiose and spectacularly fearless, and reigned as a benevolent autocrat. Absent the autocrat part—the Founders were understandably far less admiring of the Roman emperor—the same could be said of George Washington. In a world alternatively stalked by spectacular ideologues and warriors, weak monarchs and iron-fisted dictators, utopian prime ministers and would-be totalitarians, not to mention great emperors and even greater empresses, Washington would, in the long lens of history, eclipse them all. Diderot would call the 1790s "the age of liberty." This could not have happened without Washington.

If in his own country, many under their breath would soon mutter that Washington was the most infuriating, he was also the wisest. On the level of custom, he showed Americans, and the world, how a leader of a great republic should live; he invested new meanings in how a leader of a great republic should act; and he pointed the way toward independence. On the level of pragmatism, he steered America away from the world's furies that could have easily undermined the young nation—his hands-on management of the Nootka Sound crisis with the British and Spanish by today's Vancouver was one sign of this; his management of the fissures brought on by France was another. At his worst, he was maddening to his detractors. At his best, he became a living legend in the last decade of the century that birthed him. The poet Joel Barlow would go so far as to proclaim that Washington was "the greatest man on earth."

But as Washington's first term was coming to a close in 1792, the legend was slowly fading, and with it, so was the unity that all Americans had believed was crucial to the republic. The problem was in his own first family—Thomas Jefferson and Alexander Hamilton—even as the seeds of internal discord were also being laid abroad.

∞

WITH THE PRESIDENT'S blessing, Alexander Hamilton had won all the early battles over the federal government: on funding, on assumption, on taxes, and on the national bank. But as the full dimensions of Hamilton's program crystallized, the nation's two towering humanists, Jefferson and Madison, increasingly came to see Hamilton as a menace to the American experiment to be thwarted and, if necessary, destroyed. To them, Congress was supposed to be the leading branch of government, the guardian of popular liberty that would prevent the restoration of British tyranny or the reinvigoration of the monarchy. The dilemma for Jefferson, and it was a dilemma, was that he was Washington's secretary of state, and Madison was a close Washington ally. Nevertheless, their mounting fear led them to unite their disparate forces into an organized nucleus that first called themselves the "republican interest," then the "republican party," and soon after that, the "Republican party." In a word, they were becoming an opposition party, the first formal one in America's history.

To our lights, nothing could have been more inevitable or logical than the creation of a political party, which today is seen as integral to the healthy functioning of American civic life. But not so for the Founders. To them, the notion of parties, or "factions" as they were then called, was invariably a term of derision: Painful memories lingered of how Whigs and Tories had reflected the deep and bitter splits in America in 1776, while Federalists and anti-Federalists had represented bitter differences over the Constitution in 1787. Washington himself would refer to the "horrid enmities" and "frightful despotism" perpetrated by "party dissension." Parties were thus seen as corrupt, as a "fatal disease" for popular government, and as "monarchical" vestiges with no legitimate place in a genuine republic.

But a new reality was dawning.

∞

WITHIN MONTHS, JEFFERSON and Madison had irrevocably changed their minds about their alliances, and with it, the history of American civilization. Drawing support largely from Southern congressmen and planters, Southern Methodists and Baptists, and small farmers—but also anti-federalists and Northerners too—Jefferson and Madison (reportedly the first to use the word "party") came to believe that their strength lay with the people at large. Their allies rallied accordingly. At the same time, the followers of Hamilton—often tied with New England, banking interests, merchant concerns, the Atlantic Seaboard, and Episcopalians—gradually appropriated the name Federalists, which had been used earlier by advocates of the Constitution in 1787. Unlike the Republicans, who shrewdly alluded to the ancient Roman republic, for the Federalists, the executive branch was to be the chief engine of government.

Ironically, both sides denied involvement in parties; both sides accused the other of hypocrisy and perfidy; both sides saw themselves as the true heirs to the Founders; and both sides harbored lurid fantasies that the other would exert permanent control over the nation—which would only mount in intensity as they watched the Revolution in France devour first the monarchists, then the reformers, then the Girondins, then the ultra-radicals, and then the Jacobins themselves. As in France, evil conspiracies were rampant. As in France, so were paranoia and a cult of personality. And as time progressed, as in France, so did the acrimony; hatred would not be too strong a word. What the Americans lacked was perspective— the very idea of a loyal opposition did not exist, nor did the notion of an orderly transfer of power from one party to the next. And what they were saddled with was the present: a landscape defined by increasing bloodshed and chaos in Europe. In this sense, Washington, and also Hamilton, would come to view the criticism aimed at his administration as first reprehensible, then disloyal, then treasonous itself. But given the international climate of the day, it would have been almost unthinkable for either man to do otherwise.

So as the Washington administration wore on, the idyllic notion that America would be free of partisan groupings was dying. Soon it would explode altogether.

This was apparent in 1792 as a war—a newspaper war—sprouted throughout the nation.

∞

ACROSS THE FORMER thirteen colonies, the reading public was surging. In contrast to Russia and even France, Americans were a literate people, more so than any country in the world. In universities, lecture halls were crowded, growing libraries increasingly acquired well-worn works of renowned scholars, and Americans of all stripes—merchants, lawyers, doctors, tradesmen, tanners, and bankers—awaited any one of a dozen papers that flourished. Meanwhile, one paper stood above all: In 1791 the newspaper of choice was the *Gazette of the United States.* Edited by a former Boston schoolmaster, John Fenno, it covered national and international politics. Unabashedly partisan—all newspapers of the day were—it was also a quasi-official organ of the federal government, solidly aligned with Washington and Hamilton. The challenge then for Jefferson and Madison, and indeed for all Republicans, was how to counter Hamilton without being seen as direct critics of Washington, who was still viewed as untouchable; their other challenge, of course, was how to rebut Fenno's newspaper—which they saw as being in the hip pocket of resurgent "Toryism" and "aristocracy."

These were quarrelsome days. Scheming and plotting, Jefferson and Madison came up with the idea of recruiting the noted poet Philip Freneau as the republican answer to Fenno. Freneau had attended Princeton with James Madison, and was as talented a poet as the country possessed ("By midnight moons, o'er moistening dew;/In habit for the chase betrayed,/ The hunter still the deer pursues,/The hunter and the deer, a shade!"). The two breakfasted with Freneau in New York, urging him to move to Philadelphia, where he would be set up as a translator in the State Department. Arguably, it was doubly unthinkable—unthinkable because Freneau knew only one foreign language, which was spoken by many Americans, French; and unthinkable because Jefferson, a cabinet member, was now putting an enemy of Washington's on the government payroll to assault Jefferson's own administration. But Jefferson's followers, like him, were worried men. And as such, Freneau was the perfect riposte to the Federalist Fenno. His sentiments were republican—he had once been captured by the British and spent six tense weeks aboard a prison ship—words were his sword, and an acerbic wit was his rapier: Thus, King George III was "the Caligula of Great Britain," thus his rival Fenno was a "vile sycophant," thus France's Louis Capet "has lost his caput."

On October 31, 1791, the first issue of the *National Gazette* was published. It was licentious—signed articles were rare—it was fierce as well as literate, and it became the foremost Republican organ in America. As Attorney General Edmond Randolph soon boasted, Freneau was a genius and his newspaper was the "antidote" to the "doctrines and discourses circulated in favor of monarchy and aristocracy."

The fact that attacks were anonymous, or under a pseudonym, allowed the discourse to be biting, and it was. The brutal tone of these newspapers meant that virtually anything could be said, anything alleged, without any recourse. Monstrous portraits were sketched and unsubstantiated gossip peddled—this, of course, was the very kind of poison that the radical Jean-Paul Marat had used with such devastating effect in France. Freneau ominously warned that George Washington wanted to turn "limited Republican government" into an "unlimited hereditary one"; to deepen the wound, each morning he had three copies sent to Washington himself. The two sides fired away at each other all throughout the summer and fall. Under pen names, Hamilton caustically derided Jefferson as an "intriguing incendiary," a "cowardly assassin" given to "striking in the dark," and he warned that Jefferson was plotting to become "the head of a party . . . upon the ruins of National Authority." But Freneau's *National Gazette* meant to score a few bruises of its own; it counterattacked, assailing the Federalists as the "monarchical party," the "monocrats," and the "monied aristocracy." Day after day, the partisanship intensified. And at this stage, the journalistic frenzy was spinning out of control. By 1792, with Washington's first term coming to an end, political life in the young republic threatened to collapse into little more than searing critiques and unmediated malice.

Arguably, the caliber of the leading figures—Hamilton, Jefferson, Madison, and of course Washington—was unsurpassed in America's history—yet their acrimony toward one another was also unsurpassed, except perhaps by the French. How to understand the eerie spectacle of idealism and vilification, selflessness and selfishness, soaring treatises and vulgar epithets? We tend to imagine the disagreements of the Founders in lofty forums, like the Philadelphia convention or in *Federalist*-style essays. Nothing could be further from the truth. This was not an age when men of mild or sweet disposition rose to power; a savage streak was almost a prerequisite. True, personal jealousies, between Jefferson and Hamilton, Hamilton and Adams, Adams and Jefferson, and Madison and Hamilton, surely played a role; so did sheer competitive urges—the desire by each to leave his indelible mark on the policy and polity of the new nation. But these explanations only go so far. The French were fond of speaking of America as a "Virgin state," and here the deeper explanation resides: Both sides saw each other as dire threats to the still embryonic United States—to its very existence.

So, in the summer of 1792, amid this supercharged atmosphere, Jefferson decided to write a remarkable letter to Washington in which he openly spoke of Hamilton's scheming to turn the country into "monarchy" and urged the president to remove the treasury secretary from the cabinet. From there, the tensions only escalated. Washington in turn would ask Jefferson to moderate his own attacks against Hamilton—and here the commander-in-chief was singularly unsuccessful; Jefferson instead shouted back that Hamilton was a "tissue of machinations against the liberty of the country." Meanwhile, informed of the criticisms against him, Hamilton told Washington to sack Jefferson lest "the energy of the government" be destroyed. In August, Washington also pleaded with Hamilton to stop attacking Jefferson, but Hamilton actually redoubled his efforts. Drained by the open warfare and running animosity between his own prodigal sons, Washington literally seemed to tire before his cabinet's eyes.

He planned not to run for a second term.

∞

GEORGE WASHINGTON WAS sixty-one. Beaten down by fatigue and fighting, not to mention illness, his voice was now hollow and he was plagued by rheumatism. His complexion was pale, his upper jaw was in constant pain, and his vigorous physique had sagged. He complained of memory loss and failed hearing, and he struck observers as cadaverous, irritable, in short, a weary old man. And the fresh allure of the presidency had long since evaporated. For four years, he had worked extensive hours and assiduously sought to maintain an idealized image of the citizen-president hovering above partisanship. Yet this proved impossible when his own secretary of state privately derided him as a conceited, closed-minded man, manipulated by flattery or easily swayed by poor advice, and followers clapped in enthusiastic agreement. The epithets against him were unremitting. The whispers were too. By summer's end, he had had enough. He asked Madison to draft a "valedictory address" to present to the American people. The aging patriarch prepared to retire.

∞

FOR ONCE, HOWEVER, there was unity. Astounded at the thought of losing Washington, longtime advisors such as Robert Morris and Henry Knox begged him to stay on. And what of the two warring princes, Jefferson and Hamilton? They preferred to see Washington live through a second term rather than to have the office be seized by someone else. Jefferson, who planned to retire himself, pleaded with the president in a voice thick with emotion, that "North and South will hang together, if you hang on." Hamilton also implored Washington to remain at the helm. Sounding much like Jefferson, he warned that the Union might not survive without Washington to bind its divisions. Hamilton, of course, had the clearest view inside the reticent president's mind: He thus played upon Washington's desire to be loved (should the Union fail, he told him, the fickle public might blame him for its woes); he tweaked his vanity (to resign would prove to be "critically hazardous to your own reputation"); and above all else, he appealed to his patriotism—that the president's retirement would constitute the "greatest evil that could befall the country."

But the laconic Washington would not tip his hand. While weighing his decision, he returned twice to Mount Vernon, something he had not done in the three previous years. Yet in September, he realized that his attempts to calm the partisan waters had failed, utterly and totally; Jefferson refused to restrain Freneau, and the president conclusively learned that Hamilton was contributing vitriolic essays to Fenno's newspaper. Gradually, Washington began to accept that only his continued service could preserve the fledgling Union. He put Madison's valedictory in a drawer and kept it there.

With a combination of unease and dread, he stood for election a second time. For a second time, he was unanimously elected. And for a second time, he took the oath of the presidency. On March 4, 1793.

∞

THAT DAY, HE donned a black silk suit and, just before noon, set out in a handsome carriage, clacking along Philadelphia's rutted streets to Congress Hall. How different this was from the earlier acclaim he had enjoyed: the ecstatic crowds and leaping porpoises that welcomed him in New York when he took office four years earlier, or the women weeping in the streets when he arrived in Philadelphia to attend the Constitutional Convention in 1787. But today, at his own request, there were no parades of elaborate floats or polished militias; the ceremonies were closed to the public. Greeted by a small, happy crowd, he managed a wave and a bow and hurried inside the hall, where he delivered a 135-word Inaugural Address. Half as long as the Gettysburg Address and four times shorter than Lincoln's Second Inaugural, it was the shortest ever given. He was back home by one o'clock.

On the eve of his second term, anxious over what lay ahead, Washington had confessed his misgivings at "the prospect of commencing another tour of duty." He was prescient. Fisher Ames noted that the Republicans now "thirst for vengeance" and the president was "not to be spared." But this was the least of it. Much of Europe was at war over France, and the prospect of America being drawn into the conflict was more real than ever; at a minimum, it was potentially disastrous for American commerce. The only thing that Washington didn't anticipate was the degree to which the foreign hostilities would intensify the already growing divisions in the United States, with unpredictable results—perhaps leading to anarchy, violence, or yet another revolution, akin to the murderous infighting now sweeping France. Nor could he foresee that his second term would be almost wholly consumed by conflict, at home and abroad.

Soon after his inauguration, he left Philadelphia for the burgeoning spring of Mount Vernon, his third trip home in less than a year. Washington needed a break. If fate smiled, he hoped that he might absent himself from the capital for the entire sweltering summer, tend to his beloved estate, and return in the autumn when Congress reconvened. Yet he had been home less than one week when the news arrived by express that King Louis XVI had been guillotined and that in response Great Britain had gone to war with France. At once, Washington hurriedly started north for Philadelphia.

∞

FRANCE HAD CHANGED almost overnight, and so, it seemed, had America. As historian John Ferling notes, in these uncertain, gray days of March, most Americans still reveled in the achievements of the French Revolution. Were they not children of France, owing their independence to them? And had they not engaged in their own revolution in 1776? Just as they had helped make the world anew—that was Paine's soaring phrase—so they believed that the French would continue the struggle worldwide. France's heroes were America's heroes: There was Washington's surrogate son, the Marquis de Lafayette; there was the brilliant essayist Thomas Paine, now a member of the French Assembly; there was Condorcet, who had befriended Franklin, penned beautiful words, and lovingly espoused liberty; there was Admiral de Grasse, whose ships had elevated the patriots in their time of dire need; and above all, there was King Louis XVI, who had come to

America's defense against British tyranny. Indeed many Americans ardently and proudly believed that the American experience had inspired the revolutionaries in France; just as many saw the Declaration of the Rights of Man and Citizen as echoing the spirit of the Declaration of Independence. Even a skeptical Hamilton admitted that the exhilarating early events in Paris had captivated him every bit as much as the heady days of Lexington and Concord. So when the Bastille fell, Americans wept. When the National Assembly was created, Americans beamed with pride. When in October 1792 Hamilton and Washington and Madison were made honorary French citizens, Americans thrilled that France was a perfect mirror of America. And when France repulsed the Prussian and Austrian invaders at Valmy, spontaneous rejoicing erupted in the United States at the belief that the great Revolution had been saved.

But France was not a replica of America, as America's founding spirits were to soon realize—Lafayette fled the Revolution in fear for his very life and was locked up in an Austrian prison; shivering with a fever that nearly killed him, Paine was slated for execution; Condorcet soon took poison to escape the Jacobin ax; and Admiral de Grasse was decapitated. The Declaration of the Rights of Man was suspended while the king, whose birthday was an American holiday, had gone to the guillotine. And the French Revolution had devolved into yet another scurrilous great power war, pitting all of Europe against one another—and, as it turned out, Americans against Americans too.

∞

ALL THROUGHOUT 1793, reports of the revolutionary Terror and wanton mayhem spread across the Atlantic: the grisly news of the beheading of the king, the bleak reports of the compulsory de-Christianization of France, and the chilling account of the queen's bloody demise. As the list of atrocities lengthened, Americans were forced to contemplate the fruits of their own revolution, as well as their own future. While it cemented the two political parties as never before, it also opened up vast fissures between them, leading many Americans of all stripes to doubt their own country—and one another.

Ultimately, Republicans heartily embraced the spirit of Jacobinism; Jefferson expressed their sympathies best: "Rather than see [the Revolution] fail," he enthused, "I would have seen half the earth desolated." This war fever was infectious: When Robespierre announced the Revolution to be "the most beautiful . . . that has ever honored humanity," many Americans agreed. Actually, Madison hoped France would rebuke Britain while Jefferson longed to have tea in London with victorious French generals. For their part, Hamilton and the Federalists viewed an ever more militant France with growing alarm, and instead felt a common kinship with Britain. "We think in English," Hamilton privately assured a British diplomat. More than that, they thought like the English. They mocked the titles of "Citizen" or "Citizeness" now the rage in Paris and flinched at France's adoption of the workingman's sansculottes. Their disdain even extended to the new French practice of embracing one another, what some Americans ridiculed as "hugging and rugging . . ."

It would be hard to imagine America more divided than at that moment: Where Hamilton saw a vast nightmare descending across the Atlantic, a once-promising revolution giving way to indiscriminate bloodshed and authoritarianism, Jefferson remarked, "Were there but an Adam and Eve left in every country, and left free, it would be better." Where John Adams warned that "Danton, Robespierre, Marat, etc are furies [and] Dragon's teeth have been sown in France and will come up monsters," Madison described the Revolution as "wonderful in its progress and . . . stupendous in its consequences." Where Washington was sickened by the king's flight, and even more so by his death, and so was the American minister Gouverneur Morris, Jefferson asserted that Louis was amenable to punishment "like other criminals" and James Monroe opined that the execution was merely an incident "to a much greater cause."

∞

For Washington, this was America slipping into a Dantean hell.

Once more, he sought to hold himself above the fray. But unlike with economic policy, where the president allowed Hamilton to take the lead, when it came to America's security he was decisively engaged. That he would firmly stamp his own imprint on foreign policy should not have been surprising. For four decades, he had been involved in statecraft and warfare, and he had proven his courage in not one but two great conflicts; at various points in his life he had befriended Englishmen and Frenchmen— or worked against them. At various points, he also had their blood on his hands—or their amity. Yet he no longer had any need of conflict to burnish either his reputation or his ego. Moreover, he could not shake the thought that America had spent one-half of thirty years at war between 1754 and 1783, or that in 1792 America had no standing army and no navy; and he winced at the possibility of jeopardizing the fragile gains that the nation had made since 1789. If lured into Europe's wars, he felt there was precious little to gain and almost everything to lose.

Nor did he suffer from emotional attachments to any European nation—not Britain, not France, not any country. It's not that he hated an entire continent. But he had never so much as set foot on European soil, and Virginia had been home to the Washingtons since the age of France's Sun King. Having thrown his lot in with the patriots early on, he was unabashedly American, and hence instinctively wary of foreign powers. It could perhaps be said that he was little different from the chauvinistic monarchs and leaders of Europe: Pitt the Younger, Catherine the Great, the late French king, the deceased Austrian ruler; where they ultimately acted according to a shifting balance of power, which is to say, their self-interest, so did he (or in his words, "our own interest"). But that interest told him: Stay out. To be sure, it was a roll of the dice. But as a gamble, it also proved to be correct.

Meaning to practice what he preached, Washington was second to no monarch in ruthlessly acting on that self-interest. He may have been the president of a young and weak country, but he thought large and thought cleverly. Thus, though America was bound to its one ally, France, when he learned that France was preparing for a naval assault on Spanish New Orleans and was proposing that the United States help by simultaneously invading Spanish Florida, Washington spurned even the mere suggestion of an American role; at the same time he shrewdly instructed Jefferson to be prepared to extract concessions from the Spanish crown as a reward for American disengagement. And while the Franco-American treaty of 1778 committed the United States to aid France during war, to do so, Washington knew, would risk new hostilities with Great Britain, or perhaps even the monarchical coalition, the last thing he wanted. As with everything else at this stage, his cabinet was split: Hamilton and Knox advised the president to suspend the treaty, while Jefferson and Randolph urged that the pacts not be annulled.

In the end, Washington left the treaties alone, but ignored their more onerous provisions—presaging what John F. Kennedy would do almost two hundred years later with the two conflicting Russian communiqués during the Cuban missile crisis. And to ensure there was no doubt about America's stand, he convened his cabinet and, after days of hot debate, engineered consensus. It wasn't easy and wasn't without a price; but in the end, he extracted unanimous support for a hardheaded policy of American neutrality.

Actually, the policy never used the word "neutral," as is commonly supposed, instead speaking of a "friendly and impartial" America. Still, released to the world as an executive proclamation on April 22, 1793, it was henceforth dubbed the Neutrality Proclamation. Once more, a sulking Jefferson had lost in the councils of the cabinet, prompting Madison to ask indignantly why Washington should fear "the success of liberty in another country." It was a good question, but one for which Washington had a good answer. Self-interest.

But all was not resolved. The young country faced one more immediate challenge. The French Republic was sending its first minister to Philadelphia. Should Washington receive him? Hamilton, arguing with some sophistry, insisted that the treaty had been made with Louis XVI and his heirs, not with the revolutionary government, said no. Predictably, Jefferson, took the opposite tack and said yes. This time Washington sided with his secretary of state, and by doing so, set the modern practice that nations recognize nations, not just particular governments. It was farsighted and far-reaching.

Yet Washington couldn't have known that—nor could he have glimpsed the rapidly unfurling consequences of his decision, consequences that he would soon come to rue. For unknown to the president, from this point forward, his receipt of the new ambassador would not only dangerously threaten to undermine American neutrality, but raise the terrifying likelihood of an uprising in his very midst.

∞

THE EXTRAORDINARY CATARACT of events began inauspiciously enough. The new French minister arrived on April 8, 1793, sailing not into Philadelphia, as one might have thought, but into Charleston, South Carolina. The minister was apparently sidestepping standard protocol. Tactically, it was an electrifying maneuver; politically, it was sheer genius. Stepping onto American soil, he paused, almost dumbfounded. The Americans were actually cheering him. He was greeted with an astonishing reception by an enormous, clapping throng of South Carolinians—hats were tossed into the air while the "Marseillaise" was sung, men hugged each other as one "*citoyen*" to another, and a thundering cheer after cheer rolled across the city. As it turned out, this was the first of many pro-French mass demonstrations across the Eastern Seaboard. An unknowing observer would have mistaken the emissary for a beloved liberator freeing a once conquered people or a Roman general heroically returning from great victories in the far-off provinces. He wasn't; he was instead one of the stars of the French foreign service, Edmond Charles Edouard Genet—now Citizen Genet—the one-time royalist diplomat who had replaced the Comte de Ségur in Russia. There he had embraced republicanism, aroused Catherine the Great's ire, was banished from her court and, ejected from Russia upon Louis's execution, had returned as a hero to a France in revolt—and was now posted to America.

A dwarfish man with reddened cheeks and dark red hair, he had flashing eyes, undeniable good looks, and an even more undeniable vivacity. Genet is often depicted as rash and blundering, but this overstates his tactics and understates the times. He was insightful: With uncanny prescience his dispatches from St. Petersburg anticipated the great Russian Revolution to come in the twentieth century. He had considerable authority: He was speaking for one of the two most powerful countries on the planet, whose dominions spread across the globe and whose armies would soon appear invincible. He was well connected, counting Condorcet, Paine, Brissot, Danton, and Madame Roland as his friends in Paris. And though only thirty, he was brilliant, he was ambitious, and he had arrived with a deadly agenda.

Genet intended to ask for an early payment of the American debt, but that was just for starters. His larger plan was a wholly different matter—extending the French empire of liberty. From American soil, he planned to hire secret agents to strike against the Spanish and British in North America, not just at sea, but by land. Enlisting Jefferson as an accomplice—it didn't take much, notwithstanding the new policy of neutrality—he dispatched a French botanist, André Michaux, to meet with the Kentucky governor; actually, the "botany" part was a ruse. Michaux would arm Kentuckians in a dirty little private war—an invasion—by frontiersmen against Spanish Louisiana; with some luck, France would then pry New Orleans loose in the name of the Franco-American alliance. Genet, with the connivance of his government, harbored other dark plots as well, including outfitting American privateers who would then harass British merchant ships. Perhaps his most ambitious plan, however, was a 1,600-man army he began to gather on American soil to invade Spanish-held Florida. Making a mockery of U.S. sovereignty, this raised the stakes to an entirely different level. Informed of these events, President Washington was aghast; unless Genet were reined in, he would set the United States on a collision course against both Britain and Spain, pitting American against American in the process.

The president's fears were well founded; Hamilton himself growled that Genet's actions were the "height of arrogance." But Genet had little reason to second-guess his actions, for if he was an impetuous rebel, he was in good company. To the administration's utter dismay, much of America was siding with the French envoy. Already the South Carolina governor had given his permission to begin arming privateers. Genet also had held several conversations with Jefferson that confirmed his conviction that the fraternal solidarity uniting the world's two revolutionary peoples was thicker than any presidential proclamation; and he felt even more buoyed by the mass demonstrations that followed him over the course of nearly six weeks as he triumphantly made his way to the capital. The same could further be said for the "Democratic" societies that began sprouting in city after city as Genet inched northward to Philadelphia; they were in part modeled on the old Sons of Liberty, which helped fuel the American Revolution, and in part on the Jacobin societies in France, which were helping drive the French Revolution.

As it happened, Genet's successes only stoked his appetite. Wherever he went, greasing palms of supporters at each stage, it was a spectacle never seen before on American soil: Banquets were held, cannons were fired, and French flags waved. One writer rhapsodized, "It is beyond the power of figures or words to express the hugs and kisses [they] lavished on him." Indeed, it was as if borders or nations suddenly no longer mattered; what mattered was revolutionary fraternity. With each day, then, Genet himself seemed to grow braver and bolder and more audacious; and in Paris itself, the revolutionary government was becoming ever more contemptuous of the Americans—Jean Antoine Joseph Fauchet would sneeringly tell the Committee of Public Safety: "The original heroes of America who still survive, vegetate in an old age forgotten."

Meanwhile in the capital, the Federalists were in a panic.

∞

THEY SHOULD HAVE been. While Genet traveled, his ship, the *Embuscade*, had captured a British vessel, the *Grange*, and ostentatiously hauled it to Philadelphia. In doing so, he had openly flouted America's neutrality policy; Europe's dreaded war had now indisputably arrived on America's shores. But most Americans didn't seem to mind. Arriving in port, the British ship was greeted by thousands of people, crowding and jamming the wharves; Jefferson showed up too. Genet's timing was uncanny. When the British colors were seen reversed and the French flag snapped in the wind, the crowd burst into peals of exultation. Jefferson waxed to Madison that Genet "offered everything and asks nothing." In truth, the seized British ship was just the preamble; when Citizen Genet himself appeared on May 16, 1793, the outpouring of affection prompted days of celebration. Volleys of artillery rattled into the skies while French and American flags flew in unison across the capital. Liberty caps were passed around while songs commemorating the French Revolution were sung. Banquets were held while toasts were made daily, not to Washington, but to Marat and Robespierre; even the Pennsylvania governor raised his glass to the "ruling powers" in France.

But there was a darker side to this rampant enthusiasm. John Adams would later recall his fears of a deadly reprise of the bloody actions against King Louis XVI, only this time in the American capital itself. "You certainly never felt the terrorism excited by Genet," he later wrote, "when ten thousand people in the streets of Philadelphia, day after day, threatened to drag Washington out of his house and effect a revolution in the government or compel it to declare war in favor of the French Revolution." Terrified, the vice president ordered a cache of arms smuggled through the back lanes from the War Office to his home in case major violence broke out.

Was this one of those mysterious historical moments in which events acquire a momentum all their own and begin to exert an irresistible pressure? Genet, in capturing the popular imagination, was becoming the voice of millions, but the president was not about to see his own administration subverted. Washington, impatient and incensed, now brusquely stood his ground; when he received Genet, he bowed in his most detached, stilted manner. Hanging bravely behind him were the paintings of the late Louis XVI and the deposed queen, Marie Antoinette.

At the same time, the president held continuous meetings of his cabinet to furiously discuss the French situation. By this point, Washington badly needed to stanch the bleeding. Once considered invincible, he was now seen by many as a leader of a lost, outdated cause—Anglomania. His regal ways had alienated the Republicans, his lofty manner had offended a retinue of his top advisors, and his foreign policy—failing decisively to stand up for France—had proved contentious. The cabinet meetings were increasingly bitter, even cutthroat—and they got nowhere. Sensing this, Genet responded by thumbing his nose at the president in every possible way. He knew, or believed he knew, that he could appeal over the head of the chief executive to the American people themselves, thus intensifying the affections that knit the two republics together. And he knew, as the Republicans told him and as Jefferson confirmed, that the romance of the French Revolution was ineluctably linked with America's.

Was it? With Americans increasingly choking on one another's vitriol and the French revolutionary behemoth now solidly in their midst, Washington began to feel the unremitting strain. For ten days, the president suffered from lingering fevers; his aides whispered to one another that he looked unwell. They observed that he was wounded by the constant barrage of press attacks and couldn't help but notice that there were long pauses in cabinet meetings. And they recoiled when Washington unleashed his notorious temper: "By God," Washington sputtered in one meeting, he'd rather "be in his grave" than in his "present situation!"

That was a luxury he didn't have; of course, one imagines King Louis XVI may well have thought the same thing as his intimates were butchered around him and he was himself forcibly dragged from Versailles to Paris. So by late June, this was a watershed for the shaky young republic. No crisis of these proportions had taken place in the first administration. At no point before had violence and insurrection seemed such a distinct possibility. At no time since 1787 were the ties that bound Americans at such odds with the ideology that now separated them. And rarely had Washington seemed to be at such a breaking point. Twisting the noose, the *National Gazette* unleashed a broadside, half satire, half not, titled, "The funeral Dirge of George Washington." In it, a "tyrannical" president, like ex-King Louis, was executed by guillotine.

The president could be silent or he could watch his country be rent asunder, and for Washington this was no choice. Increasingly, this looked like a gathering, perfect storm. That spring had witnessed the creation of the Committee of Public Safety in France, and in June, the ruling Girondist faction was purged and placed under house arrest by the more radical Jacobins. And while the United States was a minor, immature power on the edges of empire—an afterthought for Danton and Robespierre—with each hour, it was rapidly becoming a tool for the machinations of the French colossus. At the same time, Republicans in America toasted the Jacobins and displayed models of the guillotine at their festivities. That July and August, it would now fall to a handful of men to navigate this crisis: an increasingly inflamed president, an outraged secretary of the treasury, and, ironically, a secretary of state who identified as much with Genet's cause as with the commander-in-chief he served.

Jefferson, it turned out, would be crucial to this whole drama.

∞

EVEN IN 1792, the secretary of state was one of America's greatest men. He was a beautiful dreamer, a visionary, and a romantic: When his wife, Martha, died in 1782, he wrapped a lock of her hair with a scrap of paper containing an excerpt from the couple's favorite novel and stashed the token in his desk. And he never remarried. But though he plunged into a near paralyzing stupor at her loss, he found distraction in the institutions and ideas that he eloquently championed, ideas that would ultimately carry him away from home, out of domestic seclusion, and on into history as the conscience of the country. He lived on a little mountain, set off and remote, high atop an inhospitable wilderness rise; it was his hermitage, or retreat—he called it Monticello. But there, his dinner parties reverberated like "Greek colloquia," and inside his walls percolated the ideas that helped forge a country and change the world. By any criteria, few men, even among the Founders, brought such a rare blend of prodigious vitality and formidable intellect to American life as Thomas Jefferson. And as the Genet affair intensified, few men would determine the fate of the adolescent republic as he would. Washington knew this. Hamilton knew this. Madison knew this. And Genet did too.

Born into the Virginia aristocracy, Jefferson came from the same background as George Washington, was related to such exalted Virginia gentry families as the Randolphs and the Marshalls, and, with his folded arms and regal air, his melancholy spirits and expressive eyes, he had an elevated aura not unlike Washington himself. Trained as a lawyer, he mastered Latin and Greek, learned several European languages, and absorbed enough philosophy, science, and mathematics that he was chosen to head the American Philosophical Society; he was a self-trained paleontologist to boot. But it was his contributions to the nation that made him so monumental. During the seething decade between the Stamp Act agitation and the Boston

Tea Party, many able pens and probing minds had set out constitutional solutions for British America's dilemma. Yet it was Jefferson who, in 1774, cut to the heart of the entire debate confronting the colonies in one searing treatise—*Summary View of the Rights of British America,* a compilation that interwove the primacy of individual liberty with the notion of popular sovereignty.

His finest hour came two years later in 1776, after the first shots of civil war had been fired. Jefferson was named to a distinguished committee of five appointed by the Continental Congress to consider a draft declaration of independence, "in case the Congress agreed thereto." It did. Though only thirty-three, and despite the fact that other members included such eminent figures as Benjamin Franklin, John Adams, Roger Sherman, and Robert Livingston, Jefferson was appointed chairman. He rapidly rose to the task with his electrifying draft, produced in a handful of weeks, which was adopted almost in its entirety—with a little crucial help from Franklin. Its riveting sentence said it all—"We hold these truths to be self-evident, that all men are created equal, that they are endowed by their Creator with certain inalienable rights, that among these are life, liberty, and the pursuit of happiness." The Declaration served as a national birth certificate and helped rouse the fledgling rebellion; more than that, it eventually became the classic statement of America's purpose and enduring idea. No wonder the French ardently enlisted Jefferson's help when they drew up the Declaration of the Rights of Man, and that in another turbulent time, the Declaration weighed heavily upon Abraham Lincoln's thoughts as he delivered the Gettysburg Address; arguably, Jefferson's thirty-five magic words could be the most important in American history—and quite possibly of modern human history.

Yet for all his inventive, ebullient, and surprising mind, Jefferson, who was elected to Virginia's House of Burgesses at the age of just twenty-three, who would go on to become governor of Virginia, America's minister to France, its first secretary of state, later its second vice president, then twice its president, was a bewildering grab bag of traits that would cast an ambiguous shadow over his country for generations to come. His soul was in conflict all his life. He protested that he had no wish to form a political party ("If I could go to heaven but with a party, I would not go there at all"), but in Washington's administration he did precisely that. He was, more than most of his peers, a democrat who boasted that he would "always have a jealous care of the right of election by the people," yet he vigorously opposed direct election to the Senate on the grounds that "a choice by the people themselves is not generally distinguished for its wisdom." He loved the people, and spoke in their name often, yet he disparaged "the swinish multitude." He was the man who sought to undermine the Alien and Sedition Acts and quipped, "Were it left to me to decide whether we should have government without newspapers or newspapers without a government, I should not hesitate for a moment to prefer the latter," yet he privately urged state officials to press seditious libel charges against editors unfriendly to his presidency. He was elusive too: A pacifist by temperament, he went to war against Muslim pirates with a brand new fighting force—the U.S. Marines. And he was a fighter: After George Washington was elected, he fenced peevishly with Hamilton over economic policies and foreign affairs, and waged a secret campaign against the president, even as he shrank from personal controversy whenever it touched him. A passionate idealist, he was a fervent hater, hating the Federalists, hating the speculators, hating the moneymen and merchants, hating industrialists without distinction. And though he lived in a deeply religious land, he detested clergymen too.

But one strain that defined Jefferson the man would ultimately come to cleave the nation that he had so assiduously helped to construct. Few championed personal liberty as he did, yet few reaped more from human bondage. The lengthening shadows of slavery filtered into every nook of Jefferson's long, illustrious life, until the day he died in his bed at eighty-three. He despised slavery, cursed it (under his breath, that is), and in vain tried to curb it, both publicly and in his own personal sphere; but he owned from 100 to 200 slaves, comprising some three generations of human souls, almost certainly fathered children by one of them, and bought, sold, and bred slaves his entire life. At every step, his disgust with slavery ("the dreadful firebell in the night") was marked by hesitation and temporizing. Still, he wanted no less than outright abolition, and few Quakers, abolitionists, manumission societies, or even Hamilton himself, would ever argue more forcefully or more thoroughly against the "peculiar institution." In his *Notes on the State of Virginia*, he bitterly denounced slavery in the harshest terms, insisting that it was not simply an economic evil (it crippled "industry"), but a moral one as well that debased the slave owner even more than the slave ("the most unremitting despotism on the one part, and degrading submissions on the other"). But as soon as his *Notes* was completed, he sought to avoid wide circulation—he didn't want fellow Virginians to read his scathing observations—and he refused to have his name embossed on the title page.

Early in his career, he boldly drafted a law for gradual emancipation, but then he never presumed to introduce it; when his blistering attack upon the slave trade in the Declaration was struck out, he did not quarrel. In his twilight years as an elder statesman of America, when slavery underwent a terrifying acceleration in the South, his commanding voice could have profoundly strengthened the emancipationist cause; but he instead merely sighed, "we have the wolf by the ears, and can neither hold it nor let it go."

At Monticello, the story was no less ambiguous. Upon his return from Paris after a five-year absence, his slaves eagerly collected about him at the base of the mountain, shouting and cheering and "kissing his hands and feet," some crying, some laughing. There is good reason to believe this affection was genuine. But in running his plantation his overseer severely flogged slaves, and more than once Jefferson was seen angrily shaking the whip. And two years later, this prophet of equality would mortgage all his slaves—men, women, and children—to pay for the opulent, French-inspired renovation of his home; when strapped for funds, he even paid his hired workers in slaves, as he did with his carpenter, Reuben Perry. Perhaps most troubling of all, while some of Jefferson's neighbors invoked his ringing phrases to justify the liberation of their own chattel ("the glorious and memorable revolution"), Jefferson never emancipated his own slaves, not even Sally Hemings—not in his lifetime, as did his young neighbor Edward Coles, the future Illinois governor, nor in his will, as George Washington did, nor even as did the French revolutionaries, whom he so admired.

Yet if slavery was the earthquake reverberating across Thomas Jefferson's world, his views on the Union were just as shattering. All during the late 1700s, and later as president, Jefferson relentlessly pushed to expand America; it has been said that his mind encompassed the entire continent, and in an epoch of imperialism, he was perhaps the greatest empire builder of them all. From the start of the Revolution, he envisaged the United States as one vast body, stretching from sea to sea. Here, civilization would brilliantly sweep across, from east to west, "like a cloud of light."

But what kind of America? What kind of civilization? For all of Jefferson's love of the United States—there was no fiercer patriot—on one issue he would be torn all his life. Until his dying days, he regularly propounded local self-government above all else, supporting states' rights against the

Union, county rights against the states, township rights against the county, and private rights against all. He pushed for the Tenth Amendment to the Constitution, limiting federal power in favor of the states. He would one day soberly warn that "a single consolidated government would become the more corrupt government on earth." And when during his own presidency New England darkly threatened secession ("scission") over his economic and political stances, Jefferson mourned it as an "incalculable evil," yet he answered tepidly: "Whether we remain in our confederacy, or break into Atlantic and Mississippi confederacies, I do not believe very important to the happiness of either part." And he added this: ". . . separate them if it be better." True to his divided nature and divided philosophy, he was himself of two minds.

There was one issue, however, on which there was no debate: the relationship of the French Revolution to the American. In 1792 the fledgling American republic was already a country of boundless opportunities but also worrying contradictions: A nation born by a daring act of rebellion and secession, it had just changed its government and was still in a fluid political situation; a country that was marked by astonishing growth and breathtaking vibrancy, it existed in a world of powerful, aggressive regimes and had little military power to speak of; an expanding country, it was neither a single nation nor did its own people agree on the meaning of what type of society it should be; the first free and republican society in the modern world, it already beckoned as a utopia for the average man, yet the country itself was a furnace of seething differences and smoldering enmities. And as Jefferson and Americans looked not westward, past the Alleghenies to the sprawling Great Plains and the Rocky Mountains, but eastward, to Europe, the perennial question remained whether America could or should seal itself off from the tumult of the rest of the globe, as George Washington hoped, or whether it was by definition intertwined with the great ideological battle assaulting the Old World.

Here, Jefferson suffered little doubt. His heritage was Virginian, his abiding love was for America, but his heart was cosmopolitan and his soul still largely identified with revolutionary France.

∞

WAS THIS EMPATHY? A terrible blind spot? Or was Jefferson uniquely a master of grasping the panorama of his time? The question is ultimately irresolvable. We do know that Jefferson saw America and France not simply as formal allies, but as ideological comrades in a global struggle. We also know that he saw the survival of republicanism in one as dependent upon its survival in the other, and that he believed the crowned heads of Europe knew this all too well; arguably, he was perhaps the first believer in the domino theory of international relations. Even before he departed Paris, Jefferson lived in fear that the conservative powers, especially Great Britain, would unite to destroy France; and if this were to happen, it would mean the end of republicanism in America too; only the success "of their government," he insisted firmly, "would stay up our own." As U.S. minister in Paris, Jefferson traveled widely through Europe, as well as throughout France itself; he spent time in French salons, met the philosophes, and watched monarchy in action. Never a friend of royalty, he was even less so now: He derided monarchy as the "curse of existence that must be guarded against at all costs"; he lamented that all kings do is make wars and give away sinecures; and he scoffed that aristocracy were "wolves over sheep," and the sheep, naturally, were the people.

The people. They meant everything to Jefferson. When the French Revolution began, he was there, rejoicing that the royal authority had lost much of its power; he of course advised Lafayette in writing the Declaration of the Rights of Man. Before he departed France and set sail for Virginia, Jefferson throbbed with hope that republicanism would take root in France much as it had in America ("the nation has been awakened by our revolution" he joyously wrote, "the essential principles [are] ours"), and then spread magnificently to England and across Europe. His initial exuberance was understandable; all Americans felt it. But then events took a murderous turn: corpses swinging from lamps, mutilated bodies dragged through streets, the suspension of the monarchy and the subsequent execution of the king, and, of course, the Terror. While Hamilton worried about the carnage—even before the king was killed, the September massacres cured him of his "goodwill" toward the revolutionaries ("a struggle for liberty is in itself respectable and glorious when conducted with magnanimity and justice")—and John Adams lamented that Americans were "blind" and "undistinguishing" toward the bloodthirsty mayhem, the reaction of Jefferson was, at first, studied indifference. "I have observed," he said confidently, "the mobs with my own eyes." Then it was mock laughter. "The cutting off heads is become so much a la mode that one is apt to feel of a morning whether their own is on their shoulders." Then disbelief: "I will agree to be stoned as a false prophet if all does not end well in this country." In truth, for the secretary of state this was the price of liberty. So he initially discounted the violence as propaganda, then trivialized it, then when the facts were unshakable, maintained "my fortune had been singular to see in the course of fourteen years two such revolutions as were never seen before." In the end, he counseled that the "liberty of the whole earth was depending on the issue of the contest."

Did it? That of course was the problem that haunted and divided all Americans, no less than it did the French themselves. In retrospect, Jefferson's stay in France radicalized his views, leaving him with the idea—future presidents Madison, Monroe, and Jackson shared this belief as well—that any aristocratic or monarchical sympathies in America were a terrifying slippery slope to the undoing of the spirit of 1776. "I am really astonished at the blind spirit of party which has seized on the soul of this Jefferson," Adams complained. "There is not a Jacobin in France more devoted to faction." And the fact was that the Francophile Jefferson also loathed all things British.

It further meant that when Jefferson was instructed by George Washington to tell Genet to stop outfitting privateers, most of the cabinet was also arrayed against him. As was always the case, it was Alexander Hamilton's voice that was the loudest.

∞

NEVER HAVE TWO cabinet members been so gifted and been so mercilessly at war with each other—not in the Civil War, not in World War II, not in the Cold War, not in the Iraq War in the twenty-first century—and their conflict produced the two opposite ends of the political spectrum that define American politics to this day. Early on, the two men socialized on easy terms, with little inkling that they were fated to become mortal foes. It didn't last long. Once in the Washington administration, they were like fierce, dueling satellites rotating around the aging presidential warrior. It is one of history's curiosities that both Jefferson and Hamilton had lost their fathers before becoming adults, one by death, the other by abandonment, and both looked to the childless Washington as a kind of surrogate. Jefferson had once seen himself as the acknowledged heir, but the upstart Hamilton, bursting with natural gifts and great precocity, quickly replaced him. For Jefferson, it had to have stung. Until he had moved into Washington's circle, Jefferson had never known what it was like to be eclipsed by a rival younger than himself—the treasury secretary was fourteen years his junior.

Washington had once wept unashamedly at his dinner table when describing his affection for the young French nobleman, the Marquis de Lafayette; it would be hard to imagine him doing the same for either Jefferson or Hamilton, but if he were to weep for anyone, it would have been the treasury secretary. The reasons are easy to understand. Hamilton had an impressive military record, a mind that worked with dazzling speed, and great political promise. Though he could be overbearing and arrogant, he could be equally graceful and debonair; and he brilliantly meshed with Washington—there is little doubt the two often thought alike. And where the ultra-private and aloof Jefferson preferred the quiet company of scientists and republicans, always despising the speculators and bankers, Hamilton, dapper and relaxed, glided among the wealthy merchants of New York and Philadelphia as if he had eaten off fine china plate from earliest childhood, when of course he hadn't; he loved to talk about money and finance, and did so masterfully; and he commanded great respect because he, perhaps more than anyone else in the cabinet or beyond, had Washington's ear, and hence, more influence than any man in the United States save the president himself.

Hamilton's was a uniquely American story. Born on Nevis in the Caribbean, he had spent a tortured adolescence on St. Croix; and he lacked one thing that would stalk him all his life: legitimacy. His mother was Rachel Faucette, and her abusive husband was Johann Michael Lavien—Hamilton would later write it as Lavine. At the age of twenty-one, she left Lavien and set up house with an itinerant Scotsman called James Hamilton, a drifter and failure; he also became Hamilton's father. In 1759 Lavien sued for divorce, which was granted—but Danish law in the Caribbean would not allow Rachel to remarry. Then in 1765 James himself drifted away; thus the specter of bastardy and uncertainty and abandonment hovered over Hamilton all his life. His mother died when he was just thirteen as well, leaving him orphaned; equally tragically, his thirty-two-year-old legal guardian would soon thereafter commit suicide. Even his birthdate was in doubt; he thought he was born in 1757, but just as likely it was 1755. Small, blue-eyed, with hair as red as Jefferson's (some accounts erroneously portrayed him as a swarthy Creole), he was brilliant, brash, and charming. To his critics, he was vilified as a "Bonaparte," or "a speck more obscure than Corsica," or a "bastard brat of a Scotch peddler," or that "Scottish

Creolian of Nevis," or simply that "Creole bastard." But as an orator, he could speak extemporaneously for hours. As a writer, he could crank out 10,000-word memos overnight; and as a figure on the world stage he was one of a kind: He had the superhuman energy of a Potemkin, the worldliness and panache of a Mirabeau, a genius that more than rivaled Voltaire's or Montesquieu's, and a humanitarian impulse that recalled a William Wilberforce. Talleyrand, no stranger to greatness, called him one of the greatest men of the age. Indeed, Hamilton had an intensity and frankness about him which both made him broadly admired and genuinely feared; unlike the courtly Jefferson, who mumbled speeches in a soft, inaudible voice, Hamilton never left anyone guessing as to where he stood.

Ironically, where Jefferson was a magnificent populist, Hamilton was alternately derided as pseudo-aristocrat (he married Jane Schyler, a daughter of one of America's wealthiest families), a crypto-monarchist, and an elitist; but as biographer Ron Chernow points out, he was also a passionate abolitionist with a far more expansive vision than his peers. Jefferson's heart beat to Rousseauian scenes of an agrarian America studded with sturdy yeoman farmers; by contrast, Hamilton loathed poverty and loved opportunity and devoted his life to enlarging economic prospects for all Americans—whether through trade, commerce, banks, stock exchanges, factories, or corporations; he wanted not an aristocracy, as was often alleged, but a meritocracy. If Jefferson was America's Charles James Fox—England's libertarian romantic who waxed rhapsodic about the tree of liberty—Hamilton was America's William Pitt, the high-powered financial statesman, cool, pragmatic, relentless in looking to the future. Arguably, Hamilton, a great Anglophile and hater of the French Revolution, was more of a revolutionary than Jefferson. Here, Hamilton's checkered past served him well: A descendant of the stew of buccaneering and skullduggery in the West Indies, where it was every man for himself, Hamilton was destined to be the father of American capitalism.

True, he could never match the soaring poetry or inspirational elegance of Jefferson, but who could? The secretary of state had a matchless array of stylistic devices at his disposal, including parallelism, assonance, alliteration, the use of one-syllable words, pealing flourishes, or, when needed, brevity. But this thirty-five-year-old wunderkind became a juggernaut that dominated Washington's administration and in countless ways defined America. He replaced all others in Washington's affections—and took liberties that Washington would not have tolerated in anyone else. As a result, though he never became president, Hamilton's fingerprints are on every facet of modern American life: the Constitution, the *Federalist Papers* (which he wrote the bulk of), a strong presidency, an independent judiciary, a powerful federal government, big cities, capital markets, Wall Street, the legal profession, the newspaper profession (he started the *New York Post*), a vigorous military, and, remarkably, that is not an exhaustive list. For this reason, Jefferson was wildly jealous of Hamilton, and for the same reason described Hamilton as a "colossus" and "host unto himself." Hamilton was both and more.

∞

BUT WHILE HAMILTON had the self-reliant reflexes of someone who lived by his wits, he, like Jefferson, was not without weakness or conflicting impulses. Indeed, the inner Hamilton defied simple categorization: Passionate, headstrong, ferocious, he was a born doer and a born thinker; he was also inexhaustible. His enemies labeled him a reactionary, and perhaps in the crudest sense he was. Yet there is no denying that he was a man of the future; perhaps unique among the Founding Fathers, he would have felt equally at home in the twenty-first century as in his. Still, for all his genius, and the word fits, and all his contributions, he was impulsive, or rash, or both. How else to explain that he stumbled into a searing sex scandal when he knew better, which could have wrecked his loving marriage and surely may have cost him the presidency? Or why else, even in an age of honor, would a man who had so little to gain by dueling—he had even lost a much-loved son to a duel—agree to a pistol match against Aaron Burr?

In the end, the Hamilton-Jefferson feud took on epic proportions; Hamilton was at the height of his powers, and Jefferson was not. The two argued endlessly and furiously, but almost never reasonably. Thus, the government was divided at its birth. In the spirit of the times, there were soon rival sects, the Hamiltonians and the Jeffersonians; both movements were too powerful to be suppressed. But each opposite camp tried mightily, and their tirades led to recriminations, then to vitriol. In this, they were only following their spiritual masters. In 1792 Hamilton vilified the secretary of state not only as a frozen carcass and poisonous snake, but also as an intriguer ("the *most* intriguing man in the United States"), an ambitious revolutionary ("hiding under the garb of Quaker simplicity"), a man of violent passions, and a promoter of "national insignificance." Not to be outdone, in 1793 Jefferson enlisted the likes of such statesmen as Madison,

Monroe, and Attorney General Edmund Randolph to castigate Hamilton. But while these darts hit their target, none of them seemed to stick. Echoing the Girondins' attempted impeachment of Marat, Jefferson then sought to disgrace Hamilton before the entire nation: In one last desperate attempt, he secretly drafted a series of anti-Hamiltonian resolutions for Congressman William Branch Giles, including one that accused Hamilton of "maladministration of his duties," which called for him to be "removed" from office; the resolution was voted down and it was Jefferson who, sighing that "the motion of my blood no longer keeps time with the tumult of the world," left the cabinet in defeat later that year.

But it was on the matter of France where their hostilities most threatened to spin out of control—and where the stakes were the greatest. For months now, the country had been suffocated by mutual suspicion. Jefferson was convinced that Hamilton was "bewitched" by the British model of governance, while Hamilton mocked the secretary of state's "womanish attachment" to France and thought him a gullible apologist for the Revolution's indiscriminate violence. In truth, though he never lived there, Hamilton knew a thing or two about the French: Descended from French Huguenots on his mother's side, he was fluent in French and had served as Washington's liaison with Lafayette and other French aristocrats who rallied to the American patriots. Even at the outset, Hamilton had a foreboding about the events rocking Paris, writing about the "vehement character" of the French people and the "reveries" of their utopianism that would spell disaster. As the Revolution hurtled forward, and with it, the carnage, any ambivalence he may have had rapidly dissipated. Shortly thereafter, Hamilton sounded remarkably like Catherine the Great in his assessments: Before Napoleon seized control, Hamilton predicted that France, after "wading through seas of blood," may find herself "the slave of some victorious ... Caesar."

So by the time Genet arrived, it was no surprise that Hamilton was characteristically scathing toward the Frenchman—and could hardly contain his disgust with Jefferson and his supporters in cabinet meetings.

∞

IN LATE JUNE came the confrontation. Hamilton lambasted Genet to his face, tensely telling him that France was the belligerent in Europe; unaffected, Genet then remarkably lectured Hamilton on the niceties of American government: The administration, he averred, rather timelessly, was misusing executive power and usurping the powers of Congress. For his part, Jefferson, while agreeing with Genet, begged the diplomat to exercise more restraint. Incautiously, Genet did not, thus burning his closest ally in the government.

At this stage, however, Genet was convinced that discretion no longer mattered. In July he outfitted the privateer *La Petite Democrate* in Philadelphia. Hearing this, Washington was outraged; in a white-hot fury he wrote to Jefferson: "Is the minister of the French Republic to set the Acts of this Government at defiance, *with impunity*, and then threaten the executive with an appeal to the people? What must the world think . . . ?" It was the right question.

Actually, Washington had already deemed that privateers in American ports were to be forcibly seized, but the cabinet was unwilling to enforce its own policy and use force against the continent's superpower. Genet may have run afoul of Washington, but Washington still thought twice of running afoul of Robespierre and Danton. Yet to allow all this to happen under Washington's indignant nose? The cabinet met continually throughout July and early August. Hamilton was near apoplectic—use force, he argued; publish Genet's insolent correspondence; accuse him of meddling in domestic politics; in one meeting, he railed against the Frenchman for forty-five minutes straight. Meanwhile, he continued to see Jefferson as a French stooge. The president felt the tug of this imbroglio too, exploding in a rage at yet another meeting. Philosophically he agreed with Hamilton, but practically he remained concerned about offending the French juggernaut.

The eminent Washington was now weary of constant struggle, hoarse from the polemics, and exhausted by the day-to-day demands. To the president, many of his dilemmas were incomprehensible: In his eyes, he had sought merely to offer reasonable advice to unreasonable ideologues—his Republican detractors—and to secure America's independence. But once the battle lines were drawn, at times it seemed every man's hand was against him, and sound policy was hostage to one side or the other, sometimes to both. More ominous still, reason was suspect, and common sense was painted as treachery.

By August, Washington was angry, desperate, and disgusted. But he was also a realist, and a shrewd one at that; this time, he chose to split the difference between those who hated the French and those who didn't, smartly demanding Genet's recall instead. In a flash, the flimflam was over. By now, a sickened and humiliated Jefferson—he was suffering from his legendary migraines—was reluctantly ready to wash his hands of the whole matter; "I saw the necessity of quitting a wreck," he confided to a friend. His concerns were less over what Genet was doing—that he could tolerate—but over his heavy-handed, boorish tactics that had become politically intolerable. Madison, for one, was crestfallen, saying such an action would give great pain to those "enlightened friends" of both the American and French Revolutions.

On the world stage, however, this was a stunning moment for the infant republic. For some time Washington had worried that the United States, vulnerable and widely held in contempt, was destined to fall prey to "every nation"—of course, there was no more chilling testament to the peril of American weakness than the slow, inexorable dismantlement of the once great Polish nation, which had been mercilessly carved up by Russia and Prussia a mere eight months earlier. And American opinion remained dangerously divided. In a very real sense then, the dimensions of the Genet crisis were staggering. But Washington had deftly navigated its shoals; despite the sudden rise of partisanship, the struggling center ultimately held. War had been avoided and so had domestic violence; and America showed it was no one's pawn.

Still, there was fallout. Jefferson would soon leave the government; this Washington could live with. But in a subtle way, he was now reminded that America could scarcely act with impunity—even if it were in the right. For in turn, the French sent a small but unmistakable signal of their own, demanding—and receiving—the simultaneous recall of America's minister, Washington's good friend, Gouverneur Morris.

Actually, in Paris, Morris, no less than Genet, had done more than his share of partisan meddling: He had advised the king and plotted with royal advisors and foreign diplomats, including the Russian minister Jean Simolin, to save the ill-fated monarch; he also had received 1 million livres of the king's money. In the spring of 1793, he had provoked the French government still further by sheltering aristocrats. No wonder that in the streets by the Tuileries, fervent Jacobins openly hissed at the American. But having taken his stand, George Washington knew better than to blindly charge into the cannon's mouth: To appease the French, he replaced Morris with an emissary more sympathetic to the revolution—another earnest Francophile, the Virginian James Monroe. As for Genet? He simply stayed put in Philadelphia. Summoned home to a sure execution—he was a Girondist, and the Jacobins were now in control—he instead married the daughter of the New York governor, George Clinton.

In hindsight, as the interfaith conflict in America grew, so did Washington's nerves. But even with the Genet matter resolved, Washington could hardly sit back and congratulate himself. In France, he knew, the legislature had bullied and finally overthrown the king, while the political clubs and their riotous supporters had browbeaten the legislature, and now the whole country was in anarchy. Was this an omen for the United States as well? Increasingly, the stoic president harbored fears that it was. For one thing, much as in France, democratic societies continued to bubble up across the country. And much as Russia's empress had come to fear the contagion of the French Revolution, the American president would as well—even if both would deeply resent such a coupling. For another thing, the first tentacles of what looked like a full-scale insurrection were now unfurling in western Pennsylvania, in the frontiers of every state south of the Potomac, and also in what would be today's Ohio.

It became known as the Whiskey Rebellion.

∞

THE FIRST TASTE of open rebellion occurred in Washington's own backyard. In city after city, Democratic societies, having taken a page from their French brethren, mushroomed across the country. In Baltimore and Philadelphia, in Kentucky and western Pennsylvania, in New York and Vermont and Wythe County, Virginia—some fifty all told—they protested Hamiltonian economics, an invasive executive branch, national taxation, the "rich and the well born," policies that assaulted their vision of a revolutionary America, and a neutrality policy that they saw as rabidly anti-French and pro-British. Above all, they were galvanized by the French Revolution as well as continuing gratitude to France for its assistance during the War of American Independence. As one Democratic society member enthusiastically wrote: "We were strangers and ye took us in; we were naked and ye clothed us; we were pursued by robbers and ye put arms into our hands for defense . . . by thy assistance do we enjoy all our present blessings." Jibed another society member, "If all tyrants united against free people, should not all free people unite against tyrants? Yes! Let us unite with France and stand or fall together." Or as one Pittsburgh paper simply wrote: "France's friendship was of the utmost importance . . ." During these hidebound days, the societies sang the same hymn and in chorus: Their enemies were aristocrats, monarchists, Great Britain, and, more often than not, the Washington administration; and their undying friend was France. Genet may have lost his post, but his ardor and influence lived on.

By contrast, Alexander Hamilton saw these societies in deeply conspiratorial terms: to him, they were monstrous creations of Genet or scheming secret French agents in America, and not the spontaneous groundswell of an early democratic American sentiment, which they in fact were. Was this, then, an early paranoia, akin to the red scares of the Cold War—or autocratic sentiment, as critics alleged, akin even to Catherine the Great's reactionary impulses in Russia? Or conversely, as seen in the context of the day, was it legitimate? As it turned out, while agreeing fully with their sentiments, even Jefferson and Madison initially looked askance at the societies; no less than Hamilton and Washington, they believed that the political class should steer the populace and establish the agenda. This was not as surprising as it may seem; just as "factions" were at first seen as creatures to be dreaded, so was popular democracy—and what more vivid proof of this was there than the butchery playing out in the streets of Paris? As one Vermont assemblyman confessed to Hamilton, the societies were a movement of "self-created societies and clubs"—self-created being an epithet denoting democracy—and he added that the societies had been formed for the purpose of "influencing, or dictating to, constitutional bodies." So were they merely "butchers, tinkers, broken hucksters, and trans-Atlantic traitors," as one observer tartly quipped? Or was this France all over again?

For Washington and Hamilton, the signs arriving from the distant towns in Pennsylvania were anything but reassuring. A new day was indeed coming.

∞

THE FIRST SPARK of the assault on national unity flared in the West and South over taxes. Frontier farmers all across the land revolted against Alexander Hamilton's whiskey levy. They lived in towns literally hewn from the wilderness and in constant jeopardy, if not from wild animals, then from Indians. But one thing they couldn't countenance was taxation, even with distant representation. Since 1791 the protests had continued episodically, and some eighteen months after the whiskey tax went into effect, not a cent of revenue had been collected in western Pennsylvania. By 1793, seeing themselves as shackled by far-off government, while being denied any of its benefits, these backwoods farmers angrily warned Philadelphia: "We can never be taught to submit." Tough words, and they meant them. Soon, this language was echoed in Maryland, Virginia, North Carolina, and Kentucky. In return, Washington displayed a tough side too, issuing a strident proclamation denouncing the lawbreakers and promising that the government statutes would be enforced. Still, for a time, a very brief time, tempers cooled. But as larger events buffeted the government in Philadelphia—Genet's recall, Europe's war, France's ever more violent revolution—the bands of disgruntled citizens felt emboldened and further enraged.

Now the Westerners did far more than refuse to pay the infernal whiskey tax; blackening their faces or streaking their chests with clay, they blithely shot at revenue officers who would collect it, they tarred and feathered and whipped federal officials who sought to change their minds, they kidnapped and terrorized U.S. marshals, and they set fire to the homes of George Washington's hated representatives; they even shot up signs with Washington's own likeness. Thus began the rebellion in full force. In the weeks and months that followed, the Whiskey Rebels raised liberty poles bearing inflammatory slogans, passed out petitions of angry resistance, and formed urgent committees of correspondence modeled on their revolutionary predecessors. These were the toughest and hardest of Westerners: farmers and laborers, hunters and Indian fighters, expert woodsmen and sharpshooters; most were disillusioned war veterans as well. Then, after a serious military encounter in July 1794 between frontiersmen and federal representatives—there was a two-day gun battle with the federal excise inspector—war fever began to sweep the country.

By August the call went out, and a clutch of Whiskey Rebels began to gather. Soon they were 7,000 strong at Braddock's Field, roughly the same number that the British had under arms when they defeated Washington's more numerous forces at the battle of Brandywine. Dressed for war, they hoisted their own flag (with six stripes), whooping and howling and openly flirting with independence. They began to erect mock guillotines and set up their own extra-legal courts. Taking over the local militia, they then enthusiastically praised Robespierre and brashly defied the federal government to come after them. Meanwhile, with the historic assault on the Bastille in mind, the rebels spoke of attacking the government garrison in Pittsburgh to seize weapons. A local politician, Hugh Henry Brackenridge, frantically asked President Washington: "Should an attempt be made to suppress these people?"

Another anxious witness had a different, more ominous take: "I am afraid," he asserted to the president, "the question will not be, whether you will march to Pittsburgh, but whether they will march to Philadelphia."

∞

WITH SUCH FAINT tremors do upheavals begin. The French Revolution began not with the storming of the Bastille, but with a spontaneous conflagration between a mob and the king's soldiers on the Day of Tiles; the multiyear Pugachev Rebellion in Russia began not with combat between the empress's armies and the rebels, but with a meeting in the remote Crimean steppes and an assault on a nobleman's estate; and the American Revolution began with a few charismatic hotheads and committees of correspondence. Well aware of all this, Washington was unwilling to take his chances. He had long worried that with "the mere touch of a feather" Westerners might be driven from the Union; at this stage, he fretted about that and more.

Washington now had to assess the magnitude of the threat. The issues were complex and puzzling. Was this simply a protest movement—or a separatist rebellion? Was it a dramatic sequel to the Shays rebels in Massachusetts almost a decade earlier—or a sister movement of the Jacobins? In the confusion of the day, no one could say for sure. But with France now extending its moral and fraternal support to revolutionaries across Europe, even as far away as Ireland and Poland, the president could not dismiss the possibility that the Whiskey Rebels would accept French sponsorship. Early in August, he collected the opinions of his cabinet: The attorney general, fearing that the upheaval was perhaps intended to "overthrow ... the federal government," suggested sending a commission to western Pennsylvania to judge the temper of the countryside; Hamilton, supported by War Secretary Henry Knox, instead wanted a more muscular response: to send out the militias of neighboring states—in short, a decisive show of force. "The very existence of government," he warned, "demands this course." New Secretary of State Edmund Randolph fretted that sending troops would only unify and intensify the opposition, while Chief Justice John Jay worried that warlike preparations would end up inflaming matters further. For now, Washington imaginatively followed a middle course: preparing to summon the state militias while sending out commissioners to demonstrate that the government was not acting rashly. "The public mind," he declared to his advisors, "[must] be satisfied that all other means have failed" before resorting to military might.

In August, Pennsylvania officials met privately with the president, hoping to head off a disaster in their state. And Washington did his part as well: The administration had already addressed some flaws in the original tax law, lowered the duty, and now it dispatched the emissaries into the West to appeal for obedience and peace. But these measures did nothing to assuage the tempers flaring. Federalists in the capital excoriated the rebels as an "ignorant herd," as "rabble" and "restless sons of anarchy." For their part, convinced that these taxes were every bit as illegitimate as those levied by the British ministry against the colonials, or the French king against the peasantry, the rebels now dug in their heels. These were no minor contretemps. "We will defeat the first army that comes over the mountain," boasted one rebel, "and take their arms and baggage." Would they? Even former U.S. senator William Maclay darkly averred that "war and bloodshed are the most likely consequence of all this."

In early September, the commissioners reported back that moderates had won a close vote in one meeting of the rebels, although there was still talk of erecting guillotines; to the administration, then, it remained a dangerous mob. Hamilton whispered in Washington's ear that if the rebellion was not quashed, and quickly, it would spread.

Anxious but not panicky, Washington convened a tumultuous eight-hour marathon session of his cabinet. He was by no means the last president to be in such a dilemma—assessing just when all diplomatic options had been exhausted—but he was the first. Having tried negotiation, Washington saw little alternative to his next measure, rattling his saber. Yet ironically, in doing so, he would sound much like the reviled King George III in 1775—or Empress Catherine in the early stages of the Pugachev insurrection. Proclaiming that an "open rebellion" existed—along with the Civil War, this was the only time in American history—the president authorized the use of force to suppress the insurrection. Some twelve thousand militiamen from Pennsylvania and three other states were ordered to crush the rebellion—more Americans than had defeated the British at Yorktown. And at Hamilton's recommendation, Washington himself led the army as it rode west—the only time a sitting president has actually assumed personal command of forces heading into battle. Hamilton came too.

But for all his outward presence and composure, inside Washington felt a stab of fear. And the fear was largely visceral. Privately he was convinced that the Democratic societies feeding this miasma were the product of a cabal of insurgents looking to effect a "revolution in government" and to destroy the Union. "I consider this insurrection," he would angrily rumble, "as the first formidable fruit of the Democratic Societies." This was the French Revolution in miniature—and he wasn't about to become Louis XVI as events spun out of control.

∞

HIS MIND WAS made up and his conscience at peace. To the banging of drums and tramping of the march, a "grave" and "austere" Washington reviewed his army of volunteer troops hastily assembled to quell the rebellion. He went as far as east of the Alleghenies—to Carlisle—then returned to the capital for an uneasy wait. Hamilton continued over the mountains, but as a sign of the president's commitment to this struggle, command of the expedition was turned over to the famous soldier-politician, Virginia governor Lighthorse Harry Lee—Robert E. Lee's father and the man who would memorably eulogize the president upon his death. So it now seemed that America was no longer immune to the revolutionary virus wandering across the world: As in Russia, as in France, and as in their former British nation, it was now to be countrymen shedding blood against countrymen. It was to be the United States' first war.

Adding to the general unease, the readiness of Washington's troops was still questionable. Lacking a standing army, he was forced to hastily cobble together the militias, hastily because no one wanted to allow the rebellion to fester over the winter. So in late October, having crossed the mountains and endured rain, snows, and bitter weather, the newly constituted federal militias converged on Pittsburgh. Back in Philadelphia, Washington, with the look of both predator and prey, now perceived his very government to be at stake: It would be the will of the majority or the tyranny of the minority. Still, Washington once again sought to avert the impending apocalypse, offering amnesty to all rebels who signed an oath to obey the laws of the federal government. And if they did not? Lee was instructed to confine the attacks to "armed opponents" of the laws, but Louis XVI's dismal French experience had underscored the need to have just the right commander. Was Lee the right man to quell the rebellion? Even for Washington and his supporters, that too remained an open issue; Lee was known as a born fighter, but he was also a touch ruthless (he once beheaded a deserter) and fiercely energetic. And one other question lingered. Would the militias stay loyal to the government—or would they desert over to the other side, as both French and Russian troops had done in similar situations?

Worried Hamilton: "I trembled every moment lest a great part of the militia should take it into their heads to return home rather than go forward."

∞

THESE QUESTIONS WERE answered quickly. Once Lee's men arrived, few rebels appeared; the twin approach of amnesty on the one hand and the imposing show of force on the other had worked, and, remarkably, the insurrection petered out. The leaders of the rebellion fled down the Ohio River, heading for Spanish Louisiana. Lee, however, took no chances; he seized 150 prisoners—many were ripped from their homes in the dark of night—charged them with having committed "atrocities," and kept them locked up in vile conditions until they could be questioned. Those to be tried were then force-marched to Philadelphia through the first icy blasts of the bitter winter cold. Shortly before Christmas, the army reached the capital, and thus began the final, grim procession of the defeated rebels. As the bound prisoners, chilled in their threadbare clothes and sadly dragging their feet, were paraded through the streets, merchant vessels fired salutes and a band began to play. In the end, only two men were ever convicted, one an imbecile, the other deranged. Having weathered the storm, a triumphant Washington pardoned both.

∞

THE CONVENIENT VERSION of history takes the simple resolution of this whole matter for granted, that of course all neatly tumbled into place like a well-rehearsed melody; but in countless ways, nowhere had danger and expectation been held in a more delicate balance than during the Whiskey Rebellion. This historic episode was as much a defining crossroad for the United States as the Constitutional Convention, the debate over the Bill of Rights, or the election of George Washington. Hoping for the best, the Founding Fathers had left the question of America's perpetuity to posterity;

but now, with the poisonous French Revolution raging, with Republicans seething and the Federalists vengeful, with the Whiskey Rebels openly flirting with armed rebellion, and with the enduring tension between the spirit of 1776 and the accommodations of 1787 that would not go away, the country could readily have veered in one direction or another. In France, Louis XVI had hesitated, the nation then chose violence, and that is what it got. In Russia, Catherine had used massive force against Pugachev but then abandoned all hopes of liberality. In America, Washington acted decisively but with nuance, and the country got coalitions and politics.

That choice, and its consequences, would do much to set the tone for the new nation—and arguably, never again would the United States be the same.

∞

IN HIS STATE of the Union Address, Washington, feet solidly planted, swept his eyes across the room with a challenging glance and defended his decision to employ military force, placing the blame for the insurrection squarely on the Democratic societies—or that was how it sounded to his critics. For the Republicans, this was a low point in the infant country's history; while the House openly praised Washington for crushing an insurgency that had jeopardized the Constitution, privately, many felt his words smacked of imperial menace. By attacking the societies, Madison charged that Washington was playing a "most dangerous game"—seeking to muzzle legitimate dissent as well as to circumscribe free press and free speech, both cornerstones of the republic. It was, he also maintained, "the greatest error of [Washington's] political life." Back at Monticello, a retired Jefferson shook his head in bitter agreement. In his eyes, the president was a fundamentally different man. Washington had once been "the head of a nation," Jefferson hissed; now he had become simply "the head of a party"—the Federalist Party.

The House itself refused to condemn the Democratic societies, as far as it ever would go in repudiating Washington. Did the Republicans misconstrue Washington's speech? Yes—he never actually disputed the right of aggrieved citizens to protest, only their right to threaten the survival of the nation. But in the fanatical atmosphere surrounding the crisis, that was not what Washington's enemies heard. Henceforth, the Republicans irrevocably vowed to think and act for themselves.

For all the promise that had dawned with the new decade, now the carnage of rebellion was taking its toll across the globe. There was the toll in deaths: In France, the king and queen were dead, the Terror was in full gear, and nobles and peasants alike were being capriciously dragged off to the guillotine or murdered in fields and alleyways by the regime's representatives. There was the toll in geopolitics: All of Europe was now mired in a murderous war—and the war would soon embrace not only Europe but the corners of Asia, Africa, the Middle East, and North and South America. There was the toll on freedom: The constitution was suspended in France, in England habeas corpus was suspended (and the Seditious Meeting Act was being drafted), and in Russia Catherine the Great now ruthlessly repressed her own people, suppressed the Poles, and abandoned all pretenses of reform. And there was the personal toll: The empress's health was dangerously declining, and an exhausted William Pitt the Younger, one of Britain's most majestic prime ministers, was reeling from the toils of war; he would soon die at only forty-seven. And in America, there was the toll on the sixty-two-year-old Washington as well.

The president looked and acted old. The awesome fire that had fueled his passions as a young patriot had been reduced to embers. How different this was from the robust Washington memorialized in song and legend, or, for that matter, in his own mind. Upon retirement, and before the presidency, he was always a commanding physical presence; in the same breath, he always cherished the privacy that allowed him to think and the solitude that allowed him to be outdoors. At Mount Vernon there was always a kind of ritualistic unfolding of the day, from the first gray moments of dawn to the long olive twilight, a veritable private pageant for the president. He would typically awake at five o'clock with the rising of the sun, powder his hair, don his clothes, and eat a light breakfast of soft corncakes buttered and coated with honey. From this moment on, he was the playwright, producer, director, stage manager, and, of course, hero of the day's performance. He often spent six hours on horseback supervising a small army of carpenters and painters, examining roof repairs or hedgerows, insisting that drainage ditches be widened or lengthened, dealing with poachers on his property, and asking after his slaves. At dinner he led his guests —a parade of Americans, Frenchmen, eminent dignitaries, monarchists, nobility, democrats, warriors, thinkers, Poles, Austrians, and Britons—to the piazza overlooking the Potomac; there he strode up and down, holding forth on every subject from plow designs to crop rotation schemes, from Hessian flies to his collection of medals, from the French Revolution to the dismal fate of Poland. Among them were many celebrities, but none more so than he. To the surprise of his guests, he loosened up after several glasses of champagne or wine—his one vice. As dusk fell, leaving his visitors, he retired to his study; there he rifled through his mail sorted into piles, diligently wrote letters or read any one of the ten newspapers neatly stacked for him by servants. And on quiet evenings, he ruminated about the "calm lights" of "mild philosophy."

Yet the world was now anything but calm. It had not become just dangerous but shrill; and an embattled Washington was now regular bloodsport for his critics. No longer a tribune to the whole nation, he was often heckled or mocked behind his back—or, heresy of heresies, in the open. Unflatteringly, Thomas Jefferson compared him to an aging "captain in his cabin," dozing while the country was being driven into "an enemy's port." Editorials derided him as a "tyrannical monster" and spoke of "the loathings of [his] sick mind." Even Washington himself confessed that he was "worn away" in mind as well as in body—and more than ever, he talked of "the short time" he had to stay here. But there would be no respite. In 1794 and 1795 Washington had one more crisis to face—among the greatest of his presidency. As with the French imbroglio, it raised the ugly specter of division and the terrible possibility of war. More so than even the French imbroglio, it would be Washington's most beleaguered hour.

It would also be one of his finest.

∞

AS THE MID-1790S approached, the fate of most Western nations seemed increasingly intertwined: France to England, England to Russia, Russia to Poland, Poland to France and America, and America to England and France. As if the looming problems with France weren't enough—Genet may have been cashiered but the aftermath of his escapades was hardly over—Washington was confronted with another disaster abroad, this time with Great Britain. In December 1793 British cruisers began seizing American merchant vessels in the Caribbean; hundreds were taken on the flimsy grounds that they carried French contraband, whether they actually did or not. In the United States this quickly became a casus belli: Americans had already claimed the right as neutrals to carry noncontraband goods to and from the ports of belligerents, and the mood of the country was already anti-British, which didn't help matters any. Nor did the fact that the British still treated the Americans as colonial subjects in other ways too. They agitated the Ohio Creek Indian tribes against the Americans, and they refused to remove British troops stationed in their forts on American soil, as if the Revolutionary War had never ended. So within a matter of months, the situation had become untenable.

In April 1793, as the House of Representatives debated ways to entice England to lift its economic restrictions against the Americans, news of the seizures fueled an overwhelming demand for much stronger and more virulent anti-British measures. Already seduced by Genet and the romance of the French Revolution, the country was now swept by war hysteria against the British: Volunteer defense companies sprang up, while mobs raced into the streets and harassed English seamen; meanwhile, pro-British Americans were tarred and feathered. To Washington, this was insanity: The country was steadily slipping into a war it could not win, or even if it somehow could or did win, it would prove to be a Pyrrhic victory. There was no greater example of this than the ruthless carving up of Poland and the earlier ruin of Corsica: One newspaper had aptly warned about "the partition spirit of the times," while another observer had noted, "After we had weakened ourselves with mutual devastation, we could expect no better fate than that of Poland, to be distributed as appendages to the sovereigns of Europe." Could Poland's tragic demise also become America's? Unless Washington acted, he knew events could become dangerously unmanageable. Hastily, Washington tapped the chief justice, John Jay, to negotiate an accord with London.

But the effort seemed stillborn even before it began. True, Jay was a seasoned diplomat and one of America's most eminent statesmen: Secretary of foreign affairs under the Articles of Confederation and envoy to Spain during the Revolution, he also was an author of the *Federalist Papers*. But simply negotiating with the British was unpopular at home—the young Americans had a clearly inflated sense of their ability even then—and Jay was widely criticized for his pro-British bias. Still, his mandate from Washington was broad: Much the way Franklin and other diplomats had played the European powers off each other in the War of Independence, Jay's instructions were to convince the English that unless they made concessions, they could not count on continued American neutrality with France. Jay was thus also instructed to consult with Russia, Sweden, and Denmark about the possibility of an armed neutrality agreement to bring pressure on England, not dissimilar from what Russia's Catherine had shrewdly formed during the Revolutionary War. It was a good plan. But it went nowhere, as Jay's luck appeared over even before he began. He was undercut by none other than Hamilton himself; somewhat foolishly, Hamilton persuaded Washington to decline an alliance of neutrals—though Denmark and Sweden were interested—on the grounds that it would antagonize the British. Seeking to create a favorable climate in London, Hamilton weakened Jay's delicate position even further by informing the British minister in America of Washington's decision.

Nevertheless, in England Jay did the best he could, and on March 7, 1795, the finished treaty reached Philadelphia. By now, the results were almost predictable; the British gave little and received much. Even many Federalists were unhappy. Jay did manage to secure the removal of British troops from the frontier—this was a plus—and committed the British to arbitrate American claims of compensation for cargoes confiscated on the seas, but only with onerous restrictions. Beyond that, the treaty was seemingly a near disaster, at least on the surface. It accepted British economic superiority; it tolerated British naval supremacy; it failed to secure compensation to American slave owners for slaves kidnapped by the British during the Revolution; it refused to give any guarantee against the British navy's practice of stopping American vessels to impress seamen; and it decidedly tilted away from France and toward Britain.

Of course, this all but ensured the opposition of Jefferson, Madison, and the Republicans, as well as a divisive debate. Yet, on balance, Washington was disposed to seeing the treaty ratified. To his secretary of state he muttered that he was "not favorable to it," but here he was being less than candid. He believed the treaty enshrined diplomatically what was already a fact militarily and commercially; he also believed that Jay probably got the most favorable terms he could, or as Hamilton later explained, "it is not for young and weak nations to attempt to enforce . . . pretensions of equivocal validity." And most importantly, it offered virtually the only hope of preventing a conflagration with Britain. The president decided to submit the treaty, and with the zeal of a convert, sought its ratification.

Fearing a premature hardening of opposition, Washington smartly sought to keep the terms secret until he could present them at a special session of the Senate called for June 8. Yet by the time the Senate met, the air of mystery had only sparked Republican fears and the contents of the treaty had already leaked; rumors of the provisions produced widespread public opposition and outright hostility, and that hostility only stiffened and hardened as further details became known. When the president dined alone with John Adams to enlist his support, his vice president worried, "I see nothing but a dissolution of government and immediate war."

Still, Washington was not without chits for the upcoming vote, and his qualified optimism was born out. The Senate, after deleting one clause concerning trade with the West Indies, accepted the treaty by the barest of margins: exactly the two-thirds majority required and strictly along party lines—but enough to put the issue to rest. Or did it? Actually, the fight was hardly over. To the president's growing dismay, as the treaty made its way to him for his signature, the outside furor mounted. The press denounced Jay, criticized the treaty, derided the Senate, and in a constant drumbeat, reserved some of its most trenchant words for Washington himself. One Virginia editor actually suggested a toast for a "speedy death to General Washington." Meanwhile, when the press wasn't sticking its finger in Washington's eye, popular meetings were. Across the country—in Boston, Philadelphia, New York, and countless other cities—they screeched until their voices were hoarse for Washington to reject the treaty, while in Manhattan, seven thousand Republicans, stretching from Broad Street to Wall Street, noisily marched against it. And day after day letters poured in condemning the pact as a deal with the British "Satan."

Then the opposition got truly ugly. Jay's treaty, and his effigy, were burned up and down the entire Eastern Seaboard. Rioters in Philadelphia, clogging the avenues, broke windows in the houses of the British ambassador and a Federalist senator. In New York, Alexander Hamilton was pelted with stones. And John Adams was stunned to see the presidential mansion surrounded from morning to evening by protesters repeating the same stinging calls, a deafening refrain chanted over and over again in an ever-escalating crescendo, demanding war with England, cursing Washington (a "horrid blasphemer"), and calling for the success of the French patriots; marchers even impaled the treaty on a pole and carried it to the home of the French ambassador. The vitriol was unrelenting: A pale and utterly depleted Washington was compared unfavorably to King Louis XVI, that is, when he wasn't being compared to Catherine the Great's old nemesis, the grand sultan of the Ottoman Empire, "shut up in his apartment"—or when it wasn't snidely said that he was acquainted only with the "seraskier[s]" or "mufti[s]," and not with the people.

And the storm refused to abate or die. By some estimates, hundreds of articles and many tens of thousands of words on the treaty, much of it mere diatribe, spewed forth throughout that summer and fall. Time and again, Washington had straddled the middle ground: It was uncomfortable, it was tenuous, and invariably doomed. What fleeting moments of cooperation he had with his critics became less and less. He clashed continuously with his friend, the brilliant, diminutive Madison, and he was forced to maneuver nonstop against his once surrogate son, Jefferson. The hostility was often implacable—and the abuses and intrigues were equally unrelenting. The time for holding out an olive branch to his political adversaries was over, and they felt the same way. Now, to an unprecedented degree, both parties sought to enlist public support. To an equally unprecedented degree, both sides knew that the Jay treaty reverberated with cataclysmic options for war or peace, open markets or economic stagnation, and perhaps for even the success of their sister revolution in France and the future of democracy.

And once again, the cold, dark battlefields of Europe were proving to be crucial in shaping American politics. Faced with this dismal situation, the president weighed his next moves: Yes, he believed that the treaty avoided a popular but misguided war with a stronger power; yes, he believed it preserved economic relations with the United States' major trading partner. And yes, the cabinet agreed too; only Secretary of State Edmund Randolph was against it. But wracked by the public criticism, Washington suddenly hesitated. For the first time in his career, he flinched.

Around this time the British minister handed Oliver Wolcott, Washington's new treasury secretary, some intercepted dispatches from the outgoing French minister which suggested that Randolph had turned over state secrets to the French for money; in a word, that Randolph was a traitor. Hearing about this, the president moved decisively—Randolph had to go. His blood up, flanked by other aides, on August 19 he dramatically confronted Randolph, a much beloved advisor in the cabinet and a friend of two decades. Here, behind closed doors, Washington was at his coldest and most malicious; meanwhile, with his head bowed, Randolph, reeling from the vehemence of the accusations, defended himself as best he could—and he was not without justification. In retrospect, it is almost certain that all he was guilty of was little more than what would one day be an old Washington, D.C., game—boasting of his own influence and diminishing the president's discretion. More pointedly, it is also certain that Randolph was set up by the British; simply put, the Americans had been duped. The president should have known better. Nevertheless, each word that Randolph uttered on his behalf proved more disastrous than the next. And while Washington listened much, he heard little. Irate, he accepted Randolph's resignation on the spot. The die was cast; the next day Washington signed, and thus, in his mind, finalized the acrimonious Jay treaty.

Far from being over, though, the affair continued to heat up. Jefferson, who bellowed that the treaty created a neocolonial status for the United States within the British empire, now protested that it was a treaty of the "Anglomen" and England against the legislature and the "people of the United States." And once the spring of 1796 rolled around, the machinery of opposition kicked into a higher gear. Contorting the Constitution, Jefferson argued that the House possessed a sovereign power over treaties. His House allies even requested that Washington hand over all confidential documents related to the negotiations, which Washington soberly rejected as a "dangerous precedent." It was. Yet nothing was sacrosanct anymore, and the Federalists and Republicans were squaring off like angry Jacobins and Girondists in the French Assembly. Snubbed by the president, Madison and House Republicans fought back, pledging to sabotage the treaty by starving it of the funds necessary for its implementation. The two sides headed for an April showdown as the debate in the House moved forward.

Washington was never enough of a politician not to personalize things: In a sense, it was always about *my* policies, *my* motives, *my* goals. The toll the fight took on him in these tension-ridden days was evident, in his temper tantrums, in the bouts of impatience or flashes of arrogance, or in the lines crevassing his face and the spark missing from his eyes. But the president, itching for success, kept plugging away. And on the opposite side, so did Madison.

With all the cunning he could muster, the House leader prowled the House chamber, bent arms, relentlessly finagled votes, corralled wavering members, and actively managed the floor debate. He had the votes—by his own count, twenty to spare. For Washington, the situation was perilous; as John Jay had warned: "If this treaty fails, I despair of another." A desperate Washington, aided by Hamilton and others, fought back by rushing a defense of the treaty into print. And this was a watershed of sorts for the young republic: a titanic battle over who would run foreign policy—the executive branch or the legislature, the president or the House leader. The president was determined to win, but so was the tireless Madison. At the end of April came the vote—on a second roll call Washington eked out a slim margin of three, 51–48—but it was enough to settle the matter for all time.

Yet no one escaped intact. Madison was shaken and fatigued, and his friendship with Washington—he had once been among his most trusted confidants—was completely destroyed. Jefferson, who had orchestrated a whispering campaign that Washington was a senile old bumbler, would also watch his relationship with the president wither away. Partisanship, once to be avoided at all costs, now was fixed forever in the American landscape. And a haggard Washington, despite a grand victory—he triumphantly called this one of those "great occasions" in his presidency—had lost any pretense of being the "singular figure" to unify all Americans. For some time now he had suffered acute back pain, biting facial neuralgia, and nagging rheumatism; even chewing his food was near torturous. But nothing hurt him more than having lost his luster as being indispensable to the nation. And so he made a momentous decision. He categorically decided to forgo a third term.

Finally, he was going home.

∞

But Washington did not wish to simply fade away. He had given a magnificent address when he left the army in 1783, but this would be an even more difficult challenge: It was to be the capstone of a lifetime of service to his countrymen and his last major service to the country. And it would also carry a fatherly warning of the looming dangers for the nation.

The president prevailed one last time on Hamilton. Sending him a first draft in his own words, along with Madison's original text from four years earlier, he entreated his old protégé to prepare his remarks, in "plain style." Hamilton knew all his favorite aphorisms, his prejudices, and even his anachronistic phrases; indeed, by now whatever he wrote would invariably ring with Washington's voice. Over the course of that summer, Hamilton wrote several drafts, and the president made dozens of small changes, tightening and clarifying, including in the final draft. Once finished, the president preferred Hamilton's version "greatly"—but even if the words were largely Hamilton's, the ideas were distinctly Washington's. The Farewell Address was then delivered, not in person, but in print, appearing simultaneously in two Philadelphia newspapers on September 19, 1796, then across the country. It was a heartfelt message reflecting on the wisdom he had garnered in his nearly two decades of public service, and it marked his final thoughts on the true meaning of the American Revolution.

It was also vintage Washington. He broke the news that he would soon be leaving office, explaining that he originally had hoped to retire in 1793 but remained at the helm because of the emerging foreign crisis. His theme was national unity: transforming a weak confederation into a wealthy and powerful nation capable of maintaining its sovereignty against all rivals. He thus stressed the necessity of Union—nothing was dearer to his heart—touched on the virtues of citizenship and eloquently upon "the love of liberty," and asked his readers to look beyond narrow regional or state ties to see themselves as an entirely new breed, "The name of AMERICAN, which belongs to you in your national capacity . . ." Perhaps his most trenchant commentary addressed the rush of events abroad. He boldly asserted that lasting "independence" must always be America's preeminent concern: also as such, he warned against the "insidious wiles of foreign influence" as well as indulging in "a passionate attachment" to France and "habitual hatred" of England. Lastly, he noted that it was imperative that the United States "steer clear of permanent alliances."

Yet for all his thoughtful words and gentle admonitions, the Farewell Address was also a party document scrupulously woven to justify Washington's policies and Federalist ideology. With some overstatement, when it spoke about "ill-founded jealousies and false alarms" of factious men, or the "small but artful and enterprising minority," it meant the Jeffersonians. When it advised the people that the power and responsibility belonged "to your Representatives" and praised "the real patriots," he meant the Federalists. And when he denounced those who sought to "impair the energy of the system," he was defending the federal government rather than the states. Finally, to his critics he had all but solidly aligned America with monarchical Britain and unflinchingly repudiated the long love affair with revolutionary France.

The address's initial reception was decidedly mixed. No surprise, after it appeared, the country remained mired in continuing feuds and stalled in endless bickering. Many Republicans treated the address as little more than a campaign treatise, and Madison condescendingly asserted that Washington was "compleatly [sic] in the snares" of the Federalists. As it was, it scarcely affected the upcoming election, or, for that matter, any issues in the short term. But ultimately, the address, dignified, firm, and a touch combative, would be one of America's most revered documents, forever fixed in the nation's memory; speaking of independence abroad and unity at home, it was read annually before Congress and guided the nation for a century and a half until the brimstone of two world wars made it impossible for the United States to steer clear of Europe.

In one manner it did have an immediate impact. With the publication of the Farewell Address, a congressman commented that it was a signal, "like the dropping of a hat," for party racers to start. It was. And near Christmas Eve, the election for the second president of the United States was over: John Adams, with seventy-one electoral college votes, secured a narrow victory; Jefferson, with sixty-eight electoral votes, would be the next vice president. And an outgoing Washington was suddenly a changed man. True, his face was now the portrait of exhaustion, and he looked far older than most people had remembered. Also true, he still felt the victim of gross misrepresentations and distortions. But once March 4, 1797, drew near, he again sensed that he had "the approving voice" of the country, and at long last could say he was delighted that the "curtain of my political life is about to drop."

It did, quickly. For Washington, the last weeks were spent navigating a whirlwind of lavish dinners, ceremonial receptions, "ladies' concerts," a birthday party of 1,200, and dances in his honor. Then, on his final full day in office, he threw a midafternoon dinner for department heads and diplomats. The crowd, glowing, buzzing, proud and expectant, collected in anticipation. At the meal's conclusion, Washington stood to offer a rare personal toast, delivering a public epitaph on his eight years as president; even today, his words are simple yet elegant, and laced with no small measure of emotion. "This is the last time," he said, "I shall drink to your health as a public man." A sober hush descended over the crowd, and tears of emotion streamed down the cheeks of the British minister's wife. It was a stunning moment, for the world had never seen anything like it: Upon his own volition Washington was neither going into exile nor house arrest, as was so often the norm for deposed monarchs, or to the blade or the chopping block, as was also the norm—but into retirement.

∞

THE NEXT DAY, John Adams was sworn in. It was a day of conspicuous simplicity; assuming the mantle of a private citizen, Washington walked the three blocks to the inauguration from his residence. Nonetheless, people lined the avenues, hoisting flags and humming patriotic songs. With mechanical precision, Washington and then John Adams entered the packed chamber. Thomas Jefferson was there as well. The crowd gasped, loud cheers went up, and many in the audience erupted into tears, profoundly uneasy that Washington was exiting civic life for all time. All time—that was a frightening concept for countless Americans. For this was the man who had stymied the British at Yorktown, who inspired the magnificent assemblage at the Constitutional Convention, who gave meaning to the term Founding Fathers, and who had carried the country through eight long years. And now, in fits and starts, there were those who proudly called themselves Americans, and now the American flag laid its own claim, flying indisputably in homes across the nation. All largely due to Washington. Yet when the incoming president seated himself next to Washington, he felt the outgoing president's uncharacteristic exhilaration. Said Adams, "I heard him say, 'Ay! I am fairly out and you fairly in! See which of us will be the happiest.' "

That night, hundreds of residents attended one last supper. "Washington's March" was played, and Charles Wilson Peale's painting of Washington on horseback was unveiled. Soon thereafter, Washington, with quiet dignity and self-possession, rode out into the streets, to his home in Mount Vernon, pondering the still uncertain prospects for his nation that he had so desperately sought to steer.

∞

As 1796, one of the great watersheds of history, drew to a close, along with it a darkened ray began to drop and spread across the globe. More than most could remember, it was a time of great foreboding. And arguably, no facet of this period was more important or interesting than what could best be called the transition of power. Catherine the Great, dominatrix of West and East, was now dead, and in defiance of her wishes, her son, Paul, had assumed power; he would not last long. Europe was a land tearing itself apart, Poland was extinguished, the Islamic Ottoman Empire was again growing restive, and revolutionary France was pounding the hardened fist of its *levée en masse* across the continent. In Paris, where the gutters of blood had finally dried up, a new regime had come to power in a coup, the Directory; yet it would be no less immune to upheavals than its ill-fated revolutionary predecessors.

The Old World, witness to a thousand years of power and peril, knew all too well the deadly fruits of crusade and conflict. Over the next four years, as the body counts mounted and more graves were dug, where would it all lead—republicanism or reactionary autocracy? Democracy or enlightened despotism? Revolution or counterrevolution? Curiously, only a young United States, weak, insignificant, and straddling the world's margins, had seemed initially to escape the ugly rules of the rest of the globe. What could have been more incongruous? Nevertheless, to the Founders' dismay, or at least to some of them, America was now deeply torn. To their dismay, in the wake of the global upheaval, it was often a nation at war with itself. And to their disgust, as it ventured farther away from the writing of the Constitution, the country was reinventing itself with each crisis it confronted—with wholly unpredictable consequences.

Yet for all its domestic dissent, there remained one crucial difference: Never did matters plunge out of control. Only in the United States was there a peaceful, constitutional transition of power, from the towering figure of the age, George Washington—though arguably few knew it at the time—to John Adams. It was an unprecedented event. But it was also an expedition into the unknown. Still, even as America wrestled mightily with the questions of what its own revolution stood for, even as many Americans saw themselves as one with their brother revolutionaries in Europe, the young country earnestly sought to write itself not in bloodshed, but politics, and that notion would prove to be as revolutionary as any other idea or movement or army sweeping the globe. Lord Chatham once said, "I love the Americans because they love liberty." In the end, the Americans decided that they did too.

Haunting dilemmas, however, remained: Could America divorce itself from the rest of the world and continue its bold experiment? Or would the superpowers of the day, whether the revolutionary republic of France or the monarchy of Great Britain, engulf America into the mighty vortex of their rage? The next few years would tell the tale. Throughout the millennia, such brief periods had been enough to form deadly alliances, start and finish wars, unseat kingdoms and topple great dynasties, plunge countries into unmitigated chaos, and swallow shaky republics. Indeed, there was one sobering fact: the bleak maxim that in matters of politics and war, history could abruptly shift and turn, that the exhilaration of one day could dissolve into disaster or tragedy by the next.

As the decade marched forward, this was precisely what a struggling America, under a new and untested president, and still calculating its national survival, would confront.

CHAPTER 12

AMERICA

✦

A GREAT EMPIRE AND little minds," Edmund Burke had said in 1775, "go ill together."

Implausible as the notion might have seemed then, he could well have been describing a young, beleaguered United States in 1797. The presidency, still a fragile enterprise, represented a staggering responsibility, and George Washington, a legendary figure, was still impossible to replace in America's affections. Yet against all odds, after eight years, the country remained drenched in revolutionary mythology; now, as the first president rode south, the second, John Adams, of no little mind himself, wasted scant time in assuming power. The challenge for Adams, of course, was immeasurable. For starters, on a personal level, Adams remained eclipsed by his predecessor. On a political level, Republicans and Federalists now seemed to be speaking different languages, and so did the majority of Americans, who increasingly lined up behind one party or another. The fledgling government was thus not a place of respectful debate and polite unity but a grim frontier, separating two ideologically distinct worlds. And on the level of national survival, the relentless tread of history could not be avoided: The nation was again assaulted by the murderous struggles of war and peace—once more in the guise of the armies of revolutionary France.

∞

A FRANCE REBORN was the crusading vixen on the world stage, and by the summer of 1796 the continent had entered a dangerous new phase. All of Europe shook as French forces relentlessly pushed toward Austria, breathtakingly seized most of Italy, set up a client nation called the Cisalpine Republic, and for good measure, absorbed the Papal States of Bologna and Ferrara—the pope was held a virtual prisoner in his Vatican apartment— along with the Duchy of Modena and later of Mantua. So mighty were France's victories that it was no longer the pariah of Europe; strength had trumped ideology. With Napoleon Bonaparte encamped elsewhere, Bourbon Spain preferred to keep it that way, as did the Holy Roman Empire. So the Spanish made peace with France, Austria prepared to, and while only England lay beyond its talons, even the British had opened discussions with the Directory—this, despite the fact that the French supported an Irish rebellion against their English masters. Nor was America immune from the Directory's wrathful gaze. They interpreted America's treaty with Britain as a threat to France, or at least as an affront. And so U.S.-French ties quickly began to unravel, and the storm clouds of war again increasingly hovered in the air.

As quick to anger as they were to insult, in a reprise of Edmond Genet, the Directory had dispatched a new emissary, Pierre Adet, who had openly interjected himself into the 1796 American election; this time he sought to swing it away from Adams and toward Thomas Jefferson. Actually, he came close to succeeding. Eschewing official entertainments in Philadelphia, Adet instead busied himself in manipulating American opinion. Thus, he published a series of menacing essays against America, tartly instructing all true friends of France ("a government terrible to its enemies but generous to its allies") to wear the tricolor cockade—and many Americans did. Then he hotly labeled Jay's treaty as an outright alliance with Great Britain, setting the stage for a diplomatic break. And ultimately he announced the suspension of relations between Paris and Philadelphia. Too, like Genet before him, he sought to split Republican from Federalist, American from American, and citizens from their government: "Oh America," he preened, "you who have so often flown to death and to victory with French soldiers . . . Let your government return to itself." When his silver tongue didn't work, he used intimidation. Far from being clumsy, he actually succeeded in helping to sway Pennsylvania toward Jefferson, thereby diminishing Adams's victory. Just as ominously, the French also announced they would seize American ships carrying goods bound for British ports; within weeks, virtually all American ships became fair game. And salting the wound further, Adet even joined Benjamin Franklin Bache—Benjamin Franklin's grandson—in a widely disseminated pamphlet attacking Washington himself.

Where was all this heading? Alexander Hamilton had always believed foreign influence to be the "Grecian horse" of republics, and by this stage, one had to wonder if he was correct. Washington thought so. "It remains to be seen," he wrote, "whether our country will stand upon independent ground." Independent ground. Few men believed in it more than the shaky new occupant of the presidency. Yet few men had the odds so seemingly stacked against them. So as the presidency fell to John Adams—he had won by a mere three votes—it also fell to him to guide America through a world now indisputably at war.

∞

IN EVERY AGE there are certain articles of faith that society accepts unquestioningly, with or without evidence, often, indeed, in the face of inconvenient facts. The faith may be religious, moral, or political. For America in 1797 it was political—a belief, however inflated or audacious, in its special place in the world. But history can never be so easily led into any one dock. The difficulty, invariably, is that one may steer for the wrong dock, or the tide may angle you into a different waterway altogether, or there may be oars in the water pulling at cross-purposes. Within weeks of assuming the presidency, this would be Adams's fate, as he found himself enveloped in a web of deceit—abroad, at home, and in his own councils of government.

It is hard for Americans, two centuries later, to appreciate the enormity of the change from George Washington to John Adams. Washington, unanimously elected president twice, was everything Adams was not, and Adams, a complex amalgam of character, was anything but a reincarnation of Washington. Where the Virginian defined the dignified manner of a public servant, Adams was often wild with anger, hot-tempered, even a bit flighty. Where Washington was reticent, Adams, whether out of intention or impulse, or both, always told you what he thought. Where Washington was stolid and meticulous, Adams was impatient and argumentative; some of this was due partly to style—ironically, the two men could say the same thing, yet what came out of Washington's mouth as being circumspect or exacting came out of Adams's as strident demagoguery. In the same breath, while both men were prickly, an action by Washington appeared as evidence of strength and leadership, while the same action by Adams was curmudgeonly and choleric. In a sense, Adams couldn't win. Roused to indignation, Washington got angry; Adams was instead seen as petty—or paranoid. Washington appeared statesmanly; Adams was crotchety. And where Washington was self-effacing, Adams was wild and melodramatic; Franklin once called him "absolutely out of his senses." In truth, Washington thoroughly knew who he was and was comfortable in his own skin, while Adams was chronically insecure and eclectic; all his life he was a gadfly. He didn't want to be. Still, despite his enormous achievements, too often he felt marooned or depressed or irrelevant, when of course he was anything but.

Yet in his own way, Adams was confident of his own uniqueness, enough so that he could not commit himself to any one paragon or ideology; hence, unlike a Hamilton or a Jefferson, he would never be a party man. Actually, while his flaws were numerous, his strengths were also considerable. He may well have had more integrity than Washington; he was as much a witty conversationalist as Hamilton; he was as warmhearted as Franklin; and he had a first-class mind that arguably equaled Jefferson's and Madison's. More than two centuries hence, one is still struck by his pounding energy, his ceaseless volatility, his dogged determination, his refreshing frankness, and his utter lack of humility. Still, for all this, he never looked apologetic. Nor should he have been. In truth, he was joyously human, a touch anachronistic, endearingly temperamental, capable of willful misjudgment but also of uncommon vision. While he did not dwarf all those around him—Washington did, so did Franklin and Jefferson—he nonetheless remained one of the most benevolent statesmen of his time and one of the most gifted. As it turned out, however, in the long reach of events, the impact of the Adams era was decidedly mixed. He kept the peace but inadvertently presided over the slow death of his own party. Aristotle once wrote, "History is what Alcibiades did and suffered." But while the Alcibiades of the ancient world routed the Spartans, in the end he was dismissed, fled to Asia Minor, and was murdered by Spartan agents. Tragedy is the waning shadow always cast, sooner or later, by towering heroism; in the course of his four hidebound years as president, this would be the lot for Adams as well.

He had come a long way. Unlike Jefferson and Madison, who grew up in the refined social strata of Virginia's grandest households, he was raised with few patrician comforts. Life was hardscrabble from the start. Born in 1738 in Braintree, Massachusetts, he was reared in a modest home: His father, his idol, scraped out a living making shoes in the winter and laboring as a farmer in the summer. But what the family lacked in finances they made up in ancestry: From beginning to end, Adams was a child of Massachusetts, a son of New England, and pure American—his father, his grandfather, his great-grandfather, and great-great-grandfather all came from Braintree, too. His family, riding the wave of the great Puritan migration, dated back to the 1630s—in the reign of King Charles I. Taught first at home and then in a neighbor's kitchen, Adams began his formal education at the age of six, and by fifteen he entered Harvard, studying under the distinguished astronomer John Winthrop; he was the first in his family to attend college. His lifelong creed became honesty, independence, and love of country, and his rise was meteoric: He graduated, briefly taught school, turned to law, and married well: Abigail Smith, a minister's daughter who loved books and had a keen feel for politics. Their marriage, spanning almost five decades, was tender and loving and enduring; she was also a great advantage for his political career. As president, arguably his best personal asset was his captivating spouse. And despite long separations—he was often abroad or away in government—and her ill health—she suffered from crippling rheumatoid arthritis and at one point almost died—each maintained an abiding love for the other.

Early on, through a steely drive and unceasing ambition, Adams forged his own unique trail. He made a name for himself in the 1770 Boston massacre, not for prosecuting but for *defending* the British soldiers accused of killing the five colonists. At the same time, he fought the hated Stamp Act. Thus, even then people were often not sure what to make of him: His temper was sharp and merciless; he was quick to be insulted; and he was iconoclastic. Yet by the time relations with Britain were souring, he had emerged as one of America's foremost statesmen—elected to the Massachusetts Assembly, then the Continental Congress, he quickly became one of the colonists' leading orators and impassioned voices for independence; he once bellowed: "Swim or sink, live or die, survive or perish, [I am] with my country," magnificent words that more than rivaled Patrick Henry's. From 1776 onward, Adams was seemingly everywhere: He served on an astounding ninety congressional committees, ably chairing twenty-five of them. Arguably he was the greatest talent scout in the nation's history, tapping Jefferson to write the Declaration of Independence and nominating Washington to be the commander of the Continental Army. And he was one of the revolution's most esteemed political philosophers, drafting the widely emulated Massachusetts constitution and publishing a pamphlet, *Thoughts on Government*, that became a basis for other state constitutions.

Quirky, pugnacious, outspoken, and fiercely patriotic, he also received some of the young nation's most coveted diplomatic assignments: emissary to London, commissioner to Paris, and negotiator to Amsterdam. Over and over he made epochal decisions, playing a critical role in securing Dutch loans for the struggling republic, and just as importantly, Dutch recognition, which in turn played a major part in convincing King Louis XVI and the world that America was a worthwhile gamble to back. At the same time, he mentored Francis Dana, America's emissary to Russia, who so assiduously sought Catherine the Great's aid for the colonies.

Physically, Adams was nothing to look at: a fat little man who appeared "half Washington's height," he had a round jowly face, a pale complexion, a shock of scrubby white hair, and piercing eyes. The other Founders looked the part: Washington, Jefferson, Hamilton, even the avuncular Franklin, all could walk into a room and capture the attention with their presence—or their charisma. Adams had none, so he had to rely on his wits—and his intelligence. These, at least, were considerable. It was not for nothing that Benjamin Rush said Adams possessed "more learning, probably both ancient and modern, than any man who subscribed to the Declaration of Independence." Curiously, though the twentieth century tended to overlook him, American children coming of age in the early 1800s were told that, except for Franklin, Adams was an intellectual giant among the Founding Fathers. There is a distinct ring of truth to this, and Adams's numerous writings and letters are a fantastic collection of sharp insights, powerful observations, and uncanny understanding. And there was more to him than that. In his own way, Adams was quite brave: En route to France during the War of Independence, rather than waste any time waiting out the winter for safe sail from Spain, he stubbornly took his family and his party to France by foot, an often perilous, monthlong journey by muleback simply to reach the border. Moreover, he had an uncompromising sense of morality; he was utterly opposed to slavery. Given a slave girl as a gift in 1765, he freed her immediately. And he had a fiercely idealistic streak—reading of Mary Wollstonecraft's hopes for a glorious era when "fool" and "tyrant" became synonymous, Adams penned in the book's margins, "Amen and Amen!" Finally, he was a man of genuine sentiment: Betraying his New England roots, he could, and did, ruminate lovingly about seeing his "grass and blossoms and corn"—and above all, "my books."

But too often his own volcanic opinions, and his vanities, got the better of him. Jefferson, with little overstatement, noted that Adams hated Franklin, hated Jay, hated the French, and hated the English. Routinely he was stalked by his own insecurities—he often thought the world was stacked against him—and haunted by his own inner demons; from as early as 1761, he was almost pathologically obsessed with his place in history. Over the course of his long and distinguished career, he was constantly envious of those around him: Thus, Franklin was "lazy" and overrated; thus, Washington ("Old Muttonhead") was a poseur and a not very intelligent one at that (he could "not write a sentence," Adams maliciously jibed, "without misspelling"); thus Hamilton, his archenemy, was the "incarnation of evil" and, as if that weren't enough, a "bastard brat of a Scotch peddler." When aroused, Adams's tongue was blistering: he once called Thomas Paine "a mongrel between pig and puppy, begotten by a wild boar on a butch wolf"; of France's Revolution he later wailed, "I do not know what to make of a republic of thirty million atheists." He also had a nativist streak in him, railing about the motley rabble of "saucy boys, Negroes and mulattoes, Irish teagues and outlandish jacktars." And he was always a bit of a prig: Jonathan Sewall tells us that Adams could not dance and did not drink—even George Washington did both—never gamed, was unable to make small talk with colleagues, and was incapable of flirting with women or swearing with men; in Paris of all places, he was offended by Franklin's hedonism, even as Franklin was embraced by all the French; later he condemned Hamilton's "fornications, adulteries, and his incests." Not unsurprisingly, he once confessed: "I have constantly lived in an Enemy's Country." Looking back, one wonders if that enemy country was in fact himself. Actually, one even wonders how he made it in the political arena at all; as Adams once acknowledged: "Popularity was never my mistress."

Was he aware of his weaknesses? He was. Jefferson dismissed Adams as "vain, irritable, and a bad calculator of the force and probable effect of the motives which govern men." But often it was Adams who was his own harshest critic. He constantly fretted that he was too "irascible," too "agititated in my own mind," too sensitive to "hypocracy, ingratitude, treachery." In old age—he lived to be an astonishing eighty-nine—he confessed that as a public figure, "I refused to suffer in silence." And almost pitifully, he also admitted that he "sighed, sobbed, and groaned..." Yet for his candor, it was hard not to love him—or admire him. That he cherished public service as he did, even while growing tired of the constant intrigues, the backbiting politics, and the daily pettiness, was in itself remarkable. Still, as he hopped from one elevated position to another, he fussed that the years were slipping by, that he was growing old, tarnished, and irrelevant to the service of his country. Then came the vice presidency. But for a restless Adams who obsessively cared about fame, being in Washington's shadow or presiding over the Senate did little to satisfy him.

Until he was elected president.

∞

IF EVER A man had been schooled for the presidency, it was Adams. But was he suited to lead the young republic? That was an entirely different question. His health was problematic. Throughout his decades of service, he suffered from a myriad of problems, from inner tensions to psychosomatic symptoms, from nervous tics to tremors. As a much younger man in Amsterdam, he may have suffered some form of a nervous collapse; and at the height of his early popularity in 1776, he noted that his face had gone "pale," his eyes were "weak and inflamed," and his nerves were "tremulous." Over the years, he was beset by constant headaches, fatigues, chest pains, visual impairment, night sweats, insomnia, heart palpitations, "quivering fingers," pyorrhea, skin eruptions and aching joints, memory loss and acute anxiety, melancholy, colds so severe that he thought he had yellow fever, possibly a goiter, and perhaps even Graves' disease. As he put it, "my constitution is a glass bubble." Nor did he deal well with stress—his health relapsed after the onset of war at Lexington and Concord, with the death of his daughter, when called to Paris by Vergennes to respond to a European offer to mediate the War of Independence, and in the diplomatic jockeying over the Treaty of Paris. To prevent further recurrence, he smartly changed his diet, began an exercise regime, walking several miles a day, and even worked alongside hired hands on his little farm in Quincy. But in 1797, at the age of sixty-one, he was a toothless, balding old man.

His health was one thing, his views another. Time and again he showed prescience. Time and again he evinced deep insights into the human condition. Time and again he demonstrated a profound appreciation of men and governance. When most colonists were still dreaming of the glories of the British empire, even as a young man he was convinced that America was destined to be "a mighty empire" of its own, possibly the greatest in the world. He was the first to put an independent judiciary into the Massachusetts constitution—the oldest functioning written constitution in the world—and he predicted with unerring foresight that America was in the midst of a revolution, "the most complete, unexpected, and remarkable of any in the history of the world." In the War of Independence, he rightly saw that France "was a rock" upon which the colonists could build. Yet he was never rigid; years later he was leery of a French Revolution spinning out of control—and in this sense too he was ahead of his time. True, he made his share of miscalculations. He was distrustful of capitalism, or of banks anyway ("the gangrene of avarice"), but here Adams was in the company of the rest of the Founders, except, that is, for the inimitable Hamilton.

And he had an extraordinarily developed sense of public service, not just for himself, but for others too, the notion that young men, generation after generation, should give back to their country: In contrast to European aristocracies, he venerated not land or titles or wealth, but republican duty and service to God and man as the highest ideal. And he put these ideals into practice at home—the Adamses would become the first of the great American political families, leaders in a long procession that would one day include the Lodges, Tafts, and Roosevelts; John Adams reared his own son, John Quincy Adams, to serve his nation much the way the elder Pitt had groomed his son William to assume power in Britain's Parliament. John Quincy later became president.

There was one matter, however, that would bedevil Adams all his life: his suspicion of the mob and the suspicion by others that he was a closet monarchist. There is a smattering of truth to both charges, but as in all things political, they ring only partially true. Yes, he was infatuated with Marie Antoinette and the French monarchy, once calling the king "a young and virtuous prince." Yes, he did not have natural democratic impulses and even worried about the rebel colonists themselves degenerating into a "multitude." Yes, his *Discourses on Davila* was read by many, including Thomas Paine and Thomas Jefferson, as pleading for a hereditary presidency. And yes, he believed that by necessity, America could one day lapse into monarchy; as he put it, as "an asylum against discord, sedition or civil war." But like Washington, like Hamilton, like John Marshall, he believed in a strong executive—one popularly elected. Also like them, he was an ardent nationalist; few men loved this young country more. A thoroughgoing realist, he detested cant, detested abstract ideas, and despised "ideology": After reading the French writer Destutt de Tracy, Adams once trenchantly rasped, "What was this delightful piece of French rubbish? What did 'ideology' stand for? Does it mean Idiotism? The Science of Non Compos Mentisism? The Art of Lunacy? The theory of Deliri-ism?" Typical of Adams, he also loathed the British and was firmly against close relations with America's former rulers, while still admiring their system of government. This tells us everything about him: Averse to utopian thinking, he instead believed deeply in the "science of government," the checks and balances of republics, while rejecting the fanciful notions that human nature could be perfected by democracy; for Adams, merit was always good, but pure democracy was positively dangerous. He feared that the masses ("the herd") could be every bit as tyrannical as the most despotic of kings. And what greater evidence was there than the French Revolution—reeking of "murder and massacres"—which he saw as a dangerous harbinger for what could happen in America.

For Adams, France and the Terror were no theoretical exercise. He lost one good friend, Brissot de Warville, to the guillotine, and another, the Duc de la Rochefoucauld, the philosopher and lover of liberty, was brutally massacred by a mob in Paris. Yet no less than the ardently Francophile Madison or Jefferson, he had also thought long and hard about the principles of good government. While he perhaps never equaled their ability to inject rigor or poetry into the American experiment, he was equally well read: As Americans weighed a new constitution in 1787—Adams was still abroad—he was already plunging into the dreary particulars, ancient and contemporary, of models for the fledgling United States in his three-volume *Defense of the Constitutions of the Governments of the United States.* As David McCullough reminds us, he examined modern democratic republics (the Italian commonwealth of San Marino, Biscay in Spain, the Swiss cantons), modern aristocratic republics (Nice, the Netherlands), modern monarchical and regal republics (England, Poland), and ancient democratic and aristocratic republics—Carthage, Athens, Sparta, and Rome.

Actually, he and Thomas Jefferson had enjoyed warm relations; they served together in Paris, traveled together in Europe, and upon Adams's election, Jefferson even suggested that Adams was the more deserving, chirping, "I am his junior in life, was his junior in Congress, his junior in the diplomatic line, his junior lately in our civil government." In truth, however, despite their common ties, the two had for some time been drift-ing apart. When Adams foresaw a tragic outcome for the French, Jefferson "deplored" the changes in his old comrade. By 1791, when Paine authored the *Rights of Man*, which many saw as an answer not simply to Edmund Burke but also to Adams's own ruminations on revolution, the breach had become nearly unhealable: Jefferson endorsed the tract, seen widely as a direct slap at Adams himself.

Yet in 1797, Adams was president and Jefferson his vice president.

∞

FOR A DECADE after the War of Independence, Washington, Adams, and Franklin were viewed as the great triumvirate of the American Revolu-tion. Franklin was now dead, Washington was now retired, and Adams was now president. His whole life had been leading up to this moment. Still, within weeks, he was declaring that the presidency was tantamount to a sentence of "hard labor." His pique was predictable. For all his brilliance and years in public service—he was arguably the nation's most seasoned diplomat—he had numerous handicaps to overcome. Unlike Franklin or Jefferson, he had never exercised executive power at the federal or state level. For the most part, all his positions were largely solitary diplomatic posts—or commissions or committees. He had never commanded an army the way Washington had. He was ill at ease with speeches, messages, addresses, proclamations, patronage, levees, and drawing rooms—all the trappings of power that Washington had so effectively used. And though he was the titular head of the Federalists, he dreamed of governing as a nonpartisan president—or like Viscount Bolingbroke's Patriot King, a leader beyond politics who entered the political arena from the outside and reprimanded the quarreling partisans for the good of the nation. But in doing so, Adams ignored the political sea in which he now had to swim. Not surprisingly, his own party was often arrayed against him. Federalist Congressman Theo-dore Sedgwick muttered that the "malignity of this man is . . . boundless." The Republicans were also against him. And so, it soon became clear, was his own vice president.

In the end, Adams stumbled into the worst of all worlds. An apolitical creature who philosophically eschewed coalitions, didn't embrace parties, and was initially leery of public opinion, he was often deserted by the Fed-eralists, despised by the Republicans, and governed precariously in between.

All he had was his cabinet—and even this relationship would be marred from the start.

∞

HIS FIRST STEP as president was not his most controversial, but it certainly would become his most flawed. With little precedent to guide him—there had never been a transfer of presidential power before—he was faced with the dilemma of what to do about his cabinet. Should they be retained— or replaced? On this question the Constitution was silent. Opting for a smooth transition, Adams chose to keep the core of Washington's final cabinet: Timothy Pickering, the tall, slump-shouldered, sullen New Englander, at State; James McHenry, the bland, colorless immigrant from Maryland, who was actually a surgeon by training, at War; and Oliver Wolcott Jr., the affable but relentlessly ambitious Connecticut lawyer, at Treasury. This one single measure was statesmanlike (soothing fears over an orderly transition); it was on its face logical (they had been appointed by Washington); and yet it was ultimately self-defeating. While all were loyal Federalists, they were stubbornly from the Anglophile wing of the party (called High Federalists)—and Adams was not. As it turned out, all were also far more loyal to Alexander Hamilton, whom Adams loathed. And Adams would one day come to see all three as actively betraying him.

But as Adams himself explained, "Washington had appointed them and I knew it would turn the world upside down if I removed any one of them." He added curiously: "I had then no particular objection to any of them." Still, as time went on he would alternately portray himself as their unwitting captive, or as oblivious to Hamilton's influence over them, or conversely, of knowing all along that Hamilton was tweaking the strings.

So which was it? Adams was not the hapless bystander he portrayed himself as. Where Washington had assiduously solicited his cabinet's opinions, Adams, distrustful by nature, aloof by temperament, capricious by habit, instead enlisted friends and family, and most of all Abigail, for succor and advice. Moreover, Adams frequently retreated to his home in Quincy, Massachusetts, for exceedingly long stretches of time, at first four months, then up to seven months, running the government by a curious mixture of fiat and dispatch. He was not without his reasons: One was philosophical—Adams explicitly trusted his own judgment and didn't trust that of others; in this sense, it almost didn't matter who served in his cabinet or how much time he spent with them. Others were deeply personal: Abigail's health was failing—she was often unable to travel and thus largely stayed in Massachusetts—and Adams had no great love for Philadelphia (and then Washington). Still others were practical: Yellow fever was often perilous in the capital; both Benjamin Franklin Bache and Richard Fenno would die from it. And finally, others were philosophical: The great threats to the nation Adams had warned of were sophistry, the spirit of party, and the "pestilence of foreign influence." In the quiet of Braintree, he was exempt from all of them.

In Adams's defense, he had borne great sacrifice in the past, separating himself from his cherished wife for as long as five years. Moreover, in the world as Adams saw it—more than two centuries later we of course view it differently—it was not uncommon for rulers to retreat to the country during the year. Moving her court with her, Catherine shuttled freely between her summer and winter palaces; so did Louis XVI and other French kings; and for that matter, in the United States itself, Washington often took to Mount Vernon (though never for longer than two months). Indeed, Jefferson effectively governed the Republicans from Monticello, and for that matter, Hamilton acted as a sort of shadow president over the cabinet itself from New York. Seen in this light, Adams's actions were not wholly as puzzling as they may seem today.

Yet whatever his justifications, his choice proved to be a great mistake, robbing him of the political support he would very much need as the flames of the French crisis began crackling anew; as a result, Adams often seemed fatally absent from his own administration, and when he wasn't absent, he was dealing with men who might as well have been strangers. A lifelong student of government, he should have known better. Then, as now, proximity is power; then, as now, proximity is knowledge; and then, as now, proximity helps build loyalty—or at least fear and respect. Even in Washington's administration, Adams's truancy became a question of concern ("Presuming that the vice president will have left the seat of government, for Boston," Washington once sarcastically groaned, "I have not requested his opinion to be taken."). Now it meant that the impulsive and erratic president was largely on his own, and whatever he did about his own advisors, he would get into another hateful family feud. Oliver Wolcott, Adams's closest friend in the cabinet, would one day sum up the matter this way: "Thus are the

United States governed, as Jupiter is represented to have governed Olympus. Without regarding the opinions of friends or enemies, all are summoned to hear, reverence, and obey the unchangeable fiat."

Moreover, Adams's sanctimonious ways had earned him many dedicated detractors over the years, so he needed all the friends he had. Yet his naïve conceptions—with some justification he would see them as idealistic—of dealing with his cabinet and his party would instead undermine his presidency at every step along the way.

Ironically, this became apparent even before he set foot in office.

∞

IT WAS NOW two days before the inauguration, and Congress was already clamoring for Adams to do something about revolutionary France. From all directions came appeals to clarify the country's policy, to make a statement, to announce a program, to make a grand gesture, to do something. *Do something?* What could he do? From the outset, the pressures on Adams to act were tremendous, so on March 2 and 3, even before he was officially sworn in, he ran up a trial balloon. That day, Jefferson visited Adams in the president-elect's room in the Francis Hotel; it was their first meeting in three years, and was unusually pleasant, even warm. The ice having been broken, the next day Adams called on Jefferson, thus signaling that he hoped for a government above party strife; they somberly discussed the dire situation with France. Adams said he wanted to make a "fresh start," and he disclosed that he planned to name a three-member commission to sail for Paris. Then the bombshell. While Jefferson was his first choice, would his vice president approach James Madison to also serve on the team? His reasons were good ones: The presence of such a prominent Republican could instill this diplomatic initiative with the aura of nonpartisanship, assuage French fears that the mission (by its "dignity") was a serious one, and establish a regional and political balance that could help secure domestic approval across the country. Jefferson was receptive. "I am much pleased that Mr. Adams and Mr. Jefferson lodge together," noted one observer. "The thing looks well."

But within two days, Adams's plans lay in a shambles. For starters, Jefferson informed him that Madison had refused. And told of Adams's proposed mission in the first place, Oliver Wolcott fumed that the president had consulted the leader of the opposition party before sharing his thoughts with his own advisors; in fact, Wolcott threatened to resign if such a decision were made, no doubt to be followed by the rest of the cabinet. Abigail was also against it. Even George Washington was against it. Meanwhile, Jefferson, himself reneging, had been worked over, if not by Madison, then by his own political instincts, and from that day on, all cooperation between the president and the vice president ceased.

So barely a day into his presidency, the first threads of administration policy had already unraveled. Little did Adams know that this was just the beginning. His cabinet would prove to be increasingly disloyal, yet he felt powerless; it was as though he had rejected his own instincts, covered his eyes, and waited for inept men to reform themselves so that he wouldn't have to go through the pain of sacking them. And then more shocking news: In Paris, the French Directory not only refused to recognize Charles Cotesworth Pinckney, the new American minister whom President Washington had appointed at the end of his term, but he had been summarily told to leave the country. At the same time, word soon arrived that French ships had resumed attacks on American vessels bound for the Caribbean. Meanwhile, Philadelphia was in a state of dread. The *Porcupine Gazette* announced that war with France was all but certain: "WAR," it screamed, "MAY THEREFORE BE CONSIDERED INEVITABLE." Taken together, these twin insults hurtled America's sagging relationship with its old ally into a full-fledged crisis. Adams now faced the specter of all-out conflict.

It was at this point that an already exhausted Adams responded by convening Congress for a special session in May.

∞

IN AN ATMOSPHERE of intense foreboding, Adams called his cabinet to an emergency meeting, his first. He spoke with an air of imperious solemnity, like a priest looking down on an errant flock. He knew that the fate of his party, of his own career, and most of all, of his nation, could well hinge on his decision. He did not rule out war, but his fervent hope, he whispered, was for some type of honorable diplomatic settlement. He was still wedded to dispatching a team to France, but took this moment to ask for his cabinet's advice. Would entering into negotiations, he wondered, invite shaming the nation? Yes, Pickering, Wolcott, and McHenry each thought so, and they remained opposed to a commission, insisting that the task of speaking to the French should lie solely in Pinckney's and only Pinckney's hands. Yet after a protracted back-and-forth, there was still no clear consensus. Amid the confusion, Adams, taking a page from Washington, then instructed his cabinet to submit their ideas in writing before Congress returned.

While Adams brooded over his dilemma—actually, as winds groaned and rain lashed the capital streets, he caught a "great cold" and felt unwell—Pickering and Wolcott were reporting to Hamilton on what had taken place. By week's end, Hamilton duly weighed in, and the former treasury secretary's voice was remarkably similar to Adams's. Hamilton was worried that France was nearing a decisive victory over Europe and had no wish to antagonize it further. So, he counseled, send the commissioners to Paris. However, he also cautioned that they should only do so supported by an unmistakable display of American resolve and military capability. To that end, Congress should therefore create a provisional army of 25,000 men.

In the days that followed, the lion's share of the written advice that the cabinet gave to Adams was cribbed nearly word for word from Hamilton's exhortations. Did the president see something fishy in the cabinet's new change of heart? Unlikely. In any case, Adams smartly adopted those precepts that squared with his own views and disregarded the rest. He would show the French that the United States would not be cowed and, in doing so, intended to ready the nation for the perilous moments to come. As it became apparent, in moments of crisis the president sought guidance not so much by pure reasoning but also by intuition. Yet Adams unburdened himself to his son, insisting upon his desire for peace, provided that "no stain upon honor" is exacted.

Still, he added this. "America," he bravely maintained, "is not *scared*."

Brave words indeed, for as no less than Noah Webster wrote: "The Fre. Gvt. has invaded, conquered and annexed the little, helpless republic of Geneva. She has conquered the Swiss cantons. Holland has been enslaved. Genoa and Venice annihilated." To many, France appeared almost unstoppable.

∞

ON MAY 16 Adams stood before the assembled Congress, and in a clear, controlled voice, delivered a stern message. Denouncing the French for ejecting Pinckney and stalking American ships ("inflicted a wound in the American breast"), he bleakly cautioned that the United States must never demonstrate "a colonial spirit of fear" or a "sense of inferiority." To demonstrate his conciliatory intentions, however, he insisted on his desire to preserve peace and friendship "with all nations," and thus declared his plan to send a three-member commission to Paris. But that was as far as he would go, and he solemnly swore that the country would never be the "instrument" of foreign influence. Therefore, he was proposing a two-track policy: Diplomacy undergirded by force. But rather than create a provisional army, as his cabinet wanted, Adams, a longtime naval supporter, called for the United States to build up its strength by bolstering its navy and improving the militia. Upon finishing, Adams could have relished the moment, his first major speech. Yet as he rose to leave the chamber, he nervously waited for the response. It came quickly.

The speech became a Rorschach for the nation, and that Rorschach revealed that the nation was hotly divided. The Federalists widely praised the president's message—blared one paper: Adams "shone forth" as in times of old; blared another: He manifested "humane anxiety to avert the calamities of war." It is a savage fact that his supporters also worried that the Republicans would undermine Adams with one ruinous gesture after another. "It is manifestly the design of the French party to lull the people into a fatal security," editorialized one writer, "to deaden their national energy and to defeat in this way those measures of defense which would secure to us our independence." Meanwhile, the Republicans were already expressing horror at Adams's "chauvinism." Overnight, hostile editors and politicians alike were pressuring him to tone down his actions and disparaging him for his bluster. He was compared to "George the third," accused of "gasconading like a bully" and sounding a "war-whoop," and ridiculed for "throwing the gauntlet to the most formidable power on earth." Now brilliantly coordinating the opposition, Vice President Jefferson, "much disappointed" by the tone of the speech, worried that America's defensive preparations would seem to the Directory like measures "leading directly to war." So here was the Jay treaty in reverse: Then, the Federalists had counseled a conciliatory approach; and then the Republicans had demanded a hawkish policy toward London. But now the Federalists actually hoped for war; and now the Republicans saw administration policy as the result of a cabal of High Federalists ("English in their relations and sentiments") who were making an Armageddon unavoidable.

Shortly thereafter, Adams summoned his cabinet again. His mind was made up, he told them; he would name two Southern Federalists, Charles Cotesworth Pinckney and John Marshall, and a Northerner and old friend,

Elbridge Gerry, to the commission; so there it was. Yet once more Adams was met with violent opposition, a rude awaking for a man who hoped to govern by ideas and not ideology. This time the cabinet expressly objected to Gerry. Peculiar and prickly, he was everything they were against—one of only three men who refused to sign the Constitution, an anti-Federalist during ratification, a partisan of the French Revolution, and an unpredictable politician with Republican leanings. For a second time, a bitterly disillusioned Adams swallowed his humiliation and relented. He substituted the Massachusetts Puritan Francis Dana; to be sure, it was an inspired choice: Dana was a protégé, an esteemed jurist, a tireless diplomat who had served with him in Paris and doggedly weathered great obstacles in Russia, and a much admired Federalist. Unfortunately, Dana was old and ailing, and he declined.

By the end of May, Adams had reached a decision. After having seemed so tired and unsure of himself, this time he put his foot down. He emphatically named Gerry without consulting his cabinet; to the president, they were only becoming a nest of vipers anyway. It was actually a shrewd maneuver. Jealousies and cabals were all around him, this much he sensed; and the more at odds he felt with his cabinet, the less he trusted their judgments. Now he was signaling that all debate must stop. Everyone in the administration must agree on the policy and execute it dutifully. And as the cabinet found out, Adams could be tough when pushed too far. He would stand by his oath of office and do what he thought was best for the country. He later explained that Gerry was one of the only two nonpartisan public officials in America—he, of course, was the other—and for once, other prominent Federalists, including Hamilton and Henry Knox, agreed with him, saying that the inclusion of a member of the opposition might galvanize support for the administration's efforts with France.

As his commissioners set off on their fateful mission, all Adams could do was wait. At night he would sink into a restless sleep. By day he would jot notes at his desk, write letters to his son, scan the newspapers, and throw open his door to a river of raucous humanity wandering into the President's House—ministers, foreign ministers, interviewers, ladies, congressmen, and secretaries. By this point, his nerves were gone. Many years earlier, he had once declared, "My talent, if I have one, lies in making war." Now he was hoping to make peace. "The French are no more capable of a republican government," he blustered to Gerry before his departure, "than a snowball can exist a whole week in the streets of Philadelphia under a burning sun." But privately, he was less than confident, anxiously writing to Abigail: "The times are critical and dangerous." Meanwhile, Hamilton, nurturing his own doubts about the "most horrible tyrants that ever cursed the earth," fretted that the Republicans would make America "lick the feet" of France's violent and unprincipled leaders.

In sweltering heat, the American commissioner John Marshall set sail under a pale, cloudless sky from Philadelphia. Gerry would leave from Boston. And Pinckney would meet them from his forced exile in Amsterdam. Back in Philadelphia, where sunshine bathed the streets and buildings in liquid light, a depressed Adams roamed back and forth in the President's House. Yes, the commissioners were on their way at last. He hoped to hear from them by Christmas.

They would arrive in Paris in October 1797, where, having no idea what was in store for them, they were to be greeted by the minister of foreign affairs. His was a name not to be forgotten: Charles Maurice de Talleyrand-Perigord.

∞

IN THE EARLY summer lull, Abigail Adams observed, "From every side we are in danger. We are in perils by land, and we are in perils by sea, and we are in perils by false brethren." This was certainly true. Adams, unable to relax, took all the hurts, furies, and concerns and soldiered on. He had no mandate, no loyal following in the capital, no coterie of friends in Congress. Until ships were built and men were armed and drilled, he had no military strength on land or in the oceans. And he had an undeclared war that was making a daily mockery of America. In Paris itself, the Directory was consciously carrying out what one official called "a little clandestine war." Little, perhaps, to the French juggernaut, but not so to the Americans. Actually, this last news was the most worrisome: Adams was informed that already the French had forcibly seized American trading vessels, soon to number more than 330; in one such instance, there were distressing reports of an American captain tortured with thumbscrews to make him talk. Thus, in the capital, there was little heady optimism; momentarily, those days were gone. And, unknown to Adams, there was treachery. On his own initiative, Thomas Jefferson was holding confidential talks with the French consul Joseph Letombe in Philadelphia, in which he advised the French to stall any American envoys sent to Paris. "Listen to them," the vice president exhorted, "and then drag out the negotiations at length and mollify them by [your] urbanity . . ." At the same time, he predicted that Adams would last only one term. So his message was stark. Wait Adams out.

In truth, by the time the three Americans arrived in Paris, the United States was the last thing on the Directory's mind. Under Napoleon, the French had crushed the Austrian army in Italy, turning the country into an armed camp, with cavalry, caissons, and infantry rumbling through the streets, with barracks and hospitals springing up around the cities, and with tents dotting the landscape. In early September, the Directory itself instituted a coup d'état: It canceled the results of its own 1797 election, France's first free election ever, arrested and exiled scores of deputies, together with two of its own, shut down more than forty newspapers, and purged or silenced critics of the government—including those who were friends of the United States. Of far greater concern to the young nation, however, was the fact that the new overseers of France despised this upstart America as never before. John Marshall, fearing that the already slim chances of peace between France and the United States were permanently damaged, wrote back to Philadelphia: "All power is now in the undivided possession of those who have directed against us those hostile measures." And he added this unsettling coda: "Only the Atlantic can save us." Meantime, the American diplomat in the Hague, William Vans Murray, emphatically cried that the new Directory was more ruthless than ever; as it studded the continent with vassal states, he noted, it ruled as vengefully "as Tiberius did by the Legions."

In Philadelphia, Adams slowly absorbed these brutal facts. The image of a Europe squeezed by France's fist was now a ghastly one: the long lines of wounded men marching forlornly and the ambulances clattering endlessly. The thousands of men who lay in coffins in more than a hundred cemeteries across the continent. And in their wake, columns of smoke billowing up against the sky; burned trees, broken ordnance, and dead horses littering the landscape—and of course, everywhere, the burgeoning Napoleonic armies that seemed unstoppable. With the situation thus deteriorating, it fell to the three American commissioners to avert a bloody apocalypse. So there was the handsome Marshall, witty, learned, and sure of himself. There was the pugnacious Pinckney, with his jowly mouth turned down in what seemed a permanent frown, suspicious and distrustful of all France. And there was the rawboned Gerry, the dissenter of the three, lean and forthright and partial to the French. On the shoulders of these three men rested whether a full-scale war would be detonated.

∞

AT THIS POINT the French minister Talleyrand assumed an unexpected role. By any standard, he was one of the most intriguing figures of the age. Over a remarkable forty-five-year career, Talleyrand would labor on behalf of a republic, an emperor, and three kings. With his imperial presence and stern, frowning eyes, he was one of the great diplomatic creatures in France, and soon, of all the Western world, old as well as new. In a time of towering humanists and idealists, many would misjudge him, just as many would underestimate him, and all would be wrong; he had numerous quirks and eccentricities, a number less than admirable, but was never a fool. A former royalist and bishop, he was also the ultimate survivor: Lame from the age of four, he was raised by his great-grandmother; a survivor of smallpox, he recovered unmarked; a member of the Assembly, he surreptitiously proposed his services to the king even as he became a pro-Revolutionary priest. Nothing seemed to touch him or stop him: Ordered to be arrested by the Convention, he eluded the Jacobins in London and successfully fled to America—there, he swore an oath of allegiance to the United States, was elected a member of the prestigious American Philosophical Society, and hobbled amiably down the streets of Philadelphia with his mulatto mistress on his arm and his little dog at his heel; he also speculated in bonds, futures, land, urban real estate, whatever came his way, amassing a small fortune. If survival is the test of worth, Talleyrand was the most able of all, maneuvering safely through crisis after crisis, mistress after mistress, gathering money and power at every turn; he had nine lives and bartered and sold them all. But by spring 1796, he was back in France, serving the Directory.

The Talleyrand the Americans would meet was so many things: the noble and the bishop, the constitutionalist and the revolutionary, the gentleman exile and the haughty minister, and the capitalist. Avaricious, cynical, wily, and whimsical, he had woven a powerful mystique around himself. And it turned out that his principles in public life were enriching his nation, juggling Europe's balance of power, and lining his own pockets; all he wanted, it seemed, was money, money, and more money. This he did by taking percentages and selling access. But if anything, he was also ruthlessly unsentimental. He had kept his head by living in the United States—he was a friend of both Hamilton as well as of Gouverneur Morris—and many Americans saw Talleyrand's appointment as a good omen of U.S.–French relations. They were mistaken. In truth, Talleyrand, like the Directory, didn't think much of the United States—that is, when he ruminated about it at all; he once explicitly described America as "not of great consequence," much like "Geneva," or "Genoa."

So when Talleyrand finally received the three men in October, he gave them an audience of just fifteen minutes. He was affable but curt; still, he at least promised that he would protect them from deportation by the police. But then they heard nothing from him for the next week. When he finally summoned them again, his tone grew cold as he explained the Directory was "excessively exasperated" by Adams's statements about France in his May 16 address. At that, the meeting was over. What followed in ensuing weeks was a series of intrigues, bluster, and clandestine meetings. At this stage, Talleyrand subsequently refused to meet with the Americans at all, relegating them to deal with a continuing procession of unofficial visitors—Nicholas Hubbard, Jean Conrad Hottinguer, Pierre Bellamy, and Lucien Hauteval—all who mysteriously described themselves as friends of the foreign minister. The repeated meetings—always furtive—became disconcertingly chaotic. Eventually, through his intermediaries, the foreign minister sketched his bottom line, imposing onerous and insulting demands: that President Adams retract the offensive passages from his bellicose speech; that America extend a large loan to France, $10 million to be precise, and even pay for damage inflicted on American ships by French privateers; and that a considerable *douceur* (sweetener) of some $250,000 be paid ("something for the pocket") before Talleyrand would even agree to speak with them again. All this with a straight face.

So what were the commissioners supposed to do? Grovel and pay up? Bluff? Wait and hope for a change of heart? The demands were of course preposterous, but this was, as it always had been, the prerogative of great powers in dealing with weak ones. Even Great Britain played by these audacious rules when it came to France. And in the uncertain climate of revolutionary Paris, who knew what action would avoid conflict—or precipitate it. Moreover, for the commissioners there was the matter of their own personal safety to consider.

Yet physical danger worried them less than the shabby treatment. Right was right. Wrong was wrong. Even amid the dangers, Americans were a brash and blunt people. Marshall and Pinckney were outraged; they held firm, rejecting all French claims. Once, in one meeting, when asked to pay the bribe ("It is expected . . . !"), Pinckney contemptuously answered back in indignation, "No! No! Not a sixpence!"—words that would later be immortalized in America. Afterward, Pinckney confessed that the French were attempting "to divide" the envoys, which they did with considerable success. Under French prodding and pressure, the fragile facade of American unanimity was cracking. They flattered Gerry with special attention, and over time, the commissioners were increasingly at one another's throats— Pinckney against Gerry, Marshall against Pinckney, Marshall against Gerry, with Gerry, invariably the odd man out, always counseling patience.

Then the talks took another chilling turn. When the shakedown produced no immediate results, throughout November and December the Frenchmen gravely added threats, threats that France was perfectly capable of making good on: provoking civil war and "chang[ing]" America's government, encouraging the partisans of France to rise up and begin "ravaging" America's coasts with the French navy, and, in the end, "crush[ing]" the United States.

With tones that rang equally menacing, they also reminded the Americans that they were "three thousand miles from home" and the French "were impatient." By now, even the committed Francophile Gerry had taken to sleeping with a pair of pistols under his pillow.

∞

BACK IN THE United States, in great anxiety, Adams, bags under his eyes, hair somewhat disheveled, waited for word from his commissioners. So did all of Philadelphia. Outside the President's House, Philadelphia was swarming with politicians and partisans gathering to take action—but what action? The president's son, John Quincy Adams, not prone to wild imaginings, worried that there was a French plan to send an invading army to detach the western United States and form a satellite republic like those ruthlessly being established in Europe. Congress was of the same turn of mind. So under the shadow of possible invasion, it began debating resolutions for a standing army in case of war. By July, it had approved a bill providing for naval armament; not fully what Adams had called for, but it was a start.

But the topic pervading conversation across the nation remained the fate of the commissioners in Paris. In November Adams received an ambiguous message from Pinckney, though the report was terribly confusing; feeling the increasing strain, in the last week of the month, Adams again caught cold and took to his bed for a week. From his home, he could hear crowds milling around, clamoring for further dispatches. They would have to wait. December passed and there was little word from the envoys. Then January and February. The fox was among the chickens but the farmer, out in the pasture, didn't even know he had a problem. In a state of confusion, Adams paced in his office, paced in the President's House, paced in his bedroom at night. The lack of news grew increasingly portentous as the days passed, until it appeared as if his whole grand plan had gone awry. Events seemed to be rushing to the precipice. "We are yet all in the dark respecting our envoys," Abigail Adams jotted down on February 16. A week later she nervously recorded, "Our envoys have been near six months in Paris but to this hour not a line has been received."

The delay had grown unbearable. Wild rumors swept the city; meanwhile, the capital fell into a perpetual fright. There were claims that France had already declared war on the United States, and was now preparing to seize Florida and Louisiana, even as Southerners feared that France might invade the southern coast from Santo Domingo and incite a slave insurrection—which, unbeknownst to them, Talleyrand had openly threatened. And it was even said that members of Congress were engaging in treasonable correspondence with the Directory. As it was, these conjectures merely scraped the scab to reveal a festering sore beneath.

"The period is big with events," Washington noted at Mount Vernon. But informed of the terrible silence from Paris, in early March he then urgently wrote to the secretary of war. Readying himself for disaster, he posed one frightful question.

"Are our commissioners guillotined?" he asked.

∞

IN TREPIDATION, WASHINGTON, like Adams, like all of America, would have to wait for the answer, even as Europe fell, day by day, further under the spell of France.

And that spell was dangerously deepening. As the French essayist Mallet du Pan observed morbidly, "The French republic is eating Europe leaf by leaf, like the head of an artichoke." And the man doing the eating, who ultimately would decide not just Europe's fate but, arguably, America's as well, was an individual of such remarkable talents and domineering skills that genius scarcely explains them. To the rest of the world, he was an ogre who sucked the blood of his countrymen and pillaged a continent, but to the people of France, high and low alike, he was their new savior, one of the most alluring thinkers and actors in the history of all mankind. Britain's Lord Acton put it this way: "No intellectual exercise can be more invigorating than to watch the working of the mind of Napoleon." But though some 250,000 books and booklets have been written about him, he seems repeatedly to escape our full comprehension. In the end, the questions about him are invariably the same: Was he the hero who labored to bequeath unity and law to Europe, or a monstrous dictator? A giant who modernized his nation, or a blinkered despot who made France into his own private arsenal? A visionary who brought civil equality and justice to a new order, or a terror to the patricians of the continent?

Born in Ajaccio on August 15, 1769, to a family of minor Corsican nobility, he was a short, darkish man, with Mediterranean hues and striking looks, who arguably would never have commanded anyone if dressed in civilian clothing. Ill mannered and ill tempered, he prevaricated freely; cheated at cards, often; rough-housed colleagues, pinching them by the ear; and loved to shock those around him. He was talkative, too—as it was once said, he was "no gentleman." In fact, he eluded all categorization. A child of the Enlightenment and a product of the Revolution, he was bound neither by centuries of custom, nor, for that matter, by basic morals either. Glowing with pride, he had an exalted belief in his own destiny, which only deepened and intensified as the years went on; he claimed, as many great men do, to follow his own "star." Thus he believed the world to be a protean place, ripe to be subdued by the genius of his own mind. Was this sheer hubris? Though he was not taken in by lofty philosophical sentiments, his raw, even masterful intellect invariably impressed all who met him; and at the same time, his interests were vast and wide, encompassing law and history, military science and public administration, art, astronomy, mathematics, and architecture. His mind, like his armies, was inexhaustible and disciplined; he once declared that it was like a chest of drawers, which he could open and close at will. And he was a born leader; on this matter there is complete consensus. He dazzled and captivated and overawed all those around him.

Love for him, however, or even admiration, was hardly universal. Yet while he disgusted most Americans, becoming in the 1790s the perfect epithet of "Bonaparte," some of the era's great cultural icons, including Goethe and Beethoven, were initially smitten with him. And invariably he inspired confidence by the succinctness of his orders, the resoluteness of his decisions, and his uncanny ability to grapple with the most protracted problems. And when he didn't inspire confidence, he instead aroused astonishment and fear—with his armies.

In an age of ideologues, Napoleon was also a pragmatist—or, as others have contended, a rationalist. There is considerable truth to these assertions. Where the Jacobins wanted to destroy their enemies, Napoleon as first consul would invite back all manner of exiles, from aristocratic émigrés to peasant refugees, from Vendean guerrillas to deportees of the various coups; all he asked (and received) was that they work for him and cease bickering with one another. Where revolutionaries sought to stamp out a Catholic revival, Napoleon simply coopted it—one day even signing a concordat with the Vatican. In truth, religion to him was little more than a tool: He unabashedly proclaimed himself a freethinker among the professors at the Institute in Paris, a Catholic in France, and a Muslim in Egypt, declaring himself the Sultan El Kebir. He became French to the core, but in Italy, he was one of their sons. And not unlike George Washington, he was a master at human psychology; "Every private," he once boasted, "carried a marshal's baton in his knapsack." He knew when to be silent, when to bellow, when to lead his men, and, of all things, when to abandon them. And he delighted in affirming the divinity of the people, but to his mind, the people were a power rather like Voltaire's God, who created the world but then after that never interfered in it. Actually, during the revolution he once spoke of the French crowds as "the lowest scum."

And for all this, and much much more, as did few others in his era, or really any era, Napoleon Bonaparte spectacularly entered history and legend, and has never left it. Consider this. It says something that however much one speaks of the age of Jefferson or Hamilton, or Catherine the Great or William Pitt, or Burke and Paine, one is just as likely to speak of the age of Napoleon or, in the same breath, of the Napoleonic wars.

Five of his mother's children died in childhood; Napoleon was his mother's fourth. And with his help, seven of his siblings would go on to high honors or high office as kings, queens, and advisors of state. Curiously, as a young man he was unusually taciturn; he said little and dreamed much, trusted no one, and locked himself in a world that seemed organized to torment him. He read Plutarch, studied Machiavelli and Montesquieu, and responded to the passionate prose of Rousseau. Actually, he was a bit of a romantic too, relishing poetry for the same reason "that made me delight in the murmur of the winds and waves." And at the same time he could be quite somber, even contemplating suicide as a young man ("As I must die sometime," he once suggested, "it would perhaps be better if I killed myself ").

And of course he studied in French military schools. Commissioned in the Bourbon army, in 1793 he became a fervent young Jacobin officer who helped drive the British from Toulon, and was subsequently made a brigadier general by the government of the Terror; remarkably, he was only twenty-four. Yet in the spring of 1795, he was on the outs, having been briefly jailed and again meditating on suicide; he also thought of joining the Ottoman sultan's army. Then, in one of those odd vagaries of history that thrust open a gate to the impossible, he later served the Convention by breaking up a demonstration of royalists. His life was never again the same. Sensational, notorious, cruel, and magnificent, he also burned with ambition. On March 2, 1796, he was given his chance and he seized it: When the War Ministry needed a bold and enterprising commander to lead the Army of Italy, it chose Napoleon. Seven days later he married his great love, Josephine, who had barely escaped the Terror's guillotine; she was six years his senior. And on March 11, choosing between his new love and his burning hunger for fame and glory, he left to lead his army in one of the most extraordinary campaigns in history.

His instincts for war were uncanny, his tactics, conventional and unconventional, were swashbuckling, and like few others, he grasped the inherent tempo of battle. At first his generals had reservations about him. But not for long. Employing swift marches over rough and mountainous routes, he gave his men unparalleled speed and a belief in their own supremacy. They, in return, would become powerfully devoted to him. It was needed.

Early on, braving enemy fire, he and his forces crossed the treacherous Alps, a remarkable achievement later immortalized in Gros's painting of Napoleon heroically traversing the bridge at Arcole, and drove the Austrians from the north of Italy. "Peoples of Italy," he announced with a vengeance, "the French army comes to break your chains." His own men, by now weary and wretched, were skeptical. But after a stunning series of victories in Montenotte, Millesimo, Dego, and Mondovi, the exhausted survivors of his army reached the peaks of Monte Zemoto—below them stretched the sun-soaked plains of Lombardy—and launched into a spontaneous salute to the general who had led them so brilliantly. And when his men wavered, Napoleon, eyes flashing, hair blowing in the wind, never thought twice about courting death to inspire them. More than that, he made his troops feel nearer to heaven on earth, and gave them the sense that they could walk like gods rather than crawl as men. In turn, his troops paid him back, conferring upon him the affectionate title *"le petit caporal"*—the "Little Corporal."

After storming through Italy, the Little Corporal assumed the right to make peace as well as war: He arranged a truce with the King of Naples, created the independent republics of Lombardy and of Liguria, and shamelessly plundered his conquered lands, taking precious money, precious land, and, to the revulsion of all Italians, priceless art; in the end, they could only console themselves with the pun, *"Non tutti Francesi sono ladroni, ma buona parte"* ("Not all Frenchmen are robbers, but a good part are"). Insatiable, Napoleon then lunged farther south, this time taking Pope Pius VI's Papal States; by the Treaty of Tolentino in February 1797, the pope was forced to surrender Bologna, Ferrara, Ravenna, and Ancona. Master of all northern Italy except Piedmont and Venice, Napoleon was still not contented. So he reorganized his army, again led 75,000 men across the Alps, this time through biting cold and three feet of snow, and proposed to strike at Vienna itself, the home of Haydn and Beethoven and the imperial epicenter of the assault upon the French Revolution; only the impending winter and the prospect of unfamiliar territory led him to hold back, an uncharacteristic display of hesitance. But instead, Napoleon, like Marcus Aurelius, like Caesar himself, developed a foreign policy of his own.

Wherever he marched, turmoil was left in his wake, often with a budding new order rising up against the old: Venetian cities thus revolted against Venice, Bologna against the pope, Milan against Austria, and the subjects of the Piedmont monarchy against their rulers. So he continued to stud Europe with yet more "sister republics" for France, taking the Left Bank of the Rhine, the Ionian Islands off the coast of Greece, deposing the pope and creating a Roman republic, and restructuring his other conquests into the Cisalpine Republic in the Po Valley around Milan and the Ligurian Republic around Genoa. Revolutionary republicanism, or more precisely, Napoleon's gale, raced across the mainland and the isles: Ireland fell into rebellion; the population of England became restless and there were even mutinies in the great British fleet; in Switzerland reformers "cooperated" with another invading French army to create a new Helvetic republic, thereby sending massive "indemnities" to Paris; and in the boot of Italy, a Neapolitan republic was soon to be set up. Finally, with Napoleon's blessing, it was one of his own generals, Pierre Angereau, who invaded the legislative chambers and oversaw the September 1797 coup d'état of Fructidor, in which the Directory, like the revolutionary governments before it, sank into dictatorship.

His fame now filled the world. Then, having avenged Caesar's Roman subjugation of Gaul, Napoleon set his sights on England and rushed back to Paris to ensure his many treaties were ratified by a Directory that he had helped to install. Upon his reentry in December 1797, he was treated as a conquering hero. He was sumptuously feted by Talleyrand, and amid great fanfare was greeted before the eyes of a grateful nation by the five ruling directors clad appropriately in Roman costume. Even then, the Rubicon had been crossed.

He was assigned to command an army in training to invade England, but upon reflection, he decided the time was premature. Burning with zeal, he instead turned his attention to the Orient, where he would threaten Britain's most prized colonial possession, India, with a spectacular invasion of Egypt. Wrote Napoleon, "The charm of Oriental conquest drew my thoughts away from Europe more than I should have believed possible."

But it was possible. In his hands, it seemed anything and everything was. Napoleon began at once to assemble his daring armada: 13 ships of the line, 7 frigates, 35 other warships, 130 transports, 16,000 seamen, 38,000 troops, a 300-volume library, and a bevy of renowned scientists, scholars, archaeologists, and artists who would accompany this extraordinary expedition, including even the balloonist Conde. For the time being, he labored in total secret, and the purpose of the Army of the Orient, which Napoleon merely called a "wing of the Army of England," was so well hidden that virtually all of the 54,000 men set out with no knowledge of their mission or their journey's end. The secrecy was not without reason: Ever a tactician, Napoleon meant to keep England off balance. He succeeded. The British high command was evidently convinced that the flotilla was readying to punch through Gibraltar and then join an invasion of England. They were, of course, dead wrong.

And so, on May 19, 1798, Napoleon's main fleet pulled up anchor and set off to marry the allure of antiquity with the romance of revolution.

In short, to write history anew.

∞

HAVING BEEN FOOLED by Napoleon's feints, Admiral Nelson's ships were negligent in their watch in the Mediterranean and the French handily eluded them. Then, while the general fought seasickness, he studied the Koran and set sail for Alexandria.

Approaching shore, what a sight it must have been. There, through the haze, was the ancient city, with more than five thousand years of storied history. Somewhere far off in the distance were the pyramids, testimony to the pharaohs, the magic of the deserts and the sand, and one of the most magnificent cultures ever recorded. And there in the foreground was the bustling port of Alexandria, guarded by a heavily barricaded fortress and serving as a critical gateway to the Ottoman empire. Weathering rough seas and a menacing surf, Napoleon arrived on July 1, 1798—at around the same time that America was convulsed by internal division and feared becoming France's next morsel. Once more, Napoleon put himself in harm's way, personally leading the landing party of 5,000. Without cavalry or artillery, he struck at night, losing only two hundred men and, despite house-to-house fighting, rapidly taking possession of the city. Fortified by this triumph and a handful of words in self-taught Arabic—even Catherine the Great never contemplated such a grand gesture when she toured the Crimea—Napoleon then shrewdly sat down with local elders. Pledging to respect their religion, their ideas, and their treasures, he convinced them to provide his army with laborers and supplies, as well as 1,500 camels. With that, a week later Napoleon left his ships behind and took his legions overland, across 150 miles of trackless desert. Their destination? Cairo.

His men had already braved great cold, fierce windstorms, harrowing snowstorms, high altitudes, and repeated enemy fire. But squinting into the horizon, they had never endured anything like this—such pitiless heat, such enervating thirst, such disorienting desert, or such marauding Bedouins. Napoleon sought to lift their spirits by sharing their hardships. Yet pushed to the extremes of human endurance, of agony so excruciating that only those who have comparably suffered can comprehend it, it was a nearly empty gesture. The passage was a confused, tangled path, and if Napoleon had miscalculated, if his maps or guides proved to be errant, his men might have perished there, victims of thirst and starvation—actually, a number did; many even committed suicide. Nine days later, on July 10, the gods smiled on him: He reached the Nile, enabling his men to drink, wash, and finally, to rest. But not for long. Still dreaming of conquest, the stoical Napoleon pushed his men that much harder. After five more days of cruel marching, they approached the village of Kobrakit, where they encountered an army of three thousand Mamlukes. Astride magnificent stallions, the Mamlukes were all "gleaming with gold and silver, armed with the best London carbines and pistols, and the best sabers of the east." It was a rout. Napoleon's men cut them up, and ten days after that, just eighteen miles from Cairo, the stumbling victors looked out on a prize that countless others had marveled at for centuries.

Across the mightiest of deserts, stretching to the horizons and rising out of the dust into the vast blue skies, was a sight that literally takes the breath away. In that first rapturous moment, Napoleon and his famished men, their reserves broken and overcome by emotion, surely could not speak. One imagines that the Little Corporal's private reaction was almost the same as Magellan's after crossing into the *pacifico* (Pacific Ocean): Magellan had burst into tears.

Standing there, they could glimpse the pyramids.

That evening Napoleon assembled his aides for the impending fight, and it was then that he learned an Arab army of six thousand had gathered to challenge the infidels. The next afternoon, eyes gleaming, they fell in full force upon the French invaders. Napoleon shouted to his troops, "Forty centuries have their eyes upon you!" Under a blazing sun, the two armies clashed: Cannons disemboweled men, muskets fired, fixed bayonets were thrust, and Napoleon won another great triumph, henceforth to be forever known as the Battle of the Pyramids. The Arabs knew they were beaten; the very next day the Ottoman authorities in Cairo sent Napoleon the keys of the city. It was their admission of defeat. The day after that, he strode dramatically into the ancient capital.

Smartly, unlike in Europe, however, he forbade plunder, though he did exact taxes; and he sat down with native leaders. In doing so, he declared his respect for Islamic rites, swore his fealty to Allah as the one and only God, and pleaded for Muslim help to bring wealth to Egypt. And from that day on he was the ruler of the oldest Arab civilization in the Middle East. In the weeks and months that followed, administering the country by Arab divans, or committees, he also called upon his scientists to eradicate plagues, repair canals, enhance Egyptian jurisprudence, examine fishes of the Nile and minerals of the Red Sea, and organize postal and transport services. Later he created the Institute of Egypt where he set up his scholars in a special headquarters in Cairo. And it was one of them, known simply to us as the mysterious Bouchard, who would make among the most wondrous discoveries ever recorded: the Rosetta Stone, and later, mummies themselves. As Napoleon himself reflected, "I dreamed all sorts of things, and I saw how all that I dreamed might be realized. I created a religion. I pictured myself on the road to Asia, mounted on an elephant, with a turban on my head, and in my hand a new Koran, which I should compose according to my own ideas." And this too: "I was to . . . renew my relations with *old Europe* by my conquest."

He was only twenty-nine. Swashbuckling, uncanny, sublime, he now lived like a pasha and ruled like one, outdoing even Potemkin. He negotiated with the Shah of Persia. He also took a mistress, a young Frenchwoman who had accompanied her officer husband. Yet the dream was about to burst. Learning that his wife, Josephine, had taken a lover back in Paris, he became despondent. And back in Alexandria, there was more trouble. The rest of his Army of the Orient was to unload supplies for the troops and then head for French-held Corfu. Here, suddenly, was fate. And here, just as suddenly, Napoleon could glimpse mortality. Storm clouds prevented his men's departure, so Vice Admiral François-Paul Brueys dropped sail in the Bay of Abukir—a name that would soon be known round the world.

For on the evening of July 31, 1798, Admiral Nelson of the hallowed British navy finally found the French legions. A bloody conflict broke out; fourteen English ships boldly locked horns with thirteen French ones. This momentous battle would rage into the dawn the next day, with stunning— and surprising—results that would ricochet across the globe.

∞

HALFWAY AROUND THE world in Philadelphia, after months of watching the French juggernaut in action, Adams's confidant Benjamin Rush feared that "a war with France was inevitable." If Napoleon could subjugate Europe, or, as they would learn, move his armies all the way to the Near East, surely the Directory could one day decide that America was next. But to his credit, Adams's confidence was never shaken; he still hoped for the best. Nearly six months had passed before the Americans became aware of a cold, cruel shaft of light that they had recognized as the truth. For then came electrifying news from France.

Held up by the global Franco-British war, John Marshall's account chronicling the indignities that the three commissioners had endured finally arrived, landing on the president's desk late in the day on March 4, 1798. This was the small shock. The great shock came in the contents. In fact, Secretary of State Timothy Pickering saw it first. He decoded enough of the secret dispatch to decipher the contemptible treatment of the U.S. envoys; "War," he immediately blurted out, "is made upon us." That night, Adams was roused from bed and told of the first fragmentary reports—and he agreed. As the president studied the communique by the hazy light of a fire, he rapidly scribbled notes ("unexampled arrogance" of the French; their "continuing violences . . ."). Matching Pickering's rage, he thundered that he would ask Congress for an "immediate" declaration of war on France. Yet raging in private was one thing, what he would say in public was another. Sometimes rash, Adams could also be conscientious to a fault, and here it served him well; he waited two weeks before he spoke to the Congress—it took more than a week to decode the messages in full—during which time he collected his thoughts and conferred with his cabinet. With each passing day, he weighed the stakes, and little by little his belligerency softened; the Cabinet also followed suit—even Hamilton, serving as a sort of shadow prime minister, was urging "a temperate but solemn and firm" stance, but not an outright declaration of war. Indeed, the more the president considered it, the more Adams was unsure whether the country would support hostilities against France, and in any event, he questioned whether a fragile Union could survive such a cataclysmic encounter.

So Adams tore up the first draft of his address to the country, a tempestuous, irate missive, instead substituting a more restrained text. It was at once cautious but also firm. As he appeared before Congress, his voice broke with emotion. He had been, he told the stunned members, "liberal and pacific" in sending his commissioners, but France's actions indicated that no agreement could be reached. Gathering himself for the final flourish, he thus called for a broad array of military preparations—though at the same time he still refused to sanction a provisional army. The members exploded in a roaring applause. Then word of his speech filtered out.

In the capital, the first reactions to France's audacity had been confusion and disbelief. It didn't last long. Soon thereafter Adams's tempered response— and it was tempered—was met with vehemence from the opposition. From all corners came waves of indignation. The Republicans reiterated all their old accusations: Adams was a "warmonger," pushing for a defense program that amounted to "war measures." And the Republican rumor mill labored at white heat: Jefferson, for one, went further, describing the speech as "insane" and castigating Adams as a "stalking horse" of Hamilton's, which he claimed would only drive the United States into the arms of Britain and into war with France. Madison was stronger still, savaging Adams as a "perfect Quixote," expending as much effort to get America into war as the great George Washington had spent to keep America out of it. Convinced that the president was deliberately withholding information favorable to the French, fuming Republicans became adamant that all the documents—both Adams's instructions to the commissioners as well as their reports to him—be made available at once. Their reasoning was simple. Either Adams would refuse to release them, as was the case of George Washington with the Jay treaty, which would indicate he was hiding something, or if he did turn them over, it would be a major embarrassment to the congressional war hawks.

On April 2 Congressman Giles of Virginia took to the floor and called for the reports be made public. This time Adams, so often accused of being politically inept, willingly complied, mousetrapping his opponents in the process. The very next morning the galleries were hastily cleared of visitors and the House went into a rare executive session. That was the trap. And the Republicans went for it. The scene that day was a piece of brilliant stage management. Blackening out only the names of the French agents—they were identified simply as "W , X, Y, and Z"—the documents immediately absolved the administration on charges of duplicity; the incident, which quickly became known as the XYZ affair, went out to an anxious nation, unleashing a national uproar against the French in the process. Moreover, Adams's unsuspecting adversaries now found themselves on the defensive. Jefferson immediately grasped the seriousness of the error; even countless Republicans now eagerly lined up behind the war measures. The vice president brusquely concluded that the United States was no longer "neutral and pacific." Hostilities, he came to believe in great anguish, would inexorably result from the administration's rash steps.

There was a bedrock issue here and it transcended a clash of personalities. War. This, then, became one of those inexplicable historical moments in which actions acquire a commanding pressure all their own and begin to exercise an unstoppable force.

The nation girded itself for the imminent ordeal.

∞

THE ENSUING DEBATE not only laid the outlines for national discussions in all subsequent wars—from the Mexican War to the Spanish-American War, from World War II to the wars against Iraq, but had strikingly modern echoes to events today. While the Republicans, momentarily confused and adrift, were publicly muzzled, this was not the case with their allies in the press. Almost in a chorus, many now harped on the theme that Federalists were no longer stewards of the liberties secured by the American Revolution. Thus, one Republican sarcastically protested: "*Lost. The Declaration of Rights of the Citizens of the United States.*" And the Republicans, still raging, curiously found heart in their political philosophy; answering the Federalist vilifications of democracy, one essayist heatedly wrote: a "*democrat* is one who advocates the people's rights, and a government *of* the people, or arising *out* of the people."

Throughout this extraordinary summer and into the fall the debate became increasingly passionate, even as the problems seemed to grow in complexity and frightfulness. Understandably, the boot of opposition public opinion would not rest. Their arguments now morphed and shifted. Madison came to believe that Adams, and not the Directory, was the "great obstacle" to accommodation, and grumbled that Adams had perpetrated "a libel" on the French government. His face stiff and expressionless, Jefferson argued that the whole affair was "cooked up" by John Marshall—and anyway, if there were a culprit, it was the freewheeling Talleyrand, and not the French government. In this, Jefferson, of course, was in error. Suddenly the opposition was bumbling from bad to worse; by contrast, much of the country openly reveled in the possibility of war. One New Englander summed up the giddy sentiment this way, applauding hostilities against the "factious, cut-throat, frog-eating, treaty-breaking, grace-fallen God defying devils" in Paris. Adams himself boldly told one audience—"The finger of destiny writes on the wall the word: *war!*"

If the Republicans were reeling—they were—the Federalists were a party revived. The Congress of 1798 became a war Congress, and they played up the XYZ affair for everything it was worth. Despite fears of an imminent French invasion, the call went out: Let news of the XYZ humiliation and the bold reply of Charles Cotesworth Pinckney sweep the Union. Let it sing in the newspapers, let it sing in a hundred songs and a thousand letters in every city in the nation, that this administration was not whipped and that national honor would be preserved. At Adams's prompt-ings Congress finally adopted the defensive measures he had earlier called for, and then some. By July it had created a Department of the Navy and authorized the construction or refitting of thirty-seven warships, an aston-ishing number—measures that would have been unthinkable just months before. As war hysteria grew—it didn't help matters that a French privateer even forcibly seized American vessels in full sight just outside New York harbor—Congress then went considerably further than Adams. It would soon abandon the Franco-American treaties; it halted all trade with France; and, defying all traditional fears of a standing army, it established a provi-sional military force of 12,500, to be readied "at once."

The undeclared Quasi-War was under way.

With a headiness that is hard to fathom—America, after all, was up against the globe's greatest land power—High Federalists longed in earnest for this conflict to cut down the "monster of Jacobins," to destroy the foe of "property, morality and religion." Leading the way was Congressman Fisher Ames, who had done so much to salvage the Jay treaty. "It is too late to preach peace," he said, pounding the drumbeat of hostilities. "A defen-sive war must be waged . . . that, or submission, is before us." Was it? Ames thought so: If America did not act, he thundered, the day would soon arrive when "the French tiger comes" to devour America. Yet Hamilton, hardly a picture of pusillanimity, was mouthing a different tune. He counseled a mixture of action and caution: action because the "FIVE TYRANTS of France, after binding in chains their own countrymen, after prostrating surrounding nations, and vanquishing all external resistance to the revolu-tionary despotism at home," seem resolved to decree war against all nations "not in league" with themselves. But caution because military preparation could still bring the "present despots to reason."

Could it? This, at day's end, was the question for President Adams to ponder.

∞

HE DID, OFTEN. Throughout the spring and summer, Adams fretted that war was looming. Bracing himself for the worst, Adams would sometimes wake in darkness and lie there until the first gray shafts of morning had spread throughout his room; then, tired and trembling from his chronic cares, he would pore through the latest dispatches. One in particular caught his attention: From John Quincy Adams in Berlin—his son was now emissary to Prussia—he finally heard of French intentions to attack the trans-Appalachian West (near today's Alabama and Mississippi), and sever the entire territory from the eastern half of the United States, mercilessly subjugating it much the way France had carved up Genoa, Venice, Holland, and the Swiss cantons; this, of course, would be the first step toward dismembering the United States. Adams knew the French broke many promises, but the conundrum became, did they not always make good on their threats? In one instance the president confided to the British minister that he believed war was just around the corner, which prompted the first lady to bluster that undeclared war had been flaring for *months.* So it was around this time that Abigail began hoarding household supplies for the day when the phantom of the Quasi-War dissolved into full-blown hostilities. But Adams, like Hamilton, never lost faith and never ceased praying that outright conflict might be prevented. Meanwhile, he presided over the nation's first war, his "half-war" as he put it, while pursuing a two-fisted policy.

Adams at once sought to convey the impression to Paris that the American people still felt intimately connected to the French (the first fist, opened) while at the same time would dig in and defend themselves with everything they had (the second fist, closed). He thus set out to steady Americans for a long time of travail. It couldn't have been easy: This was not an age when America's trumpet could subdue the world; far from it. But as it happened, Adams thought perfectly and spoke perfectly in the idiom of his time: the language of honor. To our ears it sounds quaint, and it perhaps was. Yet in doing so, Adams unexpectedly touched off a crest of popular passion. For once, the American people, and all of official Philadelphia, swooned over him. For once, appearing in military regalia, sword by his side, he seemed the very picture of what a great leader should be, no longer beaten down by worries about the war and doubts about himself. And for once, he exhibited an almost cocky streak, an infectious, self-confident zeal about what the future held in store. No longer was he enmeshed in hot and destructive feuds with his cabinet or his party. And just as inexplicably, he began to act like and speak like a strident partisan too: Writing back to his native Massachusetts, he spoke darkly about the duplicity of French agents and further warned against the "party in this country" devoted to the "French interest." And to young Bostonians, he provocatively wrote, *"To arms then, my young friends! TO ARMS, especially by sea!"* If this opening chord seemed unfortunate or excessively militant—or strangely evocative of France's late radical Georges Danton—the nation scarcely seemed to mind.

But while the country bristled with combativeness, President Adams could also, in a sense, scarcely believe his own good luck. As the Quasi-War heated up—the Directory continued to issue menacing statements—he called on the young United States to adopt a "warlike character"; in turn, he was warmly praised for his "fortitude" and acclaimed for his "independence." Wherever he went, he was met by huge, worshipful throngs. In New York, he received a large, boisterous welcome—young ladies threw flowers, orators sang about Adams's manhood, and bands struck up "Hail Columbia" and "Yankee Doodle." At the theater, audiences jumped to their feet and cheered. And when he called for a national day of fasting and solemnity, the churches were filled. Week after week, militia companies, merchants groups, fraternal orders, college students, and small towns sent him patriotic messages, which he answered publicly. A new song was even composed in his honor, "Adams and Liberty," by Robert Treat Paine Jr., and many came to think that Adams now rivaled Washington in the affections of his countrymen. By this point, it was hard to disagree. For his part, Adams was deeply touched by the emotions showered upon him. He almost felt serene.

It was not cost-free. America had now come to resemble revolutionary France in its halcyon days before the Terror. In Philadelphia, it was impossible to forget that one was in the capital of a nation at war. Adams's Federalist allies reveled in all the military activity that frequently clogged the streets—and so did he. Everywhere there were now black cockades in support of the administration—in the Congress, in the streets, at inns and in the theaters, in men's hats and hanging from home windows; one May afternoon, a thousand young men boisterously marched two-by-two to the President's House, humming patriotic songs and waving their hands. Adams received them decked out in a dress uniform. Headlines preached rage, nationalistic banners streamed out of houses, and leaflets and broadsheets demanded "honor" for America. In the meantime, Republicans retreated behind closed doors where they secretly sported the French tricolors, sang the Revolutionary French ditty "Ça Ira," and addressed one another as "Citizen."

The United States was again at war with itself. Men grimly crossed the streets to avoid greeting those of the rival political camp, and ladies crossed old friends off their social guest lists. Ritualistically, Republicans were now "Jacobins" and Federalists were now "Anglomen." And in city after city, paranoia marked the political landscape: Federalists feared the French, the Republicans feared the Federalists, both sides jockeyed for political advantage, and it became impossible to separate the war from domestic wrangles. Too often when men smiled, it was a vulpine smile, a smile of menace and malice, a cynical smile flashed by men who more than anything else relished revenge. Recorded one foreign observer: "Everybody was suspicious of everybody else; everywhere one saw murderous glances."

He did not exaggerate. Even in the capital itself the atmosphere had become toxic. This too was vaguely reminiscent of a throbbing Paris at the start of its revolution: Roving bands of Federalists, chanting martial songs and thumping up and down the streets, clashed violently with Republicans. Actors singing "The Marseillaise" were booed off the stage while the editor of the Republican newspaper, the *Aurora*, had his house assaulted and his windows smashed. Rumors now gathered that French saboteurs, aided by sympathetic American Jacobins, might torch the city, and John Adams stationed sentries outside the presidential residence. And in urban pockets as well as small towns, venom hung in the air; two congressmen even got into a violent skirmish on the floor of the House.

Yet Adams, at the zenith of his popularity, scarcely noticed. For all his exhilaration, he was working around the clock. And for all his triumphalism, a pale and thinner Adams was unable to distance himself from the awesome weight of the trials he now faced. Wrote Abigail, "he falls away."

It was in this context—war, violence, dissent, and personal enervation—that the young nation stumbled into the low point of Adams's presidency and one of the low points in its history.

∞

SUMMER OF 1798—"the most gloomy summer." These were Adams's words. The heat was oppressive and the flies were tormenting. But in Philadelphia, during these stifling months, the political adulation had peaked, and Adams and his allies—though it was not yet obvious—had overreached. In the lore of every nation there are scenes, phrases, and deeds that live on in the popular imagination—sometimes they are thrilling, sometimes they are disastrous. But a great event, a spectacular speech, or a legend can never be easily repeated, for part of its appeal is that it is unique. But this doesn't stop many from trying. That is why there can never be another King Arthur, another Washington, or in our own time, another Churchill. It also perhaps explains why Adams and his allies now blundered so badly.

They began with the best of intentions. Across the Eastern Seaboard, the country was gripped by martial fever. It was understandable. French cruisers were now routinely patrolling U.S. coastal waters, hovering off Long Island Sound and the Delaware Bay. So in the town of Newburyport, the local government built a twenty-gun warship to loan to the nation. Other cities quickly followed. New York raised an astonishing $80,000 or so to build a thirty-six-gun frigate; then Philadelphia matched them. And Boston outstripped them all. Meanwhile, in the capital, the quiet of the morning was every day broken by drum and fife hailing, "A Band of Brothers Joined." The frenzy so mounted that even Abigail berated Congress for dragging its heels in declaring war.

Why didn't they? Because Adams, still holding out for a peaceful resolution, had not yet asked for a declaration. A lifelong diplomat and student of history, he knew all too well that a small skirmish or a few too many words between two nations gone awry could swell into a cataclysmic upheaval, a tornado of blood and wreckage with consequences beyond prediction. So Adams stayed his hand and Congress instead did what it could—and that was to address another enemy: the enemy within. As Americans watched Napoleon's legions almost single-handedly batter and beat Europe, as predatory French ships could be spotted from American shores, much of the country was increasingly on edge about the massive influx of French émigrés in America, which now numbered more than 25,000.

To be sure, some were aristocrats who had escaped the Terror; but many, such as the refugees from slave uprisings in the Caribbean, were not. Moreover, as even a Russian seaman, whose ship had anchored in Philadelphia, noted, the French seemed to be everywhere: Who could escape the sight of French newspapers blaring the news or French booksellers peddling their goods? Or French schools teaching their own curriculum? Or French boardinghouses, which could have been nests of terror or sedition or both? Even French restaurants were suspect. Who knew, then, who among them were enemy agents—or what they might be plotting? Nor was this was idle paranoia. It was, after all, routine for all foreign governments to keep their agents in other nations, and France had already meddled in America's internal affairs, not to mention when it recruited the radical Wolfe Tone in Princeton to collude with the Directory in an invasion of Ireland. And then there were the Irish, often derided as the "Wild Irish," refugees from the recent failed rebellion, vast numbers of whom were radicals in sentiment and republicans in fact. Indeed, Timothy Pickering, the secretary of state, was so worried that he favored massive deportations of aliens. And the journalist William Cobbett spoke for many Americans when he wrote about the thousands of factious villains "vomited" from the shores of Great Britain and Ireland. This too was the summer of 1798.

In June and July, the Federalist-controlled Congress, its reputation never greater, acted decisively, passing into law a new set of bills, four all told, to ensure that foreigners or rebel sympathizers could not become a fifth column in case of war. As Europe fell into violence and republicans took to the streets, London had already embarked on similar measures, and on a far more repressive scale, so had Paris and St. Petersburg. The U.S. Congress now joined them. Representative Ames gave the rationale, with words that easily could have passed from Robespierre's lips. "Zeal," he sneered, "is better than logic." Zeal it was. The Alien Acts included a Naturalization Act, which increased the period of residence for citizenship from five to fourteen years. Then there was the Alien Friends Act, which gave the president the legal basis for expelling any foreigner he deemed "dangerous." This act was particularly favored by the secretary of state; by contrast, the Alien Enemies Act, which allowed for the removal of citizens of a country with whom the nation was at war, was the one measure enjoying broad bipartisan support. The same could not be said for the Sedition Act, which made it a federal crime to publish "false, scandalous, and malicious" writings against the government. It was passed as both a war measure and as an improvement over existing common law, which it was. Yet there was no denying that it also violated the First Amendment to the Constitution guaranteeing free speech, a crucial cornerstone of republican freedoms, or that it alarmingly stepped away from the principles of the Enlightenment and of the Revolution. Nor was there any refuting that it could—and did—constitute a fearsome weapon to muzzle Republican editors or to silence opposition press organs. Signed into law nine years to the day after the fall of the Bastille, it was the first sign that a real tempest loomed. The government now presided over a nation torn in body and soul.

Ironically, though, it was a mark of the times that such respected figures as Noah Webster, the prominent editor and author, strongly supported these measures, as did senators Theodore Sedgwick of Massachusetts and

James Lloyd of Maryland. In fact, no less than George Washington, worrying about Jacobins and disloyal citizens "poisoning the minds of our people" and "sow[ing] dissension" among them, supported them too. Yet Adams himself struck a curious stance on these four measures; he neither requested them nor did he attempt to impede their passage. Did he want to have it both ways? It's a good question. In truth, he seemed to be receptive to the acts, partly because they showed France that the United States was willing to undertake drastic measures—an edict hurled like a lance at its adversaries—and also because Adams firmly believed that French spies "swarmed our cities." Looking back, the headiness of the High Federalists was unqualified. To our modern sensibilities, it seems mindless and feckless and reactionary. But to them, the alarm, far from extreme, felt palpable and real. And likely Adams, in his tacit acquiescence, did not fully perceive these measures in terms of slippery slopes; he perceived a nation in peril.

America had already weathered an armed uprising, Shays's Rebellion, as well as foreign intervention and repeated French meddling. It is also the case that whenever war has threatened to lap up on the shores of America, or behind its lines, presidents have acted swiftly and ruthlessly. More than six decades later, during the American Civil War, Abraham Lincoln moved fiercely to contain what he called "the fire in the Rear"—habeas corpus was suspended, angry dragnets were set into motion, and some thirteen thousand citizens, mostly antiwar Democrats (including duly elected legislators), were arbitrarily rounded up and jailed for years, usually for little more than their political opinions. Later, of course, there would be Woodrow Wilson's repression of Germans in World War I and Franklin Roosevelt's internment of the Japanese in World War II, the Republican-inspired red scares of the Cold War, and the Patriot Act of the twenty-first century. Ironically, in each case, Republicans and Democrats alike followed a long, taut string of American history, stretching back to, of all presidents, John Adams.

There is no doubting that errors and excesses occurred. They did, in spades. And given how young and unformed the country still was, this was a poisonous fruit indeed. While Adams, to his credit, largely resisted the widespread clamor to run aliens out of the country, he feverishly whipped up a public frenzy by denouncing those who placed their confidence in "a foreign nation" and by enforcing the Sedition Act, thus imperiling the free press. On trumped-up, often flimsy charges, the administration indicted seventeen Republican editors; some of them were forced to pay fines, while others landed in jail. Some publications simply closed shop rather than court the law's wrath (still, not all the opposition was silenced; other Republican editors were actually emboldened and doubled their number of newspapers over the next two years). As resident Frenchmen, fearing imprisonment, signed up in droves for passage back to Europe, there were the less tangible but no less atrocious abuses: the xenophobic shouts and bigoted hatreds spewed against immigrants—said one: "All governments are to them hateful"; said another: "Like Lucifer, they carry a hell about with them in their own minds; and thus prowl from country to country." Taken as a whole, these were actions worthy of the most brutal despots and repressive regimes on the world stage.

Ironically though, despite the fears about the Union's very survival and the cascading specter of France's "contagion," Americans would come to eschew the tenor of the rest of the world—the war emergencies of revolutionary France, the turning back the clock in a once public-spirited Russia, even the mounting civil repression in enlightened Britain—and come to embrace popular assemblies and freedom of speech as the paramount bricks of a republican society. In a sense, though, Americans first had to face the devil before they averted their gaze.

By now, it seemed unavoidable. But in a curious way, this crisis would prove to be a provocative reminder of the spirit of the elusive nation. And it would be a turning point for the two political parties, as well as a turning point in the country.

∞

IN A REVOLT of conscience, a few thoughtful men, in the quietude of reflection, read the terms of the acts and were deeply troubled; even Hamilton exhorted, "Let us not establish a tyranny. Energy is a very different thing from violence." In the early days, the first doubts struggled to the surface, but it was too soon for them to coalesce. Meanwhile, the swells of admiration and praise continued to break at Adams's feet. But Jefferson, who denounced the "reign of witches," quickly sensed the coming change of public mood. And those who had cheered the vice president and the French suddenly felt vindicated. Adams and his allies, they now contended, were indeed "warring" against the "real principles" of the American people.

Suddenly Americans of all stripes increasingly wrestled less with matters of war and peace and more with the issue of civil liberties and an "arbitrary" national government that was swallowing "our public liberty."

In Philadelphia, a quivering Vice President Jefferson, unable to stomach his own government, abruptly fled the capital and moved back to Monticello. He would stay there for six months. He and the president would not exchange words for over a year.

∞

FOR ADAMS, 1798 was the most arduous and significant time for his presidency. The same could also be said for the political opposition, which now undertook concerted steps to organize. As fall swept across America, nowhere were hazard and anticipation held in more intricate balance. The political battles seemed to be without end; as it happened, so did the sickness. Yellow fever had spread through Boston, New York, Baltimore, and, most distressing of all, into Philadelphia; by September, 40,000 had fled the capital. And then the storm of political censure led to another storm, of reprisal. For back in the rolling hills of Virginia, Thomas Jefferson was conspiring.

Disgusted with the iniquity of the Federalist Congress, fed up with the folly of John Adams endangering the "World's last home of a republik," fearing that republican voices could be in permanent eclipse, he was writing letters, penning memorandums, contacting political allies—and privately drafting a set of resolutions to be introduced into the Kentucky legislature. This was his own electrifying response to the Alien and Sedition Acts. It was brazen, it was risky, and it was an act of desperation. But if the Republicans had to fight, they would. Jefferson, through Kentucky legislator John Breckenridge, introduced a resolution that asserted a "natural right" for each state to nullify the actions of the federal government. His language was extreme, but so were his concerns: At stake, he believed, was nothing less than the fate of the Constitution, the survival of "self-government," and the integrity of the Union; he even feared that a federal army might march on the South. At the same time, James Madison wrote a slightly watered-down version of his own for Virginia's legislature. In both cases, these measures were secretly authored. And in both cases, they were breathtaking reversals for these two gifted men who had done so much to write the Declaration of Independence and the Constitution.

Had their actions been uncovered, the vice president could have been brought up on sedition charges and Madison could well have been made a pariah. Yet taken together, the Virginia and Kentucky resolutions, which passed that November and December, seemed to ensure that any repressive measures put forth by the central government could be challenged. But if the Alien and Sedition Acts were a venomous fruit, so were these resolutions. In arguing that nullification was a "rightful remedy" in state and federal disputes—and in Jefferson's case he actually called for outright rebellion ("revolution and blood") if excesses could not be checked—the two men had unwittingly laid the political and conceptual foundations for national dissolution as well. If invoked, both could have been fatal to the ten-year-old Constitution and even to the very existence of the United States—and indeed, in part they were, by laying the basis for secession ("scission"), which would eventually be carried out by the Confederacy in the nightmare of the Civil War itself. But that would come years later. For the time being, the call went out for other states to speak in unison. Some, like John Randolph, the Senate candidate in Virginia, called for resistance "by force." Ultimately, however, cooler heads prevailed, and no other states adopted these radical resolutions.

But though no one quite knew it yet, the damage had been done. The gauntlet had been flung. Nor did this prevent Hamilton, intriguing more than ever, from darkly wondering if the federal government should take Spanish Florida and Louisiana—or move to put down subversion in Virginia itself; as he put it, to "put Virginia to the test . . ." How? By marching through with the provisional army that Congress had formally created—which was to be headed by a George Washington brought out of retirement, with Hamilton as second in command. Of the army itself, Abigail Adams wrote: "It is one of those strokes which the prospect and exigency of the times required." The last thing she had in mind, however, was what her husband now trembled about: an American "Bonaparte."

∞

BUT THAT WAS changing. To be sure, in these desperate days the country was in an apocalyptic mood. Dissension, mutual recrimination, and fear of civil war were the powerful undertow tugging from within. The chief target of discontent was not the French but invariably each other. And the country was surly, hostile, white-hot with grievances. Yet all this was ultimately driven by the dark underside of battle. And that, more than anything else, was the matter to which Adams turned. During the bleakest hours, Adams had refused to declare war. But as the country lurched from one crisis to another, once more he reexamined his policies.

Tired as he was, Adams pushed through a maelstrom of work that summer and fall, his days consumed with his French policy. Early that June, he leaned that Pinckney and Marshall had in fact safely left Paris in April, while Elbridge Gerry took it upon himself to remain behind. This last news, about "my ambassador," was a bitter pill to swallow. It was clear that the Directory would only negotiate with Gerry, and, to Adams's abhorrence, it was also clear that Gerry had eagerly seized upon this self-anointed role as peacemaker. Yet bit by bit, Adams was able to piece together French intentions. On June 17 John Marshall landed in New York, then made his way to Philadelphia, where he was given a hero's welcome. Tall and shrewd, Marshall confidently told the president that for all the French bellicosity, there did not need to be war; nor, as he saw it, did France even *desire* war with America. For the moment, though, Adams was still confused. As it happened, the picture remained stubbornly mixed, since France also threatened to annihilate America if Gerry did not stay behind. For America's second president, it was hard to decipher where all this was heading.

But by October, Adams was at least able to speak to Gerry directly. The meeting was not without controversy; it also proved to be an eye-opener. Gerry was suddenly a man scorned in America; upon his return, he was subjected to a fusillade of criticism—angry crowds shouted insults outside his Cambridge home; men waved their fists at him; and silence fell wherever he walked the streets. Yet now he was meeting with the president. And the more Gerry explained, the more Adams was taken in by his candor. True, Gerry was a pale, contentious man; he was also undeniably astute. And his message—that the peace feelers were genuine—reiterated Marshall's before him. In subsequent days, Adams received a similar confirmation from his son, John Quincy, as well as from Joel Barlow, a Federalist and prominent poet who had lived in Paris. In fact, Barlow added more vital information: "If the US refused these generous overtures," he said, "war of the most terrible and vindictive kind will follow." In light of this, even America's greatest warrior, George Washington himself, privately suggested to President Adams that an agreement on "terms honorable" could be negotiated. Still, for Adams it would take one more meeting to decisively tip the scales. It was his most daring consultation yet.

He conferred with a third man, George Logan, whose name today still stands for treachery with the enemy. In the end, Logan was everything that Adams despised: a Quaker, an ardent Republican, and a Francophile, all this he could stomach; but not the fact that he was also a freelancer, having conducted his own private peace mission with the French. The secretary of state had thus refused to talk with him; Washington gave him the cold shoulder; but remarkably, Adams saw him over tea. Logan's message affirmed what Adams had increasingly come to believe.

Talleyrand, and France, were at long last ready to talk.

∞

THIS WAS UNDERSTANDABLE, for that August, Admiral Nelson had decisively defeated Napoleon's men in the battle of the Nile; only two French vessels had escaped capture, which meant that the likelihood of a French invasion against the United States was now dramatically reduced. When word of the unmitigated disaster reached Napoleon in Cairo, he knew that he was now cut off by both land and sea from French assistance, and could at any moment fall prey to the whims of a restive population and an unforgiving environment. Just as quickly, the Arabs and Turks realized this too.

For Napoleon, the perils now multiplied. Almost every day the Turks, Arabs, or Mamlukes attacked their new French masters, and on October 15 the people of Cairo themselves—the so-called Arab street of today's parlance—exploded in revolt. Taken aback, Napoleon abandoned his role as the benign overlord and ordered the beheading of every armed rebel. He thought this would help staunch the bleeding. It didn't. Meanwhile the morale of Napoleon's men began to plummet.

Dire word trickled in that the sultan was preparing an army to reclaim Egypt; Napoleon resolved that he would meet the test by taking an army of thirteen thousand into Syria. This was to be his greatest challenge yet. On February 10, 1799, he captured El Arish and then crossed the Sinai Desert, where his troops once again endured unfathomable heat, unquenchable thirst, and "brackish water," when they found any at all. By this stage, his men were forced to eat dogs, monkeys, even camels, anything that they could digest. Two and a half long weeks later, they were alive, barely, and eventually made their way into Gaza, where after yet another hard battle, they finally found food: prosperous orchards of fruit. Just as quickly, Napoleon was again on the move. On March 3 he made it to the ancient port town of Jaffa—in today's Israel. The city was not only walled, but filled with a hostile populace and a daunting citadel with more than 2,700 Turks. Napoleon sought to make terms with the enemy; he was rebuffed. The date was set for an attack.

On March 7, Napoleon's men breached the wall, French troops poured in, and those who resisted were slaughtered. Then the plunder began. Napoleon tried to restore order, and prisoners were taken. In one of his most controversial decisions ever—heartless might be a better term—he raised his hands in disgust and convened a council of war: He didn't want to spare precious food and water, or, for that matter, guards to take the Turks to Cairo, nor did he want to set them free to fight another day. "What can I do with them?" he cried. Fatefully, he decided to kill the prisoners instead. That day, some 2,441, including civilians of all ages and both sexes, were murdered, the lucky ones being shot, with the brunt of them being ruthlessly bayoneted like animals to the slaughter to save on ammunition.

As the blood of the innocent stained the sand and dirt, disasters now burgeoned. Napoleon marched on, reaching the extensively fortified town of Acre on March 18; it looks much the same today as it did then. He still had visions of pushing all the way to Constantinople, and, like Catherine the Great before him, of restoring the Eastern Empire. His men dug in to set siege to the Turkish-held fortress. But aided by the English, the Muslims had ample food and matériel, and Napoleon, cut off from reinforcements and deprived of his artillery, sustained heavy losses. Two months later, reality set in. He sadly ordered a retreat from Palestine back to Egypt. Said Napoleon mournfully, "my imagination died at Acre." This was overstatement, but he now encountered a succession of setbacks. Not foreseeing the nightmare of 1812 when he would square off against Catherine's grandson, Tsar Alexander, and move on Moscow itself, he force-marched his men at a brutal clip for eleven hours a day, trekking from well to well; but the water, when they found it, was often undrinkable, or when it was, it frequently poisoned them. Now plague-stricken troops slowed the procession. Napoleon soon became desperate, asking his doctors to administer lethal doses of opium to the fatally ill. Indignant, the doctors balked, and Napoleon retracted his suggestion. So the sick were carried on horses and his officers set an example by marching on foot. Twenty-six days later, after an astonishing journey of three hundred miles, Napoleon's depleted and famished army heroically reentered Cairo, brandishing seventeen enemy flags and sixteen Turkish officers as evidence that the expedition had been a magnificent triumph. It hadn't been; but indefatigable as always, Napoleon was anything but beaten.

On July 11, a hundred vessels landed at Abukir; they were an army of Ottoman Turks raised by the sultan's grand vizier to repel the invaders from Egypt. Napoleon hurried north with his finest men and imposed so great a defeat on his attackers that many of the fighters chose to drown in the sea rather than withstand the onslaught of French cavalry. But he barely had time enough to savor his triumph when he learned from English newspapers that a Second Coalition of powers, more mighty than the first—Austria, England, Naples, Portugal, and yes, Russia and the Ottomans too—had forced the French from Germany and had reclaimed a vast swath of Italy, from the Alps to Calabria. Napoleon was aghast. He had forged one of the fiercest armies that history had every seen, swept through Europe undefeated, stamped his imprint on new republics he had created, and now this. To complete the humiliation, Russia's famed General Alexander Suvorov, that veritable "devil of a man," had decimated French armies in Italy.

France's hard-won frontiers were crumbling by the day, and the specter of these "giant barbarians," an avalanche of dreadful Russian Slavs overrunning the towns of France, was now raised. Once more France, so recently proud of its strength and its historic victories, was in a state of confusion and fear, so much so that it resembled the paranoia of 1792, which had led to the September massacres. Upon careful reflection, Napoleon knew that his absence had made all this possible. He had been too quick to hand Italy off to a French protectorate, too cavalier in ignoring England's unquenchable enmity, too insouciant in writing off Russia as being content with its eastern conquests, and too optimistic in thinking that the Ottoman empire would not want to avenge its humiliations, even at the cost of making common cause with its old enemies, including Russia itself.

So as fresh contingents of troops poured into Europe to check the French imperialist monster, the whole edifice of Napoleon's victories had collapsed, from the Rhine and the Po to Abukir and Acre. Even Malta had fallen. And now, in a humiliating checkmate, he found his decimated legions ensnared in an unfriendly kasbah—a blind alleyway—and only time would be needed to complete a final annihilation. It was at this stage that Napoleon resolved to return to France, even at the expense of shamefully deserting his brave army and leaving it to the mercies of a hostile populace and the governments of Ottoman Turkey and Egypt. "If I have the good fortune to reach France," he vowed disdainfully, "the rule of those *bavards* [babblers] will be finished." On August 23, 1799, he boarded the *Muiron*, one of the two ships that had escaped the cataclysm of Abukir, and loaded his other generals in the other surviving ship, the *Carrère*, and set sail for home. In his mind, it was now to be that same ancient choice: *aut Caesar aut nullus*—either Caesar or nobody.

Whether he could slip by the eyes and scouts of Nelson's indestructible fleet was another story. In the meantime, for the reigning Directory, having won so many victories that it had lapsed into an almost somnolent lassitude, and having now watched with each passing day as the mirage of this bright new world receded, the last thing it needed was to magnify its troubles. If this meant making peace with the Americans—for on the scale of world events the trivial United States was but a speck compared with the treasures of Europe and the Ottoman empire—then so be it.

∞

ALL THIS, OF course, would be to Adams's relief; the last thing he wanted was to confront the giant French army or face a repeat of the bleak and ghastly battlefields of Europe. Finally, on one cold November day, that prospect seemed less likely than ever. While winds howled and a driving snow fell, word seeped in of Admiral Nelson's spectacular drubbing of the French fleet in the battle of the Nile four months earlier; by now, too, three squadrons of astonishingly modern American naval vessels were fit for duty. The immediate danger thus seemed to have passed. Cried a jubilant Adams, "At present there is no more prospect of seeing a French army here than there is in heaven."

So on December 17, 1798, under a wintry sky, Adams told a joint session of Congress that the door was ajar to peace. Slowly, public passions began to cool. And on February 18, 1799, after one hundred days of agonizing uncertainty, Adams went a step further. While his cabinet had concluded that it was the French who should send a mission across the ocean, Adams bypassed them altogether, announcing that he would send William Vans Murray from The Hague to Paris, to which he then added Chief Justice Oliver Ellsworth and the Federalist governor of North Carolina, William Davie. After some back-and-forth, it was decided that they would set sail on November 15.

Overnight, the political world capsized. Of all Adams's acts, none was as unpopular as this one. High Federalists had long derided his policies as timid and indecisive. Republicans had long disparaged his administration as warmongering. And as these debacles mounted, there were deepening schisms in Adams's official family, once more among his advisors as well. Yet it was impossible to ascertain how the country would react if Adams seemed to grovel at the feet of the Directory. But as Abigail once said of Adams: "He may be torn up by the roots, or break, but he will never bend." True to himself, Adams did not waver. This, in spite of the fact that he was angrily condemned by his own party and cabinet; this, in spite of the fact that the Republicans applauded him with at best one hand.

This, despite the fact that unsettling news soon arrived from Paris of "chaos"—the Directory was breaking up.

∞

IF THE BALANCE of power was changing abroad, it was also changing closer to home. "With all my ministers against me," the president later sheepishly admitted, he was "no more at liberty than a man in prison, chained to the floor and bound hand and foot." For too long Adams had been accused of indifference to Federalist war aims, of causing friction in the cabinet and bringing on doom and drift throughout the administration. Finally, Adams had enough; he fired most of his cabinet. In typical Adams fashion it was inartfully done, leaving Pickering and McHenry unnecessarily bitter and alienated, and his party aghast. Nevertheless, when he was finished, Adams was proud of himself. At long last.

Then in March 1799, the decade in America seemed to come full circle, when force was again used in Pennsylvania to put down another budding rebellion. Stoked by Republican propaganda—it was alleged that President Adams was planning to marry off his son to a daughter of England's King George III—150 farmers armed themselves, set upon a United States marshal who had come to collect a land tax, and stormed a nearby jail holding a group of tax protesters. But unlike the Whiskey Rebellion, this uprising was not in the roughened West but in the East, in the town of Bethlehem. Led by a man named John Fries, a former militia captain who had ten children, these typically docile Pennsylvania Dutch Germans were unlikely suspects for violence, which made it all the more troubling. In any case, Adams felt that violence was violence, whether in Massachusetts, or Virginia, or Pennsylvania.

Hamilton agreed. He told the war secretary, "Beware, my dear sir, of magnifying a riot into an insurrection, by employing...inadequate force...Whenever the government appears in arms, it ought to appear like *Hercules*." It had in the Shays's and Whiskey Rebellions and it did so again. With Adams's sanction, a force commanded by a brigadier general rampaged through the "rebellious" countryside. Sixty men were arrested, and three, including Fries, were convicted of treason. They were sentenced to death by hanging. Still, in the spring of 1800, Adams showed himself to be as stubbornly independent as ever. Against the unanimous advice of his exasperated new cabinet—they thought the punishments necessary to "crush that spirit" that would overturn the government—Adams pardoned the men. He was sure Americans could take it. He was right. It was a fitting gesture; once more, Adams's own conscience had remained true.

Often throughout his presidency, Adams lurked in his home, walled in by books and memorandums, overcome by doubt. Now the president was left with two last outstanding tasks: awaiting word from his three new commissioners in Paris—and running for reelection. George Washington before him had sought to create in every American breast a republican soul and a spirit of nationalism beneath the ribs of rebellion and discord; Adams's task, by contrast, was to maintain the peace in a world at war. Did he succeed? The French, as ever, were unpredictable, turning as quickly on old friends as new ones, changing governments as rapidly as they had made them, abrogating treaties as quickly as they had signed them. So Adams as yet had no way of knowing. But this much he could know: "I congratulate you on the New Year and the New Century," Adams wrote to an old friend.

With that, for Adams, and the nation, the grand epoch of the eighteenth century came to a close.

Epilogue

The Founding

The White House

EARLY 1800.

Having recently toured in Vienna, Berlin, and Prague, the twenty-nine-year-old composer Ludwig van Beethoven is launching a public concert on a grand scale, playing the *Septet* and *First Symphony*; he is also writing a score for the ballet *The Creatures of Prometheus*. Until now, his art has sought to encompass the new spirit of humanism as captured in the works of Goethe and the ideals of the French Revolution. But his work is poised to enter a new phase. Slowly and painfully, the first symptoms of his encroaching deafness are becoming more pronounced. Tempted to take his own life, in calmer moments he is more reflective. "Ah, it seemed unthinkable for me to leave the world forever," he memorably writes to a friend. "I will seize fate by the throat."

The high-spirited young Englishwoman Jane Austen has already begun putting down her thoughts on paper. Her life was, on the surface, uneventful and serene. But in 1795 she began a novel-in-letters called "Elinore and Marianne," which will appear anonymously as *Sense and Sensibility*. Two years after that, she completes her "First Impressions," which will one day be released as *Pride and Prejudice*. Both are delicious treatments of the comedy of manners in gentry life. Yet her own life is about to take a turn for the worse: Her father will shortly retire to Bath, and Jane will have to put up with a succession of temporary lodgings or visits with relatives. Her dearest friend, Mrs. Anne Lefroy, will die suddenly, and she will agree to marry Harris Biggs-Wither, only to change her mind the next morning. She will never wed. There are some apparent consolations: Another manuscript, written under the title *Susan*, will be sold for 10 pounds. Inexplicably, however, it never appears in print.

Mary Wollstonecraft, the only daughter of Mary Wollstonecraft Godwin, is three. She will one day be best known as the author of *Frankenstein*, but arguably her mother's impact will be infinitely greater. By the time of her daughter's birth, Mary, a passionate adherent of the early French Revolution and lover of America, had joined an extraordinary circle of radicals that would include Thomas Paine, William Blake, and William Wordsworth. Her 1792 tract *A Vindication of the Rights of Woman* will one day form the basis for later pioneers of the women's rights movement, like Elizabeth Cady Stanton; in her own day, it also inspired the outspoken admiration of Abigail Adams, and her history of the French Revolution was devoured by John Adams— twice. Tragically, Mary died in 1797, eleven days after her daughter's birth; the girl will live out her life with only her mother's name to guide her.

In America, legendary frontiersman Davy Crockett is fourteen, seamstress Betsy Ross is forty-eight, painter Gilbert Stuart is forty-five, aspiring senator and orator Henry Clay is twenty-three, future war hero and president William Henry Harrison is twenty-seven, and Robert E. Lee, who will briefly taste the life of a Virginia gentleman before his father's inglorious ruin, is two.

And a thirty-three-year-old Scots-Irish lawyer named Andrew Jackson has moved back to Tennessee. One day he will become a comet unto himself, the first major figure in American politics to believe passionately and wholly in popular sentiment; as president, he will go on to create the modern Democratic Party. An ardent Francophile, he is as inscrutable as ever. His grammar and spelling are shaky; violent and temperamental, he has killed a man in a duel; and he rants constantly, calling his enemies the "Great Whore of Babylon" and delighting in the fact that his critics think he is "mad." But he is also shrewd, courteous, self-possessed, and visionary. Elected to the House as Tennessee's first congressman in 1796—there, he had been praying for a French invasion of England—he refuses to seek reelection and heads home in 1797, vowing never again to enter public life. Some think he remains destined for greatness, but others now wonder if his national career is finished. Even over a decade later, when Jackson's name is suggested for U.S. envoy to Russia, no less than Thomas Jefferson will exclaim, "Good God! He would breed you a quarrel before he had been there a month."

After a tour in Paris from which he was unceremoniously recalled by George Washington, James Monroe would again find himself back in France, where he will be startled by an offer from Talleyrand: Would the United States be interested in purchasing roughly 828,000 square miles of territory, the whole of Louisiana?

By the turn of the century, at thirty-three, John Quincy Adams is already emerging as one of America's most accomplished diplomats, having served as Francis Dana's secretary in Russia and minister to Prussia as well as Holland. He will also be selected as senator from Massachusetts, and then become minister to Russia under President James Madison. There, continuing the strange circle of history, he will frequently bump into Catherine's grandson, the dynamic new tsar.

In Britain, the earth was settling on the graves of Edward Gibbon, author of *The History of the Decline and Fall of the Roman Empire*, dead at the age of fifty-seven; sixty-eight-year-old Edmund Burke, the British statesman, friend of America and foe of the French Revolution; sixty-five-year-old Josiah Wedgwood, the brilliant porcelain maker and abolitionist; and the great Adam Smith, author of *The Wealth of Nations*. In a scant six years, Prime Minister William Pitt the Younger, exhausted by ceaseless war and turmoil, would join them amid the markers and marble tombs.

In nine years, 1809—while the age's master pamphleteer and champion of democracy, Thomas Paine, having miraculously avoided execution in France, will die unceremoniously in Greenwich Village, New York—Louis Braille will be born, and so will the writer and poet Edgar Allen Poe and the naturalist Charles Darwin; and in the poverty-ridden Kentucky backwoods, an infant boy will be named Abraham Lincoln.

As the new century dawns, one inhabitant in five of the United States is black, and far too many of them live in slavery, a regime of sorrow, degradation, unremitting toil, dreadful fear, and perpetual frustration. Remarkably, there are no slave rebellions. But Nat Turner, born a bondsman in a remote area of Virginia on a small plantation in 1800, will one day change that. Given the sign to strike in 1831—it was an eclipse of the sun—he will launch a campaign of total annihilation, murdering sixty people and inspiring terror throughout the South. For many years in black churches throughout the country, his name will be reverently whispered; moreover, the name Jerusalem will refer not only to the Bible but also covertly to the place where the daring rebel slave had met his death.

Eighteen-year-old John Calhoun has not yet embarked upon his extraordinary but ultimately tragic career in public service. A Jeffersonian Republican, a South Carolinian, and an ardent nationalist and future vice president, Calhoun will abruptly become the foremost spokesman for states' rights and the most vigorous icon of the old South until the American Civil War.

The former Marquis de Lafayette, nearly broken by the harsh years spent in an Austrian prison, is now back home in France, having settled down as a gentleman farmer.

But as the century turned, one man was noticeably absent. On December 14, 1799, came unexpected news. George Washington was dead—of "Quinsy," a streptococcus infection. He was sixty-seven.

∞

IN AN INSTANT, the mood of America changed. Wrote Abigail Adams, "A universal melancholy had pervaded all classes of people." That evening, a soft moonlight poured through the windows of Mount Vernon, where since midnight Washington's body had lain still, gently stretched before a magnificent Italian mantel. His last words were a whisper. "I am just going," he said. "Have me decently buried." He was. On December 18 a schooner maneuvered into position in the Potomac and fired minute guns, answered by the booming sounds of artillery and a funeral dirge rendered by a Masonic band. As foot soldiers formed a line, the Reverend Mr. Davis read a brief service of farewell ("I am the resurrection and the life, saith the Lord . . ."), and more guns repeated from the Potomac, echoing in the hills. Then Washington's shrouded figure was permanently laid to rest.

The news of Washington's sudden passing trickled slowly at first, then raced across the nation. Tearful Americans would never forget that moment: All throughout the country muffled bells rang and people gathered in stunned silence. In the towns, business and labor were suspended, music was played, and the Scriptures were solemnly read. In the rural areas, farmers knelt mournfully in their fields while thousands of men, women, and children stood in the rain or snow, mute and still, paying tribute to the man who was once a little-known but ambitious surveyor. Philadelphia itself heard the news on December 17. In the morning the great bells of Christ Church commenced and Congress closed its doors. While a gentle breeze caressed the sky, black was draped everywhere: in Washington's church pew, on the door of the presidential mansion, in Congress, not to mention northward to Boston Harbor and southward to Richmond, Virginia, and stretching out from Quincy, Massachusetts, and the backwoods of the Carolinas to the great city of New York. Even the nation's young army wore crepe armbands. The mourning lasted for a full ten days, until after Christmas. By now Washington himself had been interred in the family vault at Mount Vernon, but this was the nation's final tribute to its first president and its grandest hero, and it was to be done right.

To the measured beat of muted drums, a grand, subdued procession began in the capital, marching with machinelike efficiency. In slow time, starting at Congress Hall, here came the light infantry and here came the cavalry. Now came the federal leaders and the state leaders, the city magistrates and the Masons, the Supreme Court and the diplomatic corps, the Society of the Cincinnati and the assorted dignitaries, and the riderless white horse with boots reversed. The windows, and even the rooftops too, were filled with people paying their last respects. Stretching from curb to curb, the line quietly glided south on Fifth Street, then east on Walnut, then north on Fourth, then crossed over to the German Lutheran Church at Cherry Street. Several thousand mourners crept in, crowding the church to capacity. All of official Philadelphia was there: They wore black plumes and black military sashes, black epaulets and black gloves, and even carried black fans. The service would last four and a half hours. One minister that day would speak of Washington as the "savior" of the country, while General "Light Horse" Harry Lee—Robert E. Lee's father—would memorably eulogize Washington as "first in war, first in peace, first in the hearts of his countrymen." Fisher Ames, a Federalist ally, was equally perceptive, noting that Washington "changed mankind's ideas of political greatness." But ironically, it was Washington's now estranged aide, Thomas Jefferson, who would perhaps put it best. "I felt on his death, with my countrymen," he later wrote, "that verily a great man hath fallen in Israel."

And what about the nation that he loved and long saw as an "experiment"? This much could be said. Washington, who invariably had the gift of exquisite timing in life, seemed to have it in death as well. As fate would have it, his passing marked the end of one remarkable era, even as three great global events would herald the beginning of another.

∞

AGAIN. AGAIN THE world seemed to be rushing full tilt into disaster. It began in France, where change was once more brewing. Napoleon had come back from Egypt to find a nation torn between the nobles and work-ing poor, between Vendean Catholics and Jacobin atheists, between those demanding liberty and those demanding equality. No less than in the past armed gangs were fighting officials sent to conscript them for France's many wars, while municipal officers were murdered and brigands terrorized mer-chants on country roads. Every estate, every abbey, every home now faced the threat of random attack, and the nation turned to Paris in faint hope of protection. But the lawmakers had given way to the Directory, and the Directory had fallen prey to all the familiar temptations of the old, corrupt monarchy: bribery, thievery, lies, and most of all, dictatorial force.

On July 12 the Law of Hostages was passed, ordering every commune in France to compile a list of local citizens with ties to the outlawed nobil-ity and to keep them under watch. Paranoia had become the order of the day. If this wasn't bad enough, a collective exhaustion settled in. After a decade of conflict, class strife, foreign wars, political upheaval, rampaging tribunals, endless executions, and unending massacres, the heady days of the Bastille and the Rights of Man, once so filled with promise, were long gone. As it happened, nearly all of France was tired of the Revolution, and amazingly, many looked back ruefully to the "good old days" of Louis XVI. When long lines of exhausted and battered soldiers limped home from far-off battles in Italy, Austria, Germany, Switzerland, the Caribbean, and on the high seas, and eventually in the Near East, they were stunned to find despoliation in commerce, graft in industry, and unspeakable lawless-ness in the assemblies. And they asked themselves why they should give their limbs and their lives for so tarnished a dream. Nor were they alone. Even Abbé Sieyès, who ten years ago had helped spark a revolution and rouse a people with his query, "What is the Third Estate?," agreed.

By this stage, most Frenchmen had come to the conclusion that some-thing had to be done to end the chaos and give the country the security of an orderly life. Sieyès thought, as did many others, that what was needed was a "man on horseback." And when the sensational young Napoleon landed—he was still only thirty—he seemed to be that man. For three days and nights, Parisians celebrated the news of his return with exuberant sing-ing and drinking. Along towns from the coast to the capital, the people lined the way to welcome the man who seemed to embody victory; they of course had not yet heard about the cataclysm in Egypt. Reaching Paris on October 16, 1799, Napoleon rushed directly to his house, only to find that the street had been renamed in his honor: the Rue de la Victoire. For the time, however, he licked his Egyptian wounds and remained silent and somber, keeping out of the public eye. But soon his home became the headquarters of secret negotiations. He entertained visitors from left and right, pledging to the Jacobins to maintain the republic and to look after the masses, while at the same time whispering promises in the ears of Bourbon agents as well. He was proving to be as gifted a politician as he was a soldier. And shortly he was appointed commander of the Paris garrison.

After taking his oath of service at the Tuileries—he was escorted ominously by sixty officers—he prominently bellowed words worthy of Caesar himself. "What have you done with this France which I left you in its splendor? I left you peace, and I find war; I left you victories, and I find defeats! I left you millions from Italy; I find everywhere spoliation and misery." He thundered, "What have you done with the hundred thousand Frenchmen whom I knew, my companions in glory? They are dead." Briefly, Napoleon bided his time. But on November 9, 1799, he struck. One of his generals marched five hundred men near the Royal Palace, with Napoleon following, along with Abbé Sieyès and Talleyrand. Cries of protests rose up against the soldiers: "No dictatorship!" "Down with dictators!" and "We are free men here, the bayonets do not frighten us!" Yet it was too late. In the palace chambers, Napoleon now spoke abruptly, even violently. One deputy called out, "And the constitution?" Napoleon angrily shot back, "You yourselves have destroyed it. It no longer holds any man's respect." But the deputies would not be cowed as the hall resounded with "*À bas le tyran! Hors la loi!*" ("Outlaw him!")—the very chant that had precipitated the death of Robespierre. Napoleon, however, was not Robespierre. He left the hall, mounted his horse, and gave orders for the drums to sound and for the troops to scatter the recalcitrant deputies. They did, and the Grenadiers rushed in, crying "*Bravo! . . . C'est le passage du Rubicon!*"

Scarcely had truer words been spoken. As the legislators were driven from their chambers, the coup d'etat was complete. A decree was promptly passed replacing the Directory with a new form of republic, which Napoleon named the Consulate. It was headed by three consuls—Bonaparte, Sieyès, and former Jacobin Society president Pierre-Roger Ducos, with Napoleon being the first consul. Both chambers then adjourned until February 20, 1800. "Tomorrow," Napoleon rumbled triumphantly, "We shall sleep at the Luxembourg."

In the days that followed, Napoleon assured himself of a popular mandate by submitting a written constitution—actually, it was a fairy tale of parliamentary institutions—to a general referendum or "plebiscite." This was an action worthy of history's most despicable tyrants, for the voters could take it—or take nothing. They took it, by an official majority reported as 3,011,007 to 1,562. In the months and then years that followed, Napoleon thus became the last and most eminent of the world's enlightened despots. By his own admission, the revolution was now over. If the worst evils of the old regime had at least been cured, the revolutionaries' highest hopes had certainly not been accomplished. Still, within two years, Napoleon gave his people a measure of quietude. Peace was temporarily made with the papacy, Great Britain, and all the continental powers, as well as with America as ratified at the Convention of Mortefontaine in October 1800, and de facto with Russia, which withdrew from the Second Coalition in late 1799. This, though, would prove to be all illusion.

Two years later yet another plebiscite was held, and Napoleon saw himself elected consul for life. Two years after that a new constitution was again ratified, which announced that "the government of the republic is entrusted to an emperor." The Consulate became the Empire, and Bonaparte became the emperor, Napoleon I. And the emperor became the sovereign ruler of all of Europe as, once more, France flung its vast resources against all of the continent, unleashing a torrent of blood and trepidation among the monarchies from London to St. Petersburg.

∞

MEANWHILE, IN THE deepening snows of St. Petersburg, devious court machinations were under way that would more than rival those of Napoleonic France. Catherine the Great's son, Paul, who had skillfully thwarted his mother's wishes, had become tsar in 1796 upon her death. In doing so, he inherited a great empire that, for all its barbarism, had a court that continued to reach a degree of splendor second only to Versailles: French was still the spoken tongue, the ideas were those of the French aristocracy, and French nobles were almost equally at home in St. Petersburg as in Paris. And too, Italian opera was sung and applauded as properly as in Venice or Vienna, while Russian women of pedigree moved about with an ease and refinement reminiscent of the duchesses of the ancien régime. Moreover, Russia's military influence and reach were second only to France's. But now, at the apex of this lavish splendor and power sat a "madman."

Paul was nothing like his mother. He had a mania for all things Prussian, drilling the fabled regiments of Russia in the goosestep style of Frederick the Great. He was beset by morose suspicions and the dementia of absolute power; many observers then, and today, believe that he was mad, and if not mad, then the closest thing to it. Still, he was not without an attractive side: Upon becoming tsar, he magnanimously freed the radical thinkers Novikov and Radishchev, as well as Thaddeus Kosciuszko, who had gallantly fought for American independence and then for Polish freedom. He also restabilized the currency, lowered the tariffs that strangled foreign trade, and opened new canals for domestic commerce. He had a humane streak in him too: So horrified was he by the conditions in the Moscow Hospital that he ordered its prompt reorganization, thus turning it into one of the best hospitals in all of Europe.

But politically he was a bumbler. As did his father, he alienated the nobles, lost the intellectuals, and precariously estranged the army. In a few short years he also became a reactionary, forbidding the import of books published abroad and clamping down on the printing of new books in Russia. He prohibited dress that came into fashion in Europe after the French Revolution, and authorized severe penalties for rebellious serfs. And he was dangerously fickle. His foreign policy was intriguing, but in the end, it was ill-fated as well. He reached a détente with the Ottoman sultan, allowing his ships to pass through the Bosporus and the Dardanelles; he became the grandmaster of Malta; and he joined the Second Coalition, making common cause with the British, and, with Suvorov's help, ejecting the French from Naples. But then just as promptly, he developed a love affair with Napoleon and turned on the British. And all along, he was prone to fits of rage, bouts of apparent hallucinations, and a continuing stream of wrath and insults.

In Petersburg, nobles and officers united in a furtive conspiracy to depose him. They sought the consent of Paul's twenty-four-year-old son, Alexander. Gentle, dreamy, introverted, Alexander was a most unlikely Autocrat of All the Russias, let alone the man to be his father's assassin. Reared by his grandmother, Catherine herself, he was then taught for nine years by Frederic Laharpe, the Swiss republican and an enthusiastic devotee of the philosophes and later of the French Revolution. In a sense, Alexander embodied everything Catherine once had been: reformist, liberal minded, intellectual. He spoke English fluently and French perfectly—actually better than Napoleon. He read Voltaire, Rousseau, and Diderot, learned mathematics and geography, railed against "graft and embezzlement" in the Russian government, and fantasized about studying nature. He also publicly sympathized with the revolutionaries in France. In 1796 he wrote the most heretical words imaginable for a Russian duke at the decade's midpoint. "The Empire tends only toward expansion. Is it possible," he asked, "for me to administer the state, even more to reform it and to abolish the long existing evils?" He even confessed that he planned to "abdicate" and settle on the shores of the Rhine as a "private citizen."

Many years later Napoleon himself would say, "Alexander is too liberal in his vision and too democratic for his Russians . . . He would be more suited to the Parisians." Indeed, one would have expected to find Alexander quietly reading philosophy in a Paris salon or running for Congress on a Jeffersonian platform in America rather than ascending to the throne via regicide and patricide. But on March 24, 1801, that is precisely what he did.

At two o'clock in the morning a band of conspirators slipped into the Mikhailovsky Palace, overwhelmed all the guards, surrounded the struggling emperor, and choked him to death. A few hours later they notified Alexander, now implicated in the murder of his father much the way that Catherine had been implicated in the killing of her own husband, that he was tsar of Russia.

But Tsar Alexander I would become as much beloved in Russia as his fellow palace revolutionary Bonaparte was in France. And from the outset, he was as perplexing in his country as Napoleon was in his. Alexander still continued to see Laharpe as his political and philosophical mentor, and in turn, the young tsar surrounded himself with an astonishing brain trust of reform-minded and ardent young men of various nationalities, the most notable being a Polish youth named Czartoryski. He considered the many partitions of Poland as a crime and dreamed of a united Poland with himself as its constitutional king. And the new liberal tsar increasingly came to regard himself as a rival to Napoleon in shaping the future of Europe in an era of profound change. More than his contemporaries, Alexander adopted a surprisingly modern view of collective security among nations and the indivisibility of peace. He was also appalled by Napoleon's blatant imperialism and boldly declared that the issue in Europe lay unmistakably between law and force—between a common society in which the rights of each nation were secured by international consensus and an arena in which all shivered before the avariciousness of the French tyrant.

Egalitarian and sanctimonious, the young tsar nonetheless confounded and alarmed most of the diplomats of Europe, who thought they saw behind his humane and republican sentiments either a budding spokesman for Europe's "Jacobins" or the more familiar face of Russian xenophobia and duplicitous aggrandizement. In a sense, both conceptions were correct. He was alternately a prince of self-invention and a king of second acts: picturing himself a future arbiter of Europe—he would indeed be—Alexander signed a treaty with England, harbored secret designs on the Islamic Ottoman Empire, and, in eerie echoes of the Soviet-German pact of 1941, later did a stunning about-face, meeting privately with Napoleon to forge an unholy alliance designed to carve up the continent. It did not last long. He then became embroiled in a titanic war against his former ally, marked by the fabled Napoleonic march of 1812 to Moscow, and culminating in Alexander leading the Russian army across Europe, from his own capital to his foe's, where in his hour of triumph he took up residence in Talleyrand's apartment in Paris, only then to support, of all things, a Bourbon restoration.

∞

IT WAS ALMOST as if the 1790s never happened.

And it now fell to the young United States to buck the unsettling trend on the global stage toward the vicious cycle of coups and violence, hostilities and wars. But could it?

All throughout the summer and fall of 1800, as the question became who would lead America, it appeared to be touch-and-go. The contest was between President John Adams and Vice President Thomas Jefferson, between Federalists and Republicans, between those who favored a muscular national government and those who favored more limited self-government, and, as in the rest of the world, those with British sympathies and those with French sympathies. Across the country, the notion of party had indisputably seized hold—on this score there was no going back—and the campaign was rife with rancor, allegations, and counterallegations. Jefferson was alternatively decried as a hopeless visionary, a fanatic, a "half-injun, half-negro," a coward, a "violent democrat" and a vulgar demagogue, a "howling atheist" and a "bold atheist," and a schemer infatuated with the French Revolution, and therefore completely ill suited for the nation. One New York paper muttered that a Jefferson victory would mean civil war, or at a minimum, that hordes of Frenchmen and Irishmen ("the refuse of Europe") would be spit up on America's shores "whetting" their daggers and "stirring up revolution", at the same time the word went out for New Englanders to hide their family Bibles should Jefferson be elected.

For his part, Adams fared little better. He was characterized as a toothless old man, a half-frantic nag, a gross hypocrite, "an unprincipled oppressor," a Tory, a monarchist, too pro-British, not to mention a failed president ("unfit . . . for the station contemplated"), and, at the end of the day, "quite mad." His policies were wrong as well, for he gave the country the Alien and Sedition Acts, the standing army, the continual drumbeat of war, unpopular taxes, and constant feuding in the government. In a sense, however, Adams couldn't win: While the Republicans dismissed him as a warmonger, his compatriots, the more hawkish High Federalists, scorned him as weak-kneed in the face of the French menace. Was he? Only in hindsight do we see that Adams had demonstrated that through a combination of toughness and diplomacy, the actual use of military force could be forestalled. But for Adams, nothing seemed to go right. He was deserted by his cabinet, his Federalist Party was in disarray, and Hamilton would shamelessly betray him. Even Adams's bold stroke of sending out peace feelers to France was so inartfully executed that he received little political benefit—this, despite the fact that his preservation of the peace was unquestionably the supreme triumph of his presidency.

Yet the Republicans had also shrewdly positioned Jefferson as a friend of liberty, and more than that, as a "man of the people," a democrat responsive to "popular passions." True, as his critics countered, he was a patrician, a slaveholder, and an American aristocrat if ever there was one. It didn't matter. He was also sublime. And in the music of the emerging political arena, Adams was outclassed. There was no denying that in contrast to the cold realism of the Federalists, the idealistic Jefferson saw political reality through the wondrously attractive blur of his fond hopes, and he made that blur the American people's. As Fawn Brodie once noted, one of Jefferson's special qualities as a revolutionary statesman was that he could define the visionary future as it if were the living present, and do this without any sense of contradiction. By the same token, it increasingly became lost on a once rapturous American public that it was the Federalists who had bequeathed to the people a sound federal government to anchor the nation, a rule of law and the culture of capitalism to fuel the economy, a navy and a coast guard to preserve liberty, and an energetic government to respond to the fast-growing needs of the world. And while the word of Napoleon's decision to make peace with the United States had not yet filtered back—it wouldn't until after the election was held—in the end, the contest seemed to be between those who would give the country peace (Jefferson), and those who would give it war (Adams), those who would give the people a voice (Jefferson), and those who wouldn't (Adams).

When the votes were cast, Republican discipline held and Federalist discipline did not. As it happened, Adams achieved something else quite new in American history: first presidential loser. The final tally of the election gave Jefferson seventy-three electoral votes—he was tied with his Republican running mate, Aaron Burr, also with seventy-three votes, with sixty-five for Adams and sixty-four for Charles Cotesworth Pinckney. Ironically, since 1796, only one state had switched its electoral vote, but that state was the goliath New York, enough to make the difference. And the situation—a presidential tie—quickly became explosive.

In 1800, then, America now stood at its own Rubicon, no less than France had a few months before, or Russia would in a few months hence. Burr refused to step aside to be vice president, and the election was thrown to the lame-duck House of Representatives, where state delegations would vote as units. The first reactions were confusion, then disbelief. Then the politicking began. Republicans controlled eight delegations, the Federalists six, while the two remaining delegations were split. Tension and uncertainty over the Jefferson-Burr deadlock heightened by the hour; in truth, the makers of the Constitution had never anticipated that the losing party would play the role of president maker in case of a tie.

Congress met on February 11, a day that whipped the new capital of Washington with screeching winds and a blinding snowstorm. Once the Senate counted the votes, the deadlocked election was delivered to the Federalist-controlled House. The voting promptly began. By nightfall, an astounding fifteen ballots had been held, and still the election was tied: Jefferson always carried eight states, Burr six, with Maryland and Delaware split. The members briefly adjourned for dinner, and when they returned to the candlelit chamber, voting again commenced. At three A.M., the nineteenth ballot was taken, with still no change. Actually, this session lasted until nine in the morning, with delegates sleeping uncomfortably in their greatcoats on the floor. The next day saw no change either. Nor did the voting on Friday. Nor Saturday. Some ninety-six hours later, on the evening of February 14, thirty-three votes had been taken; still no winner.

The House impasse appeared unshakable. If these first votes were less than auspicious, the alarums across the nation were even more unsettling. Virulent signs of unrest increasingly multiplied—and not without some reason. In the teeth of vehement opposition, it became known that House Federalists were scheming to declare an interim president, preferably the newly appointed chief justice and former emissary to France, John Marshall. Adams, for one, did not oppose the idea—though he insisted that another election would have to follow. Republican newspapers darkly warned that the Federalists were plotting an intolerable "legislative usurpation" of the executive branch—left unsaid was that this would be hauntingly reminiscent of France. Editors and Republicans now spoke openly of civil war, not in a whisper or a hush but a roar, and indeed war seemed a distinct possibility. As crowds gathered outside the Capitol, papers reported that state militia units had been placed on alert; meanwhile, rumors swirled that men in every region were arming for battle. Were they? By now it was hard to separate rumor from fact. What was indisputable was that the governors of Virginia and Pennsylvania had begun to take secret measures to prepare their states for open conflict. Here, then, was enticement. And here, clearly, the Union was in peril.

Hamilton, seeking to play the kingmaker, threw his immense weight behind Jefferson, declaring that though "revolutionary in his notions," he was a "lover of liberty" and the preferable "choice of evils." Yet nothing budged. As the stalemate held, the reports of an interim Federalist president magnified. Jefferson himself, though outwardly "calm and self-possessed," was near panic, worrying that such a precedent would soon end "in a dictator." Smug only days before, he began to feel uneasy. By this stage, he was prepared to make a wildly generous gesture, or a completely ruinous gesture, whatever would serve his duty to country, or, as he saw it, his right. He did both. In the shadows, he connived and made deals, however tacitly stated or understood. And in the open, he met with Federalist leaders to deliver an alarming message: If an interim president were declared, he sputtered, "the Middle States would arm," and that no such usurpation would be submitted to, not "even for a single day." Then Jefferson, now looking like the Chinese god of plenty, added another threat, one that "shook" Federalists to their very depths: to call a new constitutional convention to "reorganize the government" as well as to "amend it." Ironically, where others among the Founders had resonant voices that in a crisis were dramatically heightened by their tempo, pauses, and crashing consonants, it was the lisping Jefferson, who never once gave an exciting speech, who now spoke most powerfully of all.

It was then that word filtered out that the Virginia military was planning to march on Washington. The whole experiment seemed to be unraveling. As France had demonstrated all too vividly, constitutions could be unmade as quickly as they were made, and unity could fall apart as rapidly as it was secured.

∞

IT DIDN'T. BEHIND the scenes, factions and partisans were furiously meeting. The members now felt on the threshold of something overwhelming, and they were. There were no votes on Monday. But on Tuesday, February 17, came the thirty-sixth tally—just six days after the first House vote, but seemingly an eternity. This time, it was conclusive, and it was historic. Not long after the House would adjourn on that day, tens of thousands of Belgians, Germans, Austrians, Poles, Irish, Swiss, Russians, Ottomans, Italians, and Frenchmen would, after a brief respite, again be engulfed in a frightful political storm; many would lose their lives or lose their freedoms, and for what? For nothing, save for the timeworn concepts of stability or imperialism.

As for Americans, they could not hold up the reflecting glass to the future, but one man all but chose their path. Obscure then, and obscure today, he would step forward to alter the stream of history. Fearing the onset of civil war, and to the cries of "Traitor! Traitor!" James Bayard, a Delaware Federalist, broke ranks and announced that he would abstain—actually, he submitted a blank ballot—thus throwing the election to Thomas Jefferson. Jefferson was now the third president of the United States.

"The storm is over," wrote the old firebrand Samuel Adams to Jefferson. "And we are in port." Interestingly, Jefferson had predicted a break would come on the sixth day—and it did.

By then, Jefferson had inherited a country politically divided as perhaps at no other time since the Revolution. But in the crowning triumph of this seesawing, violence-ridden age, in the decisive blow to the maxim that said the destiny of republics was inevitably to rise and fall, failing through conquest or dying through despotism, republicanism had survived its first and crucial decade in an infant America. More than that, America had accomplished one more resounding first: peacefully transferring political power from one party to another. While the hands of European and Russian monarchs and reformers alike were drenched in blood, despite all the dissension and rancor, in America they were virtually spotless.

And though it would take many years to see this in retrospect, Americans demonstrated another characteristic of their nascent country, one that was two-edged, but that also stood in marked contrast to much of the rest of the globe: on one side, a demonstrable centrist pull, a tendency to rein in anything perceived as extreme; on the other, the first sprouts of democracy. As a consequence, America was on its way to becoming a diverse and democratic country, inspiring quests for freedom around the world. And that same global stage, all too often marred by malignant hatreds and civil strife, would remind us that it most certainly did not have to be this way.

∞

ONE REMARKABLE EPOCH had come to a close and another had begun. Out of this crucible was born our modern world.

For centuries, generations had succeeded one another in a meaningless, timeless blur. Kings, emperors, and tsars ruled and died and were succeeded by the imperious voice of new kings, emperors, and tsars; wars were fought, spoils were divided, communities suffered, then recovered. But the impact on the masses was largely negligible. And too often, inertia reinforced the immobility. Just as often, far too many innovations, most of all political ones, were inconceivable or unwelcome; to suggest the possibility of one would have invited suspicion or accusation or eradication.

But then came this remarkable era, a period of perhaps the greatest galaxy of thinkers and doers in history. The list, only partial, is a long one. This was the Age of the Enlightenment—the time of Voltaire and Franklin, Rousseau and Montesquieu, of the successful Declaration of Independence and the failed Pugachev Rebellion, of the completion of the American Revolution and the onset of the French Revolution, of the last dying gasps of enlightened autocrats, the willful Catherine the Great and the hapless Louis XVI. It was the period of the bloody intersection of Christianity and Islam, from the Crimea to Constantinople to Cairo, and, most of all, of the culminating era of the 1790s: the rise of President George Washington and the first decade of the United States, the Rights of Man and the onset of the Terror, the beginning of revolutionary expansionism, and, of course, the coming of Napoleon. It was a time of dueling voices, of Paine and Burke, and Jefferson and Hamilton too, as well as of tragically silenced tongues, Kosciuszko and Novikov, and of the destruction of a whole country, Poland. And finally, it unleashed the unyielding struggle for republicanism and democracy. Monarchy, or at least auto-cratic rule, had been the linchpin of the entire political system, and when it snapped, the whole structure was ready to come down. But it would not go to its grave so willingly. To be sure, converts to republicanism were found in the unlikeliest of places, from Poland to St. Petersburg, Ireland to London, Paris to New York and Washington, and too, from Italy to Switzerland and (with qualifications) even in Vienna, as well as (also with qualifications) Selim III's seraglio in Constantinople.

Without a doubt, republican politics had help, a good deal of it. The remarkable growth of commerce and the new horizons of trade had already expanded the middle and merchant classes. These, growing in power and influence, had become exasperated with the arrogant dictates of uncaring kings and unresponsive tsars even as the supernational authority of mon-archs was being challenged by the rising gusts of nationalism, the dazzling spirit of humanism, and the burgeoning of Enlightenment thought. This last factor was perhaps the most important. It was implacable and, as it would turn out, eventually unstoppable. Yes, it was powerfully fueled by the invention of printing, the growth of literacy, the expansion of travel, and remarkably porous borders. It is also a fact that all the great figures of the day, from one end of the globe to the other, watched one another and reacted to one another—the Americans, the French, the Russians, the Ottomans all were part of one grand, interwoven tapestry—this point is a subtle one, but absolutely crucial to understand. And all these forces raised doubts, discredited ancient customs, bred skepticism, unraveled old stan-dards and gave birth to new ones, undermined the comfort and support of tradition, and, as monarchy weakened and republicanism strengthened, led to the emergence of the modern age.

It did not happen overnight. No great change does. Nor was it uni-form. This should not be surprising either. By way of analogy, during the Dark Ages, the literal interpretation of the Bible had led the Church to endorse the absurd dicta of the sixth-century monk Cosmas, who held that the world was a flat, rectangular place, surmounted by the sky, above which was heaven. In his scheme, Jerusalem was at the center of the rectangle and nearby lay the Garden of Eden, irrigated by the four rivers of Paradise. The sun, much smaller than the Earth, revolved around a conical mountain to the north. By contrast, centuries before, Homer thought the world a convex dish surrounded by the Oceanus stream. Then, four centuries after Aristotle—it would take that long—Ptolemy in the second century A.D.

shot this down—he agreed with Pythagoras that the world was a sphere. But he in turn created his own farrago of distortions, a bizarre paradigm of suns and moons all revolving around the center of the universe, the Earth; yet Ptolemy's world was no more fantastic than the whole panoply of less celestial stories that were also prevalent, of giants Gog and Magog, of jungle races with long teeth and hairy bodies, of storks who fought with pigmies and people who created their own shade by lying down and blotting out the sun with an enormous single foot. Over time, all this was proven false—Copernicus demonstrated that Ptolemy was wrong, though it would take more than three hundred years before his ideas took hold (remarkably the Vatican did not accept his views until 1838). And even in Jefferson's day some equally fantastic myths were believed—that the lost people of Israel wandered in the American West and woolly behemoths marched on the plains—until they were disproved by Lewis and Clark. This then was the story of progress, always ambling forward, in fits and starts, and never more so than in the political arena. But when that rarest of events occurred, when that universal joint was shifted, it was shifted decisively. This is what happened in the 1790s.

The picture was mixed; it is still today. The years 1787 and 1788 began as a time of boundless promise, but by 1800, great questions tragically proliferated. Was reform forever dead in Russia? Would France lapse permanently into dictatorship? Would America's republican experiment falter barely a decade after the Constitution, or had it taken root after all? The signs were not good by century's end, and for that matter, to many they had never seemed good. The Revolution in France, following on America's heels, begot a republic based on the rights of man, but the republic descended into chaos and bloodshed, and the rights of man into the waste basket, plunging France, and subsequently all of Europe, and Russia too, into an ongoing cycle of terror and violence and oppression, the remnants of which continued to plague the continent for years. Yet here, out of this decade, was the beginning of modern history's various streams: to be sure, some of its brightest strains, like democratic liberalism, but also some of its more ambiguous strains, like nationalism, and some of its darkest strains too—yes, authoritarianism, but also totalitarianism, revolutionary expansionism, and reactionary repression. Only America, with its fragile republicanism and new habit of constitutionalism, resisted.

The crowned heads of Europe and Russia thought they could ignore the United States, just as they thought they could stamp out the French

Revolution. They were wrong. Almost indefinably, the power of the monar-chical mind had been irrevocably broken. Not overnight, not fully for another century, but broken still. Its dogmatism, its infallibility, its absolute lack of ambiguity, were lost or at least fading. Slowly a new mood was tak-ing hold. And just as slowly, a resolute break was made with the past. And the result was that Europe was no longer decisively the world, and the world was no longer decisively Europe. For on the outskirts of empire was a viva-cious, flourishing America. True, America was helter-skelter, and it was raw. It was also brash, isolated, a bit cocky, and painfully unrefined, an inward-looking, negligible country at the margins. And it was filled with warts and inconsistencies as well. But by its survival in the 1790s, with its ideas and ideals intact, this proved to be America's hour. Even Napoleon on his deathbed in St. Helena admitted, "They wanted me to be another Wash-ington." Within essentially a single generation, arguably greater progress had been made politically than in all the millennia since the beginning of time. Not the rise and fall of Rome, not the Reformation, not the Renais-sance could fully compete with it. Unknown then—no one could tell how the story would turn out—American republicanism, and soon, American democracy as well, were in fact the reigning triumph of the age. In David McCullough's telling phrase, it was all "little short" of a miracle.

∞

WITH AMERICA'S SUCCESS, and the onset of the French Revolution, and Russia's harsh reaction to it, fresh questions loomed. An idea had become a fact, lodged in a country and a people. Ideas could be explained away, or replaced with other plausible ideas; they could still today. But the people of Europe were discovering a virgin political civilization that lived on the other side of the world in a young republic, thus giving the masses of France and then all of the continent hope of salvation. Where, they asked, was their paradise? They asked this in England and in Ireland, in Germany and in Italy, in Belgium and in Switzerland, and off in Poland and in Russia. The problem, of course, was that the French revolutionaries, after promising a nirvana, became murderous and bloody, and the rest of the European rul-ers, even the hallowed British, were still reactionary.

Likely, it all would have all been moot had America imitated the rest of the world, rather than one day the other way around. In the course of this tumultuous decade, America had its chances to stumble and falter. It almost did, numerous times. History should neither ignore this nor whitewash it. Yet it weathered Shays's Rebellion, the Whiskey Rebellion, and the Fries Rebellion. It survived the unexpected beginning of political parties, torn political allegiances, meddling in its internal affairs by outside powers, and a country hotly divided over its sister revolution in Paris. It groped its way through war or near war with the superpowers of the day, a political crisis with Britain and a fearful Quasi-War with France. Not only did it refuse to become a burial ground for the great powers of the age, it resisted the temptation that the French succumbed to—to treat its written constitution not as a sacred object but as a scrap of paper that could be filed away or discarded altogether. All, of course, was not perfect: For one thing, America, defying its enlightened brethren in Europe, tragically retained slavery, thus laying the seeds for a terrible civil war six decades hence. For another, it never reconciled the tension between states' rights and federal rights, keeping alive the bogeyman of secession. And it too indulged in the paranoid politics of the day, frequently treating the opposition parties as traitors, enacting the Alien and Sedition Acts, and writing the Virginia and Kentucky resolutions. But in the end, it never succumbed to organized violence, across-the-board repression, and institutionalized murder, no matter how tempting, which of course happened in civilized France and glittering Russia. In doing so, America created the first flickers of modern democratic politics. In numerous ways, just as the world knows it today.

Of course, this remains a comparative story as well. Only by seeing this era in its fullness, including contemporary leaders and thinkers across borders, can one begin to understand the totality, the sweep, the nuance, and the richness of this breathtaking and transforming period. Contrary to the way textbook history likes to tell it, the world then was far more global than we often realize. Nations, and leaders, were acutely conscious of one another. There was an uncommon fluidity of the age too, almost unheard of even today: Intellectuals, advisors, military men, and ideas freely crossed borders, changed allegiances, settled in and then moved on. Thomas Paine left England to rouse a fledgling America in 1776, then he traveled to France, joined the revolutionary Assembly, and was almost beheaded, released only due in part to pleas by James Monroe. Thomas Jefferson advised the budding French revolutionaries while Gouverneur Morris counseled the king. Voltaire hugged Franklin while lionizing his friend Empress Catherine; Rousseau found a home in Jefferson's Republican Party, and Franklin and Jefferson were studied by the Russian reformers Novikov and Radishchev. Militarily, John Paul

Jones, America's great naval hero, circumnavigated the globe and joined Catherine the Great's imperial navy to fight the Muslims, and then died a pitiful death in Paris. Another dramatic hero of the American Revolution, Polish-born Thaddeus Kosciuszko, would return home to lead a revolution that was ruthlessly crushed by Catherine's Russia. And Lafayette, after heroically serving under Washington's wing in America, fled the very French Revolution he helped start, landing in an Austrian prison, while Talleyrand, the notorious French foreign minister, would spend the bloody days of the Terror in exile in the United States, befriend Hamilton, and return home to serve the Directory (and then Napoleon), only to turn on his former American hosts.

Finally, how best to understand the wisdom and vision of the age's master spirit, George Washington, than by seeing him in the same frame as a confused or dispirited Louis XVI—or a once reformist but hardened Catherine *"la Grande"*? How best to appreciate the greatest generation of American talent in history, the Founding Brothers, Jefferson, Hamilton, Franklin, Adams, and Madison, than to see them in relation to the once idealistic, revolutionary souls of France, Mirabeau, Marat, Danton, and Robespierre—or for that matter, to the Machiavellian Russian brain trust, like the mesmerizing Potemkin? Or conversely, how best to grasp the weakness and indecision of Louis XVI than to see him in relation to the ever-steely Catherine or the ever-decisive Washington? Or how best to ponder why America's own rebellions and heated disputes were absorbed into a political process rather than resolved by the guillotine and assassination, than to see them in relation to France's Terror or to the court intrigues and mounting repression of Catherine's empire?

And how best to appreciate why republicanism succeeded in America, failed for decades in France, and never reached Russia?

∞

IT WAS THEN the decade that made the modern world. True, much remained to be done. But much had already been accomplished. To contemplate the seeds of the 1790s—of republics and revolution, of democratic hopes and liberalism, of despotism and authoritarianism, of totalitarianism and religious war too, is to begin to understand the tides that have swept the globe and battered nations, large and small, ever since. America, France, and Russia are most assuredly of their time, but their history, their modern-day founding, their latter-day echoes loudly resonate in our time as well.

∞

So as a harsh, darkening curtain was about to descend upon Napoleonic Europe, in the soft twilight that precedes the gentlest point of dusk, the new American capital was preparing for the inauguration of Thomas Jefferson the next morning. First, night fell. One by one the stars arrived in the sky like a thousand distant windows lighting up. It was March 3, 1801.

Washington had none of the luxury of Paris or the splendor of Petersburg, nor the refinement of Vienna or the culture of London, let alone the drama of Constantinople. There were no spectacular vistas or grand boulevards, no intimate squares or gentle ocean breezes. For this reason alone, Washington was perplexing to the rest of the world. It was also unique. Among all the great capitals, it was to center not around culture, but politics, and politics only.

For all its grand intentions, the city, much like the rest of the country, had only just begun, and seeing it for the first time, it was hard not to think it was a huge blunder. Planned by a Frenchman, with its diagonal streets and radiating circles, it was to reflect the traditions of neoclassical monumentality. This, at least, was the conception. The reality told a different story. Erected on a swamp and flanked by noxious flats and a stinking canal, the city emitted suffocating odors everywhere, which even wafted into the presidential quarters. Solemnly conceived, the city was, in truth, a bit of a joke. There were no schools, for a time, not a single church, and little commerce, and there were sloth and clutter. The heat was terrible, and the profusion of mosquitoes was even worse. In every direction there were not elegant women with parasols and finely suited gentlemen slipping into their carriages, but instead a plethora of workmen, discarded brick and stone, unpaved roads and unsightly shacks, and backyard privies. Even more unsightly were the slaves, on whose back much of the capital was being built, and their squalid shanties, clinging to the flanks of the finer homes.

Capitol Hill itself was a patchwork of a few nondescript stores—a printer, a tailor, a grocer—some seedy boardinghouses, and odd slabs piled by a not-yet-finished sandstone Capitol building. The vistas were filled unattractively with open fields, pigs rooting around in the dirt streets slanting off Pennsylvania Avenue, piles of rubble, rough rail fences, fetid yellow mud, and everywhere tree stumps. To the extent that there were vast, endless streets, they contained almost no buildings of any kind. A mile to the west of the Capitol, the President's House was woefully unfinished. In fact, the only department that was completed was the cramped quarters of the Treasury, a rustic, two-story brick building. And nearby Georgetown was, one observer deadpanned, "the very dirtiest hole I ever saw." Even the city's conception alone was questionable: The main buildings of the government were to be crazily spaced far apart from one another; thus the sarcastic jibe that the capital was a city "of magnificent distances."

The President's House, the largest in America and still glaringly a work in progress, conveyed nothing like the royal grandeur of France's Tuileries or the palatial magnificence of Versailles, nothing like Russia's resplendent Peterhof or Winter Palace, nothing like Berlin's exquisite Sans Souci or Vienna's Schonbrunn, or nothing like even England's more modest Blenheim Palace. There were no great halls for state receptions, or paved court-yards for troops of soldiers, throngs of retainers, or the arrival of coaches. There were no complexes of facades, terraces, balustrades, gardens, fountains, and formal vistas cascading beautifully together. There was no statuary to adorn it inside and out, no larger-than-life tapestries or magnificent mirrors to deck the interiors. In fact, the President's House still smelled of wet paint and wet plaster. Closet doors were missing and the furniture seemed out of place. For the longest time, there was only one staircase built, and it was a makeshift one at that. Thirteen fireplaces blazed at once, either to dry the paint or warm the house. And south of the mansion were sewage marshes and open drainage ditches upon which floated, as one observer later put it, "dead cats and all kinds of putridity."

Gouverneur Morris had written to a French friend, "We need only here houses, cellars, kitchens, scholarly men, amiable women, and a few other such trifles to possess a perfect city." It was hard to tell if this was bravado or prescience; most foreigners thought the former. Perhaps they could be forgiven for believing that this was not a capital of a great country, but a raw and sprawling little village, its clumps of hastily built houses divided confusingly by swamps and forests. No wonder Anthony Trollope found the city a disconcerting maze where one gets lost, as "in the deserts of the holy land," and Mary Cranch scoffed that this was a city "only so in name." No wonder cabinet secretary Oliver Wolcott was even more scathing, insisting the capital was "unfit for human habitation." No wonder diplomats quickly came to regard a tour in Washington as "a season in purgatory," and many American officials seemed to agree with them. However

exquisite in conception, in every direction, the city seemed to evoke unful-
filled promise.

∞

PROMISE, HOWEVER, WAS the word. On March 3, there were strikingly
clear skies crowded with glitter and smothered in moonlight. While it was
still dark, outgoing President John Adams managed a wan smile and eased
himself into a coach at four in the morning, departing for his home in
Quincy via Baltimore. It was the loneliest of scenes, but already quintessentially American. There was a pucker of worry over his brow and a network
of lines around his eyes. And quite by happenstance, in the same coach was
the ardent Federalist, Speaker Theodore Sedgwick; the two frequent adversaries would ride together all the way to Massachusetts.

Adams, who would live to be an astonishing eighty-nine, was bitter
and dejected and already forgotten. Sounding like so many politicians who
would follow him, he would confide to a friend that had he to do it all over
again, he would have become a shoemaker. There were consolations: he had
a new grandson by John Quincy (George Washington Adams), and by his
own admission he reveled in the start of a "new day." He soon busied himself in the hayfields of his farm, named a new puppy Juno, and intriguingly,
upon his return home—amid black skies and a torrential rain—discovered
a sealed and double-bound letter waiting for him. It was a terse, innocuous
single line from his nemesis President Jefferson, dated four days after his
inauguration. Adams opened it up, thus beginning one of the most remarkable correspondences in history.

∞

THE CAPITAL AWOKE after Adams did, a world transformed. There were
the defeated, and there were the elected, and this was to be the elected's day.
Dawn turned to day and the light grew hazy. All across Washington, the
infant capital was stirring with a new rhythm and pulse that intimated a city
of far greater magnitude than its rough facade. And all minds were preparing for the president-elect. Ceremonial artillery was moved into place, and
drum and fife units went through their last paces, soon joined by United
States marshals and the militia. At eleven A.M., Jefferson, with his deep-set
eyes and slow smile, emerged to join the procession to the Capitol. He wore

a plain suit, but conspicuously did not powder his hair or don a sword; nor did he ride in an elegant coach. This was to be a day of republican purity and equality, and, in contrast to his predecessors, he meant to set a new tone. This too was quintessentially American.

He walked, and as he did, a large crowd gathered. In the foreground was the unfinished Capitol and in the distance was the Potomac, a pale, pure blue and speckled with little boats. Europeans may have scoffed at the new capital city—as it happened, designed as much by Jefferson as by Washington—but to Jefferson it was a soothing balm to the burning eyes, a charming semi-rural landscape. As he made his way inside a packed Senate chamber, the absence of John Adams was instantly noted, and just as quickly forgotten. Elegant and soft-spoken, Jefferson was flanked by Vice President Aaron Burr and Chief Justice John Marshall—three men who profoundly disliked one another. While Jefferson had not yet been sworn in, Congress was gaveled to order, and the incoming president took his place to speak in the well. Today he was both aristocrat and democrat, and he gave a stunning oration of healing and reconciliation, crafted with great precision in a Miltonian style. And it was designed to reflect the new president's role in the history of his nation, not to mention his new political position in the Western world. As Henry Adams would later note, Jefferson had come, however implausibly, to claim the place "of an equal between Pitt and Bonaparte."

Jefferson's voice could barely be heard that day but it would continue to echo through the centuries. "We are all Republicans," he intoned almost biblically. "We are all Federalists." While admitting to anxious and awful presentiments, he then addressed himself as much to future generations as to the delegates seated before him. "Some honest men fear that a republican government cannot be strong," he said firmly, squinting into the distance. "I believe this, on the contrary, to be the strongest government on earth." Lest there were doubters, he added, "It is so. It will be so." Nor was this mere political cant or idle rhetoric. Days earlier Jefferson had tellingly written to his good friend Lafayette, "The storm we have passed through proves our vessel indestructible." And to John Dickenson he had also thundered, "Our revolution and its consequences will ameliorate the condition of man over a great portion of the globe."

∞

OUR REVOLUTION AND its consequences. In the end, on the seemingly farthest precinct of human habitation, the majesty of America or its allure would not so much be in its buildings or its architecture, or in its political programs, or one day in the vast reach of its federal government or its fearsome military, or even in any one of its leaders, but in its very underpinnings: the idea of representative government. In a sense, a very important sense, Jefferson was right, but only half so. For years and decades hence, this question of democracy would be the crux of debate not just for France or Russia, but for peoples the world over. Yet in the bloom of late spring in 1801, after the inauguration, after the streets were cleaned and the cheering crowds had gone home, it was palpable even then.

For the moment, Paris was peaceful. So too were St. Petersburg and London and Constantinople and Vienna. And Washington, too. On clear, blustery days, the belligerents who had lunged for one another's throats were now signing armistices and ratifying conventions. The guns of Europe's titanic armadas lay quiet. National flags once again snapped triumphantly in the wind, and pleasure boats cruised festively in the ocean's waters. And in the terrible evening's quiet, far from the wastelands of denuded earth that were so recently battlefields, or town squares where executioners once did their bloody work, young wives laid a loving hand on their dirt-smudged children. The sovereigns of Europe, content that all was again well in empire, talked in that most precious of phrases, peace. It would not last.

For in the end, as the Americans had already found out, the slumbering, restive masses wanted more than peace. It turns out that they always did. True, the great contest over liberty had begun in the age of the 1790s—as the eminent Welsh humanitarian Richard Price put it, "I have lived to see millions of people, indignant and resolute . . . demanding liberty with an irresistible voice." But as the 1800s began, it remained to be seen if that voice of liberty would prove to be both durable and infectious. Wordsworth himself would one day write about this most momentous of decades, "I looked for something which I could not find,/ Affecting more emotion that I felt."

But many did find it. And in the centuries to come, across the globe, many more would seek it still.

Bibliographic Notes

After six years of living in the late eighteenth century, my first debt of gratitude in undertaking this project goes to an extraordinary circle of distinguished public historians and dedicated scholars across a variety of fields whose works have educated and inspired me. Where does one begin? For my last book, suffice to say that I was fortunate to be walking in the company of the giants of the Civil War and doubted I would ever be so lucky again. Thankfully, I was wrong. The study of the age of the 1790s has produced exceptional people whose work informs every one of these pages, beginning in American history with the eminent David McCullough, Ron Chernow, Walter Isaacson, Edmund Morgan, Gordon Wood, Bernard Bailyn, Joseph Ellis, Stacy Schiff, Pauline Maier, David Hackett Fisher, and also Carol Berkin, Joyce Appleby, John Ferling, Jill Lepore, Bill Brands, Eric Foner, Richard Brookhiser, Joanne Freeman, and Richard Norton Smith; in French history, the matchless Simon Schama, Antonia Fraser, Evelyn Lever, William Doyle, J. M. Thompson, George Rudé, and George Lefebvre; in Russian history, the unsurpassed dean of Catherine scholars, Isabel de Madariaga, Simon Sebag Montefiore, John Alexander, Henri Troyat, Vincent Cronin, Norman Saul, and Mark Raef; in Ottoman history, Philip Mansel and Lord Kinross; and on a more general level, the incomparable Robert Massie, who writes about Russia but really writes about so much more; R. R. Palmer, whose books (and texts) are about France but cover the emergence of the modern world in splendid array; the peerless biographers Douglas Southall Freeman and Dumas Malone, who wrote about Washington and Jefferson, respectively, but masterfully covered the international scene abroad; Stanley Elkins and Eric McKitrick, who have written an unrivaled, seminal work on American history; Will and Ariel Durant, whose series, *The Story of Civilization*, brims in scope and detail—all are essential mainstays on which I've drawn heavily. Finally, I would be remiss if I didn't mention the countless gifted scholars in universities across the nation—there are too many to name here, but they are cited below—without whose specialized studies this book could really never have been written.

Finally, there are the actors of the era themselves, Washington and Jefferson, Catherine and Hamilton, Adams, Mirabeau, Potemkin, Paine, Danton, Morris, Napoleon, John Paul Jones, Kosciuszko, and the list goes on, who spoke with an uncommon eloquence, passion, determination and insight, and often a frailty or poignancy too, which remain almost unequaled in history and which serve as a gift for any historian or writer. Yet in a book such as this, a work of synthesis, interpretation, and analysis that stitches together a continuous narrative of a time and place that was once a complete whole but that over the years has now come down to us in fragmented segments, how to make sense of the maze of detail? Or how best to restore the age to the interconnected arena that the actors themselves saw? Or, after three years of reading, how best to draw on material culled from Washington, New York, Philadelphia, Pittsburgh, Virginia, Massachusetts, Maryland, London, Istanbul, Poland, St. Petersburg, and Paris; or from the original manuscripts and old newspapers I pored through; or the classics of the outstanding scholars listed above? Or how to make sense of some three thousand pages of handwritten notes I eventually came up with, arranged in piles, highlighted and annotated and sorted?

An almost herculean task, it began with seeing the world as the actors saw it, of listening to their voices and living in their heads. It also meant the daunting undertaking of combining disparate fields of scholarship and sources, ranging from America to Russia to France, not to mention British, Polish, and Ottoman. While almost of necessity each of these fields tends to be almost self-contained among many gifted specialists, there is an enormous value to history in combining and expanding upon these different approaches, events, and actors, particularly because the historic figures of this pivotal age were in their own right so inextricably linked.

To prevent an unusually long and unwieldy notes section—which otherwise could run several hundred pages—I have used the widespread practice of a more pared down collective reference rather than individual numeric citations. And while there is a common "text" or "body" of literature, as much as possible, however, I have sought to list the books, monographs, papers, diaries, memoirs, newspapers, and journal articles that were most influential in shaping my thinking and writing and interpretations. As such, the manner in which I have outlined things below provides a window into the construction of the book. Moreover, I have also extensively listed works that provided useful background for me and that interested readers may consult for further reference, although space limitations prevent citing every single source that I consulted. From time to time—though in the interest of length I had to resist the allure of more extensive commentary—these are accompanied by a brief discussion of the sources. I have also gone to lengths to illuminate for the reader the books and journal articles that make specific and unique contributions to our historical understanding of this era. I can further note here that from my previous book I received extensive feedback from a wide variety of readers who tell me that this sort of a section is something they greatly benefit from and enjoy, particularly because it offers a sustained look into the writing of the book.

Not unlike in my *April 1865*, my frame of reference and orientation is once again to offer a comparative story that draws upon wide and often disparate pieces of scholarship and history. And once again, drawing upon my original background in international affairs, not to mention the policy world (including advising two secretaries of defense) I have brought to bear the invaluable insights that I have gained from these arenas, which I believe can also significantly further our understanding of events here.

A final note about research. Over the years, I have often been asked if I use the Internet, and my answer was always emphatically no. Unapologetically, I have always been partial to books, paper, and pens. Near the end of this work, however, almost by happenstance, I occasionally would search for things online. To my surprise, and eventually my delight, I realized that there is now a wealth of primary sources, documents, and scholarly articles online, not only from outstanding museums, libraries, university collections, and historical associations across the nation, starting with the great Library of Congress and assisted by the National Endowment for the Humanities, but literally from across the world. It is in its own way almost revolutionary. Provided online research is used with care—in truth, the diligent historian quickly learns the same can be said for printed books and articles—researchers today and in the future will increasingly have a vast array of crucial information at their fingertips that otherwise would have taken days or weeks or months to track down. From my own experience, I can say there were some details I was able to get in a matter of hours that I estimated would otherwise have taken me up to a week. This will surely provide opportunities and approaches that have previously been denied to many scholars and open up new vistas we can only now imagine.

A number of distinguished works have been employed throughout this study. Their abbreviations are:

ALEXANDER   John T. Alexander, *Catherine the Great: Life and Legend* (Oxford University Press, 1989).

CARLYLE   Thomas Carlyle, *The French Revolution: A History* (Oxford University Press, 1989).

CTG   Isabel de Madariaga, *Catherine the Great: A Short History* (Yale University Press, 1990).

CHERNOW   Ron Chernow, *Alexander Hamilton* (Penguin, 2004).

CRONIN   Vincent Cronin, *Catherine, Empress of All the Russias* (Morrow, 1978).

DOYLE   William Doyle, *The Oxford History of the French Revolution* (Oxford University Press, 1989).

E&M   Stanley Elkins and Eric McKitrick, *The Age of Federalism* (Oxford University Press, 1993).

ELLIS, FB   Joseph J. Ellis, *Founding Brothers: The Revolutionary Generation* (Knopf, 2001).

ELLIS, HE   Joseph J. Ellis, *His Excellency George Washington* (Knopf, 2004).

FERLING   John Ferling, *A Leap in the Dark: The Struggle to Create the American Republic* (Oxford University Press, 2003).

FRASER   Antonia Fraser, *Marie Antoinette: The Journey* (First Anchor, 2001).

FREEMAN   Douglas Southhall Freeman, *Washington* (Scribner's, 1968).

GORDON   Lyndall Gordon, *Vindication: The Life of Mary Wollstonecraft* (HarperCollins, 2005).

HIBBERT   Christopher Hibbert, *The Days of the French Revolution* (Morrow, 1980).

JONES   Colin Jones, *Great Nation: France from Louis XV to Napoleon* (Penguin, 2003).

LEVER   Evelyne Lever, *Marie Antoinette: The Last Queen of France* (Farrar, Straus & Giroux, 2000).

MADARIAGA   Isabel de Madariaga, *Russia in the Age of Catherine the Great* (Yale University Press, 1981).

MANSEL   Philip Mansel, *Constantinople: City of the World's Desire, 1453–1924* (St. Martin's, 1996).

McCULLOUGH   David McCullough, *John Adams* (Simon & Schuster, 2001).

MONTEFIORE   Simon Sebag Montefiore, *Prince of Princes* (Dunne, 2001).

MORRIS   Gouverneur Morris, *The Diary and Letters of Gouverneur Morris*, 2 vols., Anne Cary Morris, ed. (Da Capo, 1970).

PALMER, *Age*   R. R. Palmer, *The Age of the Democratic Revolution: A Political History of Europe and America, 1760–1800*, 2 vols. (Princeton University Press, 1959).

PALMER, *Twelve*   R. R. Palmer, *Twelve Who Ruled: The Committee of Public Safety, During the Terror* (Oxford University Press, 1941).

P&C   R. R. Palmer and Joel Colton, *History of the Modern World to 1815* (Knopf, 1978).

RUDÉ   George Rudé, *The Crowd in the French Revolution* (Oxford University Press, 1959).

SCHAMA   Simon Schama, *Citizen: A Chronicle of the French Revolution* (Vintage, 1990).

SÉGUR   Louis-Philippe Ségur, *The Memoirs and Anecdotes of the Count de Ségur*, George Shelley, trans. (Scribner's, 1928).

THOMPSON   J. M. Thompson, *The French Revolution* (Sutton, 2001).

TROYAT   Henri Troyat, *Catherine the Great* (Dutton, 1977).

WOOD   Gordon S. Wood, *The Creation of the American Republic, 1776–1787* (University of North Carolina Press, 1969).

ZWEIG   Stefan Zweig, *Marie Antoinette* (Garden City, 1933).

REFERENCE WORKS

Adams Family Correspondence, 6 vols., L. H. Butterfield, ed. (Belknap, 1963).

John Adams, *Familiar Letters of John Adams and His Wife Abigail Adams During the Revolution with a Memoir of Mrs. Adams*, Charles Francis Adams, ed. (Books for Libraries Presses, 1970).

John Quincy Adams, *Diary of John Quincy Adams*, 2 vols., David Grayson Allen, ed. (Belknap, 1981).

Avalon Project at Yale Law School.

Daniel J. Boorstin, *The Americans: The Democratic Experience* (Random House, 1973).

Daniel J. Boorstin, *Hidden History: Exploring Our Secret Past* (Harper & Row, 1987).

Will and Ariel Durant, *Age of Louis XIV* (Simon & Schuster, 1963).

Julian Boyd, et al, eds., *The Papers of Thomas Jefferson*, 27 vols. (Princeton, 1950–).

Will and Ariel Durant, *The Age of Napoleon: A History of European Civilization from 1789 to 1815* (Simon & Schuster, 1975).

Will and Ariel Durant, *The Renaissance* (Simon & Schuster, 1953).

Will and Ariel Durant, *Rousseau and Revolution* (Simon & Schuster, 1967).

Jonathan Elliot, *The Debates in the Several State Conventions on the Adoption of the Federal Constitution as Recommended by the General Convention at Philadelphia in 1787*, 5 vols. (Franklin, 1968).

Encyclopedia Britannica, online edition.

Thomas J. Fleming, *Liberty! The American Revolution* (Viking, 1997).

Bernard Grun, *The Timetables of History* (Simon & Schuster, 1991).

Frederick Hartt, *Art: A History of Painting, Sculpture, and Architecture*, vol. 2 (Prentice-Hall, 1992).

John Jay Papers, Columbia University.

Paul Johnson, *A History of the American People* (HarperCollins, 1997).

Lafayette Collection, Cornell University.

Andrei Lobanov-Rostovsky, *Russia and Europe, 1789–1825* (Greenwood, 1968).

The Records of the Federal Convention of 1787, 4 vols., Max Farrand, ed. (Yale University Press, 1966).

The Republic of Letters: The Correspondence Between Thomas Jefferson and James Madison, 1776–1826, 3 vols., James Morton Smith, ed. (Norton, 1995).

Nicholas V. Riasonovsky, *A History of Russia* (Oxford University Press, 1963).

Donald L. Robinson, *Slavery in the Structure of American Politics, 1765–1820* (Harcourt Brace Jovanovich, 1970).

Clinton Lawrence Rossiter, *The American Quest, 1790–1860: An Emerging Nation in Search of Identity, Unity, and Modernity* (Harcourt Brace Jovanovich, 1971).

Russia and the West in the Eighteenth Century: Proceedings of the Second International Conference Organized by the Study Group on Eighteenth-Century Russia and Held at the University of East Anglia, Norwich, England, 17–22 July, 1981 (Oriental Research Partners, 1983).

Russia and the World of the Eighteenth Century: Proceedings of the Third International Conference Organized by the Study Group on Eighteenth-Century Russia and Held at Indiana University at Bloomington, USA, September 1984, R. P. Bartlett, A. G. Cross, and Karen Rasmussen, eds. (Slavica, 1988).

Paul W. Schroeder, *The Transformation of European Politics, 1763–1848* (Oxford University Press, 1994).

Supplement to Max Farrand's The Records of the Federal Convention of 1787, James H. Hutson, ed. (Yale University Press, 1987).

United States Congress, *The Debates and Proceedings in the Congress of the United States: With an Appendix, Containing Important State Papers and Public Documents, and All the Laws of a Public Nature, with a Copious Index Comprising the Period from March 3, 1789 to March 3, 1791*, 2 vols. (Gales and Seaton, 1834).

The Debate on the Constitution, 2 vols., Bernard Bailyn, ed. (Library of America, 1995).

The United States and Russia: The Beginning of Relations, 1765–1815, Nina N. Bashkina, ed. (U.S. Government Printing Office, 1980).

George Washington, *The Diaries of George Washington, 1748–1799*, 4 vols., John C. Fitzpatrick, ed. (Houghton Mifflin, 1925).

George Washington, *The Papers of George Washington: The Journal of the Proceedings of the President, 1793–1797*, Dorothy Twohig, ed., 6th rev. ed. (University Press of Virginia, 1981).

George Washington, *The Writings of George Washington from the Original Manuscript Sources, 1745–1799*, 39 vols., John C. Fitzpatrick, ed. (U.S. Government Printing Office, 1931).

Charles Whitworth, *An Account of Russia as It Was in 1710* (Strawberry Hill, 1758).

William Appleman Williams, *American-Russian Relations, 1781–1947* (Rinehart, 1952).

PRELUDE

For the Salem witch hunt and trial: See Mary Beth Norton, *In the Devil's Snare: The Salem Witchcraft Crisis of 1692* (Random House, 2002), which is the best single study. Also see Edmund S. Morgan's illuminating *The Puritan Family: Religion and Domestic Relations in Seventeenth-Century New England* (Harper & Row, 1966); John Demos, *Entertaining Satan: Witchcraft and the Culture of Early New England* (Oxford University Press, 1982); Carol F. Karlsen, *The Devil in the Shape of a Woman: Witchcraft in Colonial New England* (Norton, 1987); Larry Dale Gragg, *The Salem Witch Crisis* (Praeger, 1992); and Chadwick Hansen, *Witchcraft at Salem* (G. Braziller, 1969).

For one tale of the ravages of disease: See Jennifer Lee Carrell, *The Speckled Monster: A Historical Tale of Battling Smallpox* (Dutton, 2003); also Elizabeth A. Fenn, *Pox Americana: The Great Smallpox Epidemic of 1775–82* (Hill & Wang, 2001). **On the brutality of life before the Enlightenment:** See William Manchester, *A World Lit Only by Fire* (Little, Brown, 1992).

For France and the Age of Louis XIV: For my portrait of Louis XIV, I have most heavily relied upon four outstanding works, including Antonia Fraser, *Love and Louis XIV: The Women in the Life of the Sun King* (Doubleday, 2006); Nancy Mitford, *The Sun King* (Penguin, 1994); Robert K. Massie, *Peter the Great* (Knopf, 1980), esp.,

156–167; Will Durant and Ariel Durant, *The Age of Louis XIV* (Simon & Schuster, 1963), esp. 1–30. Also Hartt; John Laurence Carr, *Life in France Under Louis XIV* (Putnam's, 1966); G. Gooch, *Louis XV: The Monarchy in Decline* (Longmans, 1956); John B. Wolf, *Louis XIV* (Victor Gollancz, 1968); Gilette Ziegler, *At the Court of Versailles: Eye-Witness Reports from the Reign of Louis XIV* (Dutton, 1966) is quite good; A. Goodwin, ed., *The European Nobility in the Eighteenth Century* (A. & C. Black, 1953); F. J. C. Hearnshaw, ed., *Social and Political Ideas of Some Great French Thinkers of the Age of Reason* (Barnes & Noble, 1950); and Henri Martin, *The Age of Louis XIV* (Walker, Wise, 1865).

On Russia under Peter the Great: For Peter, I have drawn heavily upon the superb and richly narrated book, Massie, *Peter the Great*, which is the best single biography of Peter and which stands as a model for historical writing. Massie's book also provides an extraordinary portrait of the age, ranging from France to the Turkish sultanate to the Swedish kingdom and beyond, which has proved enormously useful to me throughout; also see Paul Bushkovitch, *Peter the Great* (Rowman and Littlefield, 2003); Lindsey Hughes, *Russia in the Age of Peter the Great* (Yale University Press, 1998) as well as his *Peter the Great: A Biography* (Yale University Press, 2002); M. S. Anderson, *Britain's Discovery of Russia, 1553–1815* (Macmillan, 1958); M. S. Anderson, *Peter the Great* (Thames & Hudson, 1978). For a rich discussion of the question of reform under Peter, see Marc Raeff, *Understanding Imperial Russia: State and Society in the Old Regime*, Arthur Goldhammer, trans. (Columbia University Press, 1984), 1–146; also Marc Raeff, *Origins of the Russian Intelligentsia: The Eighteenth-Century Nobility* (Harcourt Brace, 1966); and Marc Raeff, ed., *Peter the Great: Reformer or Revolutionary?* (Health, 1966), which is fascinating; for contemporary foreign assessments of Peter's modernization program, see Basil Dmytryshyn, *Modernization of Russia Under Peter I and Catherine II* (Wiley, 1974), esp. 16–42; Jodocus Crull, *The Ancient and Present State of Muscovy* (A. Roper, 1698); and Arthur Voyce, *Moscow and the Roots of Russian Culture* (University of Oklahoma Press, 1964). On church and religion, see Marc Raeff, *Imperial Russia 1682–1825: The Coming of Age of Modern Russia* (Knopf, 1971), esp. 123–130; G. Fedotov, *The Russian Religious Mind* (Harvard University Press, 1966). Also quite good, see H. A. L. Fischer, *A History of Europe, Vol. 1* (Fontana Library, 1960); Alexander Gordon, *History of Peter the Great* (1755); Stephen Graham, *Peter the Great* (Simon & Schuster, 1929); Ian Grey, *Peter the Great* (Lippincott, 1960); Ronald Hingley, *The Tsars: Russian Autocrats, 1533–1917* (Weidenfeld & Nicolson, 1968); L. Jay Oliva, *Peter the Great* (Prentice Hall, 1970); Bernard Pares, *A History of Russia* (Knopf, 1960); Riasonovsky; Eugene Schuyler, *Peter the Great* (Scribner's, 1884); John Stoye, *Europe Unfolding, 1648–1688* (Collins-Fontana Library, 1969); B. H. Sumner, *Peter the Great and the Emergence of Russia* (Collier, 1965). **On Streltsy revolt:** Massie, 39–49; Bushkovitch, 83–86; and Hughes, *Peter the Great*, 12–14.

On the philosophes: I have drawn upon a number of important works. See Roger Pearson, *Voltaire Almighty* (Bloomsbury, 2005); Philipp Blom, *Enlightening the World: Encyclopedie, The Book That Changed the Course of History* (Palgrave Macmillan, 2005); Henry Steele Commager, *The Empire of Reason: How Europe Imagined and America Realized the Enlightenment* (Doubleday, 1977); Thomas L. Pangle, *The Spirit of Modern Republicanism: The Moral Vision of the American Founders and the Philosophy of Locke* (University of Chicago Press, 1988); Norman Hampson, *Will and Circumstance: Montesquieu, Rousseau and the French Revolution* (Duckworth, 1983); Anne M. Cohler, Basia Carolyn Miller, and Harold Samuel Stone, eds., *Montesquieu: The Spirit of the Laws* (Cambridge Texts on the History of Political Thought, Cambridge University Press, 1989); P&C has a rich discussion on which I also drew, esp. 314–42, as well as the outstanding work of serendipity and insight, Bernard Bailyn, *To Begin the World Anew: The Genius and Ambiguities of the American Founders* (Knopf, 2003); Jean Orieux, *Voltaire* (Doubleday, 1979); Ernst Cassirer, *The Philosophy of the Enlightenment* (Princeton University Press, 1951); Ernst Cassirer, *The Question of Jean-Jacques Rousseau* (Columbia University Press, 1954); Alfred Cobban, *Rousseau and the Modern State* (Allen & Unwin, 1934); Peter Gay's wonderful, *Voltaire's Politics* (Princeton University Press, 1959); Fossey J. Hearnshaw, ed., *Social and Political Ideas*; Charles W. Hendel, *Jean-Jacques Rousseau, Moralist* (Bobbs-Merrill, 1934); Matthew Josephson, *Jean-Jacques Rousseau* (Harcourt Brace, 1932); Alfred Noyes, *Voltaire* (Sheed and Ward, 1936); Normal L. Torrey, *The Spirit of Voltaire* (Columbia University Press, 1938); Carol Blum, *Jean-Jacques Rousseau and the Republic of Virtue* (Cornell University Press, 1986); and David A. Hollinger and Charles Capper, eds., *The American Intellectual Tradition: Volume 1, 1630–1865* (Oxford University Press, 1993). For a counterbalancing viewpoint, focusing on the primacy of British thinkers, see the exceptional work by Gertrude Himmelfarb, *The Roads to Modernity: The British, French and American Enlightenments* (Vintage, 2005).

On the American Revolution: I've drawn upon a number of superb treatments from eminent scholars, including Gordon S. Wood, *The American Revolution: A History* (Modern Library, 2002); David McCullough, *1776* (Simon & Schuster, 2005); and Bernard Bailyn, *The Ideological Origins of the American Revolution* (Belknap, 1967), which remains a classic and essential contribution, as do Gordon S. Wood's exceptional *The Radicalism of the American Revolution* (Vintage, 1993), and Bailyn, *To Begin the World Anew*. On the Declaration of Independence, the best work is Pauline Maier, *American Scripture: Making the Declaration of Independence* (Vintage, 1997). Maier makes the case that the preamble to the Declaration was boilerplate Locke, secondary to the case for revolution; moreover, she makes the case that the Declaration originally spoke of "unalienable," not "inalienable," rights. Also see Bernard Bailyn, *Faces of Revolution: Personalities and Themes in the Struggle for American Independence* (Vintage, 1992); Stephen G. Kurtz and James H. Hudson, eds., *Essays on the American Revolution* (University of North Carolina Press, 1973); and Carol Berkin, *Revolutionary Mothers:*

Women in the Struggle for America's Independence (Knopf, 2005). For military aspects see Barbara Tuchman, *The First Salute: A View of the American Revolution* (Ballantine, 1989); Robert Leckie, *The Wars of America* (Harper & Row, 1968), 83–216; John W. Shy, *A People Numerous and Armed: Reflections on the Military Struggle for American Independence* (Oxford University Press, 1976); David Hackett Fischer's wonderful classic *Washington's Crossing* (Oxford University Press, 2004); Robert Leckie, *"A Few Acres of Snow": The Saga of the French and Indian Wars* (Wiley, 1999); Robert Leckie, *George Washington's War: The Saga of the American Revolution* (HarperPerennial, 1992); and Fred Anderson's excellent and sweeping *Crucible of War: The Seven Years' War and the Fate of Empire in British North America, 1754–1766* (Vintage, 2000). For diplomacy during the Revolution, the following were illuminating: Richard B. Morris, *The Peacemakers: The Great Powers and American Independence* (Harper & Row, 1965); H. M. Scott, *British Foreign Policy in the Age of the American Revolution* (Oxford University Press, 1990); Andrew Stockley, *Britain and France at the Birth of America: The European Powers and the Peace Negotiations of 1782–1783* (University of Exeter Press, 2001). I've also drawn upon P&C, 341–360. Still worthwhile are Samuel Flagg Bemis, *The Diplomacy of the American Revolution* (Indiana University Press, 1957) and his *Diplomatic History of the United States* (Holt, 1936).

On Catherine's armed neutrality and Russia's role: See Isabel de Madariaga's seminal *Britain, Russia and the Armed Neutrality of 1780: Sir James Harris's Mission to St. Petersburg During the American Revolution* (Yale University Press, 1962). For Francis Dana's too often overlooked role in Russia, see the superb biography, which I would recommend to all readers as vital to understanding the period leading up to the Revolution, W. Cresson, *A Puritan Diplomat at the Court of St. Catherine* (Longmans, Green, 1930); also see Madariaga, 80–88; and for broader context, Morris, *The Peacemakers.* On the equally overlooked American-Russian relations in the early days of the republic, see esp. the important study, Norman Saul, *Distant Friends: The United States and Russia, 1763–1867* (University Press of Kansas, 1991), 1–34. See also Clarence Augustus Manning, *Russian Influence in Early America* (Library Publishers, 1953); John R. Alden, *Stephen Sayre: American Revolutionary Adventurer* (Louisiana State University Press, 1983); Nina B. Bashkina, ed., *The United States and Russia: The Beginning of Relations, 1765–1815* (U.S. Government Printing Office, 1980); William Appleman Williams, *American-Russian Relations, 1781–1947* (Rinehart, 1952); Norman E. Saul and Richard D. McKinzie, eds., *Russian American Dialogue on Cultural Relations 1776–1914* (University of Missouri Press, 1997); Norman E. Saul, "The Beginnings of American-Russian Trade, 1763–1766," *William and Mary Quarterly* 26, no. 4 (1969): 596–600; Gordon S. Wood and Louise G. Wood, eds., *Russian-American Dialogue on the American Revolution* (University of Missouri Press, 1995); and Orville T. Murphy, *Charles Gravier, Comte de Vergennes: French Diplomacy in the Age of Revolution 1719–1787* (State University of New York Press, 1982).

On the role of France in the American Revolution: See, for starters, Lee Kennett, *The French Forces in America, 1780–1783* (Greenwood, 1977); William C. Stinchcombe, *The American Revolution and the French Alliance* (Syracuse University Press, 1969); James Breck Perkins, *France in the American Revolution* (Corner House, 1970); Elizabeth Brett White, *American Opinion of France from Lafayette to Poincaré* (Knopf, 1927); Marvin R. Zahniser, *Uncertain Friendship: American-French Diplomatic Relations Through the Cold War* (Wiley, 1975); and Thomas K. Murphy, *A Land Without Castles: The Changing Image of America in Europe, 1780–1830* (Lexington, 2001).

CHAPTER 1

For an overview of various aspects of the early republic: See R. R. Palmer's indispensable *The Age of the Democratic Revolution* (Princeton University Press, 1959), 215–238; John Ferling, *Setting the World Ablaze: Washington, Adams, Jefferson, and the American Revolution* (Oxford University Press, 2000); Bailyn, *To Begin the World Anew*; Richard B. Morris, *Seven Who Shaped Our Destiny: The Founding Fathers as Revolutionaries* (Harper & Row, 1973); Paul Johnson, *A History of the American People* (HarperCollins, 1997); Eric Foner, *Tom Paine and Revolutionary America* (Oxford University Press, 1976) remains the best on this radical for the ages; Gordon S. Wood, *The American Revolution: A History* (Modern Library, 2002), is a little gem. Also see Wood, 3–91, for background on theories of governance and republicanism. On the energy, diversity, and dynamism of early Americans, see Joyce Appleby's wonderful *Inheriting the Revolution: The First Generation of Americans* (Belknap, 2001); Adam Zamoyski, *Holy Madness: Romantics, Patriots, and Revolutionaries* (Weidenfeld & Nicholson, 1999), 6–54. Also John C. Miller, *The Federalist Era, 1789–1801* (Harper & Brother, 1960); Richard Hildreth, *The History of the United States of America*, 6 vols. (Harper & Brother, 1849–1856); essential, of course, is Alexander Hamilton, James Madison, and John Jay, *The Federalist and Anti-Federalist Papers* (Barnes & Noble, 2007); Michael Kraus, "America and the Utopian Ideal in the Eighteenth Century," *Mississippi Valley Historical Review* 22, no. 4 (1936): 487–504.

On Evacuation Day in New York and Washington's farewell to his officers: See E&M, 42, and Chernow, 185–186. I have repeatedly throughout drawn upon this magisterial biography of Hamilton, which is the best one-volume treatment to date. This book, a testament to historical writing, has also proven to be invaluable to me for its rich treatment from the revolutionary period through the 1790s; Richard Norton Smith's *Patriarch: George Washington and the New American Nation* (Houghton Mifflin, 1993), 208; Fleming, 340; Stanley Weintraub, *General Washington's Christmas Farewell: A Mount Vernon Homecoming, 1783* (Free Press, 2003); and Barnett Schecter, *The Battle for New York* (Walker, 2002).

For the role of Loyalists in the American Revolution: Still superb, by turns moving and tragic, and the best volume on the subject is Bernard Bailyn, *The Ordeal of Thomas Hutchinson: Loyalism and the Destruction of the First British Empire* (A. Lane, 1975). Of note, I originally had a much longer section on Hutchinson, whose story I found quite powerful and poignant, but to keep the narrative thread moving I opted to cut this section. However, the story of the Loyalists remains a fascinating episode in early American history. Also see Robert M. Calhoon, *The Loyalists in Revolutionary America, 1760–1781* (Harcourt Brace Jovanovich, 1973); *Reminiscences of an American Loyalist, 1738–1789*, Jonathan Boucher, ed. (Kennikat, 1967); Wallace Brown, *The Good Americans: The Loyalists in the American Revolution* (Morrow, 1969); and Sheila L. Skemp, *William Franklin: Son of a Patriot, Servant of a King* (Oxford University Press, 1990).

For the young republic floundering during the confederation period of the 1780s: See Richard Brandon Morris, *The Forging of the Union, 1781–1789* (Harper & Row, 1987); Jack N. Rakove's all-important *The Beginnings of National Politics: An Interpretive History of the Continental Congress* (Knopf, 1979); Carol Berkin, *A Brilliant Solution: Inventing the American Constitution* (Harcourt, 2002), ch. 1; Allan Nevins, ed., *The American States During and After the Revolution, 1775–1789* (Macmillan, 1924); James Roger Sharp, *American Politics in the Early Republic: The New Nation in Crisis* (Yale University Press, 1993); Chernow, ch. 7; and Robert Middlekauff, *The Glorious Cause: The American Revolution, 1763–1789* (Oxford University Press, 1982), chs. 23–24. On the lack of national unity after the revolution, see David C. Hendrickson, *Peace Pact: The Lost World of the American Founding* (University Press of Kansas, 2003), and Boorstin, *The Americans*, 327–530, which is outstanding. Also very helpful and one of my favorites, see Richard Beeman, Stephen Botein, and Edward C. Carter II, eds., *Beyond Confederation: Origins of the Constitution and American National Identity* (University of North Carolina Press, 1987), especially the illuminating article by John H. Murrin, "A Roof Without Walls: The Dilemma of American National Identity"; Samuel Eliot Morison, Henry Steele Commager, and William E. Leuchtenburg, *The Growth of the American Republic*, vol. 2 (Oxford University Press, 1980). Another thoughtful article on the fragility of young republics is Carl N. Degler, "One Among Many: The United States and National Unification," in *Lincoln, the War President: The Gettysburg Lectures*, 89–119. For a basic but helpful discussion of what constitutes a nation and the forming of a nation-state see P&C, 400, 433, 503–504. On nationalism also see O. Dann and J. R. Dinwiddy, eds., *Nationalism in the Age of the French Revolution* (London & Ronceverte, 1988). Boorstin reminds us that the American Revolution "was not the product of a nationalistic spirit," and that it lacked an "enthusiasm for the birth of a new nation"; and adds that it was "notably lacking in cultural self-consciousness and . . . any passion for national unity," Daniel J. Boorstin, *The Genius of American Politics* (University of Chicago Press,

1953), 70–73. Also on the elements of the essential amalgam for making a firm union, see Kenneth Stampp, "On the Concept of Perpetual Union," in Kenneth Stampp, *The Imperiled Union, Essays on the Background of the Civil War* (Oxford University Press, 1986), 3–36.

On Shays's Rebellion and its wider implications: The following were useful: David Szatmary, *Shays' Rebellion: The Making of an Agrarian Insurrection* (University of Massachusetts Press, 1980); Middlekauff, *The Glorious Cause,* 600–601; Chernow, 224–225; Ellis, *HE,* 172; and Fleming, 354–356, which is superb, on this and the entire period leading up to Washington's election. For the most recent scholarly treatment, see Leonard L. Richards, *Shays's Rebellion: The American Revolution's Final Battle* (University of Pennsylvania Press, 2002). Also see Taylor, *Western Massachusetts in the Revolution,* 103–167, and essays in Alfred E. Young, ed., *Beyond the American Revolution: Explorations in the History of American Radicalism* (Northern Illinois University Press, 1993); and George Connor, "The Politics of Insurrection: A Comparative Analysis of the Shays', Whiskey, and Fries Rebellion," *Social Science Journal* 29, no. 3 (1992): 259–281, which I found quite helpful. On numerous historical society websites (e.g., historyofwar.org), the reader may cull more on Benjamin Lincoln and Daniel Shays.

On the brief but important Annapolis conference: See Berkin, *A Brilliant Solution,* 24–29; Chernow, 222–224; Wood, 327, 473; E&M, 43, 102.

On the constitutional convention: Studies are voluminous. Two works on the Philadelphia Constitutional Convention that cogently synthesize a generation of scholarship are Berkin's terrific *A Brilliant Solution* and Jack Rakove's equally terrific *Original Meanings: Politics and Ideas in the Making of the Constitution* (Knopf, 1996). The best comprehensive and indispensable primary source on the convention is Max Farrand, ed., *The Records of the Federal Convention of 1787,* 4 vols. (Yale University Press, 1966), and a useful follow-up remains James H. Hutson, ed., *Supplement to Max Farrand's The Records of the Federal Convention of 1787* (Yale University Press, 1987). *The Debate on the Constitution,* 2 vols. (Viking, 1993), edited by Bernard Bailyn is another work that I leaned heavily on. Also see Wood, 309–311, 363, 376, 388–389, 535; Chernow, ch. 12; Fleming, 356–372; E&M, 43–44, 104–105, 557–560, an extraordinarily rich and landmark study, which I have drawn on throughout this book. For more, see Jonathan Elliot, *The Debates in the Several State Conventions on the Adoption of the Federal Constitution as Recommended by the General Convention at Philadelphia in 1787,* 5 vols. (Franklin, 1968); Forrest McDonald, *We the People: The Economic Origins of the Constitution* (University of Chicago Press, 1958); Middlekauff, *The Glorious Cause,* ch. 25; Joseph J. Ellis, *After the Revolution: Profiles in Early American Culture* (Norton, 1979); Forrest McDonald, *E Pluribus Unum: The Formation of the American Republic, 1776–1790* (Houghton Mifflin, 1965), is very good; John Chester Miller, *Alexander Hamilton: Portrait in Paradox* (Harper, 1959); Palmer, *Age* provides vital comparative context; Palmer, *Revolution*; Clinton Lawrence Rossiter, *1787: The Grand Convention* (Macmillan, 1966); Jack Rakove, *James Madison and the Creation of the American Republic* (Little, Brown, 1990), is quite invaluable; not to be overlooked are Richard Brookhiser, *Gentleman Revolutionary: Gouverneur Morris, the Rake Who Wrote the Constitution* (Free Press, 2003); Catherine Drinker Bowen's classic *Miracle at Philadelphia: The Story of the Constitutional Convention, May to September 1787* (Little, Brown, 1966). Michael Kammen, ed., *The Origins of the American Constitution: A Documentary History* (Penguin, 1986); and Beeman, Botein, and Carter, eds., *Beyond Confederation*, 12–113, which provide essential scholarly viewpoints.

On the often torturous and uncertain road to ratification: See Chernow, ch. 13, and Berkin, *Brilliant Solution*, ch. 8; Johnson, 190–195; Wood, 332, 341–342, 366, 533. Also see Elliot, *Debates*; Ellis, *After the Revolution*; Saul Cornell, "Aristocracy Assailed: The Ideology of Backcountry Anti-Federalism," *Journal of American History* 76 (March 1990): 148–72; and Saul Cornell, *The Other Founders: Anti-Federalism and the Dissenting Tradition in America, 1788–1828* (University of North Carolina Press, 1999). Jackson Turner Main, *The Anti-federalists: Critics of the Constitution, 1781–1790* (University of North Carolina Press, 1961), remains a useful study on the ratification struggle. Theodore Draper's "The Constitution Was Made, Not Born," in *New York Times Book Review*, October 10, 1993, review of Bernard Bailyn, ed., *Federalist and Anti-federalist Speeches, Articles and Letters During the Struggle over Ratification* (Library of America, 1993), 2 vols., is a helpful primer.

On George Washington: About Washington, arguably the most impenetrable of the Founders, I have drawn extensively upon Douglas Southhall Freeman's outstanding abridged classic, *Washington* (Simon & Schuster, 1968), as well as his majestic six-volume treatment, *George Washington*, vols. 1–6 (Scribner's, 1948–1954). To this day, it stands as the single finest work on Washington, followed closely by James Thomas Flexner's remarkable *George Washington*, 4 vols. (Little, Brown, 1965–1972). W. W. Abbott, Thilander Chase, and Dorothy Twohig, eds., *The Papers of George Washington* is indispensable; Ferling's *Setting the World Ablaze* is superb, as is the new biography by Ellis, *HE*, which perhaps stands as the best biography of Washington since Freeman, esp. 73–109, and Flexner; and also *FB*, 130–135; also excellent is Smith's *Patriarch*, which has a keen handle on foreign events; Johnson, 121–132; Edmund S. Morgan, *The Genius of George Washington* (Norton, 1980); Garry Wills, *Cincinnatus: George Washington and the Enlightenment* (Doubleday, 1984); Willard Sterne Randall, *George Washington: A Life* (Holt, 1997); and Richard Brookhiser, who has written extensively on a number of the key Founders, brings to bear rich and fascinating insights along with the additional benefit of accessibility in all his works. See here his outstanding *Founding Father: Rediscovering George Washington* (Free Press, 1996); also his *Rules of Civility: The 110 Precepts That Guided Our First President in War and Peace* (University of Virginia Press, 2003).

Don Higginbotham, *George Washington and the American Military Tradition* (University of Georgia Press, 1985), is the seminal work on Washington's military exploits; David Hackett Fischer's richly creative *Washington's Crossing* (Oxford University Press, 2004) is crucial; Robert Leckie, *"A Few Acres of Snow": The Saga of the French and Indian Wars* (Wiley, 1999); Fred Anderson, *Crucible of War: The Seven Years' War and the Fate of Empire in British North America, 1754–1766* (Vintage, 2000); and Robert Leckie, *George Washington's War: The Saga of the American Revolution* (HarperPerennial, 1992); Ségur's *Memoirs*, which straddle the worlds of the Russian and French courts, Europe as well as America, are outstanding, and show George Washington commanded enormous respect from prominent foreigners. Ségur even discussed Washington with Empress Catherine II.

CHAPTER 2

For a broad array of background: See James H. Billington, *The Icon and the Axe: An Interpretive History of Russian Culture* (Random House, 1970); Anthony Glenn Cross, *By the Banks of the Neva: Chapters from the Lives and Careers of the British in Eighteenth-Century Russia* (Cambridge University Press, 1997); Isabel de Madariaga, *Politics and Culture in Eighteenth-Century Russia: Collected Essays* (Longman, 1998); *Engraved in the Memory: James Walker, Engraver to the Empress Catherine the Great, and His Russian Anecdotes,* Anthony Cross, ed. (Berg, 1993); Carolly Erickson, *Great Catherine* (Crown, 1994); G. Gooch, *Catherine the Great, and Other Studies* (Longmans, Green, 1954); David Mark Griffiths, *Russian Court Politics and the Question of Expansionist Foreign Policy Under Catherine II, 1762–1783* (Cornell University Press, 1967); Kamenskii, *The Russian Empire in the Eighteenth Century: Searching for a Place in the World,* David Griffiths, trans. (M. E. Sharpe, 1997); Gina Kaus, *Catherine the Great* (Cassell, 1935); Charles François Philibert Masson, *The Court of St. Petersburg, Particularly Towards the End of the Reign of Catherine II and the Commencement of That of Paul I* (Arno, 1970), another one of those memoirs to be treasured; Howard I. Kushner, *American-Russian Rivalry in the Pacific Northwest, 1790–1867* (Cornell University thesis, 1970); Max M. Laserson, *The American Impact on Russia, Diplomatic and Ideological, 1784–1917* (Macmillan, 1950); Lentin, *Russia in the Eighteenth Century: From Peter the Great to Catherine the Great (1696–1796)* (Barnes & Noble, 1973); Lobanov-Rostovsky, *Russia and Europe, 1789–1825*; Philip Longworth, *The Art of Victory: The Life and Achievements of Field Marshal Suvorov (1729–1800)* (Holt, Rinehart & Winston, 1966); Zoé Oldenbourg, *Catherine the Great,* Anne Carter, trans. (Pantheon 1965); Raeff, *Understanding Imperial Russia; Russia and the West in the Eighteenth Century,* Study Group on Eighteenth-Century Russia, University of East Anglia Study; *Russia and the World of the Eighteenth Century,* Study Group on Eighteenth-Century Russia held at Indiana University at Bloomington;

Russia Under Catherine the Great, Paul Dukes, ed. (Oriental Research Partners, 1977); and *Russian-American Dialogue on Cultural Relations, 1776–1914*, Norman E. Saul and Richard D. McKinzie, eds. (University of Missouri Press, 1997).

For weather and life in Russia: See *CTG*; Massie; Hughes; W. Bruce Lincoln, *Sunlight at Midnight: St. Petersburg and the Rise of Modern Russia* (Basic, 2000).

Trip to the Crimea: This section on Catherine's Crimean trip draws extensively on Ségur, whose breadth of experience in America, as well as Russia and France, makes this indispensable reading, as well as Charles-Joseph, Prince de Ligne, *The Prince of Ligne: His Memoirs, Letters, and Miscellaneous Papers*, Katharine Prescott Wormeley, ed. and trans., 2 vols. (Brentano's, 1927); Madariaga, esp. 393–395; Alexander, esp. 256–257, and the most comprehensive overview, which I have also relied most heavily upon in this chapter, is Sebag Montefiore's magisterial *Prince of Princes: The Life of Potemkin* (Dunne, 2000); also quite helpful are Troyat, 272–288, and Cronin, 247–252. The reader may also consult the Prince de Nassau-Siegen, *Memoirs of Prince de Nassau-Siegen* (Broadway, 1927). On annexation of the Crimea, see Alan Fisher, *The Russian Annexation of the Crimea* (Cambridge, 1970). For more on Catherine's expansionist policies, see Gladys Scott Thompson, *Catherine the Great and the Expansion of Russia* (Collier, 1962); and Mansel, 406–408, which also includes quotes. For Peter's policies toward the Islamic world, see B. H. Summer, *Peter the Great and the Ottoman Empire* (Archer, 1965). On U.S.–Russian relations, see Saul, *Distant Friends*. For more on Potemkin's drive to colonize the south, see James A. Duran, "Catherine, Potemkin, and Colonization," *Russian Review* 28, no. 1 (January 1969); and Robert E. Jones, "Urban Planning and the Development of Provincial Towns in Russia, 1762–96," in *The Eighteenth Century in Russia*, J. G. Garrard, ed. (Clarendon, 1973). Masson echoes Emperor Joseph on Russia's extensive building using essentially "slaves," 78. On Islam under Catherine, see Alan Fisher, "Enlightened Despotism and Islam under Catherine II," *Slavic Review* 27 (1968).

History of Russia: For more on Russia's drive to expand, see esp. Taras Hunczak, ed., *Russian Imperialism from Ivan the Great to the Revolution* (Rutgers University Press, 1974); Norman E. Saul, *Russia and the Mediterranean, 1797–1807* (University of Chicago Press, 1970); Cynthia Hyla Whittaker, ed., *Russia Engages the World, 1453–1825* (Harvard University Press, 2003); Michael T. Florinsky, *Russia: A History and an Interpretation* (Macmillan, 1955); Leo Gershoy, *From Despotism to Revolution: 1763–89* (Harper & Row, 1944); and Charles-Augustin Sainte-Beuve, *Portraits of the Eighteenth Century*, 2 vols. (Putnam's, 1905). For more background, see Marc Raeff, *Imperial Russia 1682–1825: The Coming of Age of Modern Russia* (Knopf, 1971), esp. 39–68. On the first Russo-Turkish War, I've drawn upon the eminent dean of Catherine the Great scholars, Madariaga, 187–215, and *CTG*, 38–49.

For background on Tatars: See Alan Fisher, *The Crimean Tatars* (Stanford, 1978). I've also drawn upon Montefiore, 244–260; Madariaga, 68–69, 212, 220, 228, 242–247,

364–366; and Massie, 85–89, 137, 339–340; also François Tott's fascinating *Memoirs of Baron de Tott (1785)*, 2 vols. (G. G. J. and J. Robinson, 1785). On Prince de Ligne, see Ligne, Charles Joseph, Prince de, *Mémoires et Lettres du Prince de Ligne* (G. Crès, 1923); and O. Gilbert, *The Prince de Ligne* (McDevitt-Wilson, 1923); on the Comte de Ségur, *Memoirs*; on Lewis Littlepage, see Curtis Carroll Davis, *The King's Chevalier: A Biography of Lewis Littlepage* (Bobbs-Merrill, 1961); for more on Emperor Joseph II, see Saul K. Padover, *The Revolutionary Emperor: Joseph II* (Eyre & Spottiswoode, 1965); on King Stanislas Augustus Poniatowski, see esp. the two works by Robert Howard Lord, *The Second Partition of Poland: A Study in Diplomatic History* (AMS 1969), and Adam Zamoyski, *The Last King of Poland* (J. Cape, 1992); as well as Masson, *Court of Petersburg*, 85–88. It should be noted that when the king was in Poland, he was formally known as Stanislas-Augustus, but because the Polish king was not allowed to leave Polish territory, outside Poland he was known as Count Poniatowski. On Frederick the Great, see W. F. Reddaway, *Frederick the Great and the Rise of Prussia* (Putnam's, 1904); Nancy Mitford, *Frederick the Great* (Harper & Row, 1964), and on Miranda, see Madariaga, *Travels of General Francesco de Miranda in Russia* (London, 1950).

For more on turmoil in the Ottoman Empire: See Montefiore, particularly 562, footnote 29, which speaks to the state of war fever in Istanbul; Mansel, 200–210; Alexander, 240–41; *CTG*, 162–169. On Russian foreign policy and the Porte, see Madariaga, 377–393; Tott.

On St. Petersburg: See Lincoln. This excellent book is useful both on the history of St. Petersburg, as well as matters concerning Peter and Catherine. Beyond Dana's observations of old Muscovy, see also Francesca Wilson, *Muscovy: Russia Through Foreign Eyes, 1553–1900* (Allen & Unwin, 1970); and Masson, 77. **On Catherine:** see the three outstanding works that have proved indispensable throughout: Troyat, 140–250; Cronin, 159–259; Alexander, 3–96. In addition to the Massachusetts puritan Dana's memoirs, for other personal accounts of Russia in the eighteenth century, see Marchioness of Londonderry and H. M. Hyde, eds., *The Russian Journals of Martha and Catherine Wilmot* (Macmillan, 1934); also A. G. Cross, ed., *An English Lady at the Court of Catherine the Great: The Journal of Baroness Dimsdale, 1781* (Crest, 1989); John Parkinson, *A Tour of Russia, Siberia and the Crimea, 1792–1794*, William Collier, ed. (Cass, 1971). S. T. Aksakov, *Years of Childhood, A Russian Gentleman and a Russian Schoolboy*, J. D. Duff, trans. (Oxford University Press, 1982) is a fascinating written chronicle of Russian life seen by a boy and a youth.

CHAPTER 3

For background: See Palmer, *Age*, 3–20, 439–502; Louis Gottschalk, *Lafayette Joins the American Army* (University of Chicago Press, 1937); Louis Gottschalk, "When and Why Lafayette Became a Revolutionary," in Morris Slavin and Agnes M. Smith (eds.), *Bourgeois, Sans-culottes and Other Frenchmen: Essays of the French Revolution in Honor of John Hall Stewart* (Wilfrid Laurier University Press, 1981), 7–24; Louis Gottschalk, *Lafayette Between the American and the French Revolutions* (University of Chicago Press, 1950); Jonathan R. Dull, "Franklin in France: A Reappraisal," in *Proceedings of the Annual Meeting of the Western Society for French History* (no. 4, 1976); Philip Katz, *The Image of Benjamin Franklin in the Politics of the French Revolution 1776–1794* (Harvard University Program for Social Studies dissertation, 1986); D. M. G. Sutherland, *France 1789–1815: Revolution and Counterrevolution* (Fontana, 1984); Philip Dawson, *The French Revolution* (Prentice Hall, 1967); William Doyle, *The Ancient Regime* (Macmillan, 1986); William Doyle's vital *Origins of the French Revolution* (Oxford University Press, 1980); Susan Dunn's important *Sister Revolutions: French Lightning, American Light* (Faber & Faber, 1999); Durant, *The Age of Napoleon*; Edward E. Hale and Edward E. Hale Jr., *Franklin in France*, 2 vols. (Roberts Brothers, 1887–1888); Alfred Cobban, *The Social Interpretation of the French Revolution* (Cambridge University Press, 1999); Alfred Cobban, *Historians and the Causes of the French Revolution* (Cambridge University Press, 1958); Georges Lefebvre's classic *The Coming of the French Revolution* (Princeton University Press, 1975); Jules Michelet, *The French Revolution* (1890) (University of Chicago Press, 1967); M. J. Sydenham, *The First French Republic, 1792–1804* (University of California Press, 1973); and John Hall Stewart, *A Documentary Survey of the French Revolution* (Macmillan, 1951).

Ancien régime: Schama, 21–199. Throughout I have drawn heavily upon Schama's epic work, along with other such outstanding and powerful treatments as Doyle, 1–65; Jones, 1–363; Hibbert, 19–35; Thompson, 1–2; R. R. Palmer, *World of French Revolution* (Harper Torchbooks, 1971), 11–47; Philip Mansel, *The Court of France, 1789–1830* (Cambridge University Press, 1988); Alexis de Tocqueville, *L'Ancien Regime* (1927); and Ségur. Essays on the issue of evolving sovereignty are in the edited work of Keith Michael Baker, ed., *The Political Culture of the Old Regime* (Oxford University Press, 1987). For more on social mobility and privilege, see Guy Chaussinand-Nogaret, *The French Nobility in the Eighteenth Century: From Feudalism to Enlightenment* (trans. William Doyle, 1985); David D. Bien, "Offices, Corps, and a System of State Credit: The Uses of Privilege Under the Ancien Regime," in Keith Michael Baker (ed.), *The Political Culture of the Old Regime* (Oxford University Press, 1987), 89–114. William Doyle, *The Old European Order 1660–1800* (Oxford University Press, 1992); C. B. A. Behrens, *The Ancien Régime* (Harcourt, Brace & World, 1967); Douglas Dakin, *Turgot and the Ancien Regime in France* (Methuen, 1939); *The European Nobility in the Eighteenth Century*, A. Goodwin, ed. (A. & C. Black, 1953); G. Gooch, *Maria Theresa and Other Studies* (Longmans, Green, 1951).

Coronation of Louis XVI: Doyle, 1; Schama, 51–54; Fraser, 133–135; Lever, 545–562. Also see Keith Michael Baker's insightful essay, "French Political Thought at the Accession of Louis XVI," *Journal of Modern History* (June 1978): 279–303.

1788 problems: Jones, 364–410; Schama, 248–260; Lever, 203–204; Fraser, 269; Rudé, 10–33; Abbé Raynel; Hibbert; Arthur Young's fascinating accounts provide a poignant picture of the hardships the poor faced as a result of the terrible winter of 1788–1789, *Travels in France During the Years 1787, 1788, 1789* (Ginn, 1906), esp. 177; R. R. Palmer, *Age*, 282; Doyle, *Origins*, 66–87; An important study on finance is J. E. Bosher, *French Government Finance, 1770–1795* (Cambridge University Press, 1970). For further examination of the prerevolutionary financial crisis, see Douglas Dakin, *Turgot and the Ancien Regime in France* (London, 1939), and for background see Robert D. Harris, "French Finances and the American War, 1777–1783," *Journal of Modern History* 48, no. 2 (1976): 233–258.

Assembly of Notables: Schama, 117–118, 237–246, 253–261; Lever, 192–199; Fraser, 247–248, 254, 268; Jones, 380–385; Doyle, 70–79; Thompson, 4; Hibbert, 38–40; Carlyle, vol. I, 121–126. See the excellent Bailey Stone, *The Parlement of Paris, 1774–1789* (University of North Carolina Press, 1981) on how the judicial nobility sought to redefine sovereignty; and William Doyle's *The Parlement of Bordeaux and the End of the Old Regime 1771–1790* (St. Martin's Press, 1974) superbly covers this critical ideological development. Also see A. Goodwin, "Calonne, the Assembly of French Notables of 1787 and the Origins of the Revolte Nobilaire," *English Historical Review* (1946). On the changing attitudes of the nobility, see Guy Chaussinand-Nogaret, *The French Nobility in the Eighteenth Century: From Feudalism to Enlightenment*, William Doyle, trans. (Cambridge University Press, 1985); for Orleans's machinations during this period see esp. G. A. Kelly, "The Machine of the Duc d'Orléans and the New Politics," in *Journal of Modern History* (1979): 667–684; Vivian Gruder, "Class and Politics in the Pre-revolution: The Assembly of Notables of 1787," in Ernst Hinrichs et al., *Vom Ancien Regime* (Vandenhoeck und Ruprecht, 1978).

Day of Tiles: Schama, 272–277, is the most detailed account; also see Doyle, 83–84; Jones, 389; Stendhal's account is given in *The Life of Henry Brulard*, B. C. J. G. Knight, trans. (Minerva, 1958), 76; William Doyle, "The Parlements of France and the Breakdown of the Old Regime 1771–1788," in *French Historical Studies* (1970): 429; William Doyle, *The Parlement of Bordeaux*; Kathryn Norberg, *Rich and Poor in Grenoble 1600–1814* (University of California Press, 1985).

Necker: Schama, 88–95, 307, 345; Doyle, 62–63, 66–68, 87; Hibbert, 36–45. Robert D. Harris's superbly detailed study, *Necker and the Revolution of 1789* (University Press of America, 1986), is vital to understanding the tangled politics of 1789; Jean Egret, *Necker: Ministre de Louis XVI* (Paris, 1975); Robert D. Harris, *Necker, Reform Statesman of the Old Regime* (University of California Press, 1979); Robert D. Harris, "Necker's Compte Rendu of 1781: A Reconsideration," *Journal of Modern History* 42, no. 2 (1970): 161–183.

Reveillon riots: For best description see Schama, 326–331. I have also drawn upon Jones, 405; Thompson, 51; Rudé, 34–44; Jean Egret, "The Pre-Revolution in

Provence," in J. Kaplow, ed., *New Perspectives on the French Revolution* (Wiley, 1965); also Georges Lefebvre's masterwork, *The Great Fear of 1789: Rural Panic in Revolutionary France*, Joan White, trans. (Princeton University Press, 1973), ch. 4.

Estates-General: Jones, 405–416; Schama, 281–283, 290, 297–298, 335–347, 352; Doyle, 93–108; Hibbert, 45–60, 64; Lever: 202–207; Fraser, 261, 270–277; Carlyle, vol. I, 138–158; Zweig, 201–214. On Estates-General and National Assembly, see Thompson, 7–44. For more on the complicated question of double representation, see George Gordon Andrews, "Double Representation and Vote by Head before the French Revolution," *South Atlantic Quarterly* 26 (October 1927): 374–391. Antonia Valentin's *Mirabeau*, though almost sixty years old, remains a valuable biography, trans. E. W. Dickes (Hamish Hamilton, 1948). For Mirabeau and the Estates-General, see Guy Chaussinand-Nogaret, *Mirabeau* (Paris, 1982); Mitchell B. Garrett provides a detailed account in *The Estates-General of 1789* (Appleton-Century, 1935); for the zealous opposition of the nobles, see Daniel Wick, "The Court Nobility and the French Revolution: The Example of the Society of Thirty," in *Eighteenth-Century Studies* (1980): 263–284; Lynn Hunt, "The National Assembly," in Baker, ed., *Political Culture of the Old Regime* (Oxford University Press, 1987). For the clergy's role, see R. F. Necheles, "The Curés in the Estates General of 1789," *Journal of Modern History* (1874); and M. G. Hutt, "The Curés and the Third Estate: The Ideas of Reform in the Period 1787–89," *Journal of Ecclesiastical History* (1955 and 1957).

Sansculottes: Schama, 497–500; Rudé, 8–12, 196–199, 205–212; Carlyle, vol. I, 35–37; R. B. Rose, *The Making of the Sans-culottes: Democratic Ideas and Institutions in Paris 1789–92* (Manchester University Press, 1983); Albert Soboul, *The Parisian Sans-culottes and the French Revolution* (Oxford University Press, 1964); and Jeffrey Kaplow, *The Names of Kings: The Parisian Laboring Poor in the Eighteenth Century* (Basic, 1972), esp. ch. 7.

Tennis Court Oath: Schama, 358–361; Doyle, 105; Thompson, 19; Hibbert, 59–60; Fraser, 280–281.

Jefferson in Paris: Dumas Malone, *Jefferson and His Time*, 6 vols. (Little, Brown, 1948–1981), vol. 2, esp. 230–231; and Gilbert Chinard, "Jefferson's Influence Abroad," *Mississippi Valley Historical Review* 30, no. 2 (1943): 171–186. Jefferson's stay in Paris had a profound influence on him, and would do much to shape his actions as the U.S.-French crisis deepened from 1793 to 1799.

Bastille: Schama, 383–406; Thompson, 55–58; Doyle, 108–111; Jones, 416–420; Hibbert, 71–81; Rudé, 45–60; Fraser, 283–284; Carlyle, vol. I, 183–209; Zweig, 214–216; Jacques Godechot's *The Taking of the Bastille*, trans. Jean Stewart (Faber & Faber, 1970), is a comprehensive, often riveting story, having the additional benefit of contemporary eyewitness accounts. For the broader and increasingly dwindling military picture in the capital, see Samuel F. Scott, *The Response of the Royal Army to* *the French Revolution: The Role and Development of the Line Army* (Clarendon, 1978), esp. 46–70, as well as Alan Williams, *The Police of Paris, 1718–1789* (Louisiana State University Press, 1979). **Camille Desmoulins:** Schama, 379–82; Thompson, 54; Hibbert 65–66. **Aftermath of Bastille:** Schama, 406–414; Doyle, 112; Lever, 212–218; Fraser, 284–289; Thompson, 58–63; Hibbert, 82–83.

Louis XVI: John Hardman's richly rendered *Louis XVI* (Yale University Press, 1993) is the book that I have most heavily drawn upon. I have also extensively used Jean-Christian Petitfils, *Louis XVI* (Perrin, 2005); Jean de Viguerie, *Louis XVI: Le Roi Bienfaisant* (Rocher, 2003); John Hardman, *Louis XVI: The Silent King* (Oxford University Press, 2000); Evelyne Lever, *Louis XVI* (Fayard, 1985); John Francois Chiappe, *Louis XVI* (Libr. Academique Perrin, 1987); and Bernard Fay, *Louis XVI; or the End of a World*, Patrick O'Brian, trans. (Regnery, 1968). Also insights throughout in Schama, Fraser, Lever, and Hibbert, 19–27, 54, 61.

Marie Antoinette: I have heavily drawn here and throughout on the two superb recent biographies by Fraser and Lever, both of which prove to be invaluable in re-creating the texture and dilemma of the times as well, and which are essential mainstays for my French chapters. Also extremely good, even though more than seventy years old, is Zweig, esp. 3–19 and 76–103. See further Schama; Olivier Bernier, *The Secrets of Marie Antoinette* (Doubleday, 1985) is the published letters to the queen's mother and brother; Hilaire Belloc, *Marie Antoinette* (Doubleday, Page, 1909); Andre Castelot, *Queen of France: Marie Antoinette* (Harper Brothers, 1957); Hibbert, 20–27, 35–39; and Carlyle, 322–325.

Great Fear: For a comprehensive treatment, see Georges Lefebvre's superb *The Great Fear of 1789: Rural Panic in Revolutionary France*, trans. Joan White (Princeton University Press, 1973). Schama, 429–436; Thompson, 67–80; Jones, 418–419; Hibbert, 93; Fraser, 289–290; Doyle, 133. For the murder of Foulon, see Thompson, 63.

CHAPTER 4

Treatments in a broader sweep helpful for this period include: Joanne B. Freeman, *Affairs of Honor: National Politics in the New Republic* (Yale University Press, 2001), which conveys the rich texture of politics in this era; Palmer, *Age*, 217–235; Richard Hofstadter, *The American Political Tradition and the Men Who Made It* (Knopf, 1948), particularly the essay on Thomas Jefferson; Clinton Lawrence Rossiter, *The American Quest, 1790–1860: An Emerging Nation in Search of Identity, Unity, and Modernity* (Harcourt Brace Jovanovich, 1971); William Maclay's wonderful resource, *Journal of William Maclay*, ed. Edgard S. Maclay (Appleton, 1890); Ralph Ketchum's *Presidents Above Party: The First American Presidency, 1789–1829* (University of North Carolina Press, 1984) was very helpful to me; Joyce Appleby, *Capitalism and a New Social Order* (New

York University Press, 1984); Hugh Brogan, *The Pelican History of the United States of America* (Penguin, 1985); Samuel Eliot Morison, Henry Steele Commager, and William E. Leuchtenburg, *The Growth of the American Republic, Vol. 2* (Oxford University Press, 1980); Adrienne Koch, *Jefferson and Madison: The Great Collaboration* (Knopf, 1950); Walter A. McDougall, *Freedom Just Around the Corner: A New American History 1585–1828* (HarperCollins, 2004); and Peter Kolchin, "Some Recent Works on Slavery Outside the United States: An American Perspective. A Review Article," *Comparative Studies in Society and History* 28, no. 4 (1986): 767–777. Also see the important Sean Wilentz, *Major Problems in the Early Republic, 1787–1848, Documents and Essays* (DeHeath, 1992).

For the legendary horseback ride from Mount Vernon to New York and the inauguration of Washington: See Chernow, 276–281; E&M, 45, 74; Berkin, *Brilliant Solution*, ch. 9; and Fleming, 381. Among the best treatments of the Washington presidency are: Ellis, *HE*, ch. 6; Richard Norton Smith, *Patriarch*, which is one of my other favorites, along with James Thomas Flexner, *George Washington and the New Nation, 1783–1793* (Little, Brown, 1970); Francis Franklin, *The Rise of the American Nation, 1789–1824* (International Publishers, 1943); the indispensable Edmund S. Morgan, *The Genius of George Washington* (Norton, 1980); Edmund S. Morgan, *The Meaning of Independence: John Adams, George Washington, Thomas Jefferson* (University Press of Virginia, 1976); Randall; Marcus Cunliffe, *George Washington, Man and Monument* (Little, Brown, 1958), and of course, Freeman, *Washington*.

On the creation of the first cabinet: See McDonald, 23–46; Chernow, 286–310; E&M, 52; Ellis, *HE*, 197–200; John Chester Miller, *The Young Republic, 1789–1815* (Free Press, 1970); and Ketcham, *Presidents Above Party*. **For the debate on how to address the president of the United States:** See Ellis, *HE*, 193; Chernow, 278; and E&M, 46–48. **Debate on assumption and other financial proposals of Alexander Hamilton:** Malone, *Jefferson and the Rights of Man*, vol. 2, 299–302; Ellis, *FB*, ch. 2, also describes the political bargain between Hamilton, Jefferson, and Madison regarding the assumption of state debts and the movement of the capital to the Potomac. Other accounts I have drawn upon include E&M, 156–161; Chernow, 326–331; and Forrest McDonald, *Alexander Hamilton: A Biography* (Norton, 1979) and Frank Monaghan, *This Was New York, The Nation's Capital in 1789* (Books for Libraries Press, 1970). Three excellent scholarly treatments on the fascinating episode include Jacob E. Cooke, "The Compromise of 1790," *William and Mary Quarterly* 27 (1970): 523–545; and the follow-up, Kenneth Bowling, "Dinner at Jefferson's: A Note on Jacob E. Cooke's 'Compromise of 1790,' " *William and Mary Quarterly* 28 (1971): 629–648; and Norman K. Risjord, "The Compromise of 1790: New Evidence on the Dinner Table Bargain," *William and Mary Quarterly* 33 (1976): 309–314. For more, see James E. Ferguson, *The Power of the Purse: A History of American Public Finance, 1776–1790* (University of North Carolina Press, 1961);

and Paul A. Gilje, "The Rise of Capitalism in the Early Republic," *Journal of the Early Republic*, 16/2 (Summer 1996): 159–182.

Madison: The classic and comprehensive biography on Madison remains Irving Brant, *James Madison*, 6 vols. (Bobbs-Merrill, 1941–1961). For an excellent and thoughtful treatment see Jack N. Rakove's excellent *James Madison and the Creation of the American Republic*, Oscar Handlin, ed. (Scott, Foresman/Little, Brown, 1990); James Morton Smith, ed., *The Republic of Letters: The Correspondence Between Thomas Jefferson and James Madison, 1776–1826*, 3 vols. (Norton, 1995). Three superior treatments of Madison as a political thinker are Lance Banning, *The Sacred Fire of Liberty: James Madison and the Founding of the American Republic* (Cornell University Press, 1995); Ralph Ketcham, *James Madison: A Biography* (American Political Biography Press, 2003); Marvin Meyers, ed., *The Mind of the Founder: Sources of the Political Thought of James Madison* (University Press of New England, 1981); and Drew McCoy, *The Last of the Fathers: James Madison and the Republican Legacy* (Cambridge University Press, 1989).

Slavery: The literature on slavery is extensive. It is also by turns heartrending, poignant, and powerful. I have drawn heavily on the following, starting with the best single volume on slavery throughout history, Hugh Thomas, *The Slave Trade: The Story of the Atlantic Slave Trade, 1440–1870* (Picador, 1997). Other superb works are David Brion Davis's sophisticated and comprehensive *The Problem of Slavery in Western Culture* (Cornell University Press, 1966); Ira Berlin's rich classic *Many Thousands Gone: The First Two Centuries of Slavery in North America* (Belknap, 1998); the eloquent studies by Winthrop D. Jordan, *White over Black: American Attitudes Towards the Negro, 1550–1812* (University of North Carolina Press, 1968); Edmund Morgan, *American Slavery, American Freedom: The Ordeal of Colonial Virginia* (Norton, 1975); and Jill Lepore's superb and fascinating *New York Burning: Liberty, Slavery, and Conspiracy in Eighteenth-Century Manhattan* (Knopf, 2005); Robert Harms, *The Diligent: A Voyage Through the Worlds of the Slave Trade* (Basic, 2002), brought to my attention by Richard Gilder, is absolutely riveting. See also Robin Blackburn, *The Overthrow of Colonial Slavery, 1776–1848* (Verso, 1988); Marc E. Brandon, *Free in the World: American Slavery and Constitutional Power* (Princeton University Press, 1998); Daniel P. Resnick, "The Société des Amis des Noirs and the Abolition of Slavery," *French Historical Studies* 7, no. 4 (1972): 558–569, which provides broader historical context, as does Richard Hellie, *Slavery in Russia, 1450–1725* (University of Chicago Press, 1982). Though it is beyond the scope of this book to have treated in depth, exploring the differences between the treatment of slaves in America, slaves in the French colonies, and serfs in Russia remains an important area rich for further study. What one also sees is the degree to which abolitionists of the time were acutely conscious of each other across borders and continents. Paul Finkelman, *An Imperfect Union: Slavery, Federalism, and Comity* (University of North Carolina Press, 1981). Also important, Paul Finkelman, *Slavery and the Founders: Race and Liberty in the Age of Jefferson*

(M. E. Sharpe, 2001); Fritz Hirschfeld, *George Washington and Slavery: A Documentary Portrayal* (University of Missouri Press, 1997); Barnett Hollander, *Slavery in America* (Barnes & Noble, 1964); Duncan J. MacLeod, *Slavery, Race and the American Revolution* (Cambridge University Press, 1974); Staughton Lynd, *Class Conflict, Slavery, and the United States Constitution: Ten Essays* (Bobbs-Merrill, 1968); Matthew T. Mellon, *Early American Views on Negro Slavery, from the Letters and Papers of the Founders of the Republic* (Bergman, 1969); John Kaminski, ed., *A Necessary Evil?: Slavery and the Debate over the Constitution* (Madison House, 1995); Ira Berlin and Ronald Hoffman, eds., *Slavery and Freedom in the Age of the American Revolution* (University Press of Virginia, 1983); William M. Wiecek, *The Sources of Antislavery Constitutionalism in America, 1760–1848* (Cornell University Press, 1977); Arthur Zilversmit, *The First Emancipation: The Abolition of Slavery in the North* (University of Chicago Press, 1967), 159–160 for text of the petition; and Don E. Fehrenbacher, *Slavery, Law and Politics: The Dred Scott Case in Historical Perspective* (Oxford University Press, 1981).

On slavery and the American Revolution: See Davis's important *The Problem of Slavery and the Age of Revolution*, 48–55; Gary B. Nash, *Race and Revolution* (Madison, 1990), essays in Berlin and Hoffman, eds.; Simon Schama's penetrating *Rough Crossings: Britain, the Slaves, and the American Revolution* (Ecco, 2006); Chernow, 121–122 esp. for Laurens; and Ellis, *FB*, 89. On the slavery discussions at the Constitutional Convention, see Berkin, *A Brilliant Solution*, 113–114, 214–215; Chernow is very good, 238; Lawrence Goldstone's important new study, *Dark Bargain: Slavery, Profits, and the Struggle for the Constitution* (Walker, 2005). For broader treatments, also see Donald L. Robinson, *Slavery in the Structure of American Politics, 1765–1820* (Harcourt Brace Jovanovich, 1970), 201–247; Davis, *The Problem of Slavery and the Age of Revolution*, 122–131; MacLeod, *Slavery, Race and the American Revolution*, 37–39; Finkelman, *Slavery and the Founders*, 1–33. Also two of the best scholarly treatments, Doroty Twohig, " 'That Species of Property': Washington's Role in the Controversy over Slavery," in *George Washington Reconsidered*, Don Higginbotham, ed. (University of Virginia Press, 2001), 114–138, and Henry Wiencek, *An Imperfect God: George Washington, His Slaves, and the Creation of America* (Farrar, Straus & Giroux, 2003). On the Articles of confederation and slavery, see Finkelman, *Slavery and the Founders*, 34–57. For a less critical review see William W. Freehling, "The Founding Fathers and Slavery," *American Historical Review* 77 (1972): 81–93; also Peter Onuf, "From Constitution to Higher Law: The Reinterpretation of the Northwest Ordinance," *Ohio History* 94 (1985): 5–33, and Daniel C. Littlefield, "John Jay, the Revolutionary Generation, and Slavery," *New York History* 8, no. 1: 91–132. On the slavery debate in the first Congress, see Richard S. Newman, "Prelude to the Gag Rule: Southern Reaction to Antislavery Petitions in the First Federal Congress," *Journal of the Early Republic* 16 (1996): 571–599, is the best single rendition of the debate, including quotes which I have drawn heavily upon here. Also see Ellis, *FB*, 81–101; E&M,

151; Ferling, 285–290; and Howard Ohlone, "Slavery, Economics, and Congressional Politics," *Journal of Southern History* 46 (1980): 355–360.

Ben Franklin: My favorite treatment is Walter Isaacson's extraordinary *Benjamin Franklin* (Simon & Schuster, 2003), which I've used extensively, along with Carl Van Doren's masterful classic *Benjamin Franklin* (Viking, 1938), which is rightly considered among the finest biographies. Also outstanding is H. W. Brands, *The First American: The Life and Times of Ben Franklin* (Doubleday, 2000). Stacy Schiff's *A Great Improvisation: Franklin, France and the Birth of America* (Holt, 2005) is a tour de force, beautifully written and filled with fascinating insights and is, too, the best treatment of Franklin in France. Also Edmund S. Morgan's little jewel, *Benjamin Franklin* (Yale University Press, 2002). Further, Benjamin Franklin, *The Autobiography of Benjamin Franklin* (Yale University Press, 1964); see also Esmond Wright, *Franklin of Philadelphia* (Belknap, 1986); Robert Middlekauf, *Benjamin Franklin and His Enemies* (University of California Press, 1996). Ellis, *FB*, 108–110, gives a succinct measure of Franklin's stature among contemporaries. On Franklin's petition on slavery, see E&M, 151; and Ellis, *FB*, 110–113; also Johnson, 134–140. Also see Edward E. Hale and Edward E. Hale Jr., *Franklin in France*, 2 vols. (Roberts Brothers, 1887–1888); David Freeman Hawke, *Franklin* (Harper & Row, 1976); David Schoenbrun, *Triumph in Paris: The Exploits of Benjamin Franklin* (Harper & Row, 1976); and *Benjamin Franklin: In Search of a Better World*, Page Tailbott, ed. (Yale University Press, 2005).

CHAPTER 5

For overview reading: See Eugene Anschel, *The American Image of Russia, 1775–1917* (Ungar, 1974); Thomas Andrew Bailey, *America Faces Russia: Russian-American Relations from Early Time to Our Day* (Cornell University Press, 1950); Billington; Madariaga, *Politics and Culture*; Gooch, *Catherine the Great*; Griffiths, *Russian Court Politics . . . Under Catherine II*; Kamenskii, *The Russian Empire in the Eighteenth Century*; Kaus, *Catherine the Great*; A. Lentin, *Russia in the Eighteenth Century: From Peter the Great to Catherine the Great (1696–1796)* (Barnes & Noble, 1973); Lincoln, *Sunlight at Midnight*; Lobanov-Rostovsky, *Russia and Europe, 1789–1825*; Masson, *Memoirs of the Court of Petersburg*; Marc Raeff, *Imperial Russia 1682–1825: The Coming of Age of Modern Russia* (Knopf, 1971); Raeff, *Understanding Imperial Russia; Russia Under Catherine the Great*, Paul Dukes, ed. (Oriental Research Partners, 1977); Norman E. Saul, *Russia and the Mediterranean, 1797–1807* (University of Chicago Press, 1970); and Ségur.

On the sultan's domain: See esp. Caroline Finkle, *Osman's Dream: The History of the Ottoman Empire* (Basic, 2005). I have also drawn heavily on the fascinating discussions in Massie, 529–538; Mansel, 110–132, which, though recent, remains a classic work; Patrick Balfour Kinross, *The Ottoman Centuries: The Rise and Fall of the Turkish Empire* (Morrow, 1977), 139–228, is further very authoritative; Montefiore, 215–221; and Jason Goodwin's fascinating *Lords of the Horizons: A History of the Ottoman Empire* (Picador, 1998), 185–288. See also Tott; Lavender Cassels, *The Struggle for the Ottoman Empire, 1717–1740* (John Murray, 1966); Steven Runciman, *The Fall of Constantinople, 1453* (Cambridge University Press, 1963); Stanley Lane-Poole, *The Story of Turkey* (J. M. Dent & Sons, 1895); and Bernard Lewis, *Istanbul and the Civilization of the Ottoman Empire* (University of Oklahoma Press, 1963), 51–172.

On the harem: The fullest accounts are N. M. Penzer, *The Harem: An Account of the Institution as It Existed in the Palace of the Turkish Sultans* (Harrap, 1936), and Zeynep M. Durukan, *The Harem of the Topkapi Palace* (Hilal Matbaacilik Koll, 1973). I have also drawn extensively upon Massie, Montefiore, Mansel, Kinross, Finkle, and Goodwin.

On grand viziers: See Lewis, 87–94; Mansel, 18–19, 133–162; Kinross, 254–260, 331–336, 350–352; Massie, 536; Montefiore, 215–218; and Goodwin, 221–244.

On early history of the empire: See Finkle, 1–288; Mansel, 1–162; Kinross, 23–255; Montefiore, 215–218; Massie, 529–538; and Goodwin, 12–205. Also see Hamilton Gibb and Harold Bowen, *Islamic Society and the West* (Oxford University Press, 1950); and Lewis. **On Sultan Selim III, the aggressive reformer:** See Stanford J. Shaw, *Between the Old and New: The Ottoman Empire under Selim III, 1789–1807* (Harvard University Press, 1971).

For background on Prince Potemkin: By far the best work is the masterful biographical study by Montefiore, from which I have extensively drawn for the biography of Potemkin here and throughout. He richly captures the period in Russia as well as anyone. Also excellent, see George Soloveytchik, *Potemkin: Soldier, Statesman, Lover, and Consort of Catherine of Russia* (Norton, 1947), as well as Masson, 90–94; Madariaga, 359–377; Alexander, 203–206; Troyat, 218–292, 271–318; and Cronin, 210–220, 248–266. Absolutely critical as well are the remarkable letters in Douglas Smith, ed., *Love and Conquest: Personal Correspondence of Catherine the Great and Prince Grigory Potemkin* (Northern Illinois, 2004). Also M. Raeff, "The Style of Russia's Imperial Policy and Prince G. A. Potemkin," in *Statesmen and Statecraft of the Modern West: Essays in Honor of Dwight E. Lee and H. Donaldson Jordan*, G. N. Grob, ed. (Barre, 1967). On his youth, see Montefiore, 13–62; Soloveytchik, 39–52. **On Catherine and Potemkin as lovers:** See especially Montefiore, 109–166; Soloveytchik, 78–109; Smith; Troyat, 218–292; and Cronin, 210–221. On Potemkin's day (and diet), see Montefiore, 328–347, and Masson, 108–113. On Potemkin's building and diplomacy, see Montefiore, 203, 287–301; Soloveytchik, 176–217; and Troyat, 276–288. Also consult Jones, "Urban Planning and the Development of Provincial Towns in Russia, 1762–96," in *The Eighteenth Century in Russia*. On Potemkin's final days, see Montefiore, 480–493 and 502 for "his remains"; Soloveytchik, 320–340; Troyat, 313–315. On Potemkin villages, Mansel was more skeptical of the prince's achievements, calling them "cages without birds," 71; but Montefiore has built a strong case to the contrary. For more information on the intricate court politics between Panin and Potemkin, see David L. Ransel, *The Politics of Catherinian Russia: The Panin Party* (Yale University Press, 1975).

American-Russian relations: See esp. the most detailed study in Saul, *Distant Friends*; also Howard I. Kushner, *American-Russian Rivalry in the Pacific Northwest, 1790–1867* (Cornell University thesis, 1970); Max M. Laserson, *The American Impact on Russia, Diplomatic and Ideological, 1784–1917* (Macmillan, 1950); and Phillips Russell, *John Paul Jones: A Man of Action* (Blue Ribbon, 1930).

Jones: For the two best biographies of Jones, which I have extensively used here, see Evan Thomas's superb book, *John Paul Jones: Sailor, Hero, Father of the American Navy* (Simon & Schuster, 2003); and Samuel Eliot Morison, *John Paul Jones: A Sailor's Biography* (Little, Brown, 1959); also see James A. Mackay, *I Have Not Yet Begun to Fight: A Life of John Paul Jones* (Mainstream, 1998); George Preedy, *The Life of Rear-Admiral John Paul Jones* (H. Jenkins, 1940). **For Jones in Russia:** See F. A. Golder, *John Paul Jones in Russia* (Doubleday, Page, 1927); and esp. Thomas, 267–299; Morison, 360–390; James Otis, *The Life of John Paul Jones, Together with Chevalier Jones' Own Account of the Campaign of the Liman* (Burt, 1900); also Saul and McKinzie, eds., *Russian-American Dialogue*. It will forever remain a mystery why Thomas Jefferson brokered the meeting to send Jones to Russia, but the best explanation remains the kinship developed during the Revolutionary War as a result of Catherine's League of Armed Neutrality, and the fact that these two countries were seen as flanking a corrupt old world.

For the battle of Ochakov: See esp. K. Osipov, *Alexander Suvorov* (Hutchinson, 1941), 65–78; and Montefiore, 403–414, both of whom present the most comprehensive overview. Of note, Osipov and Montefiore provide differing interpretations of Potemkin's leadership. Also outstanding is Longworth, *The Art of Victory*, 145–150. On Potemkin's many responsibilities during this battle, see Soloveytchik, 194, 298–299; for more on how war was fought during this period, see Michael Howard, *War in European History* (Oxford University Press, 1976). For further on the second Russo-Turkish War, see Madariaga, 393–413; Mansel, 409; Christopher Duffy, *Russia's Military Way to the West: Origins and Nature of Russian Military Power, 1700–1800* (Routledge & Kegan Paul, 1981). Masson notes that Voltaire was repeatedly cheering on Catherine's effort to purge the Turks from Europe, 60–61.

For background on Russian-Swedish relations: See M. Roberts, *British Diplomacy and Swedish Politics, 1758–1773* (University of Minneapolis Press, 1980); Cynthia Hyla Whittaker, ed., *Russia Engages the World, 1453–1825* (Harvard University Press, 2003); and R. Nisbet Bain, *Gustavus III*, 2 vols. (K. Paul, Trench, Trübner, 1894). Also see Alexander, 267–72, and *CTG*, 162–175.

CHAPTER 6

For background: See Rose; Sutherland; Gouverneur Morris, *The Diary and Letters of Gouverneur Morris*, 2 vols., Anne Cary Morris, ed. (Da Capo, 1970). Morris was a keen observer of France, and his letters remain indispensable reading; Richard Brookhiser, *Gentleman Revolutionary: Gouverneur Morris, the Rake Who Wrote the Constitution* (Free Press, 2003); Carlyle; Dawson, *The French Revolution*; Doyle, *Origins of the French Revolution*; Doyle; Lawrence S. Kaplan, *Jefferson and France: An Essay on Politics and Political Ideas* (Yale University Press, 1967); Ségur; and Laura Mason and Tracy Rizzo, eds., *The French Revolution: A Documents Collection* (Houghton Mifflin, 1999).

Gouverneur Morris: For Morris's trenchant and penetrating insights, see G. Morris, *Letters*; Jared Sparks, *The Life of Gouverneur Morris, with Selections from His Correspondence and Miscellaneous Papers, Detailing Events in the American Revolution, the French Revolution, and in the Political History of the United States*, 3 vols. (Gray & Bowen, 1832); Daniel Walther, *Gouverneur Morris, Witness of Two Revolutions*, trans. Elinore Denniston (Funk & Wagnalls, 1934); Henry Bertram Hill, "Gouverneur Morris on Robespierre," *Journal of Modern History* 9, no. 2 (1937): 203–205, and Brookhiser, who has written the best recent biography.

Great Fear: Schama, 429–436; Doyle, 113–116; Jones, 418–419; Hibbert, 91–92; see Lefebvre's exceptional work for full overview. On destruction of châteaux in Burgundy, see Joachim Durandeau, *Les Châteaux Brulés* (Sirodot-Carré, 1895); on destruction of Abbaye of Murdoch, see its website. **Night of August 4:** See overviews in Schama, 436–440; Doyle, 115–118; Jones, 420–421; Thompson's epic work, 82.

Rights of Man: Schama, 442–444; Jones, 420–421; Doyle, 118. **Jefferson's role:** Malone, vol. 2. **Jefferson Paris dinner:** This dinner, too often ignored, was not unlike the momentous dinner with Hamilton and Madison a year later in 1790. See Malone, vol. 2, 230–231, and Ellis, *FB*, 48–80; William Howard Adams, *The Paris Years of Thomas Jefferson* (Yale University Press, 1997); and *Thomas Jefferson: Thomas Abroad*, Douglas L. Wilson and Lucia Stanton, eds. (Modern Library, 1999).

Various pamphlets, journals, press: For this vital subject, see Jones, 437–442; Schama, 497–500; Doyle 120–121; Thompson, 106; Carlyle, vol. 1, 328–331; On the beginnings of radical pamphlet literature, see esp. the excellent and underused study by Mitchell B. Garrett, *The Estates-General of 1789* (Appleton-Century, 1935); Robert Darnton's *The Literary Underground of the Old Regime* (1982) is an outstanding study by one of our leading French historians; Boyd C. Shafer, "Bourgeois Nationalism in Pamphlets on the Eve of the French Revolution," *Journal of Modern History* (1938): 31–50; Jack Censer, *Prelude to Power: The Parisian Radical Press 1789–1791* (Johns Hopkins University Press, 1996); Simon Burrows, *French Exile Journalism and European Politics, 1792–1814* (Royal Historical Society, 2000); Hugh Gough, *The Newspaper Press in the French Revolution* (Dorsey, 1988); Mona Ozouf, "War and

Terror in French Revolutionary Discourse (1792–1794)," *Journal of Modern History* 56, no. 4 (1984): 579–597. On patriotic oratory that became so prevalent see H. Morse Stephens, *The Principal Speeches of the Statesmen and Orators of the French Revolution 1789–1795*, 2 vols. (Clarendon, 1892), esp. his introduction. For more recent studies, see Lynn Hunt's thoughtful and scholarly "The Rhetoric of Revolution" in *Politics, Culture and Class*, 19–51. Mason and Rizzo also include a number of valuable and interesting documents. For a look beyond rhetoric, see Sophia Rosenfeld's fascinating and illuminating scholarly study, *A Revolution in Language: The Problem of Signs in Late 18th Century France* (Stanford University Press, 2004).

Marat: Schama, 731–734; Hibbert, 140–142; Carlyle; and the important Louis Gottschalk, *Marat* (Haldeman–Julius, 1927).

October I Garde du Corps and Flanders regiment: Fraser, 292; Hibbert 95–96; Thompson, 92; Doyle, 121. **Attack on Versailles:** Fraser, 293–298; Lever, 222–232; Thompson, 94–96; Jones, 421–422; Schama, 456–470; Hibbert, 97–103; Doyle, 212–213; Rudé, 61–78; Carlyle, vol. I, 270–300; and Zweig, 253–265. For the specific role of women in the march and attack see Olwen Hufton, "Women and Revolution," in Douglas Johnson, ed., *French Society and the Revolution* (Cambridge University Press, 1976), 148–166. It should be noted for the reader that there are a variety of specific accounts and interpretations of the chaotic day culminating in the march on Versailles, as well as the fast-moving events resulting in the assault on the château itself. For example, some accounts describe the meeting between the flower-girl and the King as hostile. I have leaned in the direction of a more conciliatory meeting. **King back in Paris:** Schama, 468–475; Fraser, 301–306; Lever, 233–238; Thompson, 96–97; Hibbert, 104–105; Carlyle, vol. I, 303–307.

Constitutional Assembly: Doyle, 122–133; Schama, 472–499; Thompson, 101; Jones; 422–437. **Measures against the church:** Schama, 485–487, 489–491; Doyle, 132–144; Thompson, 140–154, 161, 178; Jones, 442–448; Fraser, 317–319; Hibbert 109–112, 115–118; and Louis S. Greenbaum, *Talleyrand, Statesman-Priest: The Agent General of the Clergy and the Church at the End of the Old Regime* (Catholic University Press, 1970). For more see J. McManners, *The French Revolution and the Church* (London, 1969); Timothy Tackett's excellent study, *Religion, Revolution, and Regional Culture in Eighteenth-Century France: The Ecclesiastical Oath of 1791* (Princeton University Press, 1986), as well as Albert Mathiez's overlooked *La Révolution et l'Eglise* (Armand Colin, 1910). The "reformist" clerical ideology—*L'Ecclésiastique Citoyen*—actually began before 1787; see Ruth Necheles, *The Abbé Grégoire 1787–1831: The Odyssey of an Egalitarian* (Greenwood, 1971).

Fête de la Fédération: Schama, 500–513; Fraser, 314–315; Lever, 240–241; Hibbert, 112–115; Thompson, 120; on the cult of the *bonnet rouge*, see Jennifer Harris, "The Red Cap of Liberty: A Study of Dress Worn by French Revolutionary

Partisans 1789–1794," *Eighteenth-Century Studies* (1981): 283–312. Of note, revolutionary France turned elaborate festivals into an art form. See esp. D. L. Dowd, *Pageant-Master of the Republic: Jacques Louis-David and the French Revolution* (University of Nebraska Press, 1948).

Mirabeau: Schama, 339–345, 532–543; Thompson, 28, 190–198; Fraser, 313; Lever, 238–240; Carlyle, vol. I, 441–445 and 446–454; Zweig, 277–288; Valentin, *Mirabeau;* Guy Chaussinand-Nogaret, *Mirabeau* (Paris, 1982); and Charles F. Warwick, *Mirabeau and the French Revolution* (J. Lippincott, 1905). Mirabeau's byzantine plans for saving the monarchy are detailed in Guy Chaussinand-Nogaret, ed., *Mirabeau entre le Roi et La Révolution* (Hachette, 1986).

King's routine at Tuileries: I have found particularly valuable Fraser, 302–318; Lever, 234–239; Hardman; Zweig, 266–276; Thompson, 195.

Flight to Varennes: It should be noted for the reader that there are some considerable differences among historians about the planning and actual details of the king's dash to freedom. See Schama, 554–558; Fraser, 313, 320–350; Lever, 244–264; Zweig, 288–322; Thompson, 198–208; Doyle, 150–154; Hibbert, 118–130; Jones, 445; I also relied very heavily upon Munro Price's excellent *The Road from Versailles: Louis XVI, Marie Antoinette, and the Fall of the French Monarchy* (St. Martin's, 2002), as well as Timothy Tackett, *When King Took Flight* (Harvard University Press, 2003). **Champs de Mars massacre:** Schama, 566–559; Thompson, 220; Jones 447–448; Rudé, 80–94, which remains the classic on the French "crowds"; Fraser, 350; Lever, 266; Hibbert, 134–135; G. A. Kelly, "Bailly and the Champ de Mars Massacre," in *Journal of Modern History* (1980). **Achievements of Assembly:** Doyle, 154–158; Thompson, 224–225; Hibbert, 135–139; and R. R. Palmer, *The Improvement of Humanity: Education and the French Revolution* (Princeton University Press, 1985).

Catherine and French Revolution: See the fascinating article, William L. Blackwell, "Citizen Genet and the Revolution in Russia," *French Historical Studies Journal,* 72–92, which also lays to rest the notion of Genet as simply rash and impetuous. Also see Ségur.

CHAPTER 7

For background: See Keith Baker, *Condorcet: From Natural Philosophy to Social Mathematics* (University of Chicago Press, 1975); George Rudé, *Robespierre: Portrait of a Revolutionary Democrat* (Collins, 1975); David Jordan, *The Revolutionary Career of Maximilien Robespierre* (University of Chicago Press, 1989); Rose; Sutherland; A. Owen Aldridge, *Thomas Paine's American Ideology* (University of Delaware Press, 1984); Yvon Bizardel, *The First Expatriates: Americans in Paris During the French Revolution,* trans. June Wilson and Cornelia Higginson (Holt, Rinehart and Winston, 1975); Gordon; Simon Burrows, *French Exile Journalism and European Politics, 1792–1814* (Royal Historical Society, 2000); and Georges Jacques Danton, *Oeuvres de Danton* (F. Cournol); Dawson; Ségur.

Paine and Burke: On the extraordinary clash between these two stentorian dueling voices, see Edmund Burke, *Reflections on the French Revolution & Other Essays* (Dutton, 1910); Thomas Paine, *Rights of Man* (Dent, 1951); John Keane's excellent *Tom Paine: A Political Life* (Little, Brown, 1995), which I have greatly benefited from reading; Foner, *Tom Paine and Revolutionary America*; P&C; O'Brien, *Debate Aborted 1789–91: Priestley, Paine, Burke and the Revolution in France* (Pentland, 1996); John Morley, *Burke: A Historical Study* (Knopf, 1924). **World of Revolution:** P&C; R. R. Palmer, *The World of the French Revolution* (Harper Torchbooks, 1971); Palmer, *Age*, vol. 2, esp. 3–99. **Intellectuals:** Gordon and Mary Wollstonecraft, *A Vindication of the Rights of Woman: An Authoritative Text; Backgrounds; The Wollstonecraft Debate; Criticism*, 2nd ed. (Norton, 1988).

War against Europe: P&C; Doyle, 159–173; Jones, 449–459; Thompson, 245–262, 265, 266; Geoffrey Best, *War and Revolutionary Europe 1770–1870* (London, 1982); Paddy Griffith, *The Art of War of Revolutionary France, 1789–1802* (Greenhill, 1998), is quite authoritative.

Brissot: Thompson, 231–233, 255, 257–258; Schama, 582–600; Hibbert, 136–137. For Brissot's early career, see Robert Darnton, "A Spy in Grub Street," *Literary Underground*, 41–70; also Eloise Ellery, *Brissot de Warville: A Study in the History of the French Revolution* (Houghton Mifflin, 1915). **Girondins:** Schama, 584–606; Doyle, 174; Jones, 455–456, 472–476; Thompson, 230, 319, 330–331; Hibbert, 135–140, 144. M. J. Sydenham, *The Girondins* (London, 1961) suggests the Girondins were not a cohesive party in the modern sense. But this could be said for the early American Republicans and Federalists as well. On the ever-fascinating Vergniaud, see Claude Bowers, *Pierre Vergniaud: Voice of the French Revolution* (Macmillan, 1950).

Jacobins and clubs: Doyle, 149–150; Jones, 429–431; Carlyle, vol. 1, 332–336; vol. 2, 52–55; Thompson, 108–112 (Johns Hopkins University Press, 1976); 228, 235; Hibbert, 142, 146–147; Jack Censer, *Prelude to Power: The Parisian Radical Press 1789–1791*; For further detail, see two extremely important works, Michael L. Kennedy, *The Jacobin Clubs in the French Revolution: The First Years* (Princeton University Press, 1982), and Crane Britton, *The Jacobins* (New York, 1930). On the popular societies of Paris, see Rose, *The Making of the Sans-culottes*, ch. 6. **Elections for Assembly:** Doyle, 174–176; Thompson, 227; Gottschalk, *Marat*.

Gouverneur Morris quotes: Morris, vol. 2, esp. 1–107. Presciently, Morris wrote, "the death of the King is but a forerunner to their own destructions." Also, Jared Sparks, *The Life of Gouverneur Morris*, 3 vols. (Gray & Bowen, 1832); Daniel Walther, *Gouverneur Morris, Witness of Two Revolutions*, trans. Elinore Denniston (Funk &

Wagnalls, 1934); and Henry Bertram Hill, "Gouverneur Morris on Robespierre," *Journal of Modern History* 9, no. 2 (1937): 203–205.

Foreign war: Doyle, 177–186; P&C, 378–384; Jones, 454–459; Thompson, 245, 254–270; Hibbert, 143–145; Palmer, *World*; T. C. W. Blanning, *The Origins of the French Revolutionary Wars* (London, 1986); Howard, *War in European History*; Griffith, *The Art of War*. **Declaration of Pilnitz:** Thompson, 249–250. **General Dumouriez:** Thompson, 259–262; 266. **Marseillaise:** Thompson, 279.

June 20 attack on Tuileries: Fraser, 366–370; and Lever, 274–276, are superb, as are Schama; Thompson, 272; Rudé, 97–100; Zweig, 346–349. **July 26 Duke of Brunswick declaration:** Thompson, 284; Jones, 459; Hibbert, 153. **Revolutionary commune:** Rudé, 178–209; Schama, 289–299; Doyle 187–191; Hibbert, 154.

Actions of king and queen leading up to assault: Fraser, 370–373; Lever, 277–278; Thompson, 276–277; Jones, 459–460; see also Zweig; Carlyle; Madame Jeanne-Louise-Henriette Campan (beloved lady-in-waiting to the queen, she was Genet's sister), *Memoirs of the Court of Marie Antoinette* (Groiler Society, n.d.) is particularly helpful. **August 9 attack:** Lever, 279–282; Fraser, 373–381; Doyle, 189; Jones, 460–462; Hibbert 154–161; Schama, 611–626; Thompson, 289; Rudé, 101–108; see also Zweig, 356–366, and Campan. **Swiss Guards history:** Thompson, 289; Fraser, 376.

Danton: I have relied especially upon the outstanding Norman Hampson, *Danton* (Duckworth, 1978), which also masterfully conveys the tangled politics of the revolution; also see Charles F. Warwick, *Danton and the French Revolution* (Unwin, 1909); Herman Wendel, *Danton* (Constable, 1936); Hibbert, 165–168; and Frédéric Bluche, *Danton* (Libr. Académique Perrin, 1968).

August 28 war on church: Schama, 626; Thompson, 298–303. **September massacres:** Schama, 627–639; Doyle, 191–193, Rudé, 108–127; Hibbert, 169–179; Thompson, 303–309; Carlyle, vol. 2, 144–160; Fraser, 386–390; Lever, 283–284; J. McManners, *The French Revolution and the Church* (London, 1969). **Princess de Lamballe:** Lever, 284; Fraser, 386–390; Schama, 635; Hibbert, 175–176. Also see Carlyle, Hilaire Belloc, *Marie Antoinette* (Doubleday, Page, 1909), and Albert-Emile Sorel, *La Princesse de Lamballe: Une amie de la Reine Marie Antoinette* (Hachette, 1933).

Temple: For the poignant final days in the Temple, the two most important sources are J. B. Cléry, *A Journal of the Terror*, Sidney Scott, ed. (Folio Society, 1955); for the rich original French version, see *Journal de ce qui s'est passé à la Tour du Temple pendant la captivité de Louis XVI par M. Clery, valet du chambre du roi et autres mémoires sur le Temple*, ed. Jacques Brosse (1987). Also vital is Madame Royale, *The Ruin of a Princess*, trans. Katherine Wormely (Lamb, 1912). Both of these memoirs are remarkable for their descriptive power, their surprising balance, and their ability to recapture these days and the intense emotions at play. I've also drawn significantly upon Fraser, 382–386,

390–395; Lever, 283–283, 285–286; Schama, 653–655; Hibbert, 181–184; and Zweig, 367–378.

Third Assembly election: Thompson, 309–313; Hibbert, 180–181; Michael J. Sydenham, "The Montagnards and Their Opponents: Some Considerations on a Recent Reassessment of the Conflicts in the French National Convention, 1792–93," *Journal of Modern History* 43, no. 2 (1971): 287–293. **Republic established, battle of Valmy:** P&C, 384–385; Schama, 639–641; Jones, 462–466; Thompson, 309–315; Hibbert, 179–180. **Battlefield victories:** Fall of '92, Doyle, 192–193; Jones, 467–471; Schama, 639–644; Hibbert, 193–194; Carlyle; Thompson, 316–318; Palmer, *Age.* For the most detailed recounting, see Griffith's excellent *The Art of War;* also Howard, *War in European History.* **Edict of Fraternity:** Hibbert, 193–194. **Revolutionary calendar:** Carlyle, 312–313.

Malesherbes: For a fascinating discussion on Malesherbes, see Schama, 655–666. For the destruction of Malesherbes's family, see R. R. Palmer, ed., *The Two Tocquevilles, Father and Son: Hervé and Alexis de Tocqueville on the Coming of the French Revolution* (Princeton University Press, 1987). Also Jean-Jacques Rousseau, *The Confessions and Correspondence, Including the Letters to Malesherbes,* Christopher Kelly, Roger D. Masters, and Peter G. Stillman, eds., trans. Christopher Kelly (University Press of New England for Dartmouth College, 1995).

Trial and execution of the king: Schama, 646–675; Doyle, 193–196; Hibbert, 183–189; Jones, 466–467; Carlyle, vol. 2, 225–237. Esp. for the heartrending scene of Louis XVI on his last night with his family, Fraser, 395–399, Lever, 286–288, Hardman, 223–233, and Carlyle are particularly good, as is Madame Royale; also Zweig, 378–380. The best source on the trial remains David Jordan, *The King's Trial: The French Revolution vs. Louis XVI* (University of California Press, 2004), while Michael Walzer, *Regicide and Revolution* (Cambridge University Press, 1974) is extremely useful for documentation. Alison Patrick, *The Men of the First French Republic* (Johns Hopkins University Press, 1972) delves further into the politics of the trial.

Abbé Edgeworth: For the recollections of Abbé Edgeworth, see Schama, 668–669; Gooch, *Maria Theresa;* and also see primary documents on the Internet.

CHAPTER 8

For background: See R. Nisbet Bain, *The Last King of Poland and His Contemporaries* (Arno, 1971); Václav L. Benes and Norman J. G. Pounds, *Poland* (Praeger, 1970); Billington, *Icon;* Jerome Blum, *Lord and Peasant in Russia, from the Ninth to the Nineteenth Century* (Atheneum, 1964); *The Cambridge History of Poland,* 2 vols., W. F. Reddaway, ed. (Octagon, 1971); Madariaga, *Politics and Culture;* Gooch, *Catherine the Great;*

Griffiths, *Russian Court Politics . . . Under Catherine II*; Kaus, *Catherine the Great*; Lentin, *Russia in the Eighteenth Century: From Peter the Great to Catherine the Great (1696–1796)* (Barnes & Noble, 1973); Lincoln, *Sunlight at Midnight*; Masson, *Memoirs of the Court of St. Petersburg*; Oldenbourg, *Catherine the Great*; Ségur.

On Catherine's schedule: See esp. Cronin, 196–209; and Troyat, 140–151, 172–179; also Catherine the Great, *Memoirs* (Hamish Hamilton, 1955); Montefiore, 66–68.

Siege of Ismail: I have most relied upon Montefiore, 448–452; Soloveytchik, 85–96; and Longworth, 165–174; as well as Alexander, 257–292; and Madariaga, 413–426; also Mansel, 410–413. The true death toll of Ismail may never be known. Potemkin suggests 26,000 Turks died, where the Comte de Langeron put Russian losses higher than the official 1,815 dead, at 8,000. I am going with a higher estimate that brings the total casualties up to 45,000. What is not in doubt is the savagery of the fighting, which considerably dwarfed Western European battles at the time, not to mention the American Revolutionary War.

Background on the Ottoman Empire: The three best studies are Kinross and Mansel, as well as the most recent study by Finkle, all of which are rich, superb works. Also quite good are Goodwin and Lewis. See further Cassels; Gibb and Bowen; Durukan; S. MacCoby, *The Development of Muslim Theology, Jurisprudence and Constitutional Theory* (Scribner's, 1903).

Tensions in the Ottoman provinces: See Finkle, 358–361; Mansel, 33–34; Kinross, 331–339 and 401–403; also Lewis and Goodwin; on Egypt, see Edward W. Lane, *Manners and Customs of the Modern Egyptians* (M. A. Nattali, 1846). **On Wahhabism:** See esp. Hamid Algar, *Wahhabism, A Critical Essay* (Islamic Publications, 2002).

For fascinating personal accounts: See Charles M. Doughty, *Travels in Arabia Deserta*, 2 vols. (Boni & Leveright, 1923); Robert Halsband, *The Life of Lady Mary Wortley Montagu* (Oxford University Press, 1957); Lady Mary Wortley Montague, *Letters and Works* (G. Bell & Son, 1893); Tott; also Isobel Grundy, *Lady Mary Wortley Montagu: Comet of the Enlightenment* (Oxford University Press, 1999).

For the French Revolution and Catherine: I have drawn especially on superb treatments in Troyat, 183–193, 317–322; *CTG*, 189–202; Cronin, 269–270; as well as Catherine's and Prince de Ligne's memoirs; Catherine's in particular provides context for the degree to which she turned on the French writers whom she once so ardently admired. For more detailed information on the massive influx of French émigrés, see Leonide Ignatieff, *French Émigrés in Russia, 1789–1825: The Interaction of Cultures in Times of Stress* (unpublished doctoral diss., 1963). On Nicholas I, see Raeff, *Understanding Imperial Russia*, 147–172.

For background on Nikolai Ivanovich Novikov: See W. Gareth Jones, *Nikolay Novikov, Enlightener of Russia* (Cambridge University Press, 1984); also esp. *CTG*, 92–94, 115–116, 198–200; Troyat, 33–34, 350, as well as Cronin, 225–226, 285–288; and Madariaga, 332–334, 524–538. See also M. M. Shcherbatov, *On the Corruption* *of Morals in Russia*, A. Lentin, trans. (Cambridge University Press, 1969). New light has been thrown on Russian eighteenth-century intellectual life by G. Marker in *Publishing, Printing and the Origins of Intellectual Life in Russia* (Princeton University Press, 1984); also Raeff, *Imperial*, esp. 131–158.

Princess Ekaterina Dashkova: See E. R. Dashkova, *The Memoirs of Princess Dashkova*, Kyril Fitzlyon, trans. (Duke University Press, 1995). Dashkova played a significant role in Catherine's coup in 1762. The American Philosophical Society in 2006 had a fascinating exhibit on Dashkova's Enlightenment underpinnings as well as her relationship with Benjamin Franklin on the three hundredth anniversary of Franklin's birth, which highlighted more intimate connections between Americans and Russians than is commonly believed today.

Alexander Radishchev: See David Marshall Lang, *The First Russian Radical: Alexander Radishchev, 1749–1802* (Allen & Unwin, 1959); Aleksandr Nikolaevich Radishchev, *A Journey from St. Petersburg to Moscow*, Leo Weiner, trans. (Harvard University Press, 1958); and Allen MacConnell, *A Russian Philosophe: Alexander Radishchev, 1749–1802* (Hyperion, 1981). Also see Troyat, 334, 350; Cronin, 282–285, 287, 309–310; and Madariaga, 541–547; as well as CTG, 191–196, 215.

For the critical Pugachev Rebellion: I've drawn extensively upon Montefiore, 123–135; Longworth, 99–103; Alexander, 169–170, 175–180; Madariaga, 241–255, 263–269; as well as Longworth's fine chapter "The Pugachev Revolt," in *Rural Protest*, H. A. Landsberger, ed. (Macmillan, 1974), 194–256; and Paul Dukes, ed., *Russia Under Catherine the Great* (Oriental Research Partners, 1977), esp. 111–136. For the most detailed treatments, see Alexander, *Emperor of the Cossacks: Pugachev and the Frontier Jacquerie of 1773–1775* (Coronado, 1973); Philip Longworth, *The Cossacks* (Constable, 1969); and Alexander, *Autocratic Politics in a National Crisis: The Imperial Government and Pugachev's Revolt* (Indiana University Press, 1969). On the court crisis as a result of Pugachev's successes, see Madariaga, 239–272. For more on army life and the Cossacks' concerns with conscription, also see the relevant chapters of J. L. H. Keep, *Soldiers of the Tsar* (Clarendon, 1985). On impersonators, see M. B. Petrovich, "Catherine II and a Fake Peter III in Montenegro," *Slavic Review* 14, no. 2 (April 1955); for more on social classes in Russia, see Raeff, *Imperial Russia*, esp. 103–122. For more on serfdom, for example, see Paul Dukes, *Catherine the Great and the Russian Nobility: A Study Based on the Materials of the Legislative Commission of 1767* (Cambridge University Press, 1967), 86–144.

For background on King Stanislas Augustus Poniatowski: See esp. the two works by Lord and Adam Zamoyski, *The Last King of Poland*; Masson, *Secret Memoirs of the Court of Petersburg*; see also Catherine II, *Lettres de Catherine II a Stanislas-Auguste Poniatowski, roi de Pologne (1762–1764)* (Paris, 1914).

On Thaddeus Kosciuszko and the rape of Poland: See Anthony Walton White Evans, *Memoir of Thaddeus Kosciuszko, Poland's Hero and Patriot, and Officer in the American Army* *of the Revolution and Member of the Society of the Cincinnati* (Thitchener, 1883); Julian Ursyn Niemcewicz, *Under Their Vine and Fig Tree: Travels Through America in 1797–1799, 1805, with Some Further Account of Life in New Jersey*, Metchie J. E. Budka, ed. and trans. (Grassman, 1965); James S. Pula, *Thaddeus Kosciuszko: The Purest Son of Liberty* (Hippocrene, 1999); Francis C. Kajencki, *Thaddeus Kosciuszko: Military Engineer of the American Revolution* (Southwest Polonia Press, 1998); and Szymon Askenazy, *Thaddeus Kosciuszko* ("The Polish Review" Offices, 1917). For the rape of Poland, I have drawn especially upon Longworth's excellent *Art of Victory*; Madariaga, 409–426; Alexander, 285–292; and Osipov, 97–109; see also Lord and Zamoyski, *The Last King of Poland*, esp. 326–357, 512–528. The Lord and Zamoyski books are the two best for matters concerning Poland. Also quite helpful, Jerzy Lukowski, *The Partitions of Poland: 1772, 1793, 1795* (Longman, 1999); and J. Lojek, "Catherine's Armed Intervention in Poland: Origins of the Political Decisions at the Russian Court in 1791 and 1792," *Canadian Slavic Studies* 4, no. 3 (Fall 1970), and Jerzy Jedlicki, "The Image of America in Poland, 1776–1945," *Reviews in American History* 14, no. 4 (1986): 669–686. Also see O. Halecki, *A History of Poland*, Monica M. Gardner and Mary Corbridge-Patkaniowska, trans. (Coordinating Committee of Poland's Christianity Millenium, 1966); S. Konovalov, ed., *Russo-Polish Relations: An Historical Survey* (Princeton University Press, 1945); K. Waliszewski, *Poland the Unknown* (G. Doran, 1919); and James Breck Perkins, "The Partition of Poland," *American Historical Review* 2, no. 1 (1896): 76–92.

For U.S. dispatches on war against Poland: See *Philadelphia Aurora*, 1774–1795.

CHAPTER 9

For background: See Best, *War and Revolutionary Europe*; Soboul, *Sans-culottes*; Sutherland, *France 1789–1815*; Patrice Higonnet, *Goodness Beyond Virtue: Jacobins During the French Revolution* (Belknap, 1998); Arno Mayer, *The Furies: Violence and Terror in the French and Russian Revolutions* (Princeton University Press, 2000); Robert B. Asprey, *The Rise of Napoleon Bonaparte* (Basic, 2000); Marilyn Morris, *The British Monarchy and the French Revolution* (Yale University Press, 1998); Bizardel, *The First Expatriates*; Simon Burrows, *French Exile Journalism and European Politics, 1792–1814* (Royal Historical Society, 2000); Georges Jacques Danton, *Oeuvres de Danton* (F. Cournol); Dawson, *The French Revolution*; Maurice de La Fuye, *The Apostle of Liberty: A Life of La Fayette*, trans. Emile Babeau (Thomas Yoseloff, 1956); Richard Cobb, *Reactions to the French Revolution* (Oxford University Press, 1972); Anne C. Loveland, *Emblem of Liberty: The Image of Lafayette in the American Mind* (Louisiana State University Press, 1971); Jean Orieux, *Talleyrand: The Art of Survival*, trans. Patricia Wolf (Knopf, 1974); and William Hague, *William Pitt the Younger* (Knopf, 2004).

1793 war in Europe, First Coalition: Doyle, 197–207; Thompson, 335–342; Schama, 685–689; Griffith, *The Art of War*, and Blanning, *Origins*, are quite good on the buildup to war. For British politics during this period, see esp. Hague, 59–370. **Defection of Dumouriez:** Griffith, *The Art of War;* Doyle, 227; Thompson, 343; Jones, 468; Hibbert, 194–195.

Creation of Committee of Public Safety: R. R. Palmer, *Twelve Who Ruled: The Committee of Public Safety, During the Terror* (Oxford University Press, 1941) remains the seminal work. Of note, Schama, equally seminal, presents a very different, more critical point of view. Thompson, 349–352, is superb as well; Carlyle, 128, 146, 267; Hibbert, 194–195. **Marat trial:** Gottschalk, *Marat,* 51–54; Doyle, 228–229; Schama, 716–719; Jones, 473–447; and Hibbert, 196–198. **Girondins on the run:** Thompson, 352–356; Carlyle, vol. 2, 221–233; Schama, 714–727; Jones, 472–475; and Hibbert, 199–201. Morris Slavin, *The Making of an Insurrection: Parisian Sections and the Gironde* (Harvard University Press, 1986), is an especially important study.

New constitution: Palmer, *Twelve Who Ruled;* Thompson, 357–363; Hibbert, 202; Carlyle.

Charlotte Corday/assassination of Marat: Schama, 730–748; Carlyle, vol. 2, 291–300; Thompson, 367; Hibbert, 212–214; Gottschalk, *Marat,* 55–57; Jones, 479.

Summer of '93; who were the Twelve?: Palmer, *Twelve;* Schama, 750–754; Thompson, 383–389; Jones, 471; Hibbert, 215–216; Mason and Rizzo, 190–193. For Danton's critical early role, see Hampson, *Danton,* 117–136.

Robespierre: Assessments of Robespierre vary widely, and I've drawn upon a variety of sources. For sketches, see Schama; Thompson, 389; Hibbert, 203–212. For more, see Norman Hampson, *The Life and Opinions of Maximilien Robespierre* (Duckworth, 1974); George Rudé, *Robespierre: Portrait of a Revolutionary Democrat* (Collins, 1975), presents a sympathetic portrait, where by contrast Schama and Hibbert are more critical; David Jordan, *The Revolutionary Career of Maximilien Robespierre* (Collier Macmillan 1985); Palmer's classic, *Twelve,* is also quite enlightening; as are Hilaire Belloc, *Robespierre: A Study* (Putnam's, 1928); John Laurence Carr, *Robespierre: The Force of Circumstance* (St. Martin's, 1973); John Hardman's *Robespierre* (Longman, 1999) is important, as is *Robespierre,* Colin Hayden and William Doyle, eds. (Cambridge University Press, 1999); George Rudé, ed., *Robespierre* (Prentice Hall, 1967); Friedrich Sieburg, *Robespierre the Incorruptible* (McBride, 1938); J. M. Thompson, *Robespierre,* 2 vols. (Appleton-Century, 1936); and Charles F. Warwick, *Robespierre and the French Revolution* (Jacobs, 1909).

July 23–August 29: Thompson, 390–399; Hibbert, 214–217.

War turmoil in the provinces: Schama, 690–714; 779–792; Thompson, 401–413; 417–431; Doyle, 227–243; Rudé, 128–138; Jones, 469–471; Carlyle, vol. 2, 300–304; and Charles Tilly, *The Vendée* (Harvard University Press, 1964). On the Girondins in Normandy, see Albert Goodwin, "The Federalist Movement in Caen

During the French Revolution," *Bulletin of the John Rylands Library* (1959–1960): 313–314. For other theaters of rebellion, see Alan Forrest, *Society and Politics in Revolutionary Bordeaux* (Oxford University Press, 1975); W. H. Scott, *Terror and Repression in Revolutionary Marseilles* (Macmillan, 1973); Hubert Johnson, *The Midi in Revolution* (Princeton University Press, 1986), ch. 7; M. Crook, "Federalism and the French Revolution: The Revolt of Toulon in 1793," *History* (1980): 383–397.

Hebert and carnage: Palmer, *Twelve*; Jones, 474–475; Rudé, 178–204.

Terror: R. R. Palmer, *Twelve*, esp. 43–77; P&C, 385–91; Jones, 475–483; Doyle, 243–258; Thompson, 375–401; Schama, 767–840; Carlyle; Colin Lucas, *The Structure of the Terror: The Case of Javogues and the Loire* (Oxford University Press, 1973). For further reading on the Terror, see Colin Lucas and Gwynne Lewis, *Beyond the Terror: Essays in French Regional and Social History* (Cambridge, England, 1983).

Levée en masse: Schama, 758–767; Howard, *War*, remains the preeminent work on military history, 75–93; Palmer, *Twelve*, 78–108; P&C, 389–391; Thompson, 427–428; Jones, 483–486; also see Griffith and Pope. **French victories:** Doyle, 256; Thompson, 423; Jones, 483–486; Carlyle; Griffith; Hibbert; Richard Cobb, *Reactions to the French Revolution* (Oxford University Press, 1972). **Law of Suspects:** Schama, 766; Palmer, 67; Carlyle, 360–363.

Queen's trial and death: Powerful and crucial are Fraser, 403–410; Lever, 289–305; Zweig, esp. 401–419 and 434–454; Schama, 793–800; also Thompson, 434–436; Hibbert, 221–222, Carlyle, vol. 2, 532–535. A vital source for the final months is Rosalie Lamorliere, *"Relation de Rosalie Lamorliere, servante à la Conciergerie"* (August-October, 1793). For further insight, her reminiscences can be freely found on the Internet as well. For her imprisonment and trial, Gérard Walter, *Marie-Antoinette* (Éditions du Bateau Ivre, 1948) is particularly helpful. Also Emile Campardon, *Marie-Antoinette a la Conciergerie (du 1er aout au 16 octobre 1793)*, 2nd ed., 1864. The trial proceedings were published in the *Acte d'Accusation* and the *Bulletin of the Tribunal Révolutionnaire.*

Death of Girondins: Schama, 800–805; Thompson, 370–375; Jones, 481; Carlyle, vol. 2, 326–330; Hibbert, 222–225; Sydenham, *Girondins*; also see Guy Chaussinand-Nogaret, *Madame Roland* (Seuil, 1985).

Paine in captivity: See two outstanding works, John Keane, *Tom Paine: A Political Life* (Little, Brown, 1995), 383–452, for Paine during the Terror, and Foner, *Tom Paine and Revolutionary America;* also P. O'Brien, *Debate Aborted 1789–91: Priestley, Paine, Burke and the Revolution in France* (Pentland, 1996). **Mary Wollstonecraft:** See wonderful book by Gordon, esp. 182–232.

Terror in provinces: See varying and equally powerful accounts by Schama, 779–792; Doyle, 258–266; Jones, 486–492; Thompson, 403–414; Hibbert, 225–234; Mason and Rizzo, 218–219; Carlyle, vol. 2, 343–353. Palmer, *Twelve*, 152–176, is particularly strong on Lyon. A number of important works are Griffith, *Art of War*; Peter Paret, *Internal War and Pacification: The Vendée, 1789–1796* (Princeton University

Press, 1961); Lewis Gwynne, *The Second Vendée: The Continuity of Counterrevolution in the Department of the Gard 1789–1815* (Oxford, 1978); Charles Tilly, *The Vendée* (Harvard University Press, 1964); W. H. Scott, *Terror and Repression in Revolutionary Marseilles* (London, 1973); Colin Lucas, *The Structure of the Terror: The Case of Javogues and the Loire* (Oxford University Press, 1973); Richard Cobb, *Reactions to the French Revolution*; and John Black Sirich, *The Revolutionary Committees in the Departments of France 1793–1794* (H. Ferrig, 1971). **Terror in Paris:** Hampson, *The Life and Opinions of Maximilien Robespierre*; Palmer, *Twelve*; Mason and Rizzo, 263–270; Doyle, 266–278; Thompson, 474–502; Hibbert, 225–226, 229; J. M. Thompson, *Robespierre*, 2 vols. (Appleton-Century, 1936). For more on the enforcement of the Terror, see, for example, Cobb, *The Police and the People*, and his classic work, *The People's Armies*. **Levée en masse successes:** For detail on the dictated economy, see esp. Palmer, *Twelve*, 225–253; P&C, 391; also Doyle; Griffith; Hibbert.

Danton's downfall: I have relied most heavily upon Hampson's excellent *Danton*, 117–174, which details vividly his journey to elder statesman and death; also see Jones, 492–493; Mason and Rizzo, 236–238; Schama, 807–820; Doyle, 273–275; Hibbert, 234–244; Carlyle, vol. 2, 388–389; as well as Keane, *Tom Paine*; Charles F. Warwick, *Danton and the French Revolution* (Fisher Unwin, 1909); Herman Wendel, *Danton* (Constable, 1936).

Lafayette: From Ségur to de Ligne, Paine to John Paul Jones, this age was rich with characters who crossed borders and moved from revolution to revolution. Among the most interesting was Lafayette. It is almost ironic how to this day he is far more revered in America—there is even a Lafayette Square across from the White House—than he is in France itself. For one of the most recent books on Lafa-yette and one which I have drawn on heavily, here and throughout, see Harlow G. Unger's exceptional *Lafayette* (Wiley, 2002). See also the outstanding treatments of Lafayette: Louis Gottschalk, *The Letters of Lafayette to Washington 1777–1799* (American Philosophical Society, 1976); Louis Gottschalk, *Lafayette Between the American and the French Revolutions* (University of Chicago Press, 1950); Louis Gottschalk, *Lafayette Comes to America* (University of Chicago Press, 1935); and Louis Gottschalk, "When and Why Lafayette Became a Revolutionary," in Slavin and Smith, *Bourgeois, Sans-culottes and Other Frenchmen*: 7–24; and Oliver Bernier, *Lafayette: Hero of Two Worlds* (Dutton, 1983); also Schama, who though brief offers trenchant insights, 449–451. Further see Peter Buckman, *Lafayette: A Biography* (Paddington, 1977); Maurice de La Fuye, *The Apostle of Liberty: A Life of La Fayette*, Emile Babeau, trans. (Thomas Yoseloff, 1956); Noel B. Gerson, *Statue in Search of a Pedestal: A Biography of the Marquis de Lafayette* (Dodd, Mead, 1976); Marian Klamkin, *The Return of Lafayette, 1824–1825* (Scribner's, 1975); Marquis de Lafayette, *Lafayette in the Age of the American Revolution: Selected Letters and Papers, 1776–1790*, 5 vols., Roger E. Smith, ed. (Cornell University Press, 1977); David Goldsmith Loth, *The People's General: The Personal Story of Lafayette* (Scribner's, 1951); William E. Woodward,

Lafayette (Farrar & Rinehart, 1938); Lloyd S. Kramer, *Lafayette in Two Worlds: Public Cultures and Personal Identities in an Age of Revolution* (University of North Carolina Press, 1996); *The Letters of Lafayette to Washington, 1777–1799*, Louis Gottschalk, ed.; Charlemagne Tower, *The Marquis de La Fayette in the American Revolution, with Some Account of the Attitude of France Toward the War of Independence*, 2 vols. (Da Capo, 1970); and Robert Waln, *Life of the Marquis de Lafayette: Major General in the Service of the United States of America, in the War of the Revolution* (J. Ayres, 1825).

Feast of Supreme Being: Schama, 827–836; Thompson, 497; Hibbert, 251–254; also Dowd, *Pageant-Master of the Republic.* **Law of 22 Prairial:** Thompson, 492; Hibbert, 245–249; Carlyle; Mason and Rizzo, 206–208. **Great Terror:** Schama, 836–840; Doyle, 275–278; Thompson, 474–502; Jones, 494–500; Hibbert, 246–248, 254–256; Palmer, *Twelve*; Carlyle; Thompson, *Robespierre*; Hampson, *Danton*; and Stanley Loomis, *Paris in the Terror June 1793–July 1794* (Lippincott, 1964). For more, see Greer, *The Incidence of the Terror.* Richard T. Bienvenu, *The Ninth of Thermidor: The Fall of Robespierre* (Oxford University Press, 1968), compiles key edited documents and provides critical background.

Fall of Robespierre: This is presented with great drama and a diversity of perspectives by various scholars. See Schama, 836–846; Rudé, 138–141; Jones, 498–500; Hibbert, 256–268; Doyle, 278–282; Thompson, 505–518. Palmer, *Twelve*, 305–334, vividly shows the *enragés* and Danton's downfall, the tightening of the Terror, and Robespierre's sudden end. Also see Bienvenu, *The Ninth of Thermidor*; Hampson, *The Life and Opinions*; Rudé, *Robespierre*; Jordan, *The Revolutionary Career*; Belloc, *Robespierre*; Georges Lefebvre, *The Thermidoriens*, trans. Robert Baldick (Vintage, 1966); G. Lenotre, *Robespierre's Rise and Fall*, trans. Mrs. Rodolph Stawell (Hutchinson, 1927); Francis Leary, "Robespierre: The Meaning of Virtue," *Virginia Quarterly* (Winter 1996); and G. H. Lewes, *The Life of Maximillian Robespierre: With Extracts from His Unpublished Correspondence* (Chapman and Hall, 1899).

White Terror: Hibbert, 275–288; P&C, 392. **End of Convention:** Doyle, 282–285; Hibbert, 275–288; Carlyle, vol. 2, 444–451; P&C, 392. **Directory:** Jones, 505–520; Hibbert, 291–304; Martyn Lyons, *France Under the Directory* (Cambridge University Press, 1975); Denis Woronoff, *The Thermidorean Regime and the Directory 1794–1799* (Cambridge University Press, 1984); Isser Woloch, *Jacobin Legacy: The Democratic Movement Under the Directory* (Princeton University Press, 1978); Carl Ludwig Lokke, "French Dreams of Colonial Empire Under Directory and Consulate," *Journal of Modern History* 2, no. 2 (1930): 237–250; Sutherland, *France 1789–1815*, has an exceptional discussion. **Napoleon:** Hibbert, 285, 300–304; Louis Bergeron, *France Under Napoleon*, trans. R. R. Palmer (Princeton University Press, 1981); Asprey, *The Rise of Napoleon Bonaparte*; Paul Johnson, *Napoleon* (Viking, 2002); J. M. Thompson, *Napoleon Bonaparte* (Oxford University Press, 1952); and Frank McLynn, *Napoleon: A Biography* (Jonathan Cape, 1997).

For background: See: Robert V. Allen, *Russia Looks at America: The View to 1917* (Library of Congress, 1988); R. Nisbet Bain, *The Last King of Poland and His Contemporaries* (Arno, 1971); Václav L. Benes and Norman J. G. Pounds, *Poland* (Praeger, 1970); Billington, *Icon*; Masson, *Memoirs*; William L. Blackwell, *Alexander I and Poland: The Foundations of His Polish Policy and Its Repercussions in Russia, 1801–1825* (1959); *The Cambridge History of Poland*, 2 vols., W. F. Reddaway, ed. (Octagon, 1971); J. H. Castera, *History of Catherine II* (J. Stockdale, 1800); Madariaga, *Politics and Culture*; Madariaga; *The Eighteenth Century in Russia*, J. G. Garrard, ed. (Clarendon, 1973); *Engraved in the Memory*, Anthony Cross, ed.; Michael T. Florinsky, *Russia: A History and an Interpretation* (Macmillan, 1955); Leo Gershoy, *From Despotism to Revolution: 1763–89* (Harper & Row, 1944); Griffiths, *Russian Court Politics . . . Under Catherine II*; Lincoln, *Sunlight at Midnight*; Longworth, *The Art of Victory*; Marc Raeff, *Political Ideas and Institutions in Imperial Russia* (Westview, 1994); Raeff, *Understanding Imperial Russia*.

For the best biographies on Catherine the Great: I have drawn here extensively upon several outstanding studies, which have proved invaluable to me throughout, including Alexander; Troyat; *CTG*; Cronin; as well as Smith, ed., *Love and Conquest*. For further background, see Erickson, *Great Catherine*; Gooch, *Catherine the Great*; Kaus, *Catherine the Great*; Oldenbourg, *Catherine the Great*, Anne Carter, trans. (Pantheon, 1965); K. Waliszewski, *The Romance of an Empress: The Life of Catherine II of Russia* (Appleton, 1929). Invaluable insights can also be culled from Ségur and Masson's memoirs. I have also invaluably benefited from Raeff, ed., *Catherine the Great: A Profile* (Hill & Wang, 1972), which has a series of superb essays; scholars may also want to consult *Canadian American Slavic Studies* 4, no. 3 (Fall 1970), a special issue devoted to the reign of Catherine II.

On Catherine's childhood and younger years: See first and foremost Catherine the Great, *Memoirs*, 8–174; Alexander, 3–59; Troyat, 1–106; Cronin, 17–131; and Madariaga, 1–27; as well as M. Raeff, "The Domestic Policies of Peter III and His Overthrow," *American Historical Review* 65, no. 5 (June 1970): 1289–1310; and Ekaterina Dashkova, *The Memoirs of Princess Dashkova* (Duke University Press, 1995).

On Catherine's 1762 coup: See esp. Alexander, 57–68; Madariaga, 29–31; and Troyat, 126–139; for more on the complicated factions of her court, see Ransel, esp. 38–161. For the conservative opposition to Catherine see Brenda Meehan Waters, "Russian Convents and the Secularization of Monastic Property," in *Russia and the World of the Eighteenth Century*, R. Bartlett, A. G. Cross, and Karen Rasmussen, eds. (Slavica, 1988). On the crown and nobility in Russia up to 1762, see Dukes, *Catherine*, esp. 1–37.

On the legislative commission: See Dukes, *Catherine*; Catherine the Great, *Memoirs*; Ransel, esp. 171–201; Paul Dukes, ed., *Russia Under Catherine the Great* (Oriental Research

Partners, 1977), esp. 28–88. On the attempt to limit Catherine's power in 1762, see Raeff, *Plans for Political Reform in Imperial Russia, 1730–1905* (Prentice Hall, 1966). For difficulties of translating Western political terms into Russian, see Madariaga, "Autocracy and Sovereignty," *Canadian-American Slavic Studies* 16, nos. 3–4 (Fall/Winter 1982): 369–387. For the broader social milieu, including the peasantry and serfs, see J. Blum, *Lord and Peasant in Russia from the Ninth to the Nineteenth Century* (Princeton University Press, 1961), and Madariaga's "Catherine II and the Serfs: A Reconsideration of Some Problems," in *Slavonic and East European Review* 52, no. 126 (January 1974): 34–62. Peter Kolchin in *Unfree Labor: American Slavery and Russian Serfdom* (Harvard University Press, 1987) draws some fascinating conclusions from a comparison between serfdom and slavery, which is also useful reading for the discussion on slavery in ch. 4. "The Administration of the Borderlands," part 5 of LeDonne's *Ruling Russia*, addresses the often inflammatory eastern and southern borderlands of Russia. An English translation of Catherine's *Nakaz* has been republished in W. F. Reddaway, ed., *Documents of Catherine the Great. The Correspondence with Voltaire and the Instruction of 1767 in the English text of 1768* (Cambridge University Press, 1931), reprinted in 1971. Also on Catherine and Blackstone, see M. Raeff, "The Empress and the Vinerian Professor," *Oxford Slavonic Papers* 7 (1974): 18–40.

On government achievements: See Walter J. Gleason, *Moral Idealists, Bureaucracy, and Catherine the Great* (Rutgers University Press, 1981); for more on reforms, here I have drawn upon the following. From 1775 through the 1780s, see Madariaga, 227–308, and 38–48 for central government reforms. On education, and health, Madariaga, 488–503, as well as *CTG*, 104–117; Dukes, *Catherine*, 189–217; Cronin, 160–177; Troyat, 169–192; and Janet M. Hartley, "The Boards of Social Welfare and the Financing of Catherine II's State Schools," *Slavonic and East European Review* 67, no. 2 (April 1989): 211–227. Also Masson, *Court*, esp. 76; Catherine the Great, *Memoirs*. For more on the nobility, which is crucial to understanding a number of the obstacles that Catherine faced in governing, see the excellent study by Marc Raeff, *Origins of the Russian Intelligentsia: The Eighteenth-Century Nobility* (Harcourt Brace, 1966); on the complicated politics of Catherine's court, see especially Alexander, 97–120, 163–183. For further insight, see Hartley, "The Implementation of the Laws Relating to Local Administration with Special Reference to the Guberniya of St. Petersburg" (University of London, 1980, unpublished doctoral diss.), which casts additional light on the reforms of 1775 and 1785. On local administration reforms, R. E. Jones, *Provincial Development in Russia: Catherine II and Jacob Sievers* (Rutgers University Press, 1984); John LeDonne, *Ruling Russia: Politics and Administration in the Age of Absolutism, 1762–1796* (Princeton University Press, 1984); for a different viewpoint, see Raeff, *The Well-Ordered Police State: Social and Institutional Change Through Law in the Germanies and Russia, 1600–1800* (Yale University Press, 1983). Also Hartley, "Catherine's Conscience Court—An English

Equity Court?" in *Russia and the West in the Eighteenth Century*, ed. A. G. Cross (Oriental Research Partners, 1983); and Hartley, "Philanthropy in the Reign of Catherine II: The Theory and the Practice," in *Russia in the Age of the Enlightenment—Essays for Isabel de Madariaga*, eds. Janet M. Hartley and Roger Bartlett (Macmillan, 1990). On the treatment of Jews, see J. D. Klier, *Russia Gathers Her Jews—the Origins of the "Jewish Question" in Russia 1772–1825* (Northern Illinois University Press, 1986); also Potemkin was an eminent protector of the Jews, even employing rabbis in his court and raising a battalion of "Jewish Cossacks" to fight the Muslims in the Russo-Turkish War, Montefiore, 282–284. On Russian economic history, see A. Kahan, *The Plow, the Hammer and the Knout: An Economic History of Eighteenth-Century Russia* (University of Chicago Press, 1985); and James Duran, "Catherine the Great and the Origin of the Russian State Debt," in *Russia and the World of the Eighteenth Century*. For financial history see James A. Duran, "The Reform of Financial Administration in Russia During the Reign of Catherine II," in *Canadian Slavic Studies* 4, no. 3 (Fall 1970): 485–496, and relevant section of LeDonne. On the government, see Raeff, *Imperial Russia*, 69–88.

On the philosophes and Catherine: See Gleason; also see Lobanov-Rostovsky, *Russia and Europe, 1789–1825*; Allison Blakely, "American Influences on Russian Reformist Thought in the Era of the French Revolution," *Russian Review* 52, no. 4 (1993): 451–471; Catherine the Great, *Memoirs*; Madariaga, "Catherine and the Philosophes," in *Russia and the West in the Eighteenth Century*. Also esp. *CTG*, 91–103, and Alexander, 100–102, 172–173. On censorship see K. Papmehl, *Freedom of Expression in Eighteenth-Century Russia* (Martinus Nijhoff, 1971). There is a useful anthology of Russian literature in Harold D. Segal, *The Literature of Eighteenth-Century Russia*, 2 vols. (Dutton, 1967).

On Voltaire: See Roger Pearson, *Voltaire Almighty* (Bloomsbury, 2005); Georg Brandes, *Voltaire*, 2 vols. (A. & C. Boni, 1930); James Parton, *Life of Voltaire*, 2 vols. (Houghton Mifflin, 1892).

On Diderot: See Denis Diderot, *Dialogues* (Routledge, 1927); John Morley, *Diderot* (Macmillan, 1923). Also see Philipp Blom, *Enlightening the World: Encyclopédie, the Book That Changed the Course of History* (Palgrave Macmillan, 2005).

On Baron von Grimm: See Catherine II, *Lettres de Catherine II à Grimm*, Grot, ed. (Tip Im akademīi nauk, 1878).

For more on Laharpe: See Alexander, 311; and Troyat, 261–263, 328–329; Madariaga, 568–571.

On the crushing of Poland: See Lord; see Palmer, *Age*, 411–435; Zamoyski, *The Last King of Poland;* O. Halecki, *A History of Poland*, Monica M. Gardner and Mary Corbridge-Patkaniowska, trans. (Coordinating Committee of Poland's Christianity Millennium, 1966); Lukowski; Masson, esp. 69; and Madariaga, 450–455. Also see Konovalov, ed., *Russo-Polish Relations*; Waliszewski, *Poland the Unknown*; for more on Catherine's drive to expand, see Gladys Scott Thompson, *Catherine the Great and the Expansion of Russia* (Collier, 1962); Evans, *Memoir of Thaddeus Kosciuszko*; Pula, *Thaddeus Kosciuszko.*

For Catherine's final days: I've especially drawn upon Alexander, 321–327; Troyat, 334–348; Madariaga, 574–578; and Cronin, 289–300. Also see Mary Louise Elisabeth Vigée Lebrun, *Souvenirs of Madame Vigee Le Brun* (R. Worthington, 1879), 13–14; Angelica Goodden, *The Sweetness of Life: A Biography of Elisabeth Vigée Lebrun* (André Deutsch, 1997). For critical contemporary views of Catherine's Russia, both domestic and foreign, see Basil Dmytryshyn, *Modernization of Russia Under Peter I and Catherine II* (Wiley, 1974), esp. 109–136.

For more on Paul I: See Roderick E. McGrew, *Paul I of Russia, 1754–1801* (Oxford University Press, 1992). There is some question as to whether Paul was actually mad. See Hugh Ragsdale, *Tsar Paul and the Question of Madness: An Essay in History and Psychology* (Greenwood, 1988).

CHAPTER 11

Global turmoil: For broad overviews of the big picture for this epochal period, see these grand treatments: Palmer, *Age*; P&C; and especially R. R. Palmer, *World of the French Revolution*, 218–232, for an account of how Americans feared the French Revolution as a contagion. I have also relied quite heavily on Ferling, who has a superb grasp and sure hand of these international issues (and party politics), as do McCullough and Freeman, in conveying this agitated period. Also consult Malone, vols. 2–3, and E&M. The reader can especially consult chapter 8 in E&M, along with their notes on America and France, pp. 808–825. What is evident is just how important it is to see the degree to which the politics of France's Revolution were powerfully playing out in the young United States (and for that matter, in Russia at the same time).

Washington: Freeman, *Washington.* Freeman was indispensable in shaping my thinking here, along with Bernard Fay, *George Washington: Republican Aristocrat* (Houghton Mifflin, 1931); Flexner; Ellis, *HE*; and Washington Irving, *Life of George Washington*, 5 vols. (Putnam's, 1855); Edmund S. Morgan's superb *The Genius of George Washington* (Norton, 1980); *HE*; Shelby Little, *George Washington* (Minton, Balch, 1929). Louis Martin Sears, *George Washington & the French Revolution* (Wayne State University Press, 1960), was immensely helpful in appreciating how Washington saw the French Revolution. Sears also gives a detailed rendition of how the French revolutionaries saw Washington and the state of American politics. What clearly comes across is the radical revolutionaries' frequent contempt for the young American republic, as well as a disinterest by many officials in the nuances of the American political system. McDonald's important *The Presidency of George Washington* was quite informative on how Washington set about constructing his policies; also quite important Alexander DeConde, *Entangling Alliance: Politics and Diplomacy Under George Washington* (Duke University Press, 1958); Randall, *George Washington: A Life*; Brookhiser, *Rules of Civility* (University of Virginia Press, 2003); and Johnson.

On the tumultuous birth pangs of the first American political party system and the evolution of a loyal opposition: See Joyce Appleby's rich, important, and highly textured *Capitalism and a New Social Order: The Republican Vision in the 1790s* (New York University Press, 1984); Sharp, *American Politics in the Early Republic*; and of course E&M, ch. 7; Chernow, chs. 20–22; and McCullough, ch. 8. E&M, 451–460, assesses the state of party and faction at the end of the Washington presidency. See also Chernow, 390–392; Ellis, *FB*, 186–187; Ferling, 334–354; Richard Hofstadter, *The Idea of a Party System: The Rise of Legitimate Opposition in the United States, 1780–1840* (University of California Press, 1969); and Peter W. Schramm and Bradford Wilson, eds., *American Political Parties and Constitutional Politics* (Rowman and Littlefield, 1993). A recent and indispensable book for politics of the early American republic is Freeman, *Affairs of Honor*. Also see William Nisbet Chambers, *Political Parties in a New Nation: The American Experience, 1776–1809* (Oxford University Press, 1963); Joseph Charles, *The Origins of the American Party System* (Harper & Row, 1961); Noble Cunningham, *The Jeffersonian Republicans: The Formation of Party Organization, 1789–1801* (University of North Carolina Press, 1957); Ronald Formisano, *The Transformation of Political Culture: Massachusetts Parties, 1795–1845* (Oxford University Press, 1983); Paul Goodman, *The Democratic-Republicans of Massachusetts: Politics in a Young Republic* (Harvard University Press, 1964); Paul Goodman, "The First American Party System," in William Nisbet Chambers and Walter Dean Burnham, eds., *The American Party System: Stages of Political Development* (Oxford University Press, 1967); John E. Hoadley, *Origins of American Political Parties, 1789–1803* (University of Kentucky Press, 1986); John R. Howe Jr., "Republican Thought and the Political Violence of the 1790s," *American Quarterly* 19 (Summer 1967): 148–165. For further discussion, see Carl E. Prince, *New Jersey's Jeffersonian Republicans: The Genesis of an Early Party Machine, 1789–1817* (University of North Carolina Press, 1967); Mary Ryan, "Party Formation in the United States Congress, 1789–1796: A Quantitative Analysis," *William and Mary Quarterly*, 3d ser., 18 (October 1971): 523–542; Harry M. Tinkcom, *The Republicans and Federalists in Pennsylvania, 1790–1801* (Pennsylvania Historical and Museum Commission, 1950); Patricia Watlington, *The Partisan Spirit: Kentucky Politics, 1779–1792* (Atheneum, 1972); Alfred F. Young, *The Democratic Republicans of New York: The Origins, 1763–1797* (University of North Carolina Press, 1967); and these two important works, Bernard Bailyn, *The Origins of American Politics* (Knopf, 1968), and Lance Banning, *The Jeffersonian Persuasion: Evolution of a Party Ideology* (Cornell University Press, 1978). Perhaps one of the most difficult things for the historian or reader in the twenty-first century to grasp is the extent to which there was a revulsion for political parties in the early years of the 1790s, as well as fears that parties could lead to the ruination of the young republic. By the same token, the opposition to the Federalist administration was clearly and increasingly driven to think and act like a formal party, under the leadership of Jefferson and then Madison. What remains an enduring and fascinating question is how the Federalists and Republicans set out on one distinctive path of operating, while the Jacobins, Girondins, and Monarchists set out on an entirely different path in France. Of course, many of the continuing Federalist concerns about parties in the United States were shaped by the Jacobin excesses in Paris.

Newspaper war: Donald Henderson Steward, *Opposition Press of the Federalist Period* (State University of New York Press, 1964); Eric Burns, *Infamous Scribblers: The Founding Fathers and the Rowdy Beginnings of American Journalism* (Public Affairs, 2006); E&M, 282–292; Ferling, 343–344; and Chernow, 395–408. It almost seems perplexing how political differences playing out in the newspapers could be termed a "war," but given the stakes for the young republic—its very survival, or fears of anarchism, invasion, or despotism—such terminology becomes much more comprehensible. It becomes even more comprehensible when one sees the world as the Americans did (and where another partial explanation may lie), where scribblers like Marat and others increasingly became the engine of not only the Revolution in France, but ultimately some of its most bloody excesses.

On the question of America's neutrality: E&M, ch. 8, especially 336–341. Also lively and still insightful is Miller, *The Federalist Era*, ch. 8; Chernow's treatment in ch. 23 superbly presents the spirited debate between Jefferson and Hamilton on this critical matter. Ferling has an excellent discussion on American views of the French Revolution, as well as key insights, 356–360. Also see Albert Hall Bowman, *The Struggle for Neutrality: Franco-American Diplomacy During the Federalist Era* (University of Tennessee Press, 1974); William C. Stinchcombe, *The American Revolution and the French Alliance* (Syracuse University Press, 1969); Charles Marion Thomas, *American Neutrality in 1793: A Study in Cabinet Government* (Columbia University Press, 1931); Alexander DeConde, *A History of American Foreign Policy*, 2 vols. (Scribner's, 1978); and Ralph Ketcham, "France and American Politics," *Political Science Quarterly* 78 (June 1963): 198–223. What comes across from these accounts is just how bruising these cabinet meetings were, and how inextricably tied they were with matters of war, peace, and prosperity.

Genet affair: See Ferling's richly detailed account in *A Leap in the Dark*, 360–363, 370; as well as outstanding treatments in Chernow, 437–438, and Freeman, 618–642. And for more comprehensive treatment, Harry Ammon's important *The Genet Mission* (Norton, 1973); Harry Ammon, "The Genet Mission and the Development of American Political Parties," *Journal of American History* 52 (March 1966):

725–741; Greville Bathe, *Citizen Genet, Diplomat and Inventor* (Philadelphia, 1946); William Keller, *The Frontier Intrigues of Citizen Genet* (American Historical Company, 1940). Also Richard Lowitt, "Activities of Citizen Genet in Kentucky in 1793–1794," *Filson Club Historical Quarterly* 22 (October 1948): 252–267; Frederick Jackson Turner, "The Origins of Genet's Projected Attack on Louisiana and the Floridas," *American Historical Review* 3 (July 1898): 650–671; Charles D. Hazen, *Contemporary American Opinion of the French Revolution* (Johns Hopkins Press, 1897); Richard Brookhiser, *Alexander Hamilton, American* (Free Press, 1999) is also quite good on Genet. For more on U.S.-French relations, see Susan Dunn's excellent work, *Sister Revolutions: French Lightning, American Light* (Faber & Faber, 1999). To get a full understanding of Genet, the reader should consult the outstanding article, which was crucial to my thinking, William L. Blackwell, "Citizen Genet and the Revolution in Russia," *French Historical Studies Journal*, 72–92, as well as Palmer, *Age*, and Malone's *Jefferson and His Time*, esp. vol. 3, 91–130. Blackwell persuasively portrays Genet as an enormously talented and insightful diplomat; it is a distinct aid to the historian to study his time in Russia and France, as well as in America. It is also fascinating how Genet's sister, Madame Campan, emerged as one of Marie Antoinette's closest intimates, even as Genet increasingly became an ardent revolutionary. One final point of interest is on the question of Jefferson's handling of the Genet matter. There are two distinct versions of this, one is Jefferson's and one is Genet's. See footnote 119, 818 in E&M.

Jefferson: Dumas Malone's six-volume biography of Jefferson remains the best source on Jefferson, and I especially relied upon vols. 1–3. My other favorite, which I've drawn heavily upon, is Fawn Brodie, *Thomas Jefferson: An Intimate History* (W. W. Norton, 1974), which is also quite insightful on the impact of the French Revolution on the sweep of this era. Other crucial works are Hofstadter, "Thomas Jefferson: The Aristocrat as Democrat," in *The American Political Tradition*; Joyce Appleby's subtle and elegantly rendered *Thomas Jefferson* (Times Books, 2003); Merrill D. Peterson's outstanding *The Jefferson Image in the American Mind* (Oxford University Press, 1960); Joseph Ellis, *American Sphinx*; Peter Onuf and Leonard J. Sadosky, *Jeffersonian America* (Blackwell, 2002); and for a critical look at Jefferson, Forrest McDonald, *The Presidency of Thomas Jefferson* (University Press of Kansas, 1976). Albert Jay Nock's *Jefferson* (Harcourt Brace, 1926) remains wonderful. Also see Stephen E. Ambrose, *Undaunted Courage: Meriwether Lewis, Thomas Jefferson, and the Opening of the American West* (Simon & Schuster, 1996). For a view of ambiguity in the mind of Jefferson, see Connor Cruise O'Brien, *The Long Affair: Thomas Jefferson and the French Revolution* (Sinclair-Stevenson, 1996); Richard K. Matthews, *The Radical Politics of Thomas Jefferson: A Revisionist View* (University Press of Kansas, 1984); Merrill D. Peterson, ed., *The Portable Thomas Jefferson* (Penguin, 1975). On Jefferson's position on slavery, see esp. Ellis, *American Sphinx*, 146–152; O'Brien, "Thomas Jefferson: Radical and Racist," *Atlantic*

Monthly (October 1996); and J. C. Miller, *The Wolf by the Ears: Jefferson and Slavery* (Free Press, 1977). For a brief but extremely perceptive biographical sketch of Jefferson, see Chernow, ch. 16. On the persistent question of the limits of federalism, see Richard E. Ellis, *The Jeffersonian Crisis: Courts and Politics in the Young Republic* (Oxford University Press, 1971). For Jefferson's stint in Paris, the best treatment is Malone, *Jefferson and His Time*, vol. 2. Also excellent on Paris is Ellis, *American Sphinx*, 64–117; Douglas L. Wilson and Lucia Stanton, eds., *Thomas Jefferson: Jefferson Abroad* (Modern Library, 1999); William Howard Adams, *The Paris Years of Thomas Jefferson* (Yale University Press, 1997); and Lawrence S. Kaplan, *Jefferson and France: An Essay on Politics and Political Ideas* (Yale University Press, 1967).

Hamilton: The best single biography on Hamilton is the masterful work by Chernow, which also remains a fascinating window on the revolutionary and federalist period. Also see Alexander Hamilton, *The Papers of Alexander Hamilton*, 27 vols., Harold C. Syrett, ed. (Columbia University Press, 1961–1987). Reading *The Federalist* remains the essential source on the political mind of Hamilton, but for interpretive treatments see Garry Wills, *Explaining America: The Federalist* (Doubleday, 1981), and essays of Douglas Adair, in Trevor Colbourn, ed., *Fame and the Founding Fathers* (Norton, 1974). Also quite perceptive is Brookhiser's little gem, *Alexander Hamilton*, as well as Forrest McDonald's excellent *Alexander Hamilton: A Biography* (Norton, 1979); John Chester Miller, *Alexander Hamilton: Portrait in Paradox* (Harper, 1959); and Willard Sterne Randall's first-rate *Alexander Hamilton: A Life* (HarperCollins, 2003); Johnson, 180–182.

Whiskey Rebellion: The two essential works are William Hogeland, *The Whiskey Rebellion, George Washington, Alexander Hamilton, and the Frontier Rebels Who Challenged America's Frontier Sovereignty* (Scribner's, 2006), and Thomas Slaughter, *The Whiskey Rebellion: Frontier Epilogue to the American Revolution* (Oxford University Press, 1986). Ferling, 362–375; Freeman, 651–658; E&M, 461–473; and Albert J. Beveridge, *John Marshall* (Houghton Mifflin, 1916), esp. 86–91, are excellent as well. For heated debates in Congress, see Joseph Gales, ed., *The Annals of Congress: The Debates and Proceedings in the Congress of the United States*, 18 vols. (Gales and Seaton, 1834–1856). Stephen Ambrose, *Undaunted Courage: Meriwether Lewis, Thomas Jefferson, and the Opening of the American West* (Simon & Schuster, 1996), 38–43, makes the claim that the rebellion was "the greatest threat to national unity" between the Revolutionary War and the Civil War, and the threat of secession was credible and "quite real," 53. Washington's biographer Smith, *Patriarch*, 210–226, maintains that, in President Washington's view, the whiskey insurrection put "nothing less…at stake than the survival of the central government." For further study see Jeffrey Crow, "The Whiskey Rebellion in North Carolina," *North Carolina Historical Review* 66, no. 1 (1989): 1–28; Mary Bonsteel Tachau, "The Whiskey Rebellion in Kentucky: A Forgotten Episode of Civil Disobedience," *Journal of the Early American Republic* 2, no. 3 (1982): 239–259. Other useful sources include

Leland D. Baldwin, *Whiskey Rebels: The Story of a Frontier Uprising* (University of Pittsburgh Press, 1968); Henry W. Brackenridge, *History of the Western Insurrection in Western Pennsylvania Commonly Called the Whiskey Insurrection, 1794* (W. S. Haven, 1859). For President Washington's reaction, see Richard H. Kohn, "The Washington Administration's Decision to Crush the Whiskey Rebellion," *Journal of American History* 59 (1972): 567–574; William D. Barber, "*Techy Articles of Civil Police:* Federal Taxation and the Adoption of the Whiskey Excise," *William and Mary Quarterly* 25 (January 1968): 58–84. Also see Steven R. Boyd, ed., *The Whiskey Rebellion: Past and Present Perspectives* (Greenwood, 1985); Jacob E. Cooke, "The Whiskey Insurrection: A Reevaluation," *Pennsylvania History* 3 (July 1963): 316–346; Jeffrey J. Crow, "The Whiskey Rebellion in North Carolina," *North Carolina Historical Review* 66 (January 1989); William Miller, "The Democratic Societies and the Whiskey Insurrection," *Pennsylvania Magazine of History and Biography* 62 (July 1938): 324–349; James Roger Sharp, "The Whiskey Rebellion and the Question of Representation," in Mary K. Bonsteel Tachau, "The Whiskey Rebellion in Kentucky," *Journal of the Early Republic* 2 (Fall 1982): 239–259; Claude Milton Newlin, *The Life and Writings of Hugh Henry Brackenridge* (Princeton University Press, 1932).

Democratic societies: Eugene Perry Link, *Democratic Republican Societies, 1790–1800* (Columbia University Press, 1942); Philip S. Foner, ed., *The Democratic-Republican Societies, 1790–1800* (Greenwood, 1976); Merton E. Coulter, "The Efforts of the Democratic Societies of the West to Open Navigation of the Mississippi," *Mississippi Valley Historical Review* 11 (December 1924): 376–389; Ferling, 364–365 and 373–375; for Washington's comments themselves and for stinging comments on Washington's actions, see Twohig, ed., *Papers of George Washington*. Palmer in *Age* suggests that such phenomena as the Democratic Societies were a part of a larger tableau in Europe and Russia in which peoples were wrestling both with means for greater political participation as well as legitimate political dissent.

On the Jay treaty: Ferling, 378–382, 388–393; E&M, 415–422; Freeman, 661–685; Ellis, *HE*, 226–230. For more detailed information, see Richard B. Morris, ed., *John Jay, John Jay* (Harper & Row, 1975); James Roger Sharp, *The New Nation in Crisis: American Politics in the Early Republic* (Yale University Press, 1993), which offers an excellent review of America's first controversial and widely denounced treaty. Still quite useful is Samuel Flagg Bemis, *Jay's Treaty: A Study in Commerce and Diplomacy* (Greenwood, 1975), and his related *Pinckney's Treaty: America's Advantage from Europe's Distress, 1783–1800* (Yale University Press, 1960). Also Frank Monaghan, *John Jay* (Bobbs-Merrill, 1935). For criticism of Washington, see James Tagg, *Benjamin Franklin Bache and the Philadelphia Aurora* (University of Pennsylvania Press, 1991). For further, see Jerald A. Combs, *The Jay Treaty: Political Background of the Founding Fathers* (University of California Press, 1970); Thomas Farnham, "The Virginia Amendments of 1795: An Episode in the Opposition to Jay's Treaty," *Virginia Magazine of*

History and Biography 75 (January 1967): 75–88; DeConde, *Entangling Alliance*. Also the Avalon Project at Yale. **On the firing/resignation of Edmund Randolph:** See Freeman, 672–680; Ellis, *FB*, 146–147; Ellis, *HE*, 228; E&M, 424–431; and Moncure D. Conway, *Omitted Chapters in the History Disclosed in the Life and Papers of Edmund Randolph* (Putnam's, 1888). Mary K. Bonsteel Tachau, "George Washington and the Reputation of Edmund Randolph," *Journal of American History* 73 (June 1986): 15–34; and W. C. Ford, "Edmund Randolph on the British Treaty, 1795," *American Historical Review* 12 (1907): 587–599.

On Washington's Farewell Address: I strongly encourage readers to read the address in its entirety themselves, which can be culled from the Avalon Project. Also see Ellis, *FB*, ch. 4, an outstanding treatment of the scope, meaning, and myths of this address. Also outstanding are Ferling, 395–396; Ellis, *HE*, 230–240; Chernow, 505–508; E&M, 489–528, James Thomas Flexner, *George Washington: Anguish and Farewell, 1793–1799* (Little, Brown, 1972); and Freeman, 700–702. For more scholarly treatments, see Alexander DeConde, "Washington's Farewell, the French Alliance, and the Election of 1796," *Mississippi Valley Historical Review* 43 (March 1957): 641–658; Victor H. Paltsits, ed., *Washington's Farewell Address* (New York Public Library, 1935); Matthew Spalding and Patrick J. Garrity, *A Sacred Union of Citizens: George Washington's Farewell Address and the American Character* (Rowman & Littlefield, 1996); Burton J. Kaufman, *Washington's Farewell Address: The View from the Twentieth Century* (Quadrangle, 1969); and Arthur Markowitz, "Washington's Farewell and the Historians," *Pennsylvania Magazine of History and Biography* 94 (1970): 173–191.

CHAPTER 12

Adams: The work on the second president, John Adams, is voluminous, but David McCullough's biography *John Adams* is a masterpiece, the best one-volume treatment and an example of biography at its finest. I also found it invaluable for the politics of U.S.–French relations and dissension at home (467–524), and I have drawn heavily upon McCullough in this chapter. Merrill D. Peterson's *Adams and Jefferson: A Revolutionary Dialogue* (University of Georgia Press, 1976) is outstanding; indispensable also is L. H. Butterfield, ed., *Adams Family Correspondence*, 6 vols. (Belknap, 1963); John Adams, *Familiar Letters of John Adams and His Wife Abigail Adams During the Revolution with a Memoir of Mrs. Adams*, Charles Francis Adams, ed. (Books for Libraries Presses, 1970). The older but still venerable biographies of Adams are found in Gilbert Chinard, *Honest John Adams* (Little, Brown, 1933), and Page Smith, *John Adams*, 2 vols. (Doubleday, 1962). Also uncommonly excellent, and a work I've drawn upon, is John Ferling's *John Adams: A Life* (American Political Biography, 1997). See also Stephen G. Kurtz, *The Presidency of John Adams: The Collapse of Federalism, 1795–1800* (University of

Pennsylvania Press, 1957); Manning J. Dauer, *The Adams Federalists* (Johns Hopkins University Press, 1968); Peter Shaw, *The Character of John Adams* (University of North Carolina Press, 1976); Richard Brookhiser, *America's First Dynasty: The Adamses 1735–1918* (Free Press, 2002). Two incisive appraisals of the Adams temperament are Bernard Bailyn, *Faces of Revolution: Personalities and Themes in the Struggle for American Independence* (Knopf, 1990), 3–21; and Edmund S. Morgan's wonderful, "John Adams and the Puritan Tradition," *New England Quarterly* 34 (1961): 518–529. On the political thought of John Adams, an excellent treatment is C. Bradley Thompson, *John Adams and the Spirit of Liberty* (University Press of Kansas, 1998); James H. Hutson, *John Adams and the Diplomacy of the American Revolution* (University Press of Kentucky, 1980). For further, see Anson D. Morse, "The Politics of John Adams," *American Historical Review* 4, no. 2 (1899): 292–312; John Quincy Adams, *Diary of John Quincy Adams*, 2 vols., David Grayson Allen, ed. (Belknap, 1981); Samuel Flagg Bemis, *John Quincy Adams and the Foundations of American Foreign Policy* (Knopf, 1949); Edmund S. Morgan, *The Meaning of Independence: John Adams, George Washington, and Thomas Jefferson* (University Press of Virginia, 1976); James Grant's *John Adams: Party of One* (Farrar, Straus & Giroux, 2005) is very good on economic matters; and Johnson, 141–143.

On the XYZ affair and Quasi-War with France: See first the indispensable works that heavily shaped my thinking by Alexander DeConde, *The Quasi-War: The Politics and Diplomacy of the Undeclared War with France, 1797–1801* (Scribner's, 1966), and William C. Stinchcombe, *The XYZ Affair* (Greenwood, 1980), and William C. Stinchcombe, "The Diplomacy of the WXYZ Affair," *William and Mary Quarterly*, 3rd ser., vol. 34, no. 4 (October 1977): 590–617, which are balanced, thoughtful studies, along with Peter P. Hill, "Prologue to the Quasi-War: Stresses in Franco-American Commercial Relations, 1793–96," *Journal of Modern History* 49, no. 1 (On Demand Supplement, 1977): D1039–D1069, as well as Carl Ludwig Lokke, "French Dreams of Colonial Empire Under Directory and Consulate," *Journal of Modern History* 2, no. 2 (1930): 237–250; for more, see Ferling's penetrating and detailed, " 'Father and Protector': President Adams and Congress in the Quasi-War Crisis," in Kenneth R. Bowling and Donald R. Kennon, *Neither Separate nor Equal: Congress in the 1790s* (University of Ohio Press, 2000). Beveridge, from Marshall's point of view, is fascinating, 211–228, 273–297, 338–350; also important, see James H. Hutson, *John Adams and the Diplomacy of the American Revolution* (University Press of Kentucky, 1980). I've further relied extensively upon Ellis, *HE*, 230–231; *E&M*, 508–511; and McCullough, who is extremely insightful about the episode, *JA*, 187–216, 231–243, 473–508, as well as (and along with McCullough, contains quotes found in this chapter) Ferling, who superbly captures the political dynamic, 403–425; Peter Hill, *William Vans Murray, Federalist Diplomat: The Shaping of Peace with France, 1797–1801* (Syracuse University Press, 1971); and Linda Frey and Marsha Frey, " 'The Reign of the Charlatans Is Over': The French Revolutionary Attack on Diplomatic Practice," *Journal of Modern History*

65, no. 4 (1993): 706–744. Crucial reading for this section is culled from John D. Richardson, ed., *A Compilation of the Messages and Papers of the Presidents, 1789–1907* (Bureau of National Literature, 1908). It is also extremely helpful to read about the internal politics of France's ruling Directory to get a greater handle on this entire episode; see, for example, Georges Lefebvre, *Le directoire* (A. Colin, 1971). There is also some debate as to whether France's actions toward America were a product of deliberate wrath or imperialistic designs on the young republic, or alternatively, a function of France's lack of concern with American affairs. E&M suggest that it is a bit of both, but make the point that the directors showed virtually no interest in the United States and that references to America in the Directory's minutes are perfunctory, see especially 874–875. By the same token, it is also clear from contemporary accounts at the time that France did harbor some nascent designs on America (though they were eventually overshadowed by other military ventures considered to be of far greater importance, like Napoleon's Middle East venture). It is also interesting to see that even a Russian sailor who had ported at Philadelphia at the time picked up on what he saw as France's contempt and anger toward the Americans. He explicitly made comments on Pierre-Auguste Adet's willingness to provoke "a minor crisis" because America refused to hang the Tricolor in Congress. See Glynn de V. Barratt, "A Russian View of Philadelphia, 1795–96: From the Journal of Lieutenant Iurri Lasianskii," *Pennsylvania History* 65 (Winter 1998): 62–86.

For sketches of the major actors: see E&M, 555–579; Ferling, 419–423; and Chernow, 549–552; Émile Dard, *Napoleon and Talleyrand*, Christopher R. Turner, trans. (Philip Allan, 1937); William C. Stinchcombe, "Talleyrand and the American Negotiations of 1797–1798," *Journal of American History* 62 (December 1975), and his "A Neglected Memoir by Talleyrand on French-American Relations, 1793–1797," *Proceedings of the American Philosophical Society* 121, no. 3 (1977): 195–208, 575–590; Brookhiser, *Hamilton;* Freeman, *Washington*, Palmer, *Age.*

On John Marshall: See esp. Jean Edward Smith's more recent *John Marshall: A Definer of a Nation* (Henry Holt, 1996), which is also quite good on the politics of France, and Beveridge, *Marshall*, remains an important classic to this day; Robert Lowry Clinton, *Marbury v. Madison and Judicial Review* (1989); R. Kent Newmeyer, *The Supreme Court Under Marshall and Taney* (1968); Charles F. Hobson, *The Great Chief Justice: John Marshall and the Rule of Law* (University Press of Kansas, 1996); and William F. Swindler, *The Constitution and Chief Justice John Marshall* (Dodd, Mead, 1978); also see for Talleyrand in America, Michel Poniatowski, *Talleyrand aux Etats-Unis 1794–1796* (Presses de la Cité, 1967), and Hans Huth and Wilma J. Pugh, *Talleyrand in America as a Financial Promoter: Unpublished Letters and Memoirs* (Da Capo, 1942).

Napoleon: I have drawn heavily upon Robert Asprey, *The Rise of Napoleon Bonaparte* (Basic, 2001); Emile Ludwig, *Napoleon* (Random House, 1915), which is one of my favorites; J. Christopher Herold, *The Age of Napoleon* (Harper & Row, 1963); Alistair

Horne, *Age of Napoleon* (Modern Library, 2004); and J. M. Thompson, *Napoleon* (Blackwell, 1988). His book on the French Revolution is a classic, and so is this. The following three are all quite useful on Napoleon's military campaigns: Gunther Eric Rothenberg, *Art of Warfare in the Age of Napoleon* (Indiana University Press, 1978); Gunther Eric Rothenberg, *The Napoleonic Wars* (Smithsonian Books/HarperCollins, 2006); Owen Connelly, *Blundering to Glory: Napoleon's Military Campaigns* (Rowman & Littlefield, 2006); also see Durant, *Napoleon;* Paul Johnson, *Napoleon* (Viking, 2002); P. G. Elgood, *Napoleon's Adventures in Egypt* (Oxford University Press, 1931); Alan Schom, *Napoleon Bonaparte: A Life* (HarperPerennial, 1998); and P&C, 393–418, all of which I've heavily drawn upon.

On the Alien and Sedition Acts and the Virginia and Kentucky resolutions: John Chester Miller, *Crisis in Freedom: The Alien and Sedition Acts* (Little, Brown, 1951); Philip G. Davidson, "Virginia and the Alien and Sedition Laws," *American Historical Review* 36 (January 1931): 336–342; Michael Durey, *Transatlantic Radicals and the Early American Republic* (University Press of Kansas, 1997); and James Morton Smith, *Freedom's Fetters: The Alien and Sedition Laws and American Civil Liberties* (Cornell University Press, 1966). Also see Malone, vol. 3, 379–409; Ethelbert Dudley Warfield, *The Kentucky Resolutions of 1798: An Historical Study* (Putnam's, 1887); and Stampp, *Imperiled Union,* 22–24. I have also drawn upon Ellis, *FB,* 199–201; Chernow, 573–577; McCullough, 504–505, 521; E&M, 590–593, 700–701; Ferling, 426–436; Frank Malloy Anderson, "Contemporary Opinion of the Virginia and Kentucky Resolutions," *American Historical Review* 5 (October 1899 and January 1900): 45–63, 225–52; Adrienne Koch and Harry Ammon, "The Virginia and Kentucky Resolutions: An Episode in Jefferson's and Madison's Defense of Civil Liberties," *William and Mary Quarterly* 5 (April 1948): 145–176; and Adrienne Koch and Harry Ammon, "The Grass Roots Origin of the Kentucky Resolutions," *William and Mary Quarterly* 27 (April 1970): 221–245. For more information, see Leonard Levy, *Jefferson and Civil Liberties, the Darker Side* (Belknap, 1963). To get a sense of the tenor of the time, it is interesting to note that Abigail Adams thought that the "Alien bill, Sedition bill, and a bill declaring void all our treaties and conventions with France" was among Congress's "best" actions, E&M, 878–879. To get a greater sense of the stakes raised by both the Alien and Sedition Acts and the Virginia and Kentucky Resolutions, it can be noted that the resolutions later became prominent in the secession debate that culminated in the American Civil War.

Fries Rebellion: William W. H. Davis, *The Fries Rebellion, 1798–1799* (Arno, 1969). Peter Levine, "The Fries Rebellion: Social Violence and the Politics of the New Nation," *Pennsylvania History* 40 (July 1973); and Paul Douglas Newman, *Fries's Rebellion: The Enduring Struggle for the American Revolution* (University of Pennsylvania Press, 2004). Also see E&M, 620–621, 696–700; Chernow, 578–579; and McCullough, 540.

EPILOGUE

For background: Olivier Bernier, *World of 1800* (Wiley, 2000) is a wonderful overview of this historic year. On Mary Wollstonecraft, see Gordon, Mary Wollstonecraft, *A Vindication of the Rights of Woman: An Authoritative Text; Backgrounds; The Wollstonecraft Debate; Criticism*, 2nd ed. (Norton, 1988). On Andrew Jackson, see H. W. Brands's superb *Andrew Jackson: A Life and Times* (Doubleday, 2005).

On Napoleon: P&C, esp. 397–399; Adam Zamoyski, *Moscow 1812: Napoleon's Fatal March* (HarperCollins, 2004), 20–25; Palmer, *Age*; Durant, *Napoleon*; Herold, *Age of Napoleon*; Horne, *Age of Napoleon*; Thompson, *Napoleon*; and Schom, *Napoleon*.

On Alexander: See Alan Palmer, *Alexander I, Tsar of War and Peace* (Weidenfeld & Nicolson, 1974); Curtis Cate, *War of the Two Emperors: The Duel Between Napoleon and Alexander, Russia 1812* (Random House, 1985); Hugh Seton-Watson, *The Russian Empire 1801–1917* (Oxford University Press, 1967); Zamoyski, 20–25; Troyat; Alexander.

On election of 1800: See John Ferling, *Adams Versus Jefferson: The Tumultuous Election of 1800* (Oxford, 1994); Malone, vol. 3, 484–506, and vol. 4, 3–16. It should be noted that Malone suggests Jefferson made no explicit promises to assume the presidency. My reading of the evidence is that he made implicit promises. See also McCullough, 543–564; Ferling, 452–475. For the text of Jefferson's historic first inaugural, see the Avalon Project.

On Washington, D.C., and the transition of power: see Linda Wheeler, "Bold Vision, Humble Start: In the Beginning," *Washington Post*, March 20, 2000; McCullough, 541–542, 564–566; Ferling, 477–481, 485–488; Beveridge, *Marshall*, 1–5; Ralph A. Brown, *Presidency of John Adams* (University of Kansas Press, 1975), 195–197.

Illustration Credits

Acknowledgments

A nyone who wades into the remarkable age of the 1790s immediately confronts an extraordinary collection of public historians and university scholars, as well as almost more classics than one can count. It is to this striking circle of intellectuals, whose works are the foundation of this book, that I owe my first debt. I am greatly indebted as well to the superb staffs of numerous libraries, archives, museums, and historical associations. First and foremost, to the Library of Congress, which stands as the premier resource in the world. Here, in one place, I was able to dig into material about America, France, Russia, and Europe at large. I especially want to thank James Billington, the Librarian of Congress, who so graciously provided assistance on questions concerning Russia. Timothy Robbins also was a great help.

I would also like to thank the many specialists at the Library of Congress, the National Archives, the New York Public Library, the New-York Historical Society, the Massachusetts Historical Society, Columbia University, the Historical Society of Pennsylvania, the Pittsburgh Regional History Center, the Historical Society of Frederick County, Yale University, the University of Virginia, the Archives Nationales in Paris, the Russian State Library in Moscow, and Mount Vernon. As in the past, the University of Maryland libraries were wonderful in providing assistance and dealing with special requests.

I took intensive French at Alliance Française in Washington, D.C., to develop a reading knowledge of French. Nadine Fuger was a spectacular teacher. Elena Sokolova was tenacious in helping me solve research questions in Russia. I greatly benefited from the latest forensic research on George Washington as explained by Jeff Schwartz. Mary Thompson of Mount Vernon (an historic place that is one of our nation's treasures if ever there was one), who seems to know just about everything there is to know about Washington, was particularly helpful in answering the most obscure of questions. Historian and archivist Vlad Soshnikov from the Moscow State Institution for History and

Archives was also a great help in answering my queries. Mary Habeck of Johns Hopkins SAIS helped me navigate some of the nuances of the Islamic world.

Over the six years or more that it took to write this book, I was fortunate to have a number of fellow authors, friends, lovers of history, and colleagues who provided me with a wide array of support, wise counsel, good humor, and a steady diet of encouragement. They include John Donvan, P. J. O'Rourke, Christopher Buckley, David Ignatius, Robert Dallek, Rick Kahlenberg, Frank Williams, Evan Thomas, Richard Gilder, Wayne and Catherine Reynolds, Roger and Susan Hertog, Rusty Powell, Nina Easton, Mari Will, Bruce Cole, Jon Karp, Brent Glass, Harold Holtzer, James Q. Wilson, Erich Eichman, Jim Gilliland, and my fellow preservationists at the Civil War Preservation Trust, Nancy and Mark Penn, Laura Blumenfeld, Amy and Ken Weinstein, Jim and Marilyn Denton, Mark and Margot Bisnow, David Brooks, Burnie Bond, Alice Kelley, and Byron Hollinshead, who frequently offered sagely insights into historical questions large and small. Jerry Groopman has been a cherished friend in more ways than I could recount.

My late friend Peter Jennings was always unflagging in his support and encouragement, and believed in this book as much as I do. I did a panel at the 92nd Street Y with Peter and Joe Ellis, which helped stimulate some key ideas at a critical point early on in the project. When I spoke at the Aspen Institute, Walter Isaacson and I compared notes on writing, and he gave me some superb nuggets of advice. Another great source of inspiration for me remains Doris Kearns Goodwin.

Near the book's final stages, I benefited immensely from the several talented people who read the manuscript, made suggestions, and offered insights. Carol Berkin of Baruch College, and to my mind one of the nation's finer American historians of this period, read all my American chapters and provided detailed comments. Fred Siegel, who teaches French history at Cooper Union in New York, read my chapters on France. Mark Medish, vice president for Russian studies at the Carnegie Endowment for International Peace, a Russian scholar, and the former senior director for Russian Affairs on President Bill Clinton's National Security Council, gave a careful reading to the Russian chapters. My longtime friend Michael Humphries read the manuscript from cover to cover, and so did my in-laws, Jim and Lark Wallwork, who cheerfully tackled the task twice. I also had a reader carefully pore through my manuscript, sources, and bibliographic notes to ensure as much as possible that my facts and citations were accurate.

My learned colleague at Maryland, Ira Berlin, was good enough to help when I needed to find a new research assistant. Ellen Epstein was terrific in locating excellent assistance when it was much needed. Many thanks, too, to my research assistants, who performed a variety of tasks, especially in hunting down documents and papers: Jacqueline Gerbus, Andrew Morton, Lloyd McCoy, Siafa Hage, Daniel Feith, and Don McNeilly. I particularly want to cite Eronn Strickland, whose high caliber of work has been just superb over the course of the project.

And of course, thanks to my parents, Herb and Lynn; and also Leslie and Richard; Ken and Carol; and my ever-faithful Wheaton terrier, Bogey.

Throughout, my publisher, HarperCollins, has been almost like a family to me, starting with my editor, Tim Duggan. This is our second collaboration, and he is a pleasure to work with: He brings a superb combination of talent, patience, diligence, enthusiasm, and a genuine love for the craft of writing and history, which has been invaluable, as were his edits. An author could not ask for more. I owe special gratitude to the publisher, Jonathan Burnham, who has also been extraordinarily supportive and enthusiastic. He gave the manuscript a highly detailed and thoughtful reading and made comments on tone, language, substance, and even the French. The book is much improved as a result. For copyediting I want to thank Eleanor Mikucki, David Koral, and John Jusino. Also, many thanks to Allison Lorentzen for her always cheerful assistance on the day-to-day matters of book production. My literary agent, Michael Carlisle, of Inkwell Management, has been a valued friend and agent for more than a decade and has represented this project with special flair.

For daily joy throughout this task, I have been blessed by my wonderful sons, Nathaniel, five, and Evan "B.C.," three. True, Nathaniel knows enough about history to say, "way back when, they did . . ." and he knows who George Washington is. But even more than their father, he and his little brother will be delighted at the completion of this book.

And finally, more than words can express, my most heartfelt gratitude goes to my wife, Lyric. Where do I begin? She was the former class poet at Princeton, has a graduate degree in history from Johns Hopkins, and writes a weekly column; she is also a brilliant writer. More than that, almost as much as I do, she knows what I want to say, how I want to say it, and where I wanted it to go in this book. From start to finish over half a decade, every aspect of this project bears her loving and discerning imprint. I can think of no one else whom I would have trusted to discuss the book with me, edit it with me, and to make suggestion after suggestion—most of which I took. Without her, this book would not be what it is, yet for that and far more, she is the love of my life and the most wonderful person I know. If this is my book, then it is also hers.

Index

Insights,
Interviews
& More ...

Meet Jay Winik

Carl Caruso

JAY WINIK was born on February 8, 1957. A preemie, his first weeks were spent in an incubator; he did not speak a word until the age of three.

He grew up the youngest of three boys on a suburban block in New Haven, Connecticut. They played touch football in the front yard, studiously avoiding the stone walk and the lamp post. His father sold paneling for dens and rec rooms; his mother took care of the kids. Winik's earliest literary passion was Marvel Comics. Starting at age six he would walk to the drugstore each Saturday morning to purchase shiny new editions of *Spider-man*, *X-Men*, *Avengers*, and other comic books. His artistic leanings were almost entirely visual—from a young age he painted, drew, and cartooned, making elaborate pen-and-ink creations well into his college years.

His parents believed in public schools, and Winik was one of the few children in his area to be bused to an inner-city junior high and high school. The experience taught him lifelong lessons and instilled in him a passion for giving back.

Too small to play football, he turned to tennis, paying for lessons with his allowance, playing in tournaments throughout New England, and ultimately ranking as high as in the top five in the region. Although he dreamed of turning pro, he hedged his bets by going to Yale, where he played on the varsity team. (He once turned down a request to play doubles on a Saturday morning with a kid named John McEnroe, who had not yet made his surprise run at Wimbledon. Winik declined the invitation in order to "sleep in.")

By his junior year a pro tennis career looked increasingly unlikely, so Winik joined the *Yale Daily News* and discovered that he had a talent for writing. After graduating *cum laude* from Yale, he had, as he once put it, one goal: "not to become a lawyer." He didn't. He went to England to study at the London School of Economics and received a master's degree in international relations with distinction. "Back in those days," he told C-SPAN's *Booknotes*, "it was foreign policy, the great issue of the Cold War, that was one of the hot things to do and I was interested in it. I went to the London School of Economics, got my master's in international relations . . . and it sort of opened up a whole new world for me." ▶

> " Winik was one of the few children in his area to be bused to an inner-city junior high and high school. "

Meet Jay Winik *(continued)*

He came back home hoping to work in government, but ended up entering graduate school in political science at Yale. About to begin his dissertation, he was plucked from New Haven to join the committee staff of Representative Les Aspin, the powerful chairman of the House Armed Services Committee.

Here Winik had a front-row seat to the momentous historic events of the day—the Cold War, the U.S.-Soviet standoffs, and the cruise missile debates. From there he went on to direct the first Blue-Ribbon Military Base Closure Commission, and then to work with the Senate Foreign Relations Committee. He traveled around the globe, meeting prominent dissident leaders, including Lech Walesa and Václav Havel, and numerous heads of state. His focus quickly became civil wars and regional conflicts, especially in Cambodia, which was lurching toward a second horrific genocide. Winik helped craft the United Nations Perm 5 Plan for Cambodia, which brought peace and elections. Despite fears that the aircraft could be shot down, he was aboard the first plane (a four-seater) to land in Phnom Penh since the United States severed relations in 1975.

With the Cold War over and Cambodia stabilized, Winik longed for a fresh challenge. He found it in his true passion, writing—books and history. Winik soon ventured back in time, choosing the Civil War for *April 1865* and the Revolutionary period for

66 He traveled around the globe, meeting prominent dissident leaders, including Lech Walesa and Václav Havel. 99

The Great Upheaval. In each case, he applied his real world background in international affairs to his scholarship.

He brings an athlete's discipline to his work. While writing, he hesitates even to use email, he still doesn't give out his cell phone number, and he won't carry a BlackBerry; people looking for him end up calling his wife. He eats the same breakfast each day and follows the same schedule, revising work from the previous day before sitting down to write. Except for the occasional movie, while at work he lives an almost monkish existence. (His vice is sneaking out during the spring, summer, and early fall for a periodic "nooner"—a noontime game of tennis.) Only Bogey, his Wheaten Terrier, intrudes to make his home among the books piled in the study.

His *April 1865* climbed to the top of the bestseller lists in 2001. Not least among its readers were three former presidents, as well as President George W. Bush in the tense days just after 9/11.

Deemed an "instant classic," *April 1865* is one of those rare works to have spawned a new genre of history books; it was also honored as one of the New York Public Library's 25 Books to Remember from 2001, and received the first Walt Whitman Civil War Roundtable Award.

Winik makes frequent guest appearances on television and contributes regularly to the *New York Times* and the *Wall Street Journal*'s Leisure and Arts page. He serves on a number of boards of trustees for nonprofit organizations ▶

66 His vice is sneaking out during the spring, summer, and early fall for a periodic 'nooner'—a noontime game of tennis. 99

dealing with history and education, as well as battlefield preservation.

He lives outside Washington, D.C., with his wife, the writer Lyric W. Winik, and their sons, Nathaniel, born in 2002 and Evan, born in 2004. He loves tennis, art, the "great outdoors," family adventures, and playing with his kids.

Winik is renowned for his fresh and creative approaches to history, not to mention for his unsurpassed narrative talents. All of this is on display in *The Great Upheaval*, which was a *New York Times* bestseller and received rave reviews, including by three distinguished former Pulitzer Prize winners. Not unlike with *April 1865*, it was read by the last three presidents: Bill Clinton, for one, hailed its stunning achievement for bringing to life the international setting in which America struggled to its feet, while President Bush wrote a detailed, two-page handwritten letter to Winik with his own observations. Asked about his success by Brian Lamb on C-SPAN, Winik averred, "I feel privileged to do what I do. I wouldn't change it for any job in the world."

What will his next book be? "Something different," says Winik. "Stay tuned." ∽

Writing
The Great Upheaval

PREPARING TO WRITE *The Great Upheaval* was a little like a parent preparing to have a second child.

I have often been asked how I came up with the idea to write *The Great Upheaval*, and invariably, my answer gets to the nub of what it is to be a writer. But as a matter of course, I would be anything less than candid if I didn't talk about the experience of my previous book, *April 1865*. *April 1865*, about the end of the Civil War, was a *New York Times* and number one national bestseller; it enjoyed widespread critical acclaim; and it received extensive international notoriety. Yet more importantly, somehow everything in the book seemed to click: the narrative was strong, the stories often powerful and poignant, the characters rich and complex, and the themes compelling, particularly the threat of guerrilla war in America and the profound stakes for the nation. I also heard over and over from readers that it was one of the best works of history they had "ever read."

Amy Tan once wrote that the hardest thing she has ever done was to write her follow-on book to the much acclaimed *Joy Luck Club*, and in the aftermath of *April 1865*, I pretty much felt the same way. So what to do?

The easiest thing for me to have done would have been to write another book on the Civil War. The research challenge would have been that much easier, and I probably could've written the book in ▶

66 Amy Tan once wrote that the hardest thing she has ever done was to write her follow-on book to the much acclaimed *Joy Luck Club*, and in the aftermath of *April 1865*, I pretty much felt the same way. 99

a year or two, something publishers always love. Yet the historians whom I've always most admired, from William Manchester to Barbara Tuchman, repeatedly challenged themselves by moving from one topic to another, from one time period to another, a philosophy that I wholeheartedly embrace as well. So believing that an historian has to look for new frontiers, I checked out about 30 books from the local library on a host of subjects, everything from early American history to 20th-century history. On a yellow legal pad, I scribbled 50 pages worth of notes while ruling out ideas, even as I sought to find one that I could fall in love with. Eventually I came to settle on the topic of *The Great Upheaval*—my original working title was actually "The Founding"—weaving together the interrelated stories of America, France, and Russia in the 1790s. I realized that even more so than in *April 1865*, the period encapsulated in *The Great Upheaval* included an even richer set of gigantic personalities, an even larger set of events—it is ultimately one of the most significant eras in all of human history—and an equally compelling narrative. It became a natural choice.

But then the hard work began. The book was an enormous challenge to research as well as to write: This is the first history to weave together a global narrative of this crucial period, and in the process, I had to comb documents and sources from multiple languages across three continents, as well as delving into the Islamic world. I had no model to work from, no real template. It was not surprising then that it took me years to fully understand the age; indeed, the book took some six years to finish. But if I could pull it off, I was confident that in the end what would emerge was an entirely new picture of America's formative years, not to mention an unprecedented, new picture of the modern world.

And what fun I had as a writer, for here was an exotic, often complex world that stretched from the grandeur of Versailles to candlelight dinners in Monticello, from the mysteries of the Russian court to the Byzantine inner-workings of the Sultan's world. Suddenly it dawned on me that this book would depict an unremitting life-or-death struggle between those who believed in man-made democracy or those who believed in divinely inspired autocracy, between advocates of constitutional republicanism or of Allah's law.

Of course, in doing so I knew I would be at odds with much of the conventional wisdom about America's beginnings as a nation, which had long isolated the story of America's founding decade from the rest of the globe. But the more I researched the book, the more I came to see that the world of the 1790s was stitched together in ways we can scarcely grasp, from Philadelphia to Paris to St. Petersburg and Constantinople, and that one can't fully understand America's beginnings without viewing it in the context of global events—just as the founders themselves did. ▶

> " And what fun I had as a writer, for here was an exotic, often complex world that stretched from the grandeur of Versailles to candlelight dinners in Monticello, from the mysteries of the Russian court to the Byzantine inner-workings of the Sultan's world. "

Writing *The Great Upheaval* (continued)

Perhaps my greatest excitement in unpeeling this world that has for too long been lost to history came in seeing how the great leaders of the day watched each other, reacted to each other, and responded to each other—that was how the modern world was formed. For instance, it became apparent that the Founding Fathers were all consumed by events in Europe—from the increasing anarchy and bloodshed of the French Revolution to Russia's dismemberment of the ancient Kingdom of Poland. It also became apparent to me that one simply can't appreciate America's fears of foreign invasion in the 1790s or being swallowed by a predatory European power without seeing them in relationship to Napoleon's armies that were devouring Europe "leaf by leaf," or against the backdrop of the "tidal wave" of Russian armies laying siege to Islam, Sweden and Italy. These grim examples underscored to George Washington and the young Americans the perils of weakness in the face of imperialistic European empires. It also explains the tenuousness and fragility of the young American Republic. For instance, during the Whiskey Rebellion, the rebels were toasting the bloodthirsty dictator of France, Robespierre, as well as carrying mock guillotines—this at the height of the Terror in France—and threatening to forge an allegiance with the French Revolutionary nation to boot, which could have plunged America into the same kind of chaos and violence that was at that moment devouring Europe.

In writing the book itself, one of the hardest things for me to accomplish was structuring the narrative. On one hand I wanted to stress interconnections between events taking place in the infant, tumultuous America, a chaotic Revolutionary France, and an increasingly authoritarian Russia, not to mention a savage world war that touched all four corners of the globe along with a holy war between the Islamic world and the Christian world. On the other hand, I wanted to try to re-create as much as possible the world as the actors themselves experienced it. In the end, I decided to structure the book with chapters that would alternate between events in the three different countries, though here I was not dogmatic. By the middle of the book, when the French Revolution takes on a larger significance, I decided to keep the narrative with France. In the final third of the book, I stayed almost exclusively with America. The benefit of this was that the reader, having now lived through the events of France's revolution, could much better understand the debates and crises that pervaded the second term of Washington's administration, and all of the Adams' administration. I also went to great pains to weave the critical role of Britain throughout the book.

I found some of the scenes deeply touching: the detaining of King Louis XVI on a cold, windswept day, and the heart-rending scene when he parted for the last time with his tearful family. Some of the episodes were horrifying: the guillotine ▶

> " In writing the book itself, one of the hardest things for me to accomplish was structuring the narrative. "

that worked overtime in France and the systematic murder of hundreds of thousands of people, the old and the young, rich and poor alike, thus presaging Hitler's Nazi death machine in World War II. Others were chilling: the Islamic Sultan having his infant brothers strangled, while tiny coffins lay in the next room, or Catherine the great, the idol of the philosophes, smashing her bust of Voltaire in rage or plotting the murderous overthrow of her husband. Some chapters read like a thriller even as they were remarkably illuminating: George Washington worrying that an America divided over the French Revolution could slip into the same sort of chaos and his fears that America's envoys had been guillotined in Paris. Interestingly, (unlike in France or Russia) somehow the Americans stayed true to their Constitution and resorted to politics rather than to violence.

Finally, one of the greatest surprises for me, and one of the greatest thrills for me as a writer, was to portray the American founders along with their reigning peers across the globe. This had to have been the greatest galaxy of actors ever seen on the world stage; to be sure it was the period of Washington, Thomas Jefferson, Ben Franklin and Alexander Hamilton. But it was also the period of the doomed king and queen, Louis XVI and Marie Antoinette; of the eminent thinkers Voltaire and Diderot; of Catherine the Great, Emperor Napoleon, and the bloodthirsty tyrant Maximilian

 One of the greatest surprises for me, and one of the greatest thrills for me as a writer, was to portray the American founders along with their reigning peers across the globe.

Robespierre, not to mention a host of equally fascinating secondary figures, such as Tom Paine, John Paul Jones, Prince Potemkin, Thaddeus Kosciuszko, Edmund Burke, and the list goes on.

I have often been asked who was my favorite character to write about. Catherine the Great is probably the most charismatic and interesting; I'd want to be seated next to her at a dinner party. For some 30 years she dominated the global arena, befriended the immortal French philosophes, and after decades of enlightenment, unleashed modern authoritarianism. She was also incredibly charming, brilliant, and complex, and emerged as one of my favorites. Napoleon, of course, was one-of-a-kind, and quite important historically. Getting into the head of Robespierre, who prefigured Stalin and Hitler, was quite challenging. But ultimately, it was George Washington—stilted, austere, unapproachable—who in the long reach of history, proved to be the master spirit of the age, even if his peers didn't quite see it that way back then.

So there it was. The last major hurdle for the book came in the editing process. My long-time editor and friend, Tim Duggan, thought the book was one of the finest histories he had ever read but he had one major concern: given the current realities of modern publishing, he wanted me to cut roughly 50 pages. We went back and forth over this for many months, each time with me firmly resisting him. But in the end he persuaded me that the book ▶

would not only be enhanced in the marketplace as a result, but would benefit by being a bit shorter as well. Eventually I relented, and thus began three of the hardest months I've ever encountered as an author, a painful gauntlet of making the difficult decisions as to what stays and what goes, what to trim and what to keep intact, what to rewrite all together. The process was made even more complex by the fact that every time I cut something, say early on, that section would be referenced later in the book. My brilliant wife, Lyric, who serves as my in-house editor, was invaluable in this whole process, not least of which was to serve as another bludgeon to indeed tell me that yes, a certain section "could go."

When all was said and done, Tim was right: the book was improved. This is one of the curious things about writing, that the reader will never really know what isn't there, and as such, will never miss it. That said, I have to confess that there are times in the quiet of the night, or when I'm asked questions about *The Great Upheaval* at book events, that there are two sections that I wish had made it into the book—one was a fuller narrative exposition of the history of slavery throughout the ages, and the second was a greater detailing of the Terror in France, which I found both substantively important and riveting from the perspective of the craft of writing. Still, such decisions come with the territory.

The reactions to *The Great Upheaval* have been superb. When C-SPAN's Brian

> 66 Thus began three of the hardest months I've ever encountered as an author, a painful gauntlet of making the difficult decisions as to what stays and what goes, what to trim and what to keep intact, what to rewrite all together. 99

Lamb asked me what it was like to write another book after the great success of *April 1865*, I told him that I believed this could well be "the most important book" that I would ever write. I meant it. Not unlike *April 1865*, it was a *New York Times* bestseller and received extensive critical acclaim. President Bill Clinton wrote a warm letter to me after reading the book and said he has long believed the world was as interconnected during the time of the Founders as I wrote. President George Bush sent me a two-page hand-written, detailed letter with his observations about *The Great Upheaval*; so did Harry Reid, the Democratic Senate Majority Leader. I was invited to have dinner with 25 senators, which proved to be a fascinating evening. And one night the Winik household had a treat when an unexpected call came to my house from Jamie Lee Curtis, the actress; she's a big fan of *April 1865*. She said she was thrilled to be reading my newest book. Gordon Wood, who is one of our most preeminent scholars of the Revolutionary era, gave *The Great Upheaval* just a terrific review in the *New York Review of Books*. But I've been doing this long enough to know that the lifeblood of any author is his readers, and here I cherish the many notes, e-mails, and even gifts that I received from my readers. Perhaps most special were those who said *April 1865* was one of their favorite books, but they liked *The Great Upheaval* "even more."

That's all I could have ever asked for. ∼

Author's Suggestions: Create Your Own Tour for *The Great Upheaval*

THERE IS PERHAPS no more exciting thing for the reader than getting into the hearts and heads of the actors living through the great cataclysms of the age during "the great upheaval," to literally see and feel the world as they did. With a dog-eared copy of *The Great Upheaval*, and your favorite travel guides, take your own tour of this extraordinary era. Imagine the rarefied circle of globetrotting diplomats who crossed borders, spoke in foreign tongues, and fomented revolution—often twice. Imagine, the spirit of the age as defined by the Comte de Segur, "of philosophy and freedom. It enters palaces as well as huts." Imagine the revolutionary tide that swept Europe, America and Russia in the 1790s, and the rape of Poland, the subjugation of Italy, the rise of Napoleon, the battle by the pyramids, and the chaos across the continent. Imagine the young American Republic struggling to survive, and the ancient Russian Empire laboring to stamp out the whiff of democracy sweeping the globe.

A Few Thoughts:
Visit the historic sites in Philadelphia, Washington, DC, New York, and Monticello in Virginia. Here are the places where the Constitution was formed, where President George Washington governed America for its tumultuous first eight years, where Thomas Jefferson and

Alexander Hamilton literally wrestled with each other for the future of the nation. Put yourself in their shoes and consider their audacity, heated emotions, constant fears, and hopes for the future. And whether you are reading about the rise of capitalism, the debate over the capital, the fears of foreign invasion or domestic insurrection during the Whiskey Rebellion, and finally, the possible unraveling of the country during the election of 1800, consider the tenuousness of it all.

Visit Paris, France, as well as the area that was once the seat of the Vendee Rebellion. See the grandeur of Versailles outside the city, which was once the embodiment of this great kingdom and the envy of capitals throughout the world, the very hallmark of civilization and Enlightenment thought. Picture it all crashing down, the gorgeous mirrors being smashed, the chandeliers being destroyed, and unthinkably, the king and queen being forcibly dragged to Paris. Here, take a walking tour of the old part of the city; also visit the Tuileries, and in your mind's eye listen to the historic debates about the rights of man and the revolution, stirring words that roused a nation and oppressed people everywhere. Then watch excitement fade into tragedy and then hell itself. Re-create the mobs ruling the city, the blood staining the streets, the clank of the guillotine and the roar of the mob, and the sheer terror. Listen to the sounds of a city at war, the hush on a wintry day when Louis XVI embraced his family one last time before being beheaded, and ▶

watch your own voice fall to a whisper as did Parisians, who at the height of the terror worried about every word they said. Ponder the profound difference between America under Washington and Adams, which stayed true to its constitution and successfully transferred power from one political party to the next, and France, where Napoleon came to power in a coup. Here are two strains of the modern world unfolding before you.

Now we move on to the third strain of the world, what would become modern authoritarianism. Visit St. Petersburg, Moscow, and the Crimea. A crossroads between East and West, never forget that the world here is exotic and Byzantine, mysterious and Manichaean. It was also the largest empire on the globe, and a cosmopolitan one. Remember too that Catherine the Great's court spoke not Russian but French, and was like Versailles on the Neva; for that matter, it was intimately familiar with events in America. And wherever Catherine was, whether at the Winter Palace or Peterhof or the Crimea, everything seemed to glitter. There are balls, plays, opera, and readings nightly. Traveling with ambassadors from 22 countries into Muslim heartland—listen if you will to the babble of languages being spoken, Russian, French, German, English, and Tartar—Catherine showed how cosmopolitan her country was by discussing George Washington and the American Revolution; she'd even read Ben Franklin and corresponded with Jefferson. But don't be that easily seduced. Once the accents of liberty, whether American or French, began upending the old order, Catherine became ever more reactionary, determined to resist the democratic tide. Her country became a nation at war abroad and harshly repressive at home.

This is the tumultuous world they lived in. Writing *The Great Upheaval* for six years, this is the world I lived in. This now can be the world you live in too. ⌒

Have You Read?

APRIL 1865

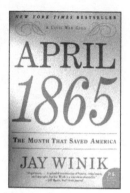

One month in 1865 witnessed the frenzied fall of Richmond, a daring last-ditch Southern plan for guerrilla warfare, Lee's harrowing retreat, and then, Appomattox. It saw Lincoln's assassination just five days later and a near-successful plot to decapitate the Union government, followed by chaos and coup fears in the North, collapsed negotiations and continued bloodshed in the South, and finally, the start of national reconciliation.

In the end, April 1865 emerged as not just the tale of the war's denouement, but the story of the making of our nation.

Jay Winik offers a brilliant new look at the Civil War's final days that will forever change the way we see the war's end and the nation's new beginning. Uniquely set within the larger sweep of history and filled with rich profiles of outsize figures, fresh iconoclastic scholarship, and a gripping narrative, this is a masterful account of the thirty most pivotal days in the life of the United States.

"Magnificent. . . . Mr. Winik is a marvelous storyteller."
— Jeff Shaara, *The Wall Street Journal*

"Brilliant. . . . Dramatic. . . . Epic. Winik's narration is masterly."
— *The New York Times Book Review*

"There probably never was a month so full of peril, so packed with action, so freighted with significance, as April 1865."
— *The Boston Globe*

9 780060 083144